Applied Mathematical Sciences

Volume 197

More information about this series at http://www.springer.com/series/34

Valery Serov

Fourier Series, Fourier Transform and Their Applications to Mathematical Physics

Valery Serov
Department of Mathematical Sciences
University of Oulu
Oulu
Finland

ISSN 0066-5452 ISSN 2196-968X (electronic)
Applied Mathematical Sciences
ISBN 978-3-319-87985-7 ISBN 978-3-319-65262-7 (eBook)
DOI: 10.1007/978-3-319-65262-7

Mathematics Subject Classification (2010): 26A16, 26A45, 35A08, 35F50, 35J05, 35J08, 35J10, 35J15, 35K05, 35K08, 35L05, 35P25, 35R30

Softcover reprint of the hardcover 1st edition 2017

Printed on acid-free paper

This Springer imprint is published by Springer Nature
The registered company is Springer International Publishing AG
The registered company address is: Gewerbestrasse 11, 6330 Cham, Switzerland

Preface

The modern theory of analysis and differential equations in general certainly includes the Fourier transform, Fourier series, integral operators, spectral theory of differential operators, harmonic analysis and much more. This book combines all these subjects based on a unified approach that uses modern view on all these themes. The book consists of four parts: Fourier series and the discrete Fourier transform, Fourier transform and distributions, Operator theory and integral equations and Introduction to partial differential equations and it outgrew from the half-semester courses of the same name given by the author at University of Oulu, Finland during 2005–2015.

Each part forms a self-contained text (although they are linked by a common approach) and can be read independently. The book is designed to be a modern introduction to qualitative methods used in harmonic analysis and partial differential equations (PDEs). It can be noted that a survey of the state of the art for all parts of this book can be found in a very recent and fundamental work of B. Simon [35].

This book contains about 250 exercises that are an integral part of the text. Each part contains its own collection of exercises with own numeration. They are not only an integral part of the book, but also indispensable for the understanding of all parts whose collection is the content of this book. It can be expected that a careful reader will complete all these exercises.

This book is intended for graduate level students majoring in pure and applied mathematics but even an advanced researcher can find here very useful information which previously could only be detected in scientific articles or monographs.

Each part of the book begins with its own introduction which contains the facts (mostly) from functional analysis used thereinafter. Some of them are proved while the others are not.

The first part, Fourier series and the discrete Fourier transform, is devoted to the classical one-dimensional trigonometric Fourier series with some applications to PDEs and signal processing. This part provides a self-contained treatment of all well known results (but not only) at the beginning graduate level. Compared with some known texts (see [12, 18, 29, 35, 38, 44, 45]) this part uses many function spaces such as Sobolev, Besov, Nikol'skii and Hölder spaces. All these spaces are introduced by special manner via the Fourier coefficients and they are used in the proofs of main results. Same definition of Sobolev spaces can be found in [35]. The advantage of such approach is that we are able to prove quite easily the precise embeddings for these spaces that are the same as in classical function theory (see [1, 3, 26, 42]). In the frame of this part some very delicate properties of the trigonometric Fourier series (Chapter 10) are considered using quite elementary proofs (see also [46]). The unified approach allows us also to consider naturally the discrete Fourier transform and establish its deep connections with the continuous Fourier transform. As a consequence we prove the famous Whittaker-Shannon-Boas theorem about the reconstruction of band-limited signal via the trigonometric Fourier series (see Chapter 13). Many applications of the trigonometric Fourier series to the one-dimensional heat, wave and Laplace equation are presented in Chapter 14. It is accompanied by a large number of very useful exercises and examples with applications in PDEs (see also [10, 17]).

The second part, Fourier transform and distributions, probably takes a central role in this book and it is concerned with distribution theory of L. Schwartz and its applications to the Schrödinger and magnetic Schrödinger operators (see Chapter 32). The estimates for Laplacian and Hamiltonian that generalize well known Agmon's estimates on the continuous spectrum are presented in this part (see Chapter 23). This part can be considered as one of the most important because of numerous applications in the scattering theory and inverse problems. Here we have considered for the first time some classical direct scattering problems for the Schrödinger operator and for the magnetic Schrödinger operator with singular (locally unbounded) coefficients including the mathematical foundations of the classical approximation of M. Born. Also, the properties of Riesz transform and Riesz potentials (see Chapter 21) are investigated very carefully in this part. Before this material could only be found in scientific journals or monographs but not in textbooks. There is a good connection of this part with Operator theory and integral equations. The main technique applied here is the Fourier transform.

The third part, Operator theory and integral equations, is devoted mostly to the self-adjoint but unbounded operators in Hilbert spaces and their applications to integral equations in such spaces. The advantage of this part is that many important results of J. von Neumann's theory of symmetric operators are collected together. J. von Neumann's spectral theorem allows us, for example, to introduce the heat kernel without solving the heat equation. Moreover, we show applications of the spectral theorem of J. von Neumann (for these operators) to the spectral theory of elliptic differential operators. In particular, the existence of Friedrichs extension for these operators with discrete spectrum is provided. Special attention is devoted to the Schrödinger and the magnetic Schrödinger operators. The famous diamagnetic inequality is proved here. We follow in this consideration B. Simon [35] (slightly different approach can be found in [28]). We recommend (in addition to this part) the reader get acquainted with the books [4, 13, 15, 24, 41]. As a consequence of the spectral theory of elliptic differential operators the integral equations with weak singularities are considered in quite simple manner not only in Hilbert spaces but also in some Banach spaces, e.g. in the space of continuous functions on closed manifolds. The central point of this consideration is the Riesz theory of compact (not necessarily self-adjoint) operators in Hilbert and Banach spaces. In order to keep this part short, some proofs will not be given, nor will all theorems be proved in complete generality. For many details of these integral equations we recommend [22]. We are able to investigate in quite simple manner one-dimensional Volterra integral equations with weak singularities in $L^\infty(a,b)$ and singular integral equations in the periodic Hölder spaces $C^\alpha[-a,a]$. Concerning approximation methods our considerations use the general theory of bounded or compact operators in Hilbert spaces and we follow mostly the monograph of Kress [22].

The fourth part, Introduction to partial differential equations, serves as an introduction to modern methods for classical theory of partial differential equations. Fourier series and Fourier transform play crucial role here too. An important (and quite independent) segment of this part is the self-contained theory of quasi-linear partial differential equations of order one. The main attention in this part is devoted to elliptic boundary value problems in Sobolev and Hölder spaces. In particular, the unique solvability of direct scattering problem for Helmholtz equation is provided. We investigate very carefully the mapping and discontinuity properties of double and single layer potentials with continuous densities. We also refer to similar properties of double and single layer potentials with densities in Sobolev spaces $H^{1/2}(S)$ and $H^{-1/2}(S)$, respectively, but will not prove any of these results, referring for their proofs to monographs [22] and [25]. Here (and elsewhere in the book) S denotes the boundary of a bounded domain in $\mathbb{R}^n$ and if the smoothness of S is not specified explicitly then it is assumed to be such that Sobolev embedding theorem holds. Compared with well known texts on partial differential equations some direct and inverse scattering problems for Helmholtz, Schrödinger and magnetic Schrödinger operators are considered in this part. As it was mentioned earlier this type of material could not be found in textbooks. The presentation in many places of this part has been strongly influenced by the monographs [6, 7, 11] (see also [8, 16, 24, 36, 40]).

In closing we note that this book is not as comprehensive as the fundamental work of B. Simon [35]. But the book can be considered as a good introduction to modern theory of analysis and differential equations and might be useful not only to students and PhD students but also to all researchers who have applications in mathematical physics and engineering sciences. This book could not have appeared without the strong participation, both in content and typesetting, of my colleague Adj. Prof. Markus Harju. Finally, a special thanks to professor David Colton from University of Delaware (USA) who encouraged the writing of this book and who has supported the author very much over the years.

Oulu, Finland
June 2017

Valery Serov

Contents

Part I
Fourier Series and the Discrete Fourier Transform

Chapter 1
Introduction

Definition 1.1. A function $f(x)$ of one variable x is said to be *periodic* with period $T > 0$ if the domain $D(f)$ of f contains $x+T$ whenever it contains x and if for every $x \in D(f)$, one has

$$f(x+T) = f(x). \tag{1.1}$$

Remark 1.2. If also $x - T \in D(f)$, then

$$f(x-T) = f(x).$$

It follows that if T is a period of f, then mT is also a period for every integer $m > 0$. The smallest value of $T > 0$ for which (1.1) holds is called the *fundamental period* of f.

For example, the functions

$$\sin\frac{m\pi x}{L}, \quad \cos\frac{m\pi x}{L}, \quad e^{i\frac{m\pi x}{L}}, \quad m = 1,2,\ldots$$

are periodic with fundamental period $T = \frac{2L}{m}$. Note also that they are periodic with common period $2L$.

If some function f is defined on the interval $[a, a+T]$, with $T > 0$ and $f(a) = f(a+T)$, then f can be *extended periodically* with period T to the whole line as

$$f(x) := f(x - mT), \quad x \in [a + mT, a + (m+1)T], \quad m = 0, \pm 1, \pm 2, \ldots.$$

Therefore, we may assume from now on that every periodic function is defined on the whole line.

V. Serov, *Fourier Series, Fourier Transform and Their Applications to Mathematical Physics*, Applied Mathematical Sciences 197,
DOI 10.1007/978-3-319-65262-7_1

We say that f is *p-integrable*, $1 \le p < \infty$, on the interval $[a,b]$ if

$$\int_a^b |f(x)|^p \mathrm{d}x < \infty.$$

The set of all such functions is denoted by $L^p(a,b)$. When $p = 1$, we say that f is *integrable*.

The following "continuity" in the sense of L^p space, $1 \le p < \infty$, holds: for every $f \in L^p(a,b)$ and $\varepsilon > 0$, there is a continuous function g on $[a,b]$ such that

$$\left(\int_a^b |f(x) - g(x)|^p \mathrm{d}x\right)^{1/p} < \varepsilon$$

(see e.g., Corollary 5.3). If f is p-integrable and g is p'-integrable on $[a,b]$, where

$$\frac{1}{p} + \frac{1}{p'} = 1, \quad 1 < p < \infty, 1 < p' < \infty,$$

then their product is integrable on $[a,b]$ and

$$\int_a^b |f(x)g(x)| \mathrm{d}x \le \left(\int_a^b |f(x)|^p \mathrm{d}x\right)^{1/p} \left(\int_a^b |g(x)|^{p'} \mathrm{d}x\right)^{1/p'}.$$

This inequality is called *Hölder's inequality* for integrals. *Fubini's theorem* states that

$$\int_a^b \left(\int_c^d F(x,y)\mathrm{d}y\right)\mathrm{d}x = \int_c^d \left(\int_a^b F(x,y)\mathrm{d}x\right)\mathrm{d}y = \int_a^b \int_c^d F(x,y)\mathrm{d}x\mathrm{d}y,$$

where $F(x,y) \in L^1((a,b)\times(c,d))$.

If $f_1, f_2, \dots, f_n$ are p-integrable on $[a,b]$ for $1 \le p < \infty$, then so is their sum $\sum_{j=1}^n f_j$, and

$$\left(\int_a^b \left|\sum_{j=1}^n f_j(x)\right|^p \mathrm{d}x\right)^{1/p} \le \sum_{j=1}^n \left(\int_a^b |f_j(x)|^p \mathrm{d}x\right)^{1/p}. \tag{1.2}$$

This inequality is called *Minkowski's inequality*. As a consequence of Hölder's inequality we obtain the *generalized Minkowski inequality*

$$\left(\int_a^b \left|\int_c^d F(x,y)\mathrm{d}y\right|^p \mathrm{d}x\right)^{1/p} \le \int_c^d \left(\int_a^b |F(x,y)|^p \mathrm{d}x\right)^{1/p} \mathrm{d}y. \tag{1.3}$$

Exercise 1.1. Prove Hölder's inequality for integrals for every $1 \le p < \infty$.

Hint. Prove first Hölder's inequality for sums, i.e.,

$$\left|\sum_{j=1}^{n} a_j b_j\right| \le \left(\sum_{j=1}^{n} |a_j|^p\right)^{1/p} \left(\sum_{j=1}^{n} |b_j|^{p'}\right)^{1/p'},$$

where $1 < p < \infty, 1/p + 1/p' = 1$, and where for $p = \infty$ (or $p' = \infty$) we consider $\max_{1\le j\le n} |a_j|$ (or $\max_{1\le j\le n} |b_j|$) instead of the corresponding sums.

Exercise 1.2. Prove (1.2) and (1.3).

Lemma 1.3. *If f is periodic with period $T > 0$ and if it is integrable on every finite interval, then*

$$\int_a^{a+T} f(x)\mathrm{d}x = \int_0^T f(x)\mathrm{d}x \tag{1.4}$$

for every $a \in \mathbb{R}$.

Proof. Let first $a > 0$. Then

$$\begin{aligned}\int_a^{a+T} f(x)\mathrm{d}x &= \int_0^{a+T} f(x)\mathrm{d}x - \int_0^a f(x)\mathrm{d}x \\ &= \int_0^T f(x)\mathrm{d}x + \left[\int_T^{a+T} f(x)\mathrm{d}x - \int_0^a f(x)\mathrm{d}x\right].\end{aligned}$$

The difference in the square brackets is equal to zero due to periodicity of f. Thus, (1.4) holds for $a > 0$.

If $a < 0$, then we proceed similarly, obtaining

$$\begin{aligned}\int_a^{a+T} f(x)\mathrm{d}x &= \int_a^0 f(x)\mathrm{d}x + \int_0^{a+T} f(x)\mathrm{d}x \\ &= \int_a^0 f(x)\mathrm{d}x + \int_0^T f(x)\mathrm{d}x - \int_{a+T}^T f(x)\mathrm{d}x \\ &= \int_0^T f(x)\mathrm{d}x + \left[\int_a^0 f(x)\mathrm{d}x - \int_{a+T}^T f(x)\mathrm{d}x\right].\end{aligned}$$

Again, the periodicity of f implies that the difference in brackets is zero. Thus the lemma is proved. □

Definition 1.4. Let us assume that the domain of f is symmetric with respect to $\{0\}$, i.e., if $x \in D(f)$, then $-x \in D(f)$. A function f is called *even* if

$$f(-x) = f(x), \quad x \in D(f),$$

and *odd* if

$$f(-x) = -f(x), \quad x \in D(f).$$

Lemma 1.5. *If f is integrable on every finite interval and if it is even, then*

$$\int_{-a}^{a} f(x)\mathrm{d}x = 2\int_{0}^{a} f(x)\mathrm{d}x$$

for every $a > 0$. Similarly, if f is odd, then

$$\int_{-a}^{a} f(x)\mathrm{d}x = 0$$

for every $a > 0$.

Proof. Since

$$\int_{-a}^{a} f(x)\mathrm{d}x = \int_{0}^{a} f(x)\mathrm{d}x + \int_{-a}^{0} f(x)\mathrm{d}x,$$

then on changing variables in the second integral we obtain

$$\int_{-a}^{a} f(x)\mathrm{d}x = \int_{0}^{a} f(x)\mathrm{d}x + \int_{0}^{a} f(-x)\mathrm{d}x.$$

The assertion of the lemma now follows from Definition 1.4. □

Definition 1.6. The notation $f(c \pm 0)$ is used to denote the *right and left limits*

$$f(c \pm 0) := \lim_{x \to c\pm} f(x).$$

Definition 1.7. A function f is said to be *piecewise continuous* (*piecewise constant*) on an interval $[a,b]$ if there are $x_0, x_1, \ldots, x_n$ such that $a = x_0 < x_1 < \cdots < x_n = b$ and

(1) f is continuous (constant) on each subinterval $(x_{j-1}, x_j), j = 1, 2, \ldots, n$,
(2) $f(x_0 + 0)$, $f(x_n - 0)$, and $f(x_j \pm 0)$, $j = 1, 2, \ldots, n-1$, exist.

Definition 1.8. A function f is said to be of *bounded variation* on an interval $[a,b]$ if there is $c_0 \geq 0$ such that for every $\{x_0, x_1, \ldots, x_n\}$ with $a = x_0 < x_1 < \cdots < x_n = b$, one has

$$\sum_{j=1}^{n} |f(x_j) - f(x_{j-1})| \leq c_0.$$

The number

$$V_a^b(f) := \sup_{x_0, x_1, \ldots, x_n} \sum_{j=1}^{n} |f(x_j) - f(x_{j-1})| \tag{1.5}$$

is called the *total variation* of f on the interval $[a,b]$. For every $x \in [a,b]$ we can also define $V_a^x(f)$ by (1.5). The class of functions of bounded variation is denoted by $BV[a,b]$.

Exercise 1.3. (1) Show that the bounded function

$$f(x) = \begin{cases} x\sin\frac{1}{x}, & x \in (0,1], \\ 0, & x = 0, \end{cases}$$

is continuous on the interval $[0,1]$ but is not of bounded variation on $[0,1]$.
(2) Show that every piecewise constant function on $[a,b]$ is of bounded variation.

Remark 1.9. This exercise shows that $C[a,b]$ and $BV[a,b]$ are not included in each other, i.e., they represent two different scales of functions.

Exercise 1.4. Prove that

(1) $V_a^x(f)$ is monotone increasing in x,
(2) for every $c \in (a,b)$, we have $V_a^b(f) = V_a^c(f) + V_c^b(f)$.

If f is real-valued, then Exercise 1.4 implies that $V_a^x(f) - f(x)$ is monotone increasing in x. Indeed, for $h > 0$ we have that

$$\begin{aligned}\left(V_a^{x+h}(f) - f(x+h)\right) - (V_a^x(f) - f(x)) &= \left(V_a^{x+h}(f) - V_a^x(f)\right) - (f(x+h) - f(x)) \\ &= V_x^{x+h}(f) - (f(x+h) - f(x)) \\ &\geq V_x^{x+h}(f) - |f(x+h) - f(x)| \geq 0.\end{aligned}$$

As an immediate consequence we obtain that every real-valued function $f \in BV[a,b]$ can be represented as the difference of two monotone increasing functions as

$$f(x) = V_a^x(f) - (V_a^x(f) - f(x)).$$

This fact allows us to define the *Stieltjes integral*

$$\int_a^b g(x)\mathrm{d}f(x), \tag{1.6}$$

where $f \in BV[a,b]$ and g is an arbitrary continuous function. The integral (1.6) is defined as

$$\int_a^b g(x)\mathrm{d}f(x) = \lim_{\Delta\to 0} \sum_{j=1}^n g(\xi_j)(f(x_j) - f(x_{j-1})),$$

where $a = x_0 < x_1 < \cdots < x_n = b$, $\xi_j \in [x_{j-1}, x_j]$, and $\Delta = \max_{1\leq j\leq n}(x_j - x_{j-1})$.

Let us introduce the *modulus of continuity* of f by

$$\omega_h(f) := \sup_{\{x\in[a,b]:x+h\in[a,b]\}} |f(x+h)-f(x)|, \quad h>0. \tag{1.7}$$

Definition 1.10. A bounded function f is said to belong to *Hölder space* $C^\alpha[a,b], 0<\alpha\leq 1$, if

$$\omega_h(f)\leq Ch^\alpha$$

with some constant $C>0$. This inequality is called the *Hölder condition with exponent* α.

Definition 1.11. A function f is said to belong to *Sobolev space* $W_p^1(a,b)$, $1\leq p<\infty$, if $f\in L^p(a,b)$ and there is $g\in L^p(a,b)$ such that

$$f(x)=\int_a^x g(t)\mathrm{d}t+C \tag{1.8}$$

with some constant C.

Definition 1.12. A function f is said to belong to *Sobolev space* $W_\infty^1(a,b)$ if there is a bounded integrable function g such that

$$f(x)=\int_a^x g(t)\mathrm{d}t+C \tag{1.9}$$

with some constant C.

Remark 1.13. Using Hölder's inequality we may conclude that

$$W_{p_1}^1(a,b)\subset W_{p_2}^1(a,b)$$

for every $1\leq p_2<p_1\leq\infty$.

Lemma 1.14. *Suppose that* $f\in W_p^1(a,b)$, $1\leq p\leq\infty$. *Then* f *is of bounded variation. Moreover, if* $p=1$, *then* f *is also continuous, and if* $1<p\leq\infty$, *then* $f\in C^{1-1/p}[a,b]$.

Proof. Let first $p=1$. Then there is an integrable function g such that (1.8) holds with some constant C. Hence for fixed $x\in[a,b]$ with $x+h\in[a,b]$ we have

$$f(x+h)-f(x)=\int_x^{x+h} g(t)\mathrm{d}t.$$

It follows that

$$|f(x+h)-f(x)|=\left|\int_x^{x+h} g(t)\mathrm{d}t\right|\to 0, \quad h\to 0,$$

since g is integrable. This proves the continuity of f. At the same time, for every $\{x_0, x_1, \dots, x_n\}$ such that $a = x_0 < x_1 < \dots < x_n = b$, we have

$$\sum_{j=1}^{n} |f(x_j) - f(x_{j-1})| = \sum_{j=1}^{n} \left| \int_{x_{j-1}}^{x_j} g(t)\mathrm{d}t \right| \leq \sum_{j=1}^{n} \int_{x_{j-1}}^{x_j} |g(t)|\mathrm{d}t = \int_a^b |g(t)|\mathrm{d}t.$$

Hence, Definition 1.8 is satisfied with constant $c_0 = \int_a^b |g(t)|\mathrm{d}t$, and f is of bounded variation.

If $1 < p < \infty$, then using Hölder's inequality for integrals we obtain for $h > 0$ that

$$|f(x+h) - f(x)| \leq \int_x^{x+h} |g(t)|\mathrm{d}t \leq \left(\int_x^{x+h} \mathrm{d}t \right)^{1/p'} \left(\int_x^{x+h} |g(t)|^p \mathrm{d}t \right)^{1/p}$$
$$\leq h^{1-1/p} \left(\int_a^b |g(t)|^p \mathrm{d}t \right)^{1/p},$$

where $1/p + 1/p' = 1$. If $p = \infty$, then $|f(x+h) - f(x)| \leq h \sup |g|$. By Definition 1.10, this means that $f \in C^{1-1/p}[a,b]$. The lemma is proved. □

Remark 1.15. Since every $f \in W_p^1(a,b)$, $1 \leq p \leq \infty$, is continuous, it follows that the constant C in (1.8)–(1.9) is equal to $f(a)$.

Definition 1.16. Two functions u and v are said to be *orthogonal* on $[a,b]$ if the product uv is integrable and

$$\int_a^b u(x)\overline{v(x)}\mathrm{d}x = 0,$$

where overline indicates the complex conjugation. A set of functions is said to be *mutually orthogonal* if each distinct pair in the set is orthogonal on $[a,b]$.

Lemma 1.17. *The functions*

$$1, \quad \sin\frac{m\pi x}{L}, \quad \cos\frac{m\pi x}{L}, \quad m = 1, 2, \dots,$$

form a mutually orthogonal set on the interval $[-L, L]$ as well as on every interval $[a, a+2L]$. In fact,

$$\int_{-L}^{L} \cos\frac{m\pi x}{L} \cos\frac{n\pi x}{L} \mathrm{d}x = \begin{cases} 0, & m \neq n, \\ L, & m = n, \end{cases} \tag{1.10}$$

$$\int_{-L}^{L} \cos\frac{m\pi x}{L} \sin\frac{n\pi x}{L} \mathrm{d}x = 0, \tag{1.11}$$

$$\int_{-L}^{L} \sin\frac{m\pi x}{L}\sin\frac{n\pi x}{L}\mathrm{d}x = \begin{cases} 0, & m \neq n, \\ L, & m = n, \end{cases} \tag{1.12}$$

and

$$\int_{-L}^{L} \sin\frac{m\pi x}{L}\mathrm{d}x = \int_{-L}^{L} \cos\frac{m\pi x}{L}\mathrm{d}x = 0. \tag{1.13}$$

Proof. By Lemma 1.3, it is enough to prove the equalities (1.10), (1.11), (1.12), and (1.13) only for integrals over $[-L,L]$. Let us derive, for example, (1.12). Using the equality

$$\sin\alpha\sin\beta = \frac{1}{2}\left(\cos(\alpha-\beta) - \cos(\alpha+\beta)\right),$$

we have for $m \neq n$ that

$$\begin{aligned}\int_{-L}^{L} \sin\frac{m\pi x}{L}\sin\frac{n\pi x}{L}\mathrm{d}x &= \frac{1}{2}\int_{-L}^{L}\cos\frac{(m-n)\pi x}{L}\mathrm{d}x - \frac{1}{2}\int_{-L}^{L}\cos\frac{(m+n)\pi x}{L}\mathrm{d}x \\ &= \frac{1}{2}\left(\frac{\sin\frac{(m-n)\pi x}{L}}{\frac{(m-n)\pi}{L}}\right)\Bigg|_{-L}^{L} - \frac{1}{2}\left(\frac{\sin\frac{(m+n)\pi x}{L}}{\frac{(m+n)\pi}{L}}\right)\Bigg|_{-L}^{L} = 0.\end{aligned}$$

If $m = n$, we have

$$\int_{-L}^{L} \sin\frac{m\pi x}{L}\sin\frac{n\pi x}{L}\mathrm{d}x = \frac{1}{2}\int_{-L}^{L} 1\mathrm{d}x - \frac{1}{2}\int_{-L}^{L}\cos\frac{2m\pi x}{L}\mathrm{d}x = L.$$

The other identities can be proved in a similar manner and are left to the reader. The lemma is proved. □

Remark 1.18. This lemma holds also for the functions $\mathrm{e}^{\mathrm{i}\frac{n\pi x}{L}}$, $n = 0, \pm1, \pm2, \ldots$, in the form

$$\int_{-L}^{L} \mathrm{e}^{i\frac{n\pi x}{L}}\mathrm{e}^{-\mathrm{i}\frac{m\pi x}{L}}\mathrm{d}x = \begin{cases} 0, & n \neq m, \\ 2L, & n = m. \end{cases}$$

Chapter 2
Formulation of Fourier Series

Let us consider a series of the form

$$\frac{a_0}{2} + \sum_{m=1}^{\infty} \left(a_m \cos \frac{m\pi x}{L} + b_m \sin \frac{m\pi x}{L} \right). \tag{2.1}$$

This series consists of $2L$-periodic functions. Thus, if the series (2.1) converges for all x, then the function to which it converges will also be $2L$-periodic. Let us denote this limiting function by $f(x)$, i.e.,

$$f(x) := \frac{a_0}{2} + \sum_{m=1}^{\infty} \left(a_m \cos \frac{m\pi x}{L} + b_m \sin \frac{m\pi x}{L} \right). \tag{2.2}$$

To determine a_m and b_m we proceed as follows: assuming that the integration can be legitimately carried out term by term (it will be, for example, if $\sum_{m=1}^{\infty}(|a_m| + |b_m|) < \infty$), we obtain

$$\begin{aligned}\int_{-L}^{L} f(x) \cos \frac{n\pi x}{L} \mathrm{d}x = \frac{a_0}{2} \int_{-L}^{L} \cos \frac{n\pi x}{L} \mathrm{d}x + \sum_{m=1}^{\infty} a_m \int_{-L}^{L} \cos \frac{m\pi x}{L} \cos \frac{n\pi x}{L} \mathrm{d}x \\ + \sum_{m=1}^{\infty} b_m \int_{-L}^{L} \sin \frac{m\pi x}{L} \cos \frac{n\pi x}{L} \mathrm{d}x\end{aligned}$$

for each fixed $n = 1, 2, \ldots$. It follows from the orthogonality relations (1.10), (1.11), and (1.13) that the only nonzero term on the right-hand side is the one for which $m = n$ in the first summation. Hence

$$a_n = \frac{1}{L} \int_{-L}^{L} f(x) \cos \frac{n\pi x}{L} \mathrm{d}x, \quad n = 1, 2, \ldots. \tag{2.3}$$

V. Serov, *Fourier Series, Fourier Transform and Their Applications to Mathematical Physics*, Applied Mathematical Sciences 197,
DOI 10.1007/978-3-319-65262-7_2

A similar expression for b_n is obtained by multiplying (2.2) by $\sin\frac{n\pi x}{L}$ and integrating termwise from $-L$ to L. The result is

$$b_n = \frac{1}{L}\int_{-L}^{L} f(x)\sin\frac{n\pi x}{L}\mathrm{d}x, \quad n = 1,2,\dots. \tag{2.4}$$

Using (1.13) we can easily obtain that

$$a_0 = \frac{1}{L}\int_{-L}^{L} f(x)\mathrm{d}x. \tag{2.5}$$

Definition 2.1. Let f be integrable (not necessarily periodic) on the interval $[-L,L]$. The *Fourier series* of f is the *trigonometric series* (2.1), where the coefficients a_0, a_m and b_m are given by (2.5), (2.3), and (2.4), respectively. In that case, we write

$$f(x) \sim \frac{a_0}{2} + \sum_{m=1}^{\infty}\left(a_m\cos\frac{m\pi x}{L} + b_m\sin\frac{m\pi x}{L}\right). \tag{2.6}$$

Remark 2.2. This definition does not imply that the series (2.6) converges to f or that f is periodic.

Definition 2.1 and Lemma 1.5 imply that if f is even on $[-L,L]$, then the Fourier series of f has the form

$$f(x) \sim \frac{a_0}{2} + \sum_{m=1}^{\infty} a_m\cos\frac{m\pi x}{L}, \tag{2.7}$$

and if f is odd, then

$$f(x) \sim \sum_{m=1}^{\infty} b_m\sin\frac{m\pi x}{L}. \tag{2.8}$$

The series (2.7) and (2.8) are called the *Fourier cosine series* and *Fourier sine series*, respectively.

If $L = \pi$, then the Fourier series (2.6) ((2.7) and (2.8)) transforms to

$$f(x) \sim \frac{a_0}{2} + \sum_{m=1}^{\infty}\left(a_m\cos mx + b_m\sin mx\right), \tag{2.9}$$

where the coefficients a_0, a_m, and b_m are given by (2.5), (2.3), and (2.4) with $L = \pi$.

There are different approaches if the function f is defined on an asymmetric interval $[0,L]$ with arbitrary $L > 0$.

(1) *Even extension.* Define a function $g(x)$ on the interval $[-L,L]$ as

$$g(x)=\begin{cases} f(x), & 0\le x\le L,\\ f(-x), & -L\le x<0.\end{cases}$$

Then $g(x)$ is even and its Fourier (cosine) series (2.7) represents f on $[0,L]$.

(2) *Odd extension.* Define a function $h(x)$ on the interval $[-L,L]$ as

$$h(x)=\begin{cases} f(x), & 0\le x\le L,\\ -f(-x), & -L\le x<0.\end{cases}$$

Then $h(x)$ is odd, and its Fourier (sine) series (2.8) represents f on $[0,L]$.

(3) Define a function $\widetilde{f}(t)$ on the interval $[-\pi,\pi]$ as

$$\widetilde{f}(t)=f\left(\frac{tL}{2\pi}+\frac{L}{2}\right).$$

If $f(0)=f(L)$, then we may extend f to be periodic with period L. Then

$$\begin{aligned} a_0(\widetilde{f}) &= \frac{1}{\pi}\int_{-\pi}^{\pi}\widetilde{f}(t)\mathrm{d}t=\frac{1}{\pi}\int_{-\pi}^{\pi}f\left(\frac{tL}{2\pi}+\frac{L}{2}\right)\mathrm{d}t=\frac{1}{\pi}\frac{2\pi}{L}\int_0^L f(x)\,\mathrm{d}x\\ &= \frac{2}{L}\int_0^L f(x)\,\mathrm{d}x := a_0(f), \end{aligned}$$

$$a_m(\widetilde{f})=(-1)^m\frac{2}{L}\int_0^L f(x)\cos\frac{2m\pi x}{L}\mathrm{d}x=(-1)^m a_m(f),$$

and

$$b_m(\widetilde{f})=(-1)^m\frac{2}{L}\int_0^L f(x)\sin\frac{2m\pi x}{L}\mathrm{d}x=(-1)^m b_m(f).$$

Hence,

$$\widetilde{f}(t)\sim\frac{a_0}{2}+\sum_{m=1}^{\infty}\left(a_m\cos mt+b_m\sin mt\right),$$

and at the same time,

$$f(x)\sim\frac{a_0}{2}+\sum_{m=1}^{\infty}(-1)^m\left(a_m\cos\frac{2m\pi x}{L}+b_m\sin\frac{2m\pi x}{L}\right),$$

where a_0, a_m, and b_m are the same and $x=\frac{tL}{2\pi}+\frac{L}{2}$.

These three alternatives allow us to consider (for simplicity) only the case of a symmetric interval $[-\pi,\pi]$ such that the Fourier series will be of the form (2.9) i.e.

$$f(x) \sim \frac{a_0}{2} + \sum_{m=1}^{\infty} (a_m \cos mx + b_m \sin mx).$$

Using Euler's formula, we will rewrite this series in the complex form

$$f(x) \sim \sum_{n=-\infty}^{\infty} c_n e^{inx}, \tag{2.10}$$

where the coefficients $c_n = c_n(f)$ are equal to

$$c_n = \begin{cases} \frac{a_n}{2} + \frac{b_n}{2i}, & n = 1, 2, \dots, \\ \frac{a_0}{2}, & n = 0, \\ \frac{a_{-n}}{2} - \frac{b_{-n}}{2i}, & n = -1, -2, \dots \end{cases} \quad \text{or} \quad \begin{aligned} a_n &= c_n + c_{-n}, \quad n = 1, 2, \dots, \\ a_0 &= 2c_0, \\ b_n &= i(c_n - c_{-n}), \quad n = 1, 2, \dots. \end{aligned} \tag{2.11}$$

The formulas (2.3), (2.4), (2.5), and (2.11) imply that

$$c_n(f) = \frac{1}{2\pi} \int_{-\pi}^{\pi} f(x) e^{-inx} dx \tag{2.12}$$

for $n = 0, \pm 1, \pm 2, \dots$. We call $c_n(f)$ the nth *Fourier coefficient* of f. It can be checked that

$$c_n(f) = \overline{c_{-n}(\overline{f})}. \tag{2.13}$$

Exercise 2.1. Prove formulas (2.10), (2.11), (2.12), and (2.13).

Exercise 2.2. Find the Fourier series of

(1) $\operatorname{sgn}(x) = \begin{cases} -1, & -\pi \le x < 0, \\ 0, & x = 0, \\ 1, & 0 < x \le \pi. \end{cases}$

(2) $|x|, -1 \le x \le 1$.

(3) $x, -1 \le x \le 1$.

(4) $f(x) = \begin{cases} 0, & -L \le x \le 0, \\ L, & 0 < x \le L. \end{cases}$

(5) $f(x) = \sin x, |x| \le 2$.

Exercise 2.3. Prove, using Part (2) of Exercise 2.2, that

$$\frac{\pi^2}{8} = \sum_{k=1}^{\infty} \frac{1}{(2k-1)^2} \quad \text{and} \quad \frac{\pi^2}{6} = \sum_{k=1}^{\infty} \frac{1}{k^2}.$$

Exercise 2.4. Suppose that

$$f(x) = \begin{cases} 1-x, & 0 \leq x \leq 1, \\ 0, & 1 < x \leq 2. \end{cases}$$

Find the Fourier cosine and sine series of $f(x)$.

Exercise 2.5. Find the Fourier series of $f(x) = \cos(x/2)$, $|x| \leq \pi$. Using this series, show that

(1) $\pi = 2 + \sum_{k=1}^{\infty} \frac{(-1)^{k+1}}{k^2 - 1/4}$;

(2) $\frac{\pi}{4} = \sum_{k=1}^{\infty} \frac{(-1)^{k+1}}{2k-1}$;

(3) $\frac{1}{2} = \sum_{k=1}^{\infty} \frac{1}{4k^2 - 1}$.

Exercise 2.6. Show that if N is odd, then $\sin^N x$ can be written as a finite sum of the form

$$\sum_{k=1}^{N} a_k \sin kx,$$

which means that this finite sum is the Fourier series of $\sin^N x$ and the coefficients a_k (which are real) are the Fourier coefficients of $\sin^N x$.

Exercise 2.7. Show that if N is odd, then $\cos^N x$ can be written as a finite sum of the form

$$\sum_{k=1}^{N} a_k \cos kx.$$

Chapter 3
Fourier Coefficients and Their Properties

Definition 3.1. A trigonometric series

$$\sum_{n=-\infty}^{\infty} c_n \mathrm{e}^{\mathrm{i}nx}$$

is said to

(1) *converge pointwise* if for each $x \in [-\pi, \pi]$ the limit

$$\lim_{N\to\infty} \sum_{|n|\le N} c_n \mathrm{e}^{\mathrm{i}nx}$$

exists,

(2) *converge uniformly* in $x \in [-\pi, \pi]$ if the limit

$$\lim_{N\to\infty} \sum_{|n|\le N} c_n \mathrm{e}^{\mathrm{i}nx}$$

exists uniformly,

(3) *converge absolutely* if the limit

$$\lim_{N\to\infty} \sum_{|n|\le N} |c_n|$$

exists, or equivalently, if

$$\sum_{n=-\infty}^{\infty} |c_n| < \infty.$$

These three different types of convergence appear frequently in the sequel, and they are presented above from the weakest to the strongest. In other words, absolute

V. Serov, *Fourier Series, Fourier Transform and Their Applications to Mathematical Physics*, Applied Mathematical Sciences 197,
DOI 10.1007/978-3-319-65262-7_3

convergence implies uniform convergence, which in turn implies pointwise convergence.

If f is integrable on the interval $[-\pi,\pi]$, then the Fourier coefficients $c_n(f)$ are uniformly bounded with respect to $n = 0, \pm 1, \pm 2, \ldots$, i.e.,

$$|c_n(f)| = \frac{1}{2\pi}\left|\int_{-\pi}^{\pi} f(x)\mathrm{e}^{-\mathrm{i}nx}\mathrm{d}x\right| \leq \frac{1}{2\pi}\int_{-\pi}^{\pi}|f(x)|\mathrm{d}x, \tag{3.1}$$

where the upper bound does not depend on n. Let us assume that a sequence $\{c_n\}_{n=-\infty}^{\infty}$ is such that

$$\sum_{n=-\infty}^{\infty} |c_n| < \infty.$$

Then the series

$$\sum_{n=-\infty}^{\infty} c_n \mathrm{e}^{\mathrm{i}nx}$$

converges uniformly in $x \in [-\pi,\pi]$ and defines a continuous and periodic function

$$f(x) := \sum_{n=-\infty}^{\infty} c_n \mathrm{e}^{\mathrm{i}nx}, \tag{3.2}$$

whose Fourier coefficients are $\{c_n\}_{n=-\infty}^{\infty} = \{c_n(f)\}_{n=-\infty}^{\infty}$. More generally, suppose that

$$\sum_{n=-\infty}^{\infty} |n|^k |c_n| < \infty$$

for some integer $k > 0$. Then the series (3.2) defines a function that is k times differentiable, with

$$f^{(k)}(x) = \sum_{n=-\infty}^{\infty} (\mathrm{i}n)^k c_n \mathrm{e}^{\mathrm{i}nx} \tag{3.3}$$

a continuous function. This follows from the fact that the series (3.3) converges uniformly with respect to $x \in [-\pi,\pi]$.

Let us consider a useful example in which Fourier coefficients are applied. If $0 \leq r < 1$, then the geometric series gives

$$\frac{1}{1 - r\mathrm{e}^{\mathrm{i}x}} = \sum_{n=0}^{\infty} r^n \mathrm{e}^{\mathrm{i}nx}, \tag{3.4}$$

and this series converges absolutely. Using the definition of the Fourier coefficients, we obtain

$$r^n = \frac{1}{2\pi}\int_{-\pi}^{\pi} \frac{\mathrm{e}^{-\mathrm{i}nx}}{1 - r\mathrm{e}^{\mathrm{i}x}}\mathrm{d}x, \quad n = 0, 1, 2, \ldots,$$

and

$$0=\frac{1}{2\pi}\int_{-\pi}^{\pi}\frac{e^{inx}}{1-re^{ix}}dx,\quad n=1,2,\ldots.$$

From the representation (3.4), we may conclude also that

$$\frac{1-r\cos x}{1-2r\cos x+r^2}=\mathrm{Re}\left(\frac{1}{1-re^{ix}}\right)=\sum_{n=0}^{\infty}r^n\cos nx=\frac{1}{2}+\frac{1}{2}\sum_{n=-\infty}^{\infty}r^{|n|}e^{inx} \tag{3.5}$$

and

$$\frac{r\sin x}{1-2r\cos x+r^2}=\mathrm{Im}\left(\frac{1}{1-re^{ix}}\right)=\sum_{n=1}^{\infty}r^n\sin nx=-\frac{i}{2}\sum_{n=-\infty}^{\infty}r^{|n|}\operatorname{sgn}(n)e^{inx}. \tag{3.6}$$

Exercise 3.1. Verify formulas (3.5) and (3.6).

Formulas (3.5) and (3.6) can be rewritten as

$$\sum_{n=-\infty}^{\infty}r^{|n|}e^{inx}=\frac{1-r^2}{1-2r\cos x+r^2}=:P_r(x) \tag{3.7}$$

and

$$-i\sum_{n=-\infty}^{\infty}\operatorname{sgn}(n)r^{|n|}e^{inx}=\frac{2r\sin x}{1-2r\cos x+r^2}=:Q_r(x). \tag{3.8}$$

Definition 3.2. The function $P_r(x)$ is called the *Poisson kernel*, while $Q_r(x)$ is called the *conjugate Poisson kernel*.

Since the series (3.7) and (3.8) converge absolutely, we have

$$r^{|n|}=\frac{1}{2\pi}\int_{-\pi}^{\pi}P_r(x)e^{-inx}dx\quad\text{and}\quad -i\operatorname{sgn}(n)r^{|n|}=\frac{1}{2\pi}\int_{-\pi}^{\pi}Q_r(x)e^{-inx}dx,$$

where $n=0,\pm1,\pm2,\ldots$. In particular,

$$\frac{1}{2\pi}\int_{-\pi}^{\pi}P_r(x)dx=1\quad\text{and}\quad\frac{1}{2\pi}\int_{-\pi}^{\pi}Q_r(x)dx=0.$$

Exercise 3.2. Prove that both $P_r(x)$ and $Q_r(x)$ are solutions of the *Laplace equation*

$$u_{x_1x_1}+u_{x_2x_2}=0$$

in the disk $x_1^2+x_2^2<1$, where $x_1+ix_2=re^{ix}$ with $0\le r<1$ and $x\in[-\pi,\pi]$.

Given a sequence $a_n, n = 0, \pm 1, \pm 2, \ldots$, we define

$$\Delta_n a = a_n - a_{n-1}.$$

Then for any two sequences a_n and b_n and integers $M < N$, the formula

$$\sum_{k=M+1}^{N} a_k \Delta_k b = a_N b_N - a_M b_M - \sum_{k=M+1}^{N} b_{k-1} \Delta_k a \tag{3.9}$$

holds. Formula (3.9) is called *summation by parts*.

Exercise 3.3. Prove (3.9).

Summation by parts allows us to investigate the convergence of a special type of trigonometric series.

Theorem 3.3. *Suppose that* $c_n > 0$*,* $n = 0, 1, 2, \ldots$*,* $c_n \geq c_{n+1}$*, and* $\lim_{n\to\infty} c_n = 0$*. Then the trigonometric series*

$$\sum_{n=0}^{\infty} c_n \mathrm{e}^{inx} \tag{3.10}$$

converges for every $x \in [-\pi, \pi] \setminus \{0\}$*.*

Proof. Let $b_n = \sum_{k=0}^{n} \mathrm{e}^{ikx}$. Since

$$b_n = \frac{1 - \mathrm{e}^{ix(n+1)}}{1 - \mathrm{e}^{ix}}, \quad x \neq 0,$$

it follows that

$$|b_n| \leq \frac{2}{|1 - \mathrm{e}^{ix}|} = \frac{1}{|\sin \frac{x}{2}|}$$

for $x \in [-\pi, \pi] \setminus \{0\}$. Applying (3.9) shows that for $M < N$ we have

$$\begin{aligned}\sum_{k=M+1}^{N} c_k \mathrm{e}^{ikx} = \sum_{k=M+1}^{N} c_k \left(\sum_{l=0}^{k} \mathrm{e}^{ilx} - \sum_{l=0}^{k-1} \mathrm{e}^{ilx} \right) &= \sum_{k=M+1}^{N} c_k \Delta_k b \\ &= c_N b_N - c_M b_M - \sum_{k=M+1}^{N} b_{k-1} \Delta_k c.\end{aligned}$$

Thus

$$\left|\sum_{k=M+1}^{N} c_k e^{ikx}\right| \le c_N|b_N| + c_M|b_M| + \sum_{k=M+1}^{N} |b_{k-1}||\Delta_k c|$$
$$\le \frac{1}{|\sin\frac{x}{2}|}\left(c_N + c_M + \sum_{k=M+1}^{N} |c_k - c_{k-1}|\right)$$
$$= \frac{1}{|\sin\frac{x}{2}|}(c_N + c_M + c_M - c_N) = \frac{2c_M}{|\sin\frac{x}{2}|} \to 0$$

as $N > M \to \infty$. This proves the theorem. □

Corollary 3.4. *Under the same assumptions as in Theorem 3.3, the trigonometric series* (3.10) *converges uniformly for all* $\pi \ge |x| \ge \delta > 0$.

Proof. If $\pi \ge |x| \ge \delta > 0$, then $\left|\sin\frac{x}{2}\right| \ge \frac{2}{\pi}\frac{|x|}{2} \ge \frac{\delta}{\pi}$. □

Theorem 3.3 implies that, for example, the series

$$\sum_{n=1}^{\infty} \frac{\cos nx}{\log(2+n)} \quad \text{and} \quad \sum_{n=1}^{\infty} \frac{\sin nx}{\log(2+n)}$$

converge for all $x \in [-\pi,\pi] \setminus \{0\}$.

Modulus of continuity and tail sum

For the trigonometric series (2.10) with

$$\sum_{n=-\infty}^{\infty} |c_n| < \infty$$

we introduce the *tail sum* by

$$E_n := \sum_{|k|>n} |c_k|, \quad n = 0,1,2,\ldots. \tag{3.11}$$

There is a good connection between the modulus of continuity (1.7) and (3.11). Indeed, if $f(x)$ denotes the series (2.10) and $h > 0$, we have

$$|f(x+h) - f(x)| \le \sum_{n=-\infty}^{\infty} |c_n||e^{inh} - 1| = \sum_{|n|h\le 1} |c_n||e^{inh} - 1| + \sum_{|n|h>1} |c_n||e^{inh} - 1|$$
$$\le h \sum_{|n|\le[1/h]} |n||c_n| + 2 \sum_{|n|>[1/h]} |c_n| =: I_1 + I_2,$$

where $[x]$ denotes the entire part of x. If we denote $[1/h]$ by N_h, then $I_2 = 2E_{N_h}$ and

$$I_1 = h \sum_{|n|\le N_h} |n|\left(E_{|n|-1} - E_{|n|}\right) = -h\sum_{l=1}^{N_h} l\,(E_l - E_{l-1}).$$

Using (3.9) in the latter sum, we obtain

$$I_1 = -h\left(N_h E_{N_h} - 0 \cdot E_0 - \sum_{n=1}^{N_h} E_{n-1}(n-(n-1))\right) = -hN_h E_{N_h} + h\sum_{n=1}^{N_h} E_{n-1}.$$

Since $hN_h = h[1/h] \leq h \cdot \frac{1}{h} = 1$, these formulas for I_1 and I_2 imply that

$$\omega_h(f) \leq 2E_{N_h} - hN_h E_{N_h} + h\sum_{n=1}^{N_h} E_{n-1} \leq 2E_{N_h} + \frac{1}{N_h}\sum_{n=0}^{N_h-1} E_n. \tag{3.12}$$

Since $E_n \to 0$ as $n \to \infty$, the inequality (3.12) implies that $\omega_h(f) \to 0$ as $h \to 0$. Moreover, if $E_n = O(n^{-\alpha})$, $n \neq 0$ for some $0 < \alpha \leq 1$, then

$$\omega_h(f) = \begin{cases} O(h^\alpha), & 0 < \alpha < 1, \\ O(h \log \frac{1}{h}), & \alpha = 1. \end{cases} \tag{3.13}$$

Here and throughout, the notation $A = O(B)$ on a set X means that $|A| \leq C|B|$ on X with some constant $C > 0$. Similarly, $A = o(B)$ means that $A/B \to 0$.

Exercise 3.4. Prove the second relation in (3.13).

We summarize (3.13) as follows: if the tail of the trigonometric series (2.10) behaves as $O(n^{-\alpha})$ for some $0 < \alpha < 1$, then the function f to which it converges belongs to the Hölder space $C^\alpha[-\pi, \pi]$.

Chapter 4
Convolution and Parseval's Equality

Let the trigonometric series (2.10) be such that

$$\sum_{n=-\infty}^{\infty} |c_n| < \infty.$$

Then the function f to which it converges is continuous and periodic. If $g(x)$ is any continuous function, then the product fg is also continuous and hence integrable on $[-\pi, \pi]$ and

$$\frac{1}{2\pi}\int_{-\pi}^{\pi} f(x)g(x)\mathrm{d}x = \frac{1}{2\pi}\sum_{n=-\infty}^{\infty} c_n(f)\int_{-\pi}^{\pi} g(x)\mathrm{e}^{inx}\mathrm{d}x = \sum_{n=-\infty}^{\infty} c_n(f)c_{-n}(g), \quad (4.1)$$

where integration of the series term by term is justified by the uniform convergence of the Fourier series. Putting $g = \overline{f}$ in (4.1) yields

$$\frac{1}{2\pi}\int_{-\pi}^{\pi} |f(x)|^2\mathrm{d}x = \sum_{n=-\infty}^{\infty} c_n(f)c_{-n}(\overline{f}) = \sum_{n=-\infty}^{\infty} |c_n(f)|^2 \quad (4.2)$$

by (2.13).

Definition 4.1. Equality (4.2) is called the *Parseval's equality* for the trigonometric Fourier series.

The formula (4.1) can be generalized as follows.

Exercise 4.1. Let a periodic function f be defined by the absolutely convergent Fourier series (2.10) and let g be integrable and periodic. Prove that

$$\frac{1}{2\pi}\int_{-\pi}^{\pi} f(x)g(y-x)\mathrm{d}x = \sum_{n=-\infty}^{\infty} c_n(f)c_n(g)\mathrm{e}^{iny}$$

V. Serov, *Fourier Series, Fourier Transform and Their Applications to Mathematical Physics*, Applied Mathematical Sciences 197,
DOI 10.1007/978-3-319-65262-7_4

and that this series converges absolutely.

A generalization of Exercise 4.1 is given by the following theorem.

Theorem 4.2. *If f_1 and f_2 are two periodic L^1 functions, then*

$$c_n(f_1 * f_2) = c_n(f_1)c_n(f_2),$$

where $f_1 * f_2$ denotes the convolution

$$(f_1 * f_2)(x) = \frac{1}{2\pi}\int_{-\pi}^{\pi} f_1(y)f_2(x-y)\mathrm{d}y \tag{4.3}$$

and where the integral converges for almost every x.

Proof. We note first that by Fubini's theorem the convolution (4.3) is well defined as an L^1 function. Indeed,

$$\begin{aligned}\int_{-\pi}^{\pi}\left(\int_{-\pi}^{\pi}|f_1(y)|\cdot|f_2(x-y)|\mathrm{d}y\right)\mathrm{d}x &= \int_{-\pi}^{\pi}|f_1(y)|\left(\int_{-\pi-y}^{\pi-y}|f_2(z)|\mathrm{d}z\right)\mathrm{d}y \\ &= \int_{-\pi}^{\pi}|f_1(y)|\left(\int_{-\pi}^{\pi}|f_2(z)|\mathrm{d}z\right)\mathrm{d}y\end{aligned}$$

by Lemma 1.3. The Fourier coefficients of the convolution (4.3) are equal to

$$\begin{aligned}c_n(f_1 * f_2) &= \frac{1}{2\pi}\int_{-\pi}^{\pi}(f_1 * f_2)(x)\mathrm{e}^{-inx}\mathrm{d}x \\ &= \frac{1}{(2\pi)^2}\int_{-\pi}^{\pi}\left(\int_{-\pi}^{\pi} f_1(y)f_2(x-y)\mathrm{d}y\right)\mathrm{e}^{-inx}\mathrm{d}x \\ &= \frac{1}{(2\pi)^2}\int_{-\pi}^{\pi} f_1(y)\left(\int_{-\pi}^{\pi} f_2(x-y)\mathrm{e}^{-inx}\mathrm{d}x\right)\mathrm{d}y \\ &= \frac{1}{(2\pi)^2}\int_{-\pi}^{\pi} f_1(y)\left(\int_{-\pi-y}^{\pi-y} f_2(z)\mathrm{e}^{-in(y+z)}\mathrm{d}z\right)\mathrm{d}y \\ &= \frac{1}{(2\pi)^2}\int_{-\pi}^{\pi} f_1(y)\mathrm{e}^{-iny}\left(\int_{-\pi}^{\pi} f_2(z)\mathrm{e}^{-inz}\mathrm{d}z\right)\mathrm{d}y = c_n(f_1)c_n(f_2)\end{aligned}$$

by Lemma 1.3. Thus, the theorem is proved. □

Exercise 4.2. Prove that if f_1 and f_2 are integrable and periodic, then their convolution is symmetric and periodic.

Exercise 4.3. Let f be a periodic L^1 function. Prove that

$$(f * P_r)(x) = (P_r * f)(x) = \sum_{n=-\infty}^{\infty} r^{|n|}c_n(f)\mathrm{e}^{inx}$$

and that $P_r * f$ satisfies the Laplace equation, i.e., $(P_r * f)_{x_1x_1} + (P_r * f)_{x_2x_2} = 0$, where $x_1^2 + x_2^2 = r^2 < 1$ with $x_1 + \mathrm{i}x_2 = re^{\mathrm{i}x}$.

Remark 4.3. We are going to prove in Chapter 10 that for every periodic continuous function f, the limit $\lim_{r\to 1-}(f * P_r)(x) = f(x)$ exists uniformly in x.

Chapter 5
Fejér Means of Fourier Series. Uniqueness of the Fourier Series.

Let us denote the *partial sum of the Fourier series* of $f \in L^1(-\pi,\pi)$ (not necessarily periodic) by

$$S_N(f) := \sum_{|n|\le N} c_n(f)\mathrm{e}^{\mathrm{i}nx}$$

for each $N = 0,1,2,\ldots$. The *Fejér means* are defined by

$$\sigma_N(f) := \frac{S_0(f)+\cdots+S_N(f)}{N+1}.$$

Writing this out in detail, we see that

$$\begin{aligned}(N+1)\sigma_N(f) &= \sum_{n=0}^{N}\sum_{|k|\le n} c_k(f)\mathrm{e}^{\mathrm{i}kx} = \sum_{|k|\le N}\sum_{n=|k|}^{N} c_k(f)\mathrm{e}^{\mathrm{i}kx}\\ &= \sum_{|k|\le N}(N+1-|k|)c_k(f)\mathrm{e}^{\mathrm{i}kx},\end{aligned}$$

which gives the useful representation

$$\sigma_N(f) = \sum_{|k|\le N}\left(1-\frac{|k|}{N+1}\right)c_k(f)\mathrm{e}^{\mathrm{i}kx}. \tag{5.1}$$

The *Fejér kernel* is

$$K_N(x) := \sum_{|k|\le N}\left(1-\frac{|k|}{N+1}\right)\mathrm{e}^{\mathrm{i}kx}. \tag{5.2}$$

V. Serov, *Fourier Series, Fourier Transform and Their Applications to Mathematical Physics*, Applied Mathematical Sciences 197,
DOI 10.1007/978-3-319-65262-7_5

The sum (5.2) can be calculated precisely as

$$K_N(x) = \frac{1}{N+1}\left(\frac{\sin\frac{N+1}{2}x}{\sin\frac{x}{2}}\right)^2. \tag{5.3}$$

Exercise 5.1. Prove the identity (5.3).

Exercise 5.2. Prove that

$$\frac{1}{2\pi}\int_{-\pi}^{\pi} K_N(x)\mathrm{d}x = 1. \tag{5.4}$$

We can rewrite $\sigma_N(f)$ from (5.1) also as

$$\sigma_N(f) = \sum_{k=-\infty}^{\infty} 1_{[-N,N]}(k)\left(1-\frac{|k|}{N+1}\right)c_k(f)\mathrm{e}^{\mathrm{i}kx},$$

where

$$1_{[-N,N]}(k) = \begin{cases} 1, & |k| \le N, \\ 0, & |k| > N. \end{cases}$$

Let us assume now that f is periodic. Then Exercise 4.1 and Theorem 4.2 lead to

$$\frac{1}{2\pi}\int_{-\pi}^{\pi} f(y)K_N(x-y)\mathrm{d}y = \sum_{k=-\infty}^{\infty} 1_{[-N,N]}(k)\left(1-\frac{|k|}{N+1}\right)c_k(f)\mathrm{e}^{\mathrm{i}kx}. \tag{5.5}$$

Exercise 5.3. Prove (5.5).

Hence, the Fejér means can be represented as

$$\sigma_N(f)(x) = (f * K_N)(x) = (K_N * f)(x). \tag{5.6}$$

The properties (5.3), (5.4), and (5.6) allow us to prove the following result.

Theorem 5.1. *Let $f \in L^p(-\pi,\pi)$ be periodic with $1 \le p < \infty$. Then*

$$\lim_{N\to\infty}\left(\int_{-\pi}^{\pi}|\sigma_N(f)(x)-f(x)|^p\,\mathrm{d}x\right)^{1/p} = 0. \tag{5.7}$$

If, in addition, f has right and left limits $f(x_0 \pm 0)$ at a point $x_0 \in [-\pi,\pi]$, then

$$\lim_{N\to\infty}\sigma_N(f)(x_0) = \frac{1}{2}\left(f(x_0+0)+f(x_0-0)\right). \tag{5.8}$$

Proof. Let us first prove (5.7). Indeed, (5.4) and (5.6) give

$$\left(\int_{-\pi}^{\pi}|\sigma_N(f)(x)-f(x)|^p\mathrm{d}x\right)^{1/p}=\left(\int_{-\pi}^{\pi}|(f*K_N)(x)-f(x)|^p\mathrm{d}x\right)^{1/p}$$
$$=\left(\int_{-\pi}^{\pi}\left|\frac{1}{2\pi}\int_{-\pi}^{\pi}K_N(y)f(x-y)\mathrm{d}y-\frac{1}{2\pi}\int_{-\pi}^{\pi}K_N(y)f(x)\mathrm{d}y\right|^p\mathrm{d}x\right)^{1/p}$$
$$=\frac{1}{2\pi}\left(\int_{-\pi}^{\pi}\left|\int_{-\pi}^{\pi}K_N(y)(f(x-y)-f(x))\mathrm{d}y\right|^p\mathrm{d}x\right)^{1/p}$$
$$\leq\frac{1}{2\pi}\left(\int_{-\pi}^{\pi}\left|\int_{|y|<\delta}K_N(y)(f(x-y)-f(x))\mathrm{d}y\right|^p\mathrm{d}x\right)^{1/p}$$
$$+\frac{1}{2\pi}\left(\int_{-\pi}^{\pi}\left|\int_{\pi\geq|y|>\delta}K_N(y)(f(x-y)-f(x))\mathrm{d}y\right|^p\mathrm{d}x\right)^{1/p}=:I_1+I_2.$$

Using the generalized Minkowski's inequality, we obtain that

$$I_1\leq\frac{1}{2\pi}\int_{|y|<\delta}K_N(y)\left(\int_{-\pi}^{\pi}|f(x-y)-f(x)|^p\mathrm{d}x\right)^{1/p}\mathrm{d}y$$
$$\leq\sup_{|y|<\delta}\left(\int_{-\pi}^{\pi}|f(x-y)-f(x)|^p\mathrm{d}x\right)^{1/p}\frac{1}{2\pi}\int_{-\pi}^{\pi}K_N(y)\mathrm{d}y$$
$$=\sup_{|y|<\delta}\left(\int_{-\pi}^{\pi}|f(x-y)-f(x)|^p\mathrm{d}x\right)^{1/p}\to 0 \tag{5.9}$$

as $\delta\to 0$, since $f\in L^p(-\pi,\pi)$. Quite similarly,

$$I_2\leq\frac{1}{2\pi}\int_{\pi\geq|y|>\delta}K_N(y)\left(\int_{-\pi}^{\pi}|f(x-y)-f(x)|^p\mathrm{d}x\right)^{1/p}\mathrm{d}y$$
$$\leq 2\left(\int_{-\pi}^{\pi}|f(x)|^p\mathrm{d}x\right)^{1/p}\frac{1}{2\pi}\int_{\pi\geq|y|>\delta}K_N(y)\mathrm{d}y, \tag{5.10}$$

since f is periodic. The next step is to note that

$$\sin^2\frac{y}{2}\geq\left(\frac{2}{\pi}\frac{|y|}{2}\right)^2\geq\left(\frac{\delta}{\pi}\right)^2$$

for $\pi\geq|y|>\delta$. That is why (5.3) leads to

$$K_N(y)\leq\frac{1}{N+1}\cdot\frac{1}{\sin^2\frac{y}{2}}\leq\frac{1}{N+1}\cdot\frac{\pi^2}{\delta^2}$$

and

$$\frac{1}{2\pi}\int_{\pi\geq|y|>\delta} K_N(y)\mathrm{d}y \leq \frac{\pi^2}{2\pi\delta^2}\cdot\frac{2(\pi-\delta)}{N+1} < \frac{\pi^2}{\delta^2(N+1)} < \frac{\pi^2}{\sqrt{N}} \to 0$$

as $N\to\infty$ if we choose $\delta = N^{-1/4}$. From (5.9) and (5.10) we may conclude (5.7). In order to prove (5.8), we use (5.4) to consider the difference

$$\begin{aligned}
&\sigma_N(f)(x_0) - \frac{1}{2}(f(x_0+0)+f(x_0-0))\\
&= \frac{1}{2\pi}\int_{-\pi}^{\pi} K_N(y)\left[f(x_0-y) - \frac{1}{2}(f(x_0+0)+f(x_0-0))\right]\mathrm{d}y\\
&= \frac{1}{2\pi}\int_{-\pi}^{\pi} K_N(y)g(y)\mathrm{d}y, \qquad (5.11)
\end{aligned}$$

where

$$g(y) = f(x_0-y) - \frac{1}{2}(f(x_0+0)+f(x_0-0)).$$

Since the Fejér kernel $K_N(y)$ is even, we can rewrite the right-hand side of (5.11) as

$$\frac{1}{2\pi}\int_0^{\pi} K_N(y)h(y)\mathrm{d}y,$$

where $h(y) = g(y)+g(-y)$. It is clear that $h(y)$ is an L^1 function. But we have more, namely,

$$\lim_{y\to 0+} h(y) = 0. \qquad (5.12)$$

Our task now is to prove that

$$\lim_{N\to\infty}\frac{1}{2\pi}\int_0^{\pi} K_N(y)h(y)\mathrm{d}y = 0.$$

We will proceed as in the proof of (5.7), i.e., we split the integral as

$$\frac{1}{2\pi}\int_0^{\pi} K_N(y)h(y)\mathrm{d}y = \frac{1}{2\pi}\int_0^{\delta} K_N(y)h(y)\mathrm{d}y + \frac{1}{2\pi}\int_{\delta}^{\pi} K_N(y)h(y)\mathrm{d}y =: I_1 + I_2.$$

The first term can be estimated as

$$|I_1| \leq \frac{1}{2\pi}\sup_{0\leq y\leq\delta}|h(y)|\int_0^{\pi} K_N(y)\mathrm{d}y = \frac{1}{2}\sup_{|y|\leq\delta}|h(y)| \to 0$$

as $\delta\to 0+$ due to (5.12). For I_2 we have

$$|I_2| \leq \frac{1}{2\pi}\left(\frac{\pi}{\delta}\right)^2 \frac{1}{N+1}\int_0^{\pi} |h(y)|\mathrm{d}y \to 0$$

as $N \to \infty$ if we choose, for example, $\delta = N^{-1/4}$. Thus, the theorem is completely proved. □

Corollary 5.2. *If f is periodic and continuous on the interval $[-\pi,\pi]$, then*

$$\lim_{N\to\infty} \sigma_N(f)(x) = f(x)$$

uniformly in $x \in [-\pi,\pi]$.

Corollary 5.3. *Every periodic L^p function, $1 \leq p < \infty$, can be approximated in the sense of L^p space by the trigonometric polynomials $\sum_{|k|\leq N} b_k \mathrm{e}^{\mathrm{i}kx}$ (which are infinitely differentiable, i.e., C^∞ functions).*

Theorem 5.4 (Uniqueness of Fourier series). *If $f \in L^1(-\pi,\pi)$ is periodic and if its Fourier coefficients are identically zero, then $f = 0$ almost everywhere.*

Proof. Since

$$\lim_{N\to\infty} \int_{-\pi}^{\pi} |\sigma_N(f)(x) - f(x)|\,\mathrm{d}x = 0$$

by (5.7), it follows that if all Fourier coefficients are zero, we have

$$\int_{-\pi}^{\pi} |f(x)|\,\mathrm{d}x = 0,$$

which means that $f = 0$ almost everywhere. This proves the theorem. □

Chapter 6
The Riemann–Lebesgue Lemma

Theorem 6.1 (Riemann–Lebesgue lemma). *If f is periodic with period 2π and belongs to $L^1(-\pi,\pi)$, then*

$$\lim_{n\to\infty}\int_{-\pi}^{\pi} f(x+z)\mathrm{e}^{-\mathrm{i}nz}\mathrm{d}z = 0 \tag{6.1}$$

uniformly in $x\in\mathbb{R}$. In particular, $c_n(f)\to 0$ as $n\to\infty$.

Proof. Since f is periodic with period 2π, it follows that

$$\int_{-\pi}^{\pi} f(x+z)\mathrm{e}^{-\mathrm{i}nz}\mathrm{d}z = \int_{-\pi+x}^{\pi+x} f(y)\mathrm{e}^{-\mathrm{i}n(y-x)}\mathrm{d}y = \mathrm{e}^{\mathrm{i}nx}\int_{-\pi}^{\pi} f(y)\mathrm{e}^{-\mathrm{i}ny}\mathrm{d}y \tag{6.2}$$

by Lemma 1.3. Formula (6.2) shows that to prove (6.1) it is enough to show that the Fourier coefficients $c_n(f)$ tend to zero as $n\to\infty$. Indeed,

$$2\pi c_n(f) = \int_{-\pi}^{\pi} f(y)\mathrm{e}^{-\mathrm{i}ny}\mathrm{d}y = \int_{-\pi+\pi/n}^{\pi+\pi/n} f(y)\mathrm{e}^{-\mathrm{i}ny}\mathrm{d}y = \int_{-\pi}^{\pi} f(t+\pi/n)\mathrm{e}^{-\mathrm{i}nt}\mathrm{e}^{-\mathrm{i}\pi}\mathrm{d}t$$

by Lemma 1.3. Hence

$$-4\pi c_n(f) = \int_{-\pi}^{\pi}\left(f(t+\pi/n)-f(t)\right)\mathrm{e}^{-\mathrm{i}nt}\mathrm{d}t. \tag{6.3}$$

If f is continuous on the interval $[-\pi,\pi]$, then

$$\sup_{t\in[-\pi,\pi]}|f(t+\pi/n)-f(t)|\to 0,\quad n\to\infty.$$

Hence $c_n(f)\to 0$ as $n\to\infty$. If f is an arbitrary L^1 function, we let $\varepsilon>0$. Then we can define a continuous function g (see Corollary 5.3) such that

V. Serov, *Fourier Series, Fourier Transform and Their Applications to Mathematical Physics*, Applied Mathematical Sciences 197,
DOI 10.1007/978-3-319-65262-7_6

$$\int_{-\pi}^{\pi} |f(x) - g(x)|\,\mathrm{d}x < \varepsilon.$$

Write

$$c_n(f) = c_n(g) + c_n(f-g).$$

The first term tends to zero as $n \to \infty$, since g is continuous, whereas the second term is less than $\varepsilon/(2\pi)$. This implies that

$$\sup_{n\to\infty}\lim |c_n(f)| \le \frac{\varepsilon}{2\pi}.$$

Since ε is arbitrary, we have

$$\lim_{n\to\infty} |c_n(f)| = 0.$$

This fact together with (6.2) gives (6.1). The theorem is thus proved. □

Corollary 6.2. *Let f be as in Theorem 6.1. If a periodic function g is continuous on $[-\pi,\pi]$, then*

$$\lim_{n\to\infty} \int_{-\pi}^{\pi} f(x+z)g(z)\mathrm{e}^{-\mathrm{i}nz}\mathrm{d}z = 0$$

and

$$\lim_{n\to\infty} \int_{-\pi}^{\pi} f(x+z)g(z)\sin(nz)\mathrm{d}z = \lim_{n\to\infty} \int_{-\pi}^{\pi} f(x+z)g(z)\cos(nz)\mathrm{d}z = 0$$

uniformly in $x \in [-\pi,\pi]$.

Exercise 6.1. Prove this corollary.

Exercise 6.2. Show that if f satisfies the Hölder condition with exponent $\alpha \in (0,1]$, then $c_n(f) = O(|n|^{-\alpha})$ as $n \to \infty$.

Exercise 6.3. Suppose that f satisfies the Hölder condition with exponent $\alpha > 1$. Prove that $f \equiv$ constant.

Exercise 6.4. Let $f(x) = |x|^\alpha$, where $-\pi \le x \le \pi$ and $0 < \alpha < 1$. Prove that $c_n(f) \asymp |n|^{-1-\alpha}$ as $n \to \infty$.

Remark 6.3. The notation $a \asymp b$ means that there exist $0 < c_1 < c_2$ such that

$$c_1|a| < |b| < c_2|a|.$$

Let us introduce for all $1 \le p < \infty$ and periodic functions $f \in L^p(-\pi,\pi)$ the L^p-*modulus of continuity* of f by

$$\omega_{p,\delta}(f) := \sup_{|h|\le\delta} \left(\int_{-\pi}^{\pi} |f(x+h) - f(x)|^p \,\mathrm{d}x \right)^{1/p}.$$

The equality (6.3) leads to

$$|c_n(f)| \leq \frac{1}{4\pi}\int_{-\pi}^{\pi} |f(x+\pi/n) - f(x)|\,\mathrm{d}x$$
$$\leq \frac{(2\pi)^{1-1/p}}{4\pi}\left(\int_{-\pi}^{\pi} |f(x+\pi/n) - f(x)|^p\,\mathrm{d}x\right)^{1/p} \leq \frac{1}{2}(2\pi)^{-1/p}\omega_{p,\pi/n}(f),$$

where we have used Hölder's inequality in the penultimate step.

Exercise 6.5. Suppose that $\omega_{p,\delta}(f) \leq C\delta^\alpha$ for some $C > 0$ and $\alpha > 1$. Prove that f is constant almost everywhere.

Hint. First show that $\omega_{p,2\delta}(f) \leq 2\omega_{p,\delta}(f)$; then iterate this to obtain a contradiction.

Suppose that $f \in L^1(-\pi,\pi)$ but f is not necessarily periodic. We can consider the Fourier series corresponding to f, i.e.,

$$f(x) \sim \sum_{n=-\infty}^{\infty} c_n \mathrm{e}^{\mathrm{i}nx},$$

where the c_n are the Fourier coefficients $c_n(f)$. The series on the right-hand side is considered formally in the sense that we know nothing about its convergence. However, the limit

$$\lim_{N\to\infty}\int_{-\pi}^{\pi}\sum_{|n|\leq N} c_n(f)\mathrm{e}^{\mathrm{i}nx}\mathrm{d}x = \int_{-\pi}^{\pi} f(x)\mathrm{d}x \tag{6.4}$$

exists. Indeed,

$$\int_{-\pi}^{\pi}\sum_{|n|\leq N} c_n(f)\mathrm{e}^{\mathrm{i}nx}\mathrm{d}x = c_0(f)\int_{-\pi}^{\pi}\mathrm{d}x + \sum_{0<|n|\leq N} c_n(f)\int_{-\pi}^{\pi}\mathrm{e}^{\mathrm{i}nx}\mathrm{d}x = 2\pi c_0(f)$$
$$= \int_{-\pi}^{\pi} f(x)\mathrm{d}x.$$

Remark 6.4 (Important properties of the Fourier series). The existence of the limit (6.4) shows us that we can always integrate the Fourier series of an L^1 function term by term.

$$\int_{-\pi}^{x}\sum_{|n|\le N} c_n(f)e^{int}dt = c_0(f)\int_{-\pi}^{x}dt + \sum_{0<|n|\le N} c_n(f)\int_{-\pi}^{x} e^{int}dt \to 2\pi c_0(f) \ldots = \int_{-\pi}^{x} f(t)dt.$$

Chapter 7
The Fourier Series of a Square-Integrable Function. The Riesz–Fischer Theorem.

The set of *square-integrable functions* $L^2(-\pi,\pi)$ is an inner product space (linear Euclidean space) equipped with the inner product

$$(f,g)_{L^2(-\pi,\pi)} = \frac{1}{2\pi}\int_{-\pi}^{\pi} f(x)\overline{g(x)}\mathrm{d}x.$$

We can measure the degree of approximation by the (square of) *mean square distance*

$$\frac{1}{2\pi}\int_{-\pi}^{\pi}|f(x)-g(x)|^2\mathrm{d}x = (f-g,f-g)_{L^2(-\pi,\pi)}.$$

In particular, if $g(x)=\sum_{|n|\le N} b_n \mathrm{e}^{\mathrm{i}nx}$ is a trigonometric polynomial, then this distance can be written as

$$\frac{1}{2\pi}\int_{-\pi}^{\pi}|f(x)|^2\mathrm{d}x + \frac{1}{2\pi}\int_{-\pi}^{\pi}|g(x)|^2\mathrm{d}x - \frac{1}{2\pi}2\mathrm{Re}\int_{-\pi}^{\pi} f(x)\overline{g(x)}\mathrm{d}x,$$

or

$$\begin{aligned}
&\frac{1}{2\pi}\int_{-\pi}^{\pi}|f(x)|^2\mathrm{d}x \\
&\quad+\frac{1}{2\pi}\int_{-\pi}^{\pi}\sum_{|n|\le N} b_n\mathrm{e}^{\mathrm{i}nx}\sum_{|k|\le N}\overline{b_k}\mathrm{e}^{-\mathrm{i}kx}\mathrm{d}x - \frac{1}{2\pi}2\mathrm{Re}\int_{-\pi}^{\pi} f(x)\sum_{|n|\le N}\overline{b_n}\mathrm{e}^{-\mathrm{i}nx}\mathrm{d}x \\
&\quad=\frac{1}{2\pi}\int_{-\pi}^{\pi}|f(x)|^2\mathrm{d}x + \frac{1}{2\pi}\sum_{|n|\le N}|b_n|^2\int_{-\pi}^{\pi}\mathrm{d}x - 2\mathrm{Re}\sum_{|n|\le N}\overline{b_n}c_n(f) \\
&=\frac{1}{2\pi}\int_{-\pi}^{\pi}|f(x)|^2\mathrm{d}x + \sum_{|n|\le N}|b_n|^2 - 2\mathrm{Re}\sum_{|n|\le N}\overline{b_n}c_n(f) + \sum_{|n|\le N}|c_n(f)|^2 - \sum_{|n|\le N}|c_n(f)|^2.
\end{aligned}$$

V. Serov, *Fourier Series, Fourier Transform and Their Applications to Mathematical Physics*, Applied Mathematical Sciences 197,
DOI 10.1007/978-3-319-65262-7_7

So

$$\frac{1}{2\pi}\int_{-\pi}^{\pi}|f(x)-g(x)|^2\mathrm{d}x=\frac{1}{2\pi}\int_{-\pi}^{\pi}|f(x)|^2\mathrm{d}x-\sum_{|n|\le N}|c_n(f)|^2+\sum_{|n|\le N}|b_n-c_n(f)|^2.$$

This equality has the following consequences:

(1) The minimum error is

$$\min_{g(x)=\sum_{|n|\le N}b_n\mathrm{e}^{inx}}\frac{1}{2\pi}\int_{-\pi}^{\pi}|f(x)-g(x)|^2\mathrm{d}x$$
$$=\frac{1}{2\pi}\int_{-\pi}^{\pi}|f(x)|^2\mathrm{d}x-\sum_{|n|\le N}|c_n(f)|^2, \quad (7.1)$$

and it is attained when $b_n=c_n(f)$.

(2) For $N=1,2,\ldots$, it is true that

$$\sum_{|n|\le N}|c_n(f)|^2\le\frac{1}{2\pi}\int_{-\pi}^{\pi}|f(x)|^2\mathrm{d}x,$$

and in particular,

$$\sum_{n=-\infty}^{\infty}|c_n(f)|^2\le\frac{1}{2\pi}\int_{-\pi}^{\pi}|f(x)|^2\mathrm{d}x. \quad (7.2)$$

This inequality is called *Bessel's inequality*.

It turns out that (7.2) holds with equality. This is *Parseval's equality* for $f\in L^2(-\pi,\pi)$, which we state as the following theorem.

Theorem 7.1. *For every periodic function $f\in L^2(-\pi,\pi)$ with period 2π, its Fourier series converges in $L^2(-\pi,\pi)$, i.e.,*

$$\lim_{N\to\infty}\frac{1}{2\pi}\int_{-\pi}^{\pi}\left|f(x)-\sum_{|n|\le N}c_n(f)\mathrm{e}^{inx}\right|^2\mathrm{d}x=0,$$

and Parseval's equality

$$\frac{1}{2\pi}\int_{-\pi}^{\pi}|f(x)|^2\mathrm{d}x=\sum_{n=-\infty}^{\infty}|c_n(f)|^2 \quad (7.3)$$

holds.

Proof. By Bessel's inequality (7.2), we have for every $f \in L^2(-\pi,\pi)$ that

$$\frac{1}{2\pi}\int_{-\pi}^{\pi}\left|\sum_{|n|\le N} c_n(f)\mathrm{e}^{\mathrm{i}nx} - \sum_{|n|\le M} c_n(f)\mathrm{e}^{\mathrm{i}nx}\right|^2 \mathrm{d}x = \sum_{M+1\le|n|\le N} |c_n(f)|^2 \to 0$$

as $N > M \to \infty$. Due to the completeness of the trigonometric polynomials in $L^2(-\pi,\pi)$ (see Corollary 5.3), we may now conclude that there exists $F \in L^2(-\pi,\pi)$ such that

$$\lim_{N\to\infty}\frac{1}{2\pi}\int_{-\pi}^{\pi}\left|F(x) - \sum_{|n|\le N} c_n(f)\mathrm{e}^{\mathrm{i}nx}\right|^2 \mathrm{d}x = 0.$$

It remains to show that $F(x) = f(x)$ almost everywhere. To do this, we compute the Fourier coefficients $c_n(F)$ by writing

$$2\pi c_n(F) = \int_{-\pi}^{\pi} F(x)\mathrm{e}^{-\mathrm{i}nx}\mathrm{d}x = \int_{-\pi}^{\pi}\left(F(x) - \sum_{|k|\le N} c_k(f)\mathrm{e}^{\mathrm{i}kx}\right)\mathrm{e}^{-\mathrm{i}nx}\mathrm{d}x$$
$$+ \sum_{|k|\le N} c_k(f)\int_{-\pi}^{\pi}\mathrm{e}^{\mathrm{i}(k-n)x}\mathrm{d}x.$$

If $N > |n|$, then the last sum is equal to $2\pi c_n(f)$. Thus, by Hölder's inequality,

$$2\pi|c_n(F) - c_n(f)| \le \int_{-\pi}^{\pi}\left|F(x) - \sum_{|k|\le N} c_k(f)\mathrm{e}^{\mathrm{i}kx}\right|\mathrm{d}x$$
$$\le \sqrt{2\pi}\left(\int_{-\pi}^{\pi}\left|F(x) - \sum_{|k|\le N} c_k(f)\mathrm{e}^{\mathrm{i}kx}\right|^2 \mathrm{d}x\right)^{1/2} \to 0$$

as $N \to \infty$, i.e., $c_n(F) = c_n(f)$ for all $n = 0, \pm 1, \pm 2, \ldots$. Theorem 5.4 (uniqueness of Fourier series) implies now that $F = f$ almost everywhere. Parseval's equality follows from (7.1) if we let $N \to \infty$. □

Corollary 7.2 (Riesz–Fischer theorem). *Suppose* $\{b_n\}_{n=-\infty}^{\infty}$ *is a sequence of complex numbers with* $\sum_{n=-\infty}^{\infty}|b_n|^2 < \infty$. *Then there is a unique periodic function* $f \in L^2(-\pi,\pi)$ *such that* $b_n = c_n(f)$.

Proof The proof is identical to that of Theorem 7.1. □

Theorem 7.3. *Suppose that* $f \in L^2(-\pi,\pi)$ *is periodic with period* 2π *and that its Fourier coefficients satisfy*

$$\sum_{n=-\infty}^{\infty} |n|^2|c_n(f)|^2 < \infty. \tag{7.4}$$

Then $f \in W_2^1(-\pi,\pi)$ with the Fourier series for $f'(x)$ given by

$$f'(x) \sim \sum_{n=-\infty}^{\infty} inc_n(f)\mathrm{e}^{inx}.$$

Proof. Since (7.4) holds, it follows that by the Riesz–Fischer theorem there is a unique function $g(x) \in L^2(-\pi,\pi)$ such that

$$g(x) \sim \sum_{n=-\infty}^{\infty} inc_n(f)\mathrm{e}^{inx}.$$

Integrating term by term, we obtain

$$\int_{-\pi}^{\pi} g(x)\mathrm{d}x = 0.$$

Let $F(x) := \int_{-\pi}^{x} g(t)\mathrm{d}t$. Then for $n \neq 0$, we have

$$\begin{aligned} c_n(F) &= \frac{1}{2\pi}\int_{-\pi}^{\pi}\left(\int_{-\pi}^{x} g(t)\mathrm{d}t\right)\mathrm{e}^{-inx}\mathrm{d}x = \frac{1}{2\pi}\int_{-\pi}^{\pi} g(t)\left(\int_{t}^{\pi}\mathrm{e}^{-inx}\mathrm{d}x\right)\mathrm{d}t \\ &= \frac{1}{2\pi}\int_{-\pi}^{\pi} g(t)\left(\frac{\mathrm{e}^{-in\pi}}{-in} + \frac{\mathrm{e}^{-int}}{in}\right)\mathrm{d}t = \frac{1}{in}\frac{1}{2\pi}\int_{-\pi}^{\pi} g(t)\mathrm{e}^{-int}\mathrm{d}t = \frac{1}{in}c_n(g). \end{aligned}$$

On the other hand, $c_n(g) = inc_n(f)$. Thus, by the uniqueness of Fourier series, we obtain that $F(x) - f(x) = \text{constant}$ almost everywhere, or $f'(x) = g(x)$ almost everywhere. This means that

$$f(x) = \int_{-\pi}^{x} g(t)\mathrm{d}t + \text{constant},$$

where $g \in L^2(-\pi,\pi)$. Therefore, $f \in W_2^1(-\pi,\pi)$ and

$$f'(x) = g(x) \sim \sum_{n=-\infty}^{\infty} inc_n(f)\mathrm{e}^{inx}.$$

This completes the proof. □

Corollary 7.4. *Under the conditions of Theorem 7.3, it is true that*

$$\sum_{n=-\infty}^{\infty} |c_n(f)| < \infty.$$

Proof. Due to (7.4) we have

$$\sum_{n=-\infty}^{\infty} |c_n(f)| = |c_0(f)| + \sum_{n\neq 0} |c_n(f)|$$
$$\leq |c_0(f)| + \frac{1}{2}\sum_{n\neq 0} |n|^2 |c_n(f)|^2 + \frac{1}{2}\sum_{n\neq 0} \frac{1}{|n|^2} < \infty,$$

where we have used the basic inequality $2ab \leq a^2 + b^2$ for real numbers a and b. □

Using Parseval's equality (7.3), we can obtain for every periodic function $f \in L^2(-\pi,\pi)$ and $N = 1,2,\ldots$ that

$$\frac{1}{2\pi}\int_{-\pi}^{\pi} \left| \sum_{|n|\leq N} c_n(f)\mathrm{e}^{inx} - f(x) \right|^2 \mathrm{d}x = \sum_{|n|>N} |c_n(f)|^2. \tag{7.5}$$

Exercise 7.1. Prove (7.5).

Using Parseval's equality again, we have

$$\begin{aligned}\frac{1}{2\pi}\int_{-\pi}^{\pi} |f(x+h) - f(x)|^2 \mathrm{d}x &= \sum_{n=-\infty}^{\infty} |c_n(f(x+h) - f(x))|^2 \\ &= \sum_{n=-\infty}^{\infty} |\mathrm{e}^{ihn} - 1|^2 |c_n(f)|^2.\end{aligned} \tag{7.6}$$

Theorem 7.5. *Suppose $f \in L^2(-\pi,\pi)$ is periodic with period 2π. Then*

$$\sum_{|n|>N} |c_n(f)|^2 = O(N^{-2\alpha}), \quad N = 1,2,\ldots, \tag{7.7}$$

with $0 < \alpha < 1$ if and only if

$$\sum_{n=-\infty}^{\infty} |\mathrm{e}^{inh} - 1|^2 |c_n(f)|^2 = O(|h|^{2\alpha}) \tag{7.8}$$

for $|h|$ sufficiently small.

Proof. From (7.6) we have for every integer $M > 0$ that

$$\frac{1}{2\pi}\int_{-\pi}^{\pi} |f(x+h) - f(x)|^2 \mathrm{d}x \leq \sum_{|n|\leq M} n^2 h^2 |c_n(f)|^2 + 4\sum_{|n|>M} |c_n(f)|^2 \tag{7.9}$$

if $M|h| \leq 1$. If (7.7) holds, then the second sum is $O(M^{-2\alpha})$. To estimate the first sum we use summation by parts. Writing

$$I_n := \sum_{|k|\le n} |c_k(f)|^2,$$

we have

$$\sum_{1\le|n|\le M} n^2|c_n(f)|^2 = M^2 I_M - 0\cdot I_0 - \sum_{n=1}^{M} I_{n-1}(n^2-(n-1)^2)$$
$$= M^2 I_M - \sum_{n=2}^{M}(2n-1)I_{n-1} - I_0.$$

By hypothesis,

$$I_\infty - I_n = O(n^{-2\alpha}), \quad n=1,2,\ldots.$$

Thus,

$$M^2 I_M - \sum_{n=2}^{M}(2n-1)\left(I_\infty + O((n-1)^{-2\alpha})\right)$$
$$= M^2\left(I_\infty + O(M^{-2\alpha})\right) - I_\infty\sum_{n=2}^{M}(2n-1) - \sum_{n=2}^{M}(2n-1)O((n-1)^{-2\alpha})$$
$$= O(M^{2-2\alpha}) + I_\infty\underbrace{\left(M^2-\sum_{n=2}^{M}(2n-1)\right)}_{=1} - \sum_{n=2}^{M}(2n-1)O((n-1)^{-2\alpha})$$
$$= O(M^{2-2\alpha}) + O(M^{2-2\alpha}) = O(M^{2-2\alpha}).$$

Exercise 7.2. Prove that

(1) $\sum_{n=1}^{M}(2n-1) = M^2$;
(2) $\sum_{n=2}^{M} O((2n-1)(n-1)^{-2\alpha}) = O(M^{2-2\alpha})$ for $0<\alpha<1$.

Combining these two estimates, we may conclude from (7.9) that there exists $C>0$ such that

$$\frac{1}{2\pi}\int_{-\pi}^{\pi}|f(x+h)-f(x)|^2\,dx \le C\left(h^2M^{2-2\alpha}+M^{-2\alpha}\right).$$

Since $0<\alpha<1$, choosing $M=[1/|h|]$ we obtain

$$\frac{1}{2\pi}\int_{-\pi}^{\pi}|f(x+h)-f(x)|^2\,dx \le C\left(h^2[1/|h|]^{2-2\alpha}+[1/|h|]^{-2\alpha}\right)$$
$$\le C\left(h^2(1/|h|)^{2-2\alpha}+(1/|h|-\{1/|h|\})^{-2\alpha}\right)$$
$$\le C\left(|h|^{2\alpha}+(1/|h|-1)^{-2\alpha}\right)$$
$$= C\left(|h|^{2\alpha}+|h|^{2\alpha}/(1-|h|)^{2\alpha}\right) \le C|h|^{2\alpha}$$

if $|h| \le 1/2$. Here $\{1/|h|\}$ denotes the fractional part of $1/|h|$, i.e., $\{1/|h|\} = 1/|h| - [1/|h|] \in [0,1)$.

Conversely, if (7.8) holds, then

$$\sum_{n=-\infty}^{\infty} (1-\cos(nh))|c_n(f)|^2 \le Ch^{2\alpha}$$

with some $C > 0$. Integrating this inequality with respect to h over the interval $[0,l]$, $l > 0$, we have

$$\sum_{n=-\infty}^{\infty} |c_n(f)|^2 \int_0^l (1-\cos(nh))\mathrm{d}h \le C \int_0^l h^{2\alpha}\mathrm{d}h,$$

or

$$\sum_{n=-\infty}^{\infty} |c_n(f)|^2 \left(l - \frac{\sin(nl)}{n}\right) \le Cl^{2\alpha+1},$$

or

$$\sum_{n=-\infty}^{\infty} |c_n(f)|^2 \left(1 - \frac{\sin(nl)}{nl}\right) \le Cl^{2\alpha}.$$

It follows that

$$Cl^{2\alpha} \ge \sum_{|n|l\ge 2} |c_n(f)|^2 \left(1 - \frac{\sin(nl)}{nl}\right) \ge \frac{1}{2} \sum_{|n|l\ge 2} |c_n(f)|^2. \tag{7.10}$$

Taking $l = 2/N$ for the integer $N > 0$ in (7.10) yields

$$\sum_{|n|\ge N} |c_n(f)|^2 \le CN^{-2\alpha}.$$

This completes the proof. □

Remark 7.6. If $\alpha = 1$, then (7.8) implies (7.7) but not conversely.

Exercise 7.3. Suppose that a periodic function $f \in L^2(-\pi,\pi)$ satisfies the condition

$$\int_{-\pi}^{\pi} |f(x+h) - f(x)|^2 \,\mathrm{d}x \le Ch^2$$

with some $C > 0$. Prove that

$$\sum_{n=-\infty}^{\infty} |n|^2 |c_n(f)|^2 < \infty$$

and therefore $f \in W_2^1(-\pi,\pi)$.

For an integrable function f periodic on the interval $[-\pi,\pi]$ let us introduce the mapping

$$f \mapsto \{c_n(f)\}_{n=-\infty}^{\infty},$$

where $c_n(f)$ are the Fourier coefficients of f. This mapping is a linear transformation. Formula (3.1) says that this mapping is bounded from $L^1(-\pi,\pi)$ to $l^\infty(\mathbb{Z})$. Here $\mathbb{Z}$ denotes all integers, and the *sequence space* $l^p(\mathbb{Z})$ consists of sequences $\{b_n\}_{n=-\infty}^{\infty}$ for which

$$\sum_{n=-\infty}^{\infty} |b_n|^p < \infty$$

if $1 \le p < \infty$ and $\sup_{n\in\mathbb{Z}} |b_n| < \infty$ if $p = \infty$.

Parseval's equality (7.3) shows that it is also bounded from $L^2(-\pi,\pi)$ to $l^2(\mathbb{Z})$. By the Riesz–Thorin interpolation theorem (Theorem 17.7), we may conclude that this mapping is bounded from $L^p(-\pi,\pi)$ to $l^{p'}(\mathbb{Z})$ for every $1 < p < 2$, $1/p + 1/p' = 1$, and

$$\sum_{n=-\infty}^{\infty} |c_n(f)|^{p'} \le c_p \left(\int_{-\pi}^{\pi} |f(x)|^p \mathrm{d}x \right)^{p'/p} < \infty.$$

Chapter 8
Besov and Hölder Spaces

In this chapter we will consider integrable 2π-periodic functions f defined via trigonometric Fourier series in $L^2(-\pi,\pi)$ as

$$f(x) \sim \sum_{n=-\infty}^{\infty} c_n(f)\mathrm{e}^{inx}, \tag{8.1}$$

where the Fourier coefficients $c_n(f)$ satisfy Parseval's equality

$$\frac{1}{2\pi}\int_{-\pi}^{\pi} |f(x)|^2 \mathrm{d}x = \sum_{n=-\infty}^{\infty} |c_n(f)|^2;$$

that is, (8.1) can be understood in the sense of $L^2(-\pi,\pi)$ as

$$\lim_{N\to\infty}\int_{-\pi}^{\pi}\left|f(x) - \sum_{|n|\le N} c_n(f)\mathrm{e}^{inx}\right|^2 \mathrm{d}x = 0.$$

We will introduce new spaces of functions (as subspaces of $L^2(-\pi,\pi)$) in terms of Fourier coefficients. The motivation of such an approach is the following: we proved (see Theorem 7.3 and Exercise 7.3) that a periodic function f belongs to $W_2^1(-\pi,\pi)$ if and only if

$$\sum_{n=-\infty}^{\infty} |n|^2 |c_n(f)|^2 < \infty.$$

This fact and Parseval's equality justify the following definitions.

V. Serov, *Fourier Series, Fourier Transform and Their Applications to Mathematical Physics*, Applied Mathematical Sciences 197,
DOI 10.1007/978-3-319-65262-7_8

Definition 8.1. A 2π-periodic function f is said to belong to the *Sobolev space*

$$W_2^\alpha(-\pi,\pi)$$

for some $\alpha \geq 0$ if

$$\sum_{n=-\infty}^{\infty} |n|^{2\alpha}|c_n(f)|^2 < \infty, \quad 0^0 := 1.$$

Definition 8.2. A 2π-periodic function f is said to belong to the *Besov space*

$$B_{2,\theta}^\alpha(-\pi,\pi)$$

for some $\alpha \geq 0$ and some $1 \leq \theta < \infty$ if

$$\sum_{j=0}^{\infty} \left(\sum_{2^j \leq |n| < 2^{j+1}} |n|^{2\alpha}|c_n(f)|^2 \right)^{\theta/2} < \infty.$$

Definition 8.3. A 2π-periodic function f is said to belong to the *Nikol'skii space*

$$H_2^\alpha(-\pi,\pi)$$

for some $\alpha \geq 0$ if

$$\sup_{j=0,1,2,\ldots} \sum_{2^j \leq |n| < 2^{j+1}} |n|^{2\alpha}|c_n(f)|^2 < \infty.$$

Definition 8.4 (See also Definition 1.10**).**

1) A 2π-periodic function f is said to belong to the *Hölder space* $C^\alpha[-\pi,\pi]$ for some noninteger $\alpha > 0$ if f is continuous on the interval $[-\pi,\pi]$, there is a continuous derivative $f^{(k)}$ of order $k = [\alpha]$ on the interval $[-\pi,\pi]$, and for all $h \neq 0$ small enough, we have

$$\sup_{x\in[-\pi,\pi]} |f^{(k)}(x+h) - f^{(k)}(x)| \leq C|h|^{\alpha-k},$$

where the constant $C > 0$ does not depend on h.

2) By the space $C^k[-\pi,\pi]$ for integer $k > 0$ we mean the set of 2π-periodic functions f that have continuous derivatives $f^{(k)}$ of order k on the interval $[-\pi,\pi]$.

Remark 8.5. We shall use later the following sufficient condition (see (3.13)): if there is a constant $C > 0$ such that for each $n = 1,2,\ldots$ we have

$$\sum_{|m|\geq n} |m|^k |c_m(f)| \leq Cn^{-\alpha} \tag{8.2}$$

with some integer $k \geq 0$ and some $0 < \alpha < 1$, then f belongs to the Hölder space $C^{k+\alpha}[-\pi, \pi]$.

The definitions 8.1–8.3 imply the following equalities and embeddings:

(1) $B^{\alpha}_{2,2}(-\pi, \pi) = W^{\alpha}_2(-\pi, \pi), \alpha \geq 0$.
(2) $W^0_2(-\pi, \pi) = L^2(-\pi, \pi)$.
(3) $B^{\alpha}_{2,1}(-\pi, \pi) \subset B^{\alpha}_{2,\theta}(-\pi, \pi) \subset H^{\alpha}_2(-\pi, \pi), \alpha \geq 0, 1 \leq \theta < \infty$.
(4) $B^0_{2,\theta}(-\pi, \pi) \subset L^2(-\pi, \pi), 1 \leq \theta \leq 2$ and $L^2(-\pi, \pi) \subset B^0_{2,\theta}(-\pi, \pi), 2 \leq \theta < \infty$.
(5) $L^2(-\pi, \pi) \subset H^0_2(-\pi, \pi)$.

Exercise 8.1. Prove embeddings (3), (4), and (5).

More embeddings are formulated in the following theorems.

Theorem 8.6. *If $\alpha \geq 0$, then*

$$C^{\alpha}[-\pi, \pi] \subset W^{\alpha}_2(-\pi, \pi)$$

and

$$C^{\alpha}[-\pi, \pi] \subset H^{\alpha}_2(-\pi, \pi).$$

Proof. Let us prove the first claim for integer $\alpha \geq 0$. If $\alpha = 0$ then

$$C[-\pi, \pi] \subset L^2(-\pi, \pi) = W^0_2(-\pi, \pi).$$

If $\alpha = k > 0$ is an integer, then Definition 8.4 implies that

$$\frac{1}{2\pi}\int_{-\pi}^{\pi} |f^{(k-1)}(x+h) - f^{(k-1)}(x)|^2 \mathrm{d}x \leq Ch^2.$$

Using Parseval's equality, we obtain

$$\sum_{n=-\infty}^{\infty} |n|^{2k-2}|c_n(f)|^2|\mathrm{e}^{\mathrm{i}nh} - 1|^2 \leq Ch^2$$

or

$$4\sum_{n=-\infty}^{\infty} |n|^{2k-2}|c_n(f)|^2 \sin^2(nh/2) \leq Ch^2.$$

It follows that

$$\sum_{|nh|\leq 2} |n|^{2k-2}|c_n(f)|^2 n^2 h^2 \leq Ch^2$$

or

$$\sum_{|n|\leq 2/|h|} |n|^{2k}|c_n(f)|^2 \leq C.$$

Letting $h \to 0$ yields

$$\sum_{n=-\infty}^{\infty} |n|^{2k} |c_n(f)|^2 \le C$$

i.e., $f \in W_2^k(-\pi,\pi)$. For noninteger α one needs to interpolate between the spaces $C^{[\alpha]}[-\pi,\pi]$ and $C^{[\alpha]+1}[-\pi,\pi]$.

Now let us consider the second claim. As above, for $f \in C^\alpha[-\pi,\pi]$, we have

$$\frac{1}{2\pi}\int_{-\pi}^{\pi} |f^{(k)}(x+h) - f^{(k)}(x)|^2 \mathrm{d}x \le C|h|^{2(\alpha-k)},$$

where $k = [\alpha]$ if α is not an integer and $k = \alpha - 1$ if α is an integer. By Parseval's equality,

$$\sum_{n=-\infty}^{\infty} |n|^{2k} |c_n(f)|^2 |\mathrm{e}^{inh} - 1|^2 \le C|h|^{2(\alpha-k)},$$

or

$$2\sum_{n=-\infty}^{\infty} |n|^{2k} |c_n(f)|^2 (1 - \cos(nh)) \le C|h|^{2(\alpha-k)}.$$

It suffices to consider $h > 0$. If we integrate the last inequality with respect to $h > 0$ from 0 to l, then

$$\sum_{n=-\infty}^{\infty} |n|^{2k} |c_n(f)|^2 \left(1 - \frac{\sin(nl)}{nl}\right) \le Cl^{2(\alpha-k)}.$$

It follows that

$$\sum_{2\le|n|l\le 4} |n|^{2k} |c_n(f)|^2 \le Cl^{2(\alpha-k)}$$

or equivalently,

$$\sum_{2\le|n|l\le 4} |n|^{2k} l^{2(k-\alpha)} |c_n(f)|^2 \le C,$$

where the constant $C > 0$ does not depend on l. Since $2(k-\alpha) < 0$, it follows that

$$\sum_{2/l\le|n|\le 4/l} |n|^{2\alpha} |c_n(f)|^2 \le C$$

for every $l > 0$. Choosing $l = 2^{-j+1}$, we obtain

$$\sum_{2^j\le|n|\le 2^{j+1}} |n|^{2\alpha} |c_n(f)|^2 \le C$$

i.e., $f \in H_2^\alpha(-\pi,\pi)$. This completes the proof. □

Theorem 8.7. *Assume that $\alpha > 1/2$ and that $\alpha - 1/2$ is not an integer. Then*

$$H_2^\alpha(-\pi,\pi) \subset C^{\alpha-1/2}[-\pi,\pi].$$

Proof. Let $k = [\alpha]$, so that $\alpha = k + \{\alpha\}$, where $\{\alpha\}$ denotes the fractional part of α. Note that in general, $0 \le \{\alpha\} < 1$ and in this theorem $\{\alpha\} \ne 1/2$. We will assume first that $\{\alpha\} = 0$, i.e., $\alpha = k$ is an integer and $k \ge 1$. If $f \in H_2^k(-\pi,\pi)$, then there is a constant $C > 0$ such that

$$\sum_{2^j \le |m| < 2^{j+1}} |m|^{2k} |c_m(f)|^2 \le C \tag{8.3}$$

for each $j = 0,1,2,\ldots$. Let us estimate the tail (8.2). Indeed, by the Cauchy–Bunyakovsky–Schwarz inequality and (8.3),

$$\begin{aligned}
\sum_{|m|\ge n} |m|^{k-1}|c_m(f)| &\le \sum_{\substack{j=j_0\\ 2^{j_0}\sim n}}^{\infty} \sum_{2^j\le|m|<2^{j+1}} |m|^{k-1}|c_m(f)| \\
&\le \sum_{\substack{j=j_0\\ 2^{j_0}\sim n}}^{\infty} \left(\sum_{2^j\le|m|<2^{j+1}} |m|^{2k}|c_m(f)|^2 \right)^{1/2} \left(\sum_{2^j\le|m|<2^{j+1}} \frac{1}{m^2} \right)^{1/2} \\
&\le \sqrt{C} \sum_{\substack{j=j_0\\ 2^{j_0}\sim n}}^{\infty} \left(\sum_{2^j\le|m|<2^{j+1}} \frac{1}{m^2} \right)^{1/2} \le \sqrt{C} \sum_{\substack{j=j_0\\ 2^{j_0}\sim n}}^{\infty} 2^{-\frac{j}{2}} \le C 2^{-\frac{j_0}{2}}.
\end{aligned}$$

Here, $2^{j_0} \sim n$ means that $2^{j_0} \le n < 2^{j_0+1}$. Therefore, we obtain

$$\sum_{|m|\ge n} |m|^{k-1}|c_m(f)| \le C n^{-1/2},$$

where the constant $C > 0$ is independent of n. This means that (see (8.2)) f belongs to $C^{k-1+1/2}[-\pi,\pi] = C^{k-1/2}[-\pi,\pi]$.

If $\alpha > 1/2$ is not an integer and $\alpha - 1/2$ is not an integer, then for $f \in H_2^\alpha(-\pi,\pi)$ we have instead of (8.3) the estimate

$$\sum_{2^j\le|m|<2^{j+1}} |m|^{2k+2\{\alpha\}}|c_m(f)|^2 \le C,$$

where $k = [\alpha], 0 < \{\alpha\} < 1$ and $\{\alpha\} \ne 1/2$. If $k = 0$, then $1/2 < \alpha = \{\alpha\} < 1$. Repeating now the above procedure, we obtain easily

$$\sum_{|m|\ge n}|c_m(f)| \le \sum_{\substack{j=j_0\\ 2^{j_0}\sim n}}^{\infty}\left(\sum_{2^j\le|m|<2^{j+1}}|m|^{2\alpha}|c_m(f)|^2\right)^{\frac{1}{2}}\left(\sum_{2^j\le|m|<2^{j+1}}|m|^{-2\alpha}\right)^{\frac{1}{2}}$$

$$\le C\sum_{\substack{j=j_0\\ 2^{j_0}\sim n}}^{\infty}\left(\sum_{2^j\le|m|<2^{j+1}}|m|^{-2\alpha}\right)^{1/2}\le C\sum_{\substack{j=j_0\\ 2^{j_0}\sim n}}^{\infty}2^{-(\alpha-1/2)j}\le Cn^{-(\alpha-1/2)},$$

i.e., we have again that $f\in C^{\alpha-1/2}[-\pi,\pi]$.

For the case $[\alpha]=k\ge 1$, α is not an integer, and $\alpha-1/2$ is not an integer, we consider two cases: $0<\{\alpha\}<1/2$ and $1/2<\{\alpha\}<1$. In the first case we have

$$\sum_{|m|\ge n}|m|^{k-1}|c_m(f)|\le$$

$$\sum_{\substack{j=j_0\\ 2^{j_0}\sim n}}^{\infty}\left(\sum_{2^j\le|m|<2^{j+1}}|m|^{2k+2\{\alpha\}}|c_m(f)|^2\right)^{1/2}\left(\sum_{2^j\le|m|<2^{j+1}}|m|^{-2-2\{\alpha\}}\right)^{1/2}$$

$$\le C\sum_{\substack{j=j_0\\ 2^{j_0}\sim n}}^{\infty}2^{-j-j\{\alpha\}}2^{j/2}\le Cn^{-1/2-\{\alpha\}}.$$

This means again that $f\in C^{k-1+1/2+\{\alpha\}}[-\pi,\pi]=C^{\alpha-1/2}[-\pi,\pi]$. In the second case, $1/2<\{\alpha\}<1$, we proceed as follows:

$$\sum_{|m|\ge n}|m|^{k}|c_m(f)|\le$$

$$\sum_{\substack{j=j_0\\ 2^{j_0}\sim n}}^{\infty}\left(\sum_{2^j\le|m|<2^{j+1}}|m|^{2k+2\{\alpha\}}|c_m(f)|^2\right)^{1/2}\left(\sum_{2^j\le|m|<2^{j+1}}|m|^{-2\{\alpha\}}\right)^{1/2}$$

$$\le C\sum_{\substack{j=j_0\\ 2^{j_0}\sim n}}^{\infty}2^{-(\{\alpha\}-1/2)j}\le Cn^{-\{\alpha\}+1/2}.$$

This means that $f\in C^{k+\{\alpha\}-1/2}[-\pi,\pi]=C^{\alpha-1/2}[-\pi,\pi]$. Hence, the theorem is completely proved. □

Corollary 8.8. *Assume that $\alpha=k+1/2$ for some integer $k\ge 1$. Then*

$$H_2^{\alpha}(-\pi,\pi)\subset C^{\beta-1/2}[-\pi,\pi]$$

for every $1/2<\beta<\alpha$.

Corollary 8.9. *Assume that $\alpha > 1/2$ and $\alpha - 1/2$ is not an integer. Then*

$$B^{\alpha}_{2,\theta}(-\pi,\pi) \subset C^{\alpha-1/2}[-\pi,\pi]$$

for every $1 \leq \theta < \infty$.

Exercise 8.2. Prove Corollaries 8.8 and 8.9.

Exercise 8.3. Prove that the Fourier series (8.1) with coefficients

$$c_n(f) = \frac{1}{|n|\log(1+|n|)}, n \neq 0, \quad c_0(f) = 1$$

defines a function from the Besov space $B^{1/2}_{2,\theta}(-\pi,\pi)$ for every $1 < \theta < \infty$, but not for $\theta = 1$.

Exercise 8.4. Prove that the Fourier series (8.1) with coefficients

$$c_n(f) = \frac{1}{|n|^{3/2}\log(1+|n|)}, n \neq 0, \quad c_0(f) = 1$$

defines a function from the Besov space $B^{1}_{2,\theta}(-\pi,\pi)$ for every $1 < \theta < \infty$, but not for $\theta = 1$.

Exercise 8.5. Consider the Fourier series (8.1) with coefficients

$$c_n(f) = \frac{1}{|n|^2\log^{\beta}(1+|n|)}, n \neq 0, \quad c_0(f) = 1.$$

Prove that

(1) $f \in H^{3/2}_2(-\pi,\pi)$ if $\beta \geq 0$
(2) $f \in W^{3/2}_2(-\pi,\pi)$ if $\beta > 1/2$
(3) $f \in C^1[-\pi,\pi]$ if $\beta > 1$ but $f \notin C^1[-\pi,\pi]$ if $\beta \leq 1$.

So the embeddings $W^{\alpha}_2(-\pi,\pi) \subset C^{\alpha-1/2}[-\pi,\pi]$ and $H^{\alpha}_2(-\pi,\pi) \subset C^{\alpha-1/2}[-\pi,\pi]$ are not valid for $\alpha - 1/2$ an integer (see Theorem 8.6).

Exercise 8.6. Let

$$\sum_{k=0}^{\infty} a^k \cos(b^k x)$$

be a trigonometric series, where $b = 2,3,\ldots$ and $0 < a < 1$. Prove that the series defines a function from $C^1[-\pi,\pi]$ if $0 < ab < 1$ and a function from the Hölder space $C^{\gamma}[-\pi,\pi]$, $\gamma < 1$ if $ab = 1$.

Exercise 8.7. Assume that $a = 1/b^2$ in Exercise 8.6. Is it true that this function belongs to $C^1[-\pi,\pi]$?

Chapter 9
Absolute Convergence. Bernstein and Peetre Theorems.

We begin by proving the equivalence between a 2π-periodic function f belonging to the Nikol'skii space $H_2^\alpha(-\pi,\pi)$ for some $0<\alpha<1$ in the sense of Definition 8.3 and the L^2 *Hölder condition* of order α, i.e.,

$$\frac{1}{2\pi}\int_{-\pi}^{\pi}|f(x+h)-f(x)|^2\mathrm{d}x \le K|h|^{2\alpha} \tag{9.1}$$

for some constant $K>0$ and for all $h\neq 0$ sufficiently small. Indeed, due to Parseval's equality we have

$$\begin{aligned}\frac{1}{2\pi}\int_{-\pi}^{\pi}|f(x+h)-f(x)|^2\mathrm{d}x &= \sum_{n=-\infty}^{\infty}|c_n(f)|^2|\mathrm{e}^{inh}-1|^2\\ &\le \sum_{|n|\le 2^{j_0}}|n|^2|h|^2|c_n(f)|^2+4\sum_{|n|\ge 2^{j_0}}|c_n(f)|^2, \end{aligned}\tag{9.2}$$

where j_0 is chosen so that $2^{j_0}\le\frac{1}{|h|}<2^{j_0+1}$. The first sum on the right-hand side of (9.2) is estimated from above as

$$\begin{aligned}|h|^2\sum_{|n|\le 2^{j_0}}|n|^2|c_n(f)|^2 &\le |h|^2\sum_{j=0}^{j_0}\sum_{2^j\le|n|<2^{j+1}}|n|^2|c_n(f)|^2\\ &\le |h|^2\sum_{j=0}^{j_0}\left(2^{j+1}\right)^{2-2\alpha}\sum_{2^j\le|n|<2^{j+1}}|n|^{2\alpha}|c_n(f)|^2\\ &\le C|h|^2\sum_{j=0}^{j_0}\left(2^{2-2\alpha}\right)^{j+1}=C|h|^2\frac{\left(2^{2-2\alpha}\right)^{j_0+2}-2^{2-2\alpha}}{2^{2-2\alpha}-1}\\ &\le C|h|^2\left(2^{j_0}\right)^{2-2\alpha}\le C|h|^2\left(\frac{1}{|h|}\right)^{2-2\alpha}=C|h|^{2\alpha}\end{aligned}$$

V. Serov, *Fourier Series, Fourier Transform and Their Applications to Mathematical Physics*, Applied Mathematical Sciences 197,
DOI 10.1007/978-3-319-65262-7_9

if $0 < \alpha < 1$. We used this condition for α because we considered a geometric sum with common ratio $2^{2-2\alpha} \neq 1$. The second sum on the right-hand side of (9.2) is estimated from above as

$$4\sum_{j=j_0}^{\infty} \sum_{2^j \le |n| < 2^{j+1}} |n|^{2\alpha}|n|^{-2\alpha}|c_n(f)|^2 \le 4\sum_{j=j_0}^{\infty} 2^{-2\alpha j} \sum_{2^j \le |n| < 2^{j+1}} |n|^{2\alpha}|c_n(f)|^2$$
$$\le C\sum_{j=j_0}^{\infty} 2^{-2\alpha j} \le C2^{-2j_0\alpha} \le C|h|^{2\alpha},$$

since $\frac{1}{|h|} \le 2^{j_0+1}$ and the criterion of Definition 8.3 is satisfied. Thus, (9.1) is proved. Conversely, if the L^2 Hölder condition (9.1) is fulfilled, then Theorem 7.5 implies for each $N = 1,2,\ldots$ the inequality

$$\sum_{|n|\ge N} |c_n(f)|^2 \le CN^{-2\alpha}$$

with the same α as in (9.1). But this leads to the inequality

$$N^{2\alpha} \sum_{N\le|n|<2N} |c_n(f)|^2 \le C,$$

where the constant C is independent of N. Thus, we obtain for every integer $N > 0$ that

$$\sum_{N\le|n|<2N} |n|^{2\alpha}|c_n(f)|^2 \le C.$$

Since N is arbitrary, we may conclude that $f \in H_2^{\alpha}(-\pi,\pi)$ for $0 < \alpha \le 1$ in the sense of Definition 8.3. Therefore, the L^2 Hölder condition (9.1) can be considered as an equivalent definition of the Nikol'skii space $H_2^{\alpha}(-\pi,\pi)$ for $0 < \alpha < 1$.

Exercise 9.1. Prove that f belongs to the Nikol'skii space $H_2^{\alpha}(-\pi,\pi)$ for every noninteger $\alpha > 0$ in the sense of Definition 8.3 if and only if the following L^2 Hölder condition holds:

$$\frac{1}{2\pi}\int_{-\pi}^{\pi} |f^{(k)}(x+h) - f^{(k)}(x)|^2 \mathrm{d}x \le K|h|^{2\alpha-2k}$$

with some constant $K > 0$ and $k = [\alpha]$.

Exercise 9.2. Prove that $f \in W_2^k(-\pi,\pi)$, $k = 1,2,\ldots$, if and only if

$$\frac{1}{2\pi}\int_{-\pi}^{\pi} |f^{(k-1)}(x+h) - f^{(k-1)}(x)|^2 \mathrm{d}x \le C|h|^2.$$

Theorem 9.1 (Bernstein, 1914). *Assume that a 2π-periodic function f satisfies the L^2 Hölder condition with $1/2 < \alpha \le 1$. Then its trigonometric Fourier series converges absolutely, i.e.,*

$$\sum_{n=-\infty}^{\infty} |c_n(f)| < \infty.$$

Proof. Since the L^2 Hölder condition (9.1) is equivalent to $f \in H_2^{\alpha}(-\pi,\pi)$ for $0 < \alpha < 1$, there is a constant $C > 0$ such that

$$\sum_{2^j \le |n| < 2^{j+1}} |n|^{2\alpha} |c_n(f)|^2 \le C$$

for each $j = 0,1,2,\ldots$. Hence we have

$$\begin{aligned}
\sum_{n=-\infty}^{\infty} |c_n(f)| &= |c_0(f)| + \sum_{j=0}^{\infty} \sum_{2^j \le |n| < 2^{j+1}} |n|^{\alpha} |c_n(f)| |n|^{-\alpha} \\
&\le |c_0(f)| + \sum_{j=0}^{\infty} \left(\sum_{2^j \le |n| < 2^{j+1}} |n|^{2\alpha} |c_n(f)|^2 \right)^{1/2} \left(\sum_{2^j \le |n| < 2^{j+1}} |n|^{-2\alpha} \right)^{1/2} \\
&\le |c_0(f)| + \sqrt{C} \sum_{j=0}^{\infty} 2^{-\alpha j} 2^{j/2} = |c_0(f)| + \sqrt{C} \sum_{j=0}^{\infty} \left(2^{-(\alpha - 1/2)} \right)^j < \infty,
\end{aligned}$$

since $\alpha > 1/2$. Thus, the theorem is proved. □

Corollary 9.2. *Theorem 9.1 holds for $C^{\alpha}[-\pi,\pi]$, $B_{2,\theta}^{\alpha}(-\pi,\pi)$, and $H_2^{\alpha}(-\pi,\pi)$ for every $\alpha > 1/2$ and $1 \le \theta < \infty$.*

Exercise 9.3. Prove this Corollary.

Theorem 9.3 (Peetre, 1967). *Assume that a 2π-periodic function f belongs to the Besov space $B_{2,1}^{1/2}(-\pi,\pi)$. Then its trigonometric Fourier series converges absolutely.*

Proof. If $f \in B_{2,1}^{1/2}(-\pi,\pi)$, then

$$\sum_{j=0}^{\infty} \left(\sum_{2^j \le |n| < 2^{j+1}} |n| |c_n(f)|^2 \right)^{1/2} < \infty. \tag{9.3}$$

Hence we have

$$\begin{aligned}
\sum_{n=-\infty}^{\infty} |c_n(f)| &= |c_0(f)| + \sum_{j=0}^{\infty} \sum_{2^j \le |n| < 2^{j+1}} |n|^{1/2} |c_n(f)| |n|^{-1/2} \\
&\le |c_0(f)| + \sum_{j=0}^{\infty} \left(\sum_{2^j \le |n| < 2^{j+1}} |n| |c_n(f)|^2 \right)^{1/2} \left(\sum_{2^j \le |n| < 2^{j+1}} |n|^{-1} \right)^{1/2}
\end{aligned}$$

$$\leq |c_0(f)| + \sum_{j=0}^{\infty} \left(\sum_{2^j \leq |n| < 2^{j+1}} |n| |c_n(f)|^2 \right)^{1/2} \left(2^{-j} 2^j\right)^{1/2}$$

$$= |c_0(f)| + \sum_{j=0}^{\infty} \left(\sum_{2^j \leq |n| < 2^{j+1}} |n| |c_n(f)|^2 \right)^{1/2} < \infty$$

due to (9.3). This proves the theorem. □

Corollary 9.4. *It is true that* $B_{2,1}^{1/2}(-\pi,\pi) \subset C[-\pi,\pi]$.

Exercise 9.4. Prove that the embedding $B_{2,\theta}^{1/2}(-\pi,\pi) \subset C[-\pi,\pi]$ does not hold for $1 < \theta < \infty$ by considering the function from Exercise 8.3.

Exercise 9.5. Prove that $H_2^{\alpha}(-\pi,\pi) \subset B_{2,1}^{1/2}(-\pi,\pi)$ if $\alpha > 1/2$.

Theorem 9.5. *Assume that a 2π-periodic function f belongs to the Sobolev space $W_p^1(-\pi,\pi)$ with some $1 < p < \infty$. Then its trigonometric Fourier series converges absolutely.*

Proof. Since $W_{p_1}^1(-\pi,\pi) \subset W_{p_2}^1(-\pi,\pi)$ for $1 \leq p_2 < p_1$, we may assume without loss of generality that $f \in W_p^1(-\pi,\pi)$ with $1 < p \leq 2$. Then there is a function $g \in L^p(-\pi,\pi)$ with $1 < p \leq 2$ such that

$$f(x) = \int_{-\pi}^{x} g(t)\mathrm{d}t + f(-\pi), \quad \int_{-\pi}^{\pi} g(t)\mathrm{d}t = 0. \tag{9.4}$$

As we know from the proof of Theorem 7.3, (9.4) leads to

$$c_n(f) = \frac{1}{in} c_n(g), \quad n \neq 0.$$

Since $g \in L^p(-\pi,\pi)$ with $1 < p \leq 2$, the results of Chapter 7 give

$$\left(\sum_{n=-\infty}^{\infty} |c_n(g)|^{p'} \right)^{1/p'} < \infty, \tag{9.5}$$

where $\frac{1}{p} + \frac{1}{p'} = 1$. The facts (9.4), (9.5) and Hölder's inequality imply that

$$\sum_{n=-\infty}^{\infty} |c_n(f)| = |c_0(f)| + \sum_{n\neq 0} \frac{1}{|n|} |c_n(g)|$$

$$\leq |c_0(f)| + \left(\sum_{n \neq 0} \frac{1}{|n|^p} \right)^{1/p} \left(\sum_{n\neq 0} |c_n(g)|^{p'} \right)^{1/p'} < \infty,$$

since $1 < p \leq 2$. This completes the proof. □

Remark 9.6. For the Sobolev space $W_1^1(-\pi,\pi)$, this theorem is not valid, i.e., there is a function f from $W_1^1(-\pi,\pi)$ with absolutely divergent trigonometric Fourier series. More precisely, we will prove in the next chapter that the function

$$f(x) := \sum_{n=1}^{\infty} \frac{\sin nx}{n\log(1+n)} \tag{9.6}$$

belongs to the Sobolev space $W_1^1(-\pi,\pi)$ and is continuous on the interval $[-\pi,\pi]$, but its trigonometric Fourier series (9.6) diverges absolutely.

The next theorem is due to Zigmund (1958–1959).

Theorem 9.7. *Suppose that* $f \in W_1^1(-\pi,\pi) \cap C^\alpha[-\pi,\pi]$ *with some* $0 < \alpha < 1$. *Then its trigonometric Fourier series converges absolutely.*

Proof. Since $f \in W_1^1(-\pi,\pi)$, it follows that f is of bounded variation. The periodicity of f implies that

$$\frac{1}{2\pi}\int_{-\pi}^{\pi} |f(x+h)-f(x)|^2 \mathrm{d}x = \frac{1}{2\pi N}\int_{-\pi}^{\pi}\sum_{k=1}^{N} |f(x+kh)-f(x+(k-1)h)|^2 \mathrm{d}x, \tag{9.7}$$

where the integer N is chosen so that $N|h| \le 1$ for $h \neq 0$ sufficiently small. We choose $N = [1/|h|]$. Since $f \in C^\alpha[-\pi,\pi]$, the right-hand side of (9.7) can be estimated as

$$\begin{aligned}
&\frac{1}{2\pi N}\int_{-\pi}^{\pi}\sum_{k=1}^{N} |f(x+kh)-f(x+(k-1)h)|^2 \mathrm{d}x \\
&\quad \le \frac{C|h|^\alpha}{2\pi N}\int_{-\pi}^{\pi}\sum_{k=1}^{N} |f(x+kh)-f(x+(k-1)h)| \mathrm{d}x \\
&\le \frac{C|h|^\alpha}{2\pi N}\left(V_{-\pi-1}^{-\pi}(f)+V_{-\pi}^{\pi}(f)+V_{\pi}^{\pi+1}(f)\right)2\pi \le \frac{C|h|^\alpha}{N} \\
&= \frac{C|h|^\alpha}{[1/|h|]} = \frac{C|h|^\alpha}{1/|h|-\{1/|h|\}} \le \frac{C|h|^\alpha}{1/|h|-1} = \frac{C|h|^{\alpha+1}}{1-|h|} \le C|h|^{\alpha+1}
\end{aligned}$$

if $|h| \le 1/2$. This inequality means (see (9.7)) that $f \in H_2^{\frac{\alpha+1}{2}}(-\pi,\pi)$ with $\frac{\alpha+1}{2} > 1/2$ for $\alpha > 0$. An application of Bernstein's theorem completes the proof of the theorem. □

Exercise 9.6. Let a (periodic) function f be defined by

$$f(x) := \sum_{k=1}^{\infty} \frac{\mathrm{e}^{ikx}}{k}.$$

Prove that the function f belongs to the Nikol'skii space $H_2^{1/2}(-\pi,\pi)$ but its trigonometric Fourier series is not absolutely convergent.

Exercise 9.7. Let a (periodic) function f be defined by the absolutely convergent Fourier series

$$f(x) := \sum_{k=1}^{\infty} \frac{\mathrm{e}^{ikx}}{k^{3/2}}.$$

(1) Show that f belongs to the Nikol'skii space $H_2^1(-\pi,\pi)$ but

$$\frac{1}{2\pi}\int_{-\pi}^{\pi} |f(x+h)-f(x)|^2 \mathrm{d}x \geq \frac{4h^2}{\pi^2}\log\frac{\pi}{|h|}$$

for $0<|h|<1$, that is, (9.1) does not hold for $\alpha = 1$.

(2) Show that f does not belong to the Besov space $B_{2,\theta}^1(-\pi,\pi)$ for any $1 \leq \theta < \infty$.

Chapter 10
Dirichlet Kernel. Pointwise and Uniform Convergence.

The material of this chapter forms a central part of the theory of trigonometric Fourier series. In this chapter we will answer the following question: to what value does a trigonometric Fourier series converge?

The *Dirichlet kernel* $D_N(x)$, which is defined by the symmetric finite trigonometric sum

$$D_N(x) := \sum_{|n|\le N} \mathrm{e}^{\mathrm{i}nx}, \tag{10.1}$$

plays a key role in this chapter. If $x \in [-\pi, \pi] \setminus \{0\}$, then $D_N(x)$ from (10.1) can be recalculated as follows. Using Euler's formula, we have

$$\begin{aligned} D_N(x) &= \sum_{n=-N}^{N} \mathrm{e}^{\mathrm{i}nx} = \mathrm{e}^{-\mathrm{i}Nx} \sum_{n=-N}^{N} \mathrm{e}^{\mathrm{i}(n+N)x} = \mathrm{e}^{-\mathrm{i}Nx} \sum_{k=0}^{2N} \mathrm{e}^{\mathrm{i}kx} \\ &= \mathrm{e}^{-\mathrm{i}Nx} \frac{1-\mathrm{e}^{\mathrm{i}(2N+1)x}}{1-\mathrm{e}^{\mathrm{i}x}} = \frac{\mathrm{e}^{-\mathrm{i}Nx} - \mathrm{e}^{\mathrm{i}(N+1)x}}{1-\mathrm{e}^{\mathrm{i}x}} \\ &= \frac{\mathrm{e}^{\mathrm{i}(N+1/2)x} - \mathrm{e}^{-\mathrm{i}(N+1/2)x}}{\mathrm{e}^{\mathrm{i}x/2} - \mathrm{e}^{-\mathrm{i}x/2}} = \frac{\sin(N+1/2)x}{\sin x/2}. \end{aligned}$$

Thus, the Dirichlet kernel equals

$$D_N(x) = \frac{\sin(N+1/2)x}{\sin x/2}, \quad x \neq 0. \tag{10.2}$$

For $x = 0$ we have

$$D_N(0) = 2N + 1 = \lim_{x\to 0} D_N(x),$$

so that (10.2) holds for all $x \in [-\pi, \pi]$.

V. Serov, *Fourier Series, Fourier Transform and Their Applications to Mathematical Physics*, Applied Mathematical Sciences 197,
DOI 10.1007/978-3-319-65262-7_10

Exercise 10.1. Prove that

(1) $\frac{1}{2\pi}\int_{-\pi}^{\pi} D_N(x)\mathrm{d}x = 1, N = 0,1,2,\ldots;$

(2) $K_N(x) = \frac{1}{N+1}\sum_{j=0}^{N} D_j(x)$, where $K_N(x)$ is the Fejér kernel (5.2)

Recall that the trigonometric Fourier partial sum is given by

$$S_N f(x) = \sum_{|n|\leq N} c_n(f)\mathrm{e}^{\mathrm{i}nx}. \tag{10.3}$$

The Fourier coefficients of $D_N(x)$ are equal to

$$c_n(D_N) = \frac{1}{2\pi}\int_{-\pi}^{\pi} \mathrm{e}^{-\mathrm{i}nx}\sum_{|k|\leq N} \mathrm{e}^{\mathrm{i}kx}\mathrm{d}x = \begin{cases} 0, & |n| > N, \\ 1, & |n| \leq N. \end{cases}$$

Hence, if f is periodic and integrable, then the partial sum (10.3) can be rewritten as (see Exercise 4.1)

$$\begin{aligned} S_N f(x) &= \sum_{n=-\infty}^{\infty} c_n(D_N)c_n(f)\mathrm{e}^{\mathrm{i}nx} = (f * D_N)(x) = \frac{1}{2\pi}\int_{-\pi}^{\pi} D_N(x-y)f(y)\mathrm{d}y \\ &= \frac{1}{2\pi}\int_{-\pi}^{\pi} D_N(y)f(x+y)\mathrm{d}y = \frac{1}{2\pi}\int_{-\pi}^{\pi} f(x+y)\frac{\sin(N+1/2)y}{\sin y/2}\mathrm{d}y. \end{aligned} \tag{10.4}$$

Exercise 10.2. Let f be the function

$$f(x) = \begin{cases} \frac{1}{2} - \frac{x}{2\pi}, & 0 < x \leq \pi, \\ -\frac{1}{2} - \frac{x}{2\pi}, & -\pi \leq x < 0. \end{cases}$$

Show that

(1) $(S_N f)'(x) = \frac{1}{2\pi}(D_N(x) - 1);$

(2) $\lim_{N\to\infty} S_N f(x) = f(x), x \neq 0;$

(3) $\lim_{N\to\infty} S_N f(0) = 0.$

Exercise 10.3. Prove that as $N \to \infty$,

$$\frac{1}{2\pi}\int_{-\pi}^{\pi} |D_N(x)|\mathrm{d}x = \frac{4\log N}{\pi^2} + O(1).$$

Since the Dirichlet kernel is an even function (see (10.2)), we can rewrite (10.4) as

$$S_N f(x) = \frac{1}{2\pi}\int_0^{\pi} (f(x+y) + f(x-y))\frac{\sin(N+1/2)y}{\sin y/2}\mathrm{d}y.$$

Using the normalization of the Dirichlet kernel (see Exercise 10.1), we have for every function $S(x)$ that

$$S_N f(x) - S(x) = \frac{1}{2\pi} \int_0^\pi (f(x+y) + f(x-y) - 2S(x)) \frac{\sin(N+1/2)y}{\sin y/2} dy. \quad (10.5)$$

Our aim is to define $S(x)$ so that the limit of the right-hand side of (10.5) is equal to zero. We will simplify the problem by splitting it into two steps. The first simplification is connected with the following technical lemma.

Lemma 10.1. *For all $z \in [-\pi, \pi]$, it is true that*

$$\left| \frac{1}{\sin z/2} - \frac{2}{z} \right| \le \frac{\pi^2}{24}.$$

Proof. First we show that

$$\left| \sin z/2 - \frac{z}{2} \right| \le \frac{|z|^3}{48} \quad (10.6)$$

for all $z \in [-\pi, \pi]$. In order to prove this inequality, it is enough to show that

$$0 < x - \sin x < \frac{x^3}{6}$$

for all $0 < x < \pi/2$. The left inequality is well known. To prove the right inequality we introduce $h(x)$ as

$$h(x) = x - \sin x - x^3/6.$$

Then its derivative satisfies

$$h'(x) = 1 - \cos x - x^2/2 = 2(\sin^2 x/2 - x^2/4) < 0$$

for all $0 < x < \pi/2$. Thus, $h(x)$ is monotonically decreasing on the interval $[0, \pi/2]$, which implies that

$$0 = h(0) > h(x) = x - \sin x - x^3/6$$

for all $0 < x < \pi/2$. This proves (10.6), which in turn yields

$$\left| \frac{1}{\sin z/2} - \frac{2}{z} \right| = \frac{2|z/2 - \sin(z/2)|}{|z||\sin(z/2)|} \le \frac{|z|^3/24}{|z||\sin(z/2)|} \le \frac{|z|^3/24}{|z||z|/\pi} \le \frac{\pi|z|}{24} \le \frac{\pi^2}{24},$$

since $|\sin z/2| \ge |z|/\pi$ for all $z \in [-\pi, \pi]$. This finishes the proof. □

As an immediate corollary of Lemma 10.1, we obtain for every periodic and integrable function f that the function

$$y \mapsto (f(x+y) + f(x-y) - 2S(x)) \left(\frac{1}{\sin y/2} - \frac{2}{y} \right)$$

is integrable on the interval $[0,\pi]$ uniformly in $x \in [-\pi,\pi]$ if $S(x)$ is bounded, i.e.,

$$\begin{aligned}\int_0^\pi |f(x+y)+f(x-y)-2S(x)| &\left|\frac{1}{\sin y/2}-\frac{2}{y}\right| \mathrm{d}y \\ &\le \frac{\pi^2}{24}\left(\int_0^\pi |f(x+y)|\mathrm{d}y+\int_0^\pi |f(x-y)|\mathrm{d}y+2\pi|S(x)|\right) \\ &\le \frac{\pi^2}{24}\left(\int_{-\pi}^\pi |f(y)|\mathrm{d}y+2\pi \sup_{x\in[-\pi,\pi]} |S(x)|\right).\end{aligned}$$

Application of the Riemann–Lebesgue lemma (Theorem 6.1) gives us that

$$\lim_{N\to\infty}\frac{1}{2\pi}\int_0^\pi (f(x+y)+f(x-y)-2S(x))\left(\frac{1}{\sin y/2}-\frac{2}{y}\right)\sin(N+1/2)y=0$$

pointwise in $x \in [-\pi,\pi]$ and even uniformly in $x \in [-\pi,\pi]$ if $S(x)$ is bounded on the interval $[-\pi,\pi]$.

Thus, we have reduced the question of pointwise or uniform convergence in (10.5) to proving that

$$\lim_{N\to\infty}\int_0^\pi (f(x+y)+f(x-y)-2S(x))\frac{\sin(N+1/2)y}{y}\mathrm{d}y=0 \tag{10.7}$$

pointwise or uniformly in $x \in [-\pi,\pi]$.

For the second simplification we consider the contribution to (10.7) from the interval $0<\delta\le y\le\pi$. Note that the function

$$\frac{f(x+y)+f(x-y)-2S(x)}{y}$$

is integrable in y on the interval $[\delta,\pi]$ uniformly in $x \in [-\pi,\pi]$ if $S(x)$ is bounded. Hence, by the Riemann–Lebesgue lemma this contribution tends to zero as $N\to\infty$. We summarize these two simplifications as

$$\begin{aligned}S_N f(x)-S(x)& \\ &=\frac{1}{\pi}\int_0^\delta (f(x+y)+f(x-y)-2S(x))\frac{\sin(N+1/2)y}{y}\mathrm{d}y+o(1)\end{aligned} \tag{10.8}$$

as $N\to\infty$ pointwise or uniformly in $x \in [-\pi,\pi]$.

Let us assume now that f is a piecewise continuous periodic function. We wish to know the values of $S(x)$ in (10.8) to which the trigonometric Fourier partial sum can converge. The second part of Theorem 5.1 shows that the Fejér means converge to

$$\lim_{N\to\infty}\sigma_N f(x)=\frac{1}{2}(f(x+0)+f(x-0))$$

for every $x \in [-\pi, \pi]$ pointwise. But since

$$\sigma_N f(x) = \frac{\sum_{j=0}^{N} S_j f(x)}{N+1},$$

$S_N f(x)$ can converge (if it converges) only to the value

$$\frac{1}{2}(f(x+0) + f(x-0)).$$

We can obtain some sufficient conditions when the limit in (10.8) exists.

Theorem 10.2. *Suppose that $S(x)$ is chosen so that*

$$\int_0^\delta \frac{|f(x+y) + f(x-y) - 2S(x)|}{y} dy < \infty \tag{10.9}$$

pointwise or uniformly in $x \in [-\pi, \pi]$. Then

$$\lim_{N \to \infty} S_N f(x) = S(x) \tag{10.10}$$

pointwise or uniformly in $x \in [-\pi, \pi]$.

Proof. The result follows immediately from (10.8) and the Riemann–Lebesgue lemma. □

Remark 10.3. If in (10.10) we have uniform convergence, then $S(x)$ must necessarily be periodic ($S(-\pi) = S(\pi)$) and continuous on the interval $[-\pi, \pi]$.

Corollary 10.4. *Suppose a periodic function f belongs to the Hölder space $C^\alpha[-\pi, \pi]$ for some $0 < \alpha \le 1$. Then*

$$\lim_{N \to \infty} S_N f(x) = f(x)$$

uniformly in $x \in [-\pi, \pi]$.

Proof. Since $f \in C^\alpha[-\pi, \pi]$, it follows that

$$|f(x+y) + f(x-y) - 2f(x)| \le |f(x+y) - f(x)| + |f(x-y) - f(x)| \le Cy^\alpha$$

for $0 < y < \delta$. This means that the condition (10.9) holds with $S(x) = f(x)$ uniformly in $x \in [-\pi, \pi]$, from which the statement of the corollary follows. □

Theorem 10.5 (Dirichlet, Jordan). *Suppose that a periodic function f is of bounded variation on the interval $[x-\delta, x+\delta]$ for some $\delta > 0$ and some fixed x. Then*

$$\lim_{N\to\infty} S_N f(x) = \frac{1}{2}\left(f(x+0)+f(x-0)\right).$$

Proof. Since f is of bounded variation, the limit

$$\lim_{y\to 0+} \frac{1}{2}\left(f(x+y)+f(x-y)\right) = \frac{1}{2}\left(f(x+0)+f(x-0)\right) =: S(x) \tag{10.11}$$

exists. For $0 < y < \delta$ we define

$$F(y) := f(x+y)+f(x-y)-2S(x),$$

where $S(x)$ is defined by (10.11). Note that $F(0) = 0$. Let us also define

$$G_N(y) := \int_0^y \frac{\sin(N+1/2)t}{t}\mathrm{d}t, \quad 0 < y \le \delta. \tag{10.12}$$

It is easy to check that

$$G_N(y) = \int_0^{(N+1/2)y} \frac{\sin\rho}{\rho}\mathrm{d}\rho, \quad 0 < y \le \delta.$$

This representation implies that

$$\lim_{N\to\infty} G_N(y) = \int_0^{\infty} \frac{\sin\rho}{\rho}\mathrm{d}\rho = \frac{\pi}{2}. \tag{10.13}$$

For fixed x we have from (10.8) and (10.12) that

$$S_N f(x) - S(x) = \frac{1}{\pi}\int_0^{\delta} F(y)G_N'(y)\mathrm{d}y + o(1), \quad N\to\infty.$$

Here integration by parts gives

$$\begin{aligned} S_N f(x) - S(x) &= \frac{1}{\pi}\left(F(y)G_N(y)|_0^{\delta} - \int_0^{\delta} G_N(y)\mathrm{d}F(y)\right) + o(1) \\ &= \frac{1}{\pi}\left(F(\delta)G_N(\delta) - \int_0^{\delta} G_N(y)\mathrm{d}F(y)\right) + o(1), \quad N\to\infty, \end{aligned} \tag{10.14}$$

where the last integral is well defined as the Stieltjes integral of the continuous function $G_N(y)$ with respect to the function of bounded variation $F(y)$. Since the limit (10.13) holds and $G_N(y)$ is continuous, we can consider the limit in (10.14) as $N \to \infty$. Hence, we obtain

$$\lim_{N\to\infty} (S_N f(x) - S(x)) = \frac{1}{\pi}\left(F(\delta)\frac{\pi}{2} - \frac{\pi}{2}\int_0^\delta \mathrm{d}F(y)\right)$$
$$= \frac{1}{2}(F(\delta) - F(\delta) + F(0)) = 0.$$

This completes the proof. □

Corollary 10.6. *If f is periodic and if f and f' are piecewise continuous, then the Fourier series of f converges to $\frac{1}{2}(f(x+0)+f(x-0))$ at all points. If in addition f is continuous on $(-\infty,\infty)$, then its Fourier series converges to $f(x)$ uniformly on $(-\infty,\infty)$.*

Corollary 10.7. *If f is periodic and belongs to the Sobolev space $W_1^1(-\pi,\pi)$, then its trigonometric Fourier series converges pointwise to $f(x)$ everywhere.*

Proof. Since $f \in W_1^1(-\pi,\pi)$, it is of bounded variation and continuous on the interval $[-\pi,\pi]$. In this case, the value $S(x)$ from (10.11) equals $f(x)$ at every point $x \in [-\pi,\pi]$. Thus,

$$\lim_{N\to\infty} S_N f(x) = f(x)$$

pointwise in $x \in [-\pi,\pi]$. □

Remark 10.8. The above proof does not allow us to conclude uniform convergence of the trigonometric Fourier series of functions from the Sobolev space $W_1^1(-\pi,\pi)$. However, uniform convergence is in fact the case, as we will prove later in this chapter.

Exercise 10.4. Show that

$$f(x) = \frac{1}{\log\frac{1}{|x|}}, \quad |x| < 1/2$$

is of bounded variation but this function does not satisfy condition (10.9) at $x = 0$.
Hint.

$$\left|\int_0^\delta \frac{1}{y\log y}\mathrm{d}y\right| = +\infty.$$

Exercise 10.5. Show that

$$f(x) = x\sin\frac{1}{x}$$

satisfies condition (10.9) at $x = 0$ but this function is not of bounded variation, see Exercise 1.3.

We can return now to the question of term-by-term integration of trigonometric Fourier series.

Theorem 10.9. *Suppose f belongs to $L^1(-\pi,\pi)$. Then*

$$\lim_{N\to\infty}\int_a^b S_N f(x)\mathrm{d}x=\int_a^b f(x)\mathrm{d}x$$

for every interval $(a,b)\subset[-\pi,\pi]$.

Proof. For a given L^1 function f (not necessarily periodic) we introduce a new function F as

$$F(x):=\int_{-\pi}^{x}(f(t)-c_0(f))\mathrm{d}t. \tag{10.15}$$

It is clear that $F(x)$ belongs to the Sobolev space $W_1^1(-\pi,\pi)$ with $F(-\pi)=F(\pi)=0$ (periodicity) and

$$F'(x)=f(x)-c_0(f).$$

This implies

$$c_n(F')=\mathrm{i}nc_n(F)=c_n(f),\quad n\neq 0,\quad c_0(F')=0.$$

Corollary 10.7 gives us that $F(x)$ has everywhere convergent trigonometric Fourier series

$$F(x)=c_0(F)+\sum_{n\neq 0}\frac{c_n(f)}{\mathrm{i}n}\mathrm{e}^{\mathrm{i}nx}. \tag{10.16}$$

In particular, for every $-\pi\le a<b\le\pi$ we have from (10.16) that

$$F(b)-F(a)=\sum_{n\neq 0}\frac{c_n(f)}{\mathrm{i}n}(\mathrm{e}^{\mathrm{i}nb}-\mathrm{e}^{\mathrm{i}na})=\sum_{n\neq 0}c_n(f)\int_a^b\mathrm{e}^{\mathrm{i}nx}\mathrm{d}x,$$

or equivalently (see (10.15)),

$$\begin{aligned}\int_{-\pi}^{b}(f(x)-c_0(f))\mathrm{d}x-\int_{-\pi}^{a}(f(x)-c_0(f))\mathrm{d}x&=\int_a^b f(x)\mathrm{d}x-(b-a)c_0(f)\\&=\sum_{n\neq 0}c_n(f)\int_a^b\mathrm{e}^{\mathrm{i}nx}\mathrm{d}x.\end{aligned}$$

Thus, we obtain finally

$$\begin{aligned}\int_a^b f(x)\mathrm{d}x=\sum_{n=-\infty}^{\infty}c_n(f)\int_a^b\mathrm{e}^{\mathrm{i}nx}\mathrm{d}x&=\lim_{N\to\infty}\sum_{|n|\le N}c_n(f)\int_a^b\mathrm{e}^{\mathrm{i}nx}\mathrm{d}x\\&=\lim_{N\to\infty}\int_a^b S_N f(x)\mathrm{d}x.\end{aligned}$$

This proves the theorem. □

Exercise 10.6. Calculate $c_0(F)$ for F defined by (10.15).

Corollary 10.10. *If $f \in L^1(-\pi,\pi)$, then the series*

$$\sum_{n\neq 0} \frac{c_n(f)}{n} \quad \textit{and} \quad \sum_{n\neq 0} \frac{c_n(f)(-1)^n}{n}$$

converge.

Proof. The result follows from (10.16). □

Corollary 10.11. *The series*

$$\sum_{n=1}^{\infty} \frac{\sin(nx)}{\log(1+n)}$$

is not the Fourier series of an L^1 function.

Proof. Let us assume to the contrary that there is a function $f \in L^1(-\pi,\pi)$ such that

$$\begin{aligned} f(x) &\sim \sum_{n=1}^{\infty} \frac{\sin(nx)}{\log(1+n)} = \frac{1}{2\mathrm{i}} \sum_{n=1}^{\infty} \frac{\mathrm{e}^{\mathrm{i}nx}}{\log(1+n)} - \frac{1}{2\mathrm{i}} \sum_{n=1}^{\infty} \frac{\mathrm{e}^{-\mathrm{i}nx}}{\log(1+n)} \\ &= \frac{1}{2\mathrm{i}} \sum_{n=1}^{\infty} \frac{\mathrm{e}^{\mathrm{i}nx}}{\log(1+n)} + \frac{1}{2\mathrm{i}} \sum_{n=-1}^{-\infty} \frac{\mathrm{sgn}(n)\mathrm{e}^{\mathrm{i}nx}}{\log(1+|n|)} = \frac{1}{2\mathrm{i}} \sum_{n\neq 0} \frac{\mathrm{sgn}(n)\mathrm{e}^{\mathrm{i}nx}}{\log(1+|n|)}, \end{aligned}$$

i.e., we have

$$c_n(f) = \frac{1}{2\mathrm{i}} \frac{\mathrm{sgn}(n)}{\log(1+|n|)}, \quad n \neq 0, \quad c_0(f) = 0.$$

Since $c_n(f) = -c_{-n}(f)$, this trigonometric Fourier series can be interpreted as the Fourier series of some odd L^1 function. Then Corollary 10.10 implies that

$$\sum_{n\neq 0} \frac{\mathrm{sgn}(n)}{2\mathrm{i}n\log(1+|n|)} = \frac{1}{\mathrm{i}} \sum_{n=1}^{\infty} \frac{1}{n\log(1+n)}$$

must be convergent. But this is not true. This contradiction proves this corollary. □

Remark 10.12. If we define the function f by the series in Corollary 10.11, then it turns out that

$$\int_{-\pi}^{\pi} f(x)\mathrm{d}x = 0, \quad \int_{-\pi}^{\pi} |f(x)|\mathrm{d}x = +\infty.$$

Recall that the Poisson kernel is equal to

$$P_r(x) = \frac{1-r^2}{1-2r\cos x + r^2}, \quad 0 \le r < 1$$

and its trigonometric Fourier series is

$$P_r(x) = \sum_{n=-\infty}^{\infty} r^{|n|} e^{inx}.$$

Corollary 10.4 shows us that this series converges to $P_r(x)$ uniformly in $x \in [-\pi, \pi]$.

Theorem 10.13. *Suppose that $f \in C[-\pi, \pi]$ is periodic. Then*

$$\lim_{r \to 1-} (P_r * f)(x) = f(x)$$

uniformly in $x \in [-\pi, \pi]$ or

$$\lim_{r \to 1-} \sum_{n=-\infty}^{\infty} r^{|n|} c_n(f) e^{inx} = f(x) \tag{10.17}$$

uniformly in $x \in [-\pi, \pi]$ even if f has no convergent trigonometric Fourier series.

Proof. Using the normalization

$$\frac{1}{2\pi} \int_{-\pi}^{\pi} P_r(x) dx = 1,$$

we have

$$\begin{aligned}(P_r * f)(x) - f(x) &= \frac{1}{2\pi} \int_{-\pi}^{\pi} P_r(y)(f(x-y) - f(x)) dy \\ &= \frac{1}{2\pi} \int_{|y| \le \delta} P_r(y)(f(x-y) - f(x)) dy \\ &+ \frac{1}{2\pi} \int_{\delta \le |y| \le \pi} P_r(y)(f(x-y) - f(x)) dy =: I_1 + I_2.\end{aligned}$$

Since f is continuous on $[-\pi, \pi]$, it follows that I_1 can be estimated as

$$\begin{aligned}|I_1| \le \sup_{x \in [-\pi,\pi], |y| \le \delta} |f(x-y) - f(x)| \frac{1}{2\pi} \int_{-\pi}^{\pi} P_r(y) dy \\ = \sup_{x \in [-\pi,\pi], |y| \le \delta} |f(x-y) - f(x)| \to 0\end{aligned}$$

as $\delta \to 0$. At the same time, I_2 can be estimated as

$$|I_2| \le 2 \max_{|x| \le \pi} |f(x)| \frac{1}{2\pi} \int_{\delta \le |y| \le \pi} P_r(y) dy.$$

For $\delta \le |y| \le \pi$ the Poisson kernel can be estimated as

$$P_r(y) = \frac{1-r^2}{1-2r\cos y + r^2} = \frac{1-r^2}{4r\sin^2 y/2 + (1-r)^2}$$
$$\le \frac{1-r}{2r\sin^2 y/2} \le \frac{1-r}{2r\delta^2/\pi^2} = \frac{\pi^2}{2}\frac{1-r}{r\delta^2}.$$

If we choose $\delta^4 = 1-r$, then I_2 is estimated as

$$|I_2| \le \frac{1}{\pi}\max_{|x|\le\pi}|f(x)|\frac{\pi^2}{2}\frac{1-r}{r\sqrt{1-r}} = \frac{\pi}{2}\max_{|x|\le\pi}|f(x)|\frac{\sqrt{1-r}}{r} \to 0$$

as $r \to 1-$. Hence, the estimates for I_1 and I_2 show that

$$\lim_{r\to 1-}((P_r * f)(x) - f(x)) = 0$$

uniformly in $x \in [-\pi, \pi]$. The equality (10.17) follows from this fact and Exercise 4.1. This completes the proof. □

We will prove now the well known *Hardy's theorem* and then apply it to the uniform convergence of the trigonometric sum $S_N f(x)$.

Theorem 10.14 (Hardy, 1949). *Let $\{a_k\}_{k=0}^{\infty}$ be a sequence of complex numbers such that*

$$k|a_k| \le M, \quad k = 0, 1, 2, \ldots, \tag{10.18}$$

where the constant M is independent of k. If the limit

$$\lim_{n\to\infty}\sigma_n := \lim_{n\to\infty}\sum_{j=0}^{n}\left(1 - \frac{j}{n+1}\right)a_j = a \tag{10.19}$$

exists, then

$$\lim_{n\to\infty}(\sigma_n - s_n) = 0,$$

where $s_n = \sum_{j=0}^{n} a_j$, i.e. also

$$\lim_{n\to\infty} s_n = a. \tag{10.20}$$

If a_k depends on x and (10.18) *holds uniformly in x and convergence in* (10.19) *is uniform, then convergence in* (10.20) *is also uniform.*

Proof. For $n < m$ it is true that

$$(m+1)\sigma_m - (n+1)\sigma_n - \sum_{j=n+1}^{m}(m+1-j)a_j = (m-n)s_n.$$

Indeed,

$$
\begin{aligned}
(m+1)\sigma_m - (n+1)\sigma_n - &\sum_{j=n+1}^{m}(m+1-j)a_j \\
&= \sum_{j=0}^{m}(m+1-j)a_j - \sum_{j=0}^{n}(n+1-j)a_j - \sum_{j=n+1}^{m}(m+1-j)a_j \\
&= \sum_{j=0}^{n}(m+1-j)a_j - \sum_{j=0}^{n}(n+1-j)a_j = \sum_{j=0}^{n}(m-n)a_j = (m-n)s_n.
\end{aligned}
$$

Therefore,

$$(m+1)\sigma_m - (n+1)\sigma_n - \sum_{j=n+1}^{m}(m+1-j)a_j - (m-n)\sigma_n = (m-n)s_n - (m-n)\sigma_n$$

or equivalently,

$$m(\sigma_m - \sigma_n) + (\sigma_m - \sigma_n) - \sum_{j=n+1}^{m}(m+1-j)a_j = (m-n)(s_n - \sigma_n),$$

i.e.,

$$\frac{m+1}{m-n}(\sigma_m - \sigma_n) - \frac{m+1}{m-n}\sum_{j=n+1}^{m}\left(1 - \frac{j}{m+1}\right)a_j = s_n - \sigma_n.$$

Let $m > n \to \infty$ be such that $m/n \sim 1+\delta$ (i.e., $\lim_{m,n\to\infty} m/n = 1+\delta$) with some positive δ to be chosen. Since σ_m is a Cauchy sequence by (10.19), it follows that

$$\frac{m+1}{m-n}(\sigma_m - \sigma_n) \to 0, \quad m > n \to \infty.$$

At the same time, the condition (10.18) implies that

$$
\begin{aligned}
\left|\frac{m+1}{m-n}\sum_{j=n+1}^{m}\left(1 - \frac{j}{m+1}\right)a_j\right| &\le \frac{m+1}{m-n}\sum_{j=n+1}^{m}\left(1 - \frac{j}{m+1}\right)\frac{M}{j} \\
&\sim M(1+1/\delta)\sum_{j=n+1}^{m}\left(\frac{1}{j} - \frac{1}{m+1}\right) \\
&= M(1+1/\delta)\left(\sum_{j=n+1}^{m}\frac{1}{j} - \frac{m-n}{m+1}\right) \\
&\le M(1+1/\delta)\left(\int_n^m \frac{1}{\xi}\mathrm{d}\xi - \frac{m-n}{m+1}\right) \\
&= M(1+1/\delta)\left(\log\frac{m}{n} - \frac{m-n}{m+1}\right)
\end{aligned}
$$

$$\begin{aligned}
&\sim M(1+1/\delta)\left(\log(1+\delta)-\frac{\delta}{1+\delta+1/n}\right)\\
&\sim M(1+1/\delta)\\
&\times\left(\delta-\delta^2/2+o(\delta^2)-\delta(1-\delta+O(\delta^2))\right)\\
&=M(1+1/\delta)\left(\delta^2/2+o(\delta^2)\right)\le 2M\delta
\end{aligned}$$

if δ is chosen small enough. Since δ is arbitrary, we may conclude that also the second term converges to zero. This proves the theorem. □

Corollary 10.15. *Suppose that $f \in C[-\pi,\pi]$ is periodic and that its Fourier coefficients satisfy*

$$|c_n(f)| \le \frac{M}{|n|}, \quad n \neq 0$$

with positive constant M that does not depend on n. Then the trigonometric Fourier series of f converges to f uniformly in $x \in [-\pi,\pi]$.

Proof. Since $f \in C[-\pi,\pi]$ is periodic, Theorem 5.1 gives the convergence of Fejér means

$$\lim_{N\to\infty} \sigma_N f(x) = f(x)$$

uniformly in $x \in [-\pi,\pi]$. Let us define $a_0 = c_0(f)$ and

$$a_k(x) = c_k(f)\mathrm{e}^{ikx} + c_{-k}(f)\mathrm{e}^{-ikx}, \quad k = 1,2,\ldots.$$

Then

$$s_n \equiv \sum_{k=0}^{n} a_k(x) = S_n f(x)$$

and

$$\sigma_n f(x) = \frac{1}{n+1}\sum_{k=0}^{n} S_k f(x) = \frac{1}{n+1}\sum_{k=0}^{n} s_k.$$

Thus, we are in the setting of Hardy's theorem, because

$$|a_k(x)| \le \frac{2M}{k}, \quad k = 1,2,\ldots.$$

Since this inequality is uniform in $x \in [-\pi,\pi]$, on applying Hardy's theorem we obtain that

$$\lim_{N\to\infty} s_N \equiv \lim_{N\to\infty} S_N f(x) = f(x)$$

uniformly in $x \in [-\pi,\pi]$. □

Corollary 10.16. *If $f \in W_1^1(-\pi,\pi)$ is periodic, then*

$$\lim_{N\to\infty} S_N f(x) = f(x)$$

uniformly in $x \in [-\pi,\pi]$.

Proof. Since $f \in W_1^1(-\pi,\pi)$ is periodic, there is $g \in L^1(-\pi,\pi)$ such that

$$f(x) = \int_{-\pi}^{x} g(t)\mathrm{d}t + f(-\pi), \quad \int_{-\pi}^{\pi} g(t)\mathrm{d}t = 0.$$

Thus, $f' = g$ and

$$c_n(g) = inc_n(f),$$

or equivalently,

$$|c_n(f)| = \left|\frac{c_n(g)}{in}\right| \leq \frac{M}{|n|}, \quad n \neq 0.$$

Due to embedding (see Lemma 1.14), the function f is continuous on the interval $[-\pi,\pi]$. Using again Hardy's theorem, we obtain

$$\lim_{N\to\infty} S_N f(x) = f(x)$$

uniformly in $x \in [-\pi,\pi]$. This completes the proof. □

Exercise 10.7. Using Theorem 10.9, prove the embedding

$$W_2^{\alpha}(-\pi,\pi) \subset W_1^1(-\pi,\pi), \quad \alpha > 1/2.$$

Remark 10.17. Corollary 10.16 and Exercise 10.7 show that for every function f from the spaces H_2^{α} and $B_{2,\theta}^{\alpha}$ with $1 \leq \theta < \infty$ and $\alpha > 1/2$, its trigonometric Fourier series converges to this function uniformly. Here one must take into account that f might be changed on a set of measure zero.

Let us return to some special trigonometric Fourier series. Namely, we consider functions $f_1(x)$ and $f_2(x)$ that are defined by the Fourier series

$$f_1(x) = \sum_{n=1}^{\infty} \frac{\sin(nx)}{n\log(1+n)} \tag{10.21}$$

and

$$f_2(x) = \sum_{n=1}^{\infty} \frac{\cos(nx)}{n\log(1+n)}. \tag{10.22}$$

These functions are well defined for all $x \in [-\pi, \pi] \setminus \{0\}$; see Theorem 3.3. In addition, $f_1(0) = 0$, whereas $f_2(0)$ is not defined, since the series (10.22) diverges at zero. We will show that $f_2(x)$ does not belong to $W_1^1(-\pi, \pi)$ but $f_1(x)$ does.

If we assume to the contrary that $f_2 \in W_1^1(-\pi, \pi)$, then its derivative f_2' has the Fourier series

$$f_2'(x) \sim -\sum_{n=1}^{\infty} \frac{\sin(nx)}{\log(1+n)}.$$

But due to Corollary 10.11 this is not a Fourier series of an L^1 function. This contradiction proves that $f_2 \notin W_1^1(-\pi, \pi)$.

Concerning the series (10.21), let us prove first that it converges uniformly in $x \in [-\pi, \pi]$, i.e., $f_1(x)$ is (at least) continuous on the interval $[-\pi, \pi]$. Indeed, by summation by parts we obtain for $0 \le M < N$ that

$$\begin{aligned}\sum_{n=M+1}^{N} \frac{\sin(nx)}{n\log(1+n)} &= \sum_{n=M+1}^{N} \frac{1}{n\log(1+n)} \left(\sum_{k=1}^{n} \sin(kx) - \sum_{k=1}^{n-1} \sin(kx) \right) \\ &= \frac{\sin(Nx)}{N\log(1+N)} - \frac{\sin(Mx)}{M\log(1+M)} \\ &- \sum_{n=M+2}^{N+1} \left(\sum_{k=1}^{n} \sin(kx) \right) \left(\frac{1}{(n+1)\log(2+n)} - \frac{1}{n\log(1+n)} \right). \end{aligned} \tag{10.23}$$

Using the calculation from Exercise 5.1, we have

$$\sum_{k=1}^{n} \sin(kx) = \frac{\cos x/2 - \cos(n+1/2)x}{2\sin x/2} = \frac{\sin(nx/2)\sin((n+1)x/2)}{\sin x/2}. \tag{10.24}$$

The first two terms on the right-hand side of (10.23) converge to zero as $N > M \to \infty$ uniformly in $x \in [-\pi, \pi]$. The sum on the right-hand side of (10.23) becomes, using (10.24),

$$\begin{aligned} &-\sum_{n=M+2}^{N+1} \frac{\sin(nx/2)\sin((n+1)x/2)}{\sin x/2} \frac{n\log\frac{2+n}{1+n} + \log(2+n)}{n(n+1)\log(1+n)\log(2+n)} \\ &= -\sum_{n=M+2}^{N+1} \frac{\sin(nx/2)\sin((n+1)x/2)}{\sin x/2} \frac{\log\frac{2+n}{1+n}}{(n+1)\log(1+n)\log(2+n)} \\ &- \sum_{n=M+2}^{N+1} \frac{\sin(nx/2)\sin((n+1)x/2)}{\sin x/2} \frac{1}{n(n+1)\log(1+n)} =: I_1 + I_2. \end{aligned}$$

Let us consider two cases: $n|x| < 1$ and $n|x| > 1$. In the first case,

$$\left| \frac{\sin(nx/2)\sin((n+1)x/2)}{\sin x/2} \right| \le \frac{n|x|/2 \cdot 1}{|x|/\pi} = \frac{\pi n}{2}.$$

Then

$$
\begin{aligned}
|I_1| &\le \frac{\pi}{2}\sum_{n=M+2}^{\left[\frac{1}{|x|}\right]} \frac{n\log(1+1/(n+1))}{(n+1)\log(1+n)\log(2+n)} < \frac{\pi}{2}\sum_{n=M+2}^{\left[\frac{1}{|x|}\right]} \frac{n\frac{1}{n}}{n\log^2 n} \\
&< \frac{\pi}{2}\sum_{n=M+2}^{\infty} \frac{1}{n\log^2 n} \to 0
\end{aligned} \tag{10.25}
$$

as $M \to \infty$ uniformly in x. In the first case, for I_2 we have, by integration by parts, that

$$
\begin{aligned}
|I_2| &\le \frac{\pi}{2}\sum_{n=M+2}^{\left[\frac{1}{|x|}\right]} \frac{n\frac{n+1}{2}|x|}{n(n+1)\log(1+n)} = \frac{\pi}{4}|x|\sum_{n=M+2}^{\left[\frac{1}{|x|}\right]} \frac{1}{\log(1+n)} \\
&< \frac{\pi}{4}|x|\int_{M+2}^{1/|x|} \frac{dt}{\log t} \\
&= \frac{\pi}{4}|x|\left(\frac{t}{\log t}\Big|_{M+2}^{1/|x|} + \int_{M+2}^{1/|x|} \frac{dt}{\log^2 t} \right) \\
&= \frac{\pi}{4}|x|\left(\frac{1/|x|}{\log(1/|x|)} - \frac{M+2}{\log(M+2)} + \int_{M+2}^{1/|x|} \frac{dt}{\log^2 t} \right) \\
&\le \frac{\pi}{4}\left(\frac{1}{\log(1/|x|)} + \frac{|x|(M+2)}{\log(M+2)} + \frac{|x|}{\log^2(M+2)}(1/|x|+M+2) \right) \\
&< \frac{\pi}{4}\left(\frac{1}{\log(M+2)} + \frac{1}{\log(M+2)} + \frac{1}{\log^2(M+2)} + \frac{1}{\log^2(M+2)} \right) \\
&\le \frac{\pi}{\log(M+2)} \to 0
\end{aligned}
$$

uniformly in x as $M \to \infty$. In the second case,

$$
\left| \frac{\sin(nx/2)\sin((n+1)x/2)}{\sin x/2} \right| \le \frac{1}{|\sin x/2|} \le \frac{\pi}{|x|} \le \pi n.
$$

Then

$$
\begin{aligned}
|I_1| &\le \pi \sum_{n=\left[\frac{1}{|x|}\right]}^{N+1} \frac{n\log(1+1/(n+1))}{(n+1)\log(1+n)\log(2+n)} < \pi \sum_{n=\left[\frac{1}{|x|}\right]}^{N+1} \frac{n\frac{1}{n}}{n\log^2 n} \\
&< \pi \sum_{n=\left[\frac{1}{|x|}\right]\ge M}^{\infty} \frac{1}{n\log^2 n} \to 0, \quad M \to \infty
\end{aligned}
$$

uniformly in x. For I_2 we have, by integration by parts,

$$
\begin{aligned}
|I_2| &\le \frac{1}{|\sin x/2|}\sum_{n=\left[\frac{1}{|x|}\right]+1}^{N+1}\frac{1}{n^2\log n}\le \pi\frac{1}{|x|}\int_{\left[\frac{1}{|x|}\right]\ge M}^{\infty}\frac{dt}{t^2\log t}\\
&= \pi\frac{1}{|x|}\left(-\frac{1}{t\log t}\Big|_{\left[\frac{1}{|x|}\right]\ge M}^{\infty}+\int_{\left[\frac{1}{|x|}\right]\ge M}^{\infty}\frac{dt}{t^2\log^2 t}\right)\\
&= \pi\left(\frac{1/|x|}{[1/|x|]\log[1/|x|]}+\frac{1}{|x|}\int_{\left[\frac{1}{|x|}\right]\ge M}^{\infty}\frac{dt}{t^2\log^2 t}\right)\\
&< \pi\left(\frac{[1/|x|]+1}{[1/|x|]\log[1/|x|]}+\frac{[1/|x|]+1}{[1/|x|]}\int_M^{\infty}\frac{dt}{t\log^2 t}\right)\\
&\le \pi\left(\frac{1+1/M}{\log M}+\frac{1+1/M}{\log M}\right)\to 0,\quad M\to\infty
\end{aligned}
\tag{10.26}
$$

uniformly in x. Finally, we may conclude that the trigonometric Fourier series (10.21) converges uniformly on the interval $[-\pi,\pi]$, and therefore it defines a continuous function $f_1(x)$. This series, as well as (10.22), can be differentiated term by term for $\pi\ge|x|\ge\delta>0$, because the series

$$\sum_{n=1}^{\infty}\frac{\cos(nx)}{\log(1+n)}$$

converges uniformly (see Corollary 3.4) for $\pi\ge|x|\ge\delta>0$. This means that for this interval, $f_1(x)$ belongs to C^1. Thus, it remains to investigate the behavior of the series (10.21) as $x\to 0+$. But the estimates (10.25)–(10.26) show us that (if we choose $M\asymp 1/x, x\to 0+$) the function $f_1(x)$ from (10.21) has the asymptotic behavior

$$f_1(x)\sim\frac{C}{\log x}. \tag{10.27}$$

It is possible to prove (see [46, Chapter V, formula (2.19)]) that the asymptotic (10.27) can be differentiated, and we obtain

$$f_1'(x)\sim-\frac{C}{x\log^2 x}.$$

This singularity is integrable at zero. Therefore, the function $f_1(x)$ belongs to $W_1^1(-\pi,\pi)$.

Exercise 10.8. Prove that the series (10.21) and (10.22) do not converge absolutely.

Chapter 11
Formulation of the Discrete Fourier Transform and Its Properties.

Let $x(t)$ be a 2π-periodic continuous signal. Assume that $x(t)$ can be represented by an absolutely convergent trigonometric Fourier series

$$x(t) = \sum_{m=-\infty}^{\infty} c_m e^{imt}, \quad t \in [-\pi, \pi], \tag{11.1}$$

where the c_m are the Fourier coefficients of $x(t)$.

Let now N be an even positive integer and

$$t_k = \frac{2\pi k}{N}, \quad k = -\frac{N}{2}, \ldots, \frac{N}{2} - 1.$$

Then $x(t_k)$ is a response at t_k, i.e.,

$$x(t_k) = \sum_{m=-\infty}^{\infty} c_m e^{i\frac{2\pi km}{N}}. \tag{11.2}$$

Since $e^{i2\pi kl} = 1$ for integers k and l, the series (11.2) can be rewritten as

$$\begin{aligned}\sum_{m=-\infty}^{\infty} c_m e^{i\frac{2\pi k}{N}(m-lN)} &= \sum_{l=-\infty}^{\infty} \sum_{-N/2 \le m-lN \le N/2-1} c_m e^{i\frac{2\pi k}{N}(m-lN)} \\ &= \sum_{l=-\infty}^{\infty} \sum_{n=-N/2}^{N/2-1} c_{n+lN} e^{i\frac{2\pi k}{N}n} \\ &= \sum_{n=-N/2}^{N/2-1} e^{i\frac{2\pi k}{N}n} \sum_{l=-\infty}^{\infty} c_{n+lN} = \sum_{n=-N/2}^{N/2-1} e^{i\frac{2\pi k}{N}n} X_n, \end{aligned} \tag{11.3}$$

where $X_n, n = -N/2, \ldots, N/2-1$ is given by

V. Serov, *Fourier Series, Fourier Transform and Their Applications to Mathematical Physics*, Applied Mathematical Sciences 197,
DOI 10.1007/978-3-319-65262-7_11

$$X_n = \sum_{l=-\infty}^{\infty} c_{n+lN}. \tag{11.4}$$

In the calculation (11.3) we have used the fact that the series (11.1) converges absolutely. Combining (11.2) and (11.3), we obtain

$$x_k := x(t_k) = \sum_{n=-N/2}^{N/2-1} X_n \mathrm{e}^{\mathrm{i}\frac{2\pi kn}{N}}. \tag{11.5}$$

The formula (11.5) can be viewed as an inverse discrete Fourier transform, and it appeared quite naturally in the discretization of a continuous periodic signal. Moreover, the formula (11.4) becomes the main property of this approach. Since

$$\sum_{n=-N/2}^{N/2-1} \mathrm{e}^{\mathrm{i}(k-m)\frac{2\pi n}{N}} = \begin{cases} 0, & k-m \neq 0, \pm N, \pm 2N, \ldots \\ N, & k-m = 0, \pm N, \pm 2N, \ldots \end{cases} \tag{11.6}$$

we solve the linear system (11.5) with respect to $X_n, n = -N/2, \ldots, N/2-1$, and obtain

$$X_n = \frac{1}{N} \sum_{k=-N/2}^{N/2-1} x_k \mathrm{e}^{-\mathrm{i}\frac{2\pi kn}{N}}. \tag{11.7}$$

Exercise 11.1. Prove (11.6) and (11.7).

In fact, the formulas (11.5) and (11.7) give us the inverse and direct discrete Fourier transforms, respectively.

Definition 11.1. The sequence $\{X_n\}_{n=-N/2}^{N/2-1}$ of complex numbers is called the *discrete Fourier transform* (DFT) of the sequence $\{Y_k\}_{k=-N/2}^{N/2-1}$ if for each $n = -N/2, \ldots,$ $N/2-1$ we have

$$X_n = \frac{1}{N} \sum_{k=-N/2}^{N/2-1} Y_k \mathrm{e}^{-\mathrm{i}\frac{2\pi kn}{N}}. \tag{11.8}$$

We use the symbol $\mathscr{F}$ for the DFT and write

$$X_n = \mathscr{F}(Y_k)_n$$

or simply $X = \mathscr{F}(Y)$.

Definition 11.2. The sequence $\{Z_k\}_{k=-N/2}^{N/2-1}$ of complex numbers is said to be the *inverse discrete Fourier transform* (IDFT) of the sequence $\{X_n\}_{n=-N/2}^{N/2-1}$ if for each $k = -N/2, \ldots, N/2-1$ we have

$$Z_k = \sum_{n=-N/2}^{N/2-1} X_n e^{i\frac{2\pi kn}{N}}. \tag{11.9}$$

We use the symbol $\mathscr{F}^{-1}$ for the IDFT and write

$$Z_k = \mathscr{F}^{-1}(X_n)_k$$

or simply $Z = \mathscr{F}^{-1}(X)$.

The properties of the DFT and IDFT are collected in the following lemmas.

Lemma 11.3. *The following equalities hold:*

(1) $\mathscr{F}^{-1}(\mathscr{F}(Y)) = Y$;
(2) $\mathscr{F}(\mathscr{F}^{-1}(X)) = X$;
(3)

$$\sum_{k=-N/2}^{N/2-1} \mathscr{F}(X_n)_k \overline{\mathscr{F}(Y_n)_k} = \frac{1}{N} \sum_{n=-N/2}^{N/2-1} X_n \overline{Y_n}.$$

Proof. Using (11.6), (11.8), and (11.9), we have

$$\begin{aligned}\mathscr{F}^{-1}(\mathscr{F}(Y))_k &= \sum_{n=-N/2}^{N/2-1} \mathscr{F}(Y_l)_n e^{i\frac{2\pi kn}{N}} = \frac{1}{N} \sum_{n=-N/2}^{N/2-1} \left(\sum_{l=-N/2}^{N/2-1} Y_l e^{-i\frac{2\pi ln}{N}} \right) e^{i\frac{2\pi kn}{N}} \\ &= \frac{1}{N} \sum_{l=-N/2}^{N/2-1} Y_l \left(\sum_{n=-N/2}^{N/2-1} e^{i\frac{2\pi n(k-l)}{N}} \right) = \frac{1}{N} Y_k N = Y_k.\end{aligned}$$

This proves part (1). Part (2) can be proved in the same manner. □

Exercise 11.2. Prove part (3) of Lemma 11.3.

Corollary 11.4 (Parseval's equality).

$$\frac{1}{N} \sum_{n=-N/2}^{N/2-1} |X_n|^2 = \sum_{k=-N/2}^{N/2-1} |\mathscr{F}(X_n)_k|^2.$$

Remark 11.5. Due to the periodicity of the complex exponential, we may extend the values of X_m, $m = -N/2, \ldots, N/2-1$ periodically to any integer by

$$X_{m+lN} = X_m, \quad l = 0, \pm 1, \pm 2, \ldots. \tag{11.10}$$

Corollary 11.6. *For a sequence* $X = \{X_n\}_{n=-N/2}^{N/2-1}$ *we define*

$$X_{\text{rev}} = \{X_{N-n}\}_{n=-N/2}^{N/2-1}.$$

Then

$$\mathscr{F}^{-1}(X) = N\mathscr{F}(X_{\text{rev}}).$$

Proof. By Definition 11.2 and the periodicity condition (11.10) we have

$$\begin{aligned}
N\mathscr{F}(X_{\text{rev}})_k &= \sum_{n=-N/2}^{N/2-1} X_{N-n} e^{-i\frac{2\pi kn}{N}} = \sum_{n=-N/2}^{N/2-1} X_{-n} e^{-i\frac{2\pi kn}{N}} = \sum_{n=N/2}^{-N/2+1} X_n e^{i\frac{2\pi kn}{N}} \\
&= \sum_{n=-N/2+1}^{N/2} X_n e^{i\frac{2\pi kn}{N}} = \sum_{n=-N/2}^{N/2-1} X_n e^{i\frac{2\pi kn}{N}} + X_{N/2} e^{i\pi k} - X_{-N/2} e^{-i\pi k} \\
&= \sum_{n=-N/2}^{N/2-1} X_n e^{i\frac{2\pi kn}{N}} = \mathscr{F}^{-1}(X)_k.
\end{aligned}$$

This concludes the proof. □

Definition 11.7. The *convolution* of discrete sequences $X = \{X_n\}_{n=-N/2}^{N/2-1}$ and $Y = \{Y_n\}_{n=-N/2}^{N/2-1}$ is defined as the sequence whose elements are given by

$$(X * Y)_k = \sum_{l=-N/2}^{N/2-1} X_l Y_{k-l}, \tag{11.11}$$

where X_n and Y_n satisfy the periodicity condition (11.10).

Proposition 11.8. *For every integer l, it is true that*

$$\sum_{m=-N/2-l}^{N/2-l-1} Y_m e^{-i\frac{2\pi nm}{N}} = \sum_{m=-N/2}^{N/2-1} Y_m e^{-i\frac{2\pi nm}{N}}.$$

Proof. The claim is trivial for $l = 0$. If $l > 0$, then

$$\begin{aligned}
\sum_{m=-N/2-l}^{N/2-l-1} Y_m e^{-i\frac{2\pi nm}{N}} &= \sum_{m=-N/2}^{N/2-1} Y_m e^{-i\frac{2\pi nm}{N}} + \sum_{m=-N/2-l}^{-N/2-1} Y_m e^{-i\frac{2\pi nm}{N}} - \sum_{m=N/2-l}^{N/2-1} Y_m e^{-i\frac{2\pi nm}{N}} \\
&= \sum_{m=-N/2}^{N/2-1} Y_m e^{-i\frac{2\pi nm}{N}} + \sum_{m=N/2-l}^{N/2-1} Y_{m-N} e^{-i\frac{2\pi n}{N}(m-N)} \\
&\quad - \sum_{m=N/2-l}^{N/2-1} Y_m e^{-i\frac{2\pi nm}{N}} = \sum_{m=-N/2}^{N/2-1} Y_m e^{-i\frac{2\pi nm}{N}}
\end{aligned}$$

due to the periodicity condition (11.10). If $l < 0$, then

$$\sum_{m=-N/2-l}^{N/2-l-1} Y_m e^{-i\frac{2\pi nm}{N}} = \sum_{m=-N/2}^{N/2-1} Y_m e^{-i\frac{2\pi nm}{N}} + \sum_{m=N/2}^{N/2-l-1} Y_m e^{-i\frac{2\pi nm}{N}} - \sum_{m=-N/2}^{-N/2-l-1} Y_m e^{-i\frac{2\pi nm}{N}}$$
$$= \sum_{m=-N/2}^{N/2-1} Y_m e^{-i\frac{2\pi nm}{N}} + \sum_{m=N/2}^{N/2-l-1} Y_m e^{-i\frac{2\pi nm}{N}}$$
$$- \sum_{m=N/2}^{N/2-l-1} Y_{m-N} e^{-i\frac{2\pi n}{N}(m-N)} = \sum_{m=-N/2}^{N/2-1} Y_m e^{-i\frac{2\pi nm}{N}}$$

due to periodicity condition (11.10). This proves the proposition. □

Corollary 11.9. *For every integer l it is true that*

$$\sum_{m=-N/2-l}^{N/2-l-1} Y_m = \sum_{m=-N/2}^{N/2-1} Y_m.$$

Lemma 11.10. *The convolution* (11.11) *is symmetric, i.e.,*

$$(X*Y)_k = (Y*X)_k$$

for every $k = -N/2,\dots,N/2-1$.

Proof. We have

$$(X*Y)_k = \sum_{l=-N/2}^{N/2-1} X_l Y_{k-l} = \sum_{j=k+N/2}^{k+1-N/2} Y_j X_{k-j}$$
$$= \sum_{j=-N/2+(k+1)}^{N/2-1+(k+1)} Y_j X_{k-j} = \sum_{j=-N/2}^{N/2-1} Y_j X_{k-j} = (Y*X)_k$$

by Corollary 11.9. □

Lemma 11.11. *For each $n = -N/2,\dots,N/2-1$ it is true that*

(1) $\mathscr{F}(X*Y)_n = N\mathscr{F}(X)_n\mathscr{F}(Y)_n$;
(2) $\mathscr{F}^{-1}(X*Y)_n = \mathscr{F}^{-1}(X)_n\mathscr{F}^{-1}(Y)_n$.

Proof. Using (11.11), we have

$$\mathscr{F}(X*Y)_n = \frac{1}{N}\sum_{k=-N/2}^{N/2-1}(X*Y)_k e^{-i\frac{2\pi kn}{N}} = \frac{1}{N}\sum_{l=-N/2}^{N/2-1} X_l \sum_{k=-N/2}^{N/2-1} Y_{k-l} e^{-i\frac{2\pi kn}{N}}$$

$$= \frac{1}{N} \sum_{l=-N/2}^{N/2-1} X_l \sum_{m=-N/2-l}^{N/2-l-1} Y_m \mathrm{e}^{-\mathrm{i}\frac{2\pi n}{N}(m+l)}$$

$$= \frac{1}{N} \sum_{l=-N/2}^{N/2-1} X_l \mathrm{e}^{-\mathrm{i}\frac{2\pi nl}{N}} \sum_{m=-N/2-l}^{N/2-l-1} Y_m \mathrm{e}^{-\mathrm{i}\frac{2\pi nm}{N}}. \tag{11.12}$$

Proposition 11.8 allows us to rewrite (11.12) as

$$\mathscr{F}(X*Y)_n = \frac{1}{N} \sum_{l=-N/2}^{N/2-1} X_l \mathrm{e}^{-\mathrm{i}\frac{2\pi nl}{N}} \sum_{m=-N/2}^{N/2-1} Y_m \mathrm{e}^{-\mathrm{i}\frac{2\pi nm}{N}} = N\mathscr{F}(X)_n \mathscr{F}(Y)_n.$$

Part (2) is proved in a similar manner. □

Corollary 11.12. *For each $n = -N/2,\dots,N/2-1$ it is true that*

$$\mathscr{F}^{-1}(X \cdot Y)_n = \frac{1}{N}\left(\mathscr{F}^{-1}(X) * \mathscr{F}^{-1}(Y)\right)_n,$$
$$\mathscr{F}(X \cdot Y)_n = \left(\mathscr{F}(X) * \mathscr{F}(Y)\right)_n,$$

where $X \cdot Y$ denotes the sequence $\{X_k \cdot Y_k\}_{k=-N/2}^{N/2-1}$.

Proof. Lemmas 11.3 and 11.11 imply that

$$\left(\widetilde{X} * \widetilde{Y}\right)_n = \mathscr{F}^{-1}\left(\mathscr{F}\left(\widetilde{X} * \widetilde{Y}\right)\right)_n = N\mathscr{F}^{-1}\left(\mathscr{F}(\widetilde{X}) \cdot \mathscr{F}(\widetilde{Y})\right)_n.$$

Setting $\widetilde{X} := \mathscr{F}^{-1}(X)$ and $\widetilde{Y} := \mathscr{F}^{-1}(Y)$, we obtain easily from the latter equality that

$$N\mathscr{F}^{-1}(X \cdot Y)_n = \left(\mathscr{F}^{-1}(X) * \mathscr{F}^{-1}(Y)\right)_n.$$

The second part is proved in a similar manner. □

Let us return to the continuous signal $x(t)$, $t \in [-\pi, \pi]$, which is represented by an absolutely convergent trigonometric Fourier series (11.1). Formula (11.4) allows us to obtain

$$\sum_{n=-N/2}^{N/2-1} |X_n - c_n| = \sum_{n=-N/2}^{N/2-1} \left| \sum_{l=-\infty}^{\infty} c_{n+lN} - c_n \right| = \sum_{n=-N/2}^{N/2-1} \left| \sum_{l \neq 0} c_{n+lN} \right|$$
$$\leq \sum_{n=-N/2}^{N/2-1} \sum_{l \neq 0} |c_{n+lN}| \leq \sum_{|\nu| \geq N/2} |c_\nu|. \tag{11.13}$$

Similarly, we have

$$\left|\mathscr{F}^{-1}(X_n)_k - \sum_{n=-N/2}^{N/2-1} c_n e^{i\frac{2\pi kn}{N}}\right| = \left|\sum_{n=-N/2}^{N/2-1} (X_n - c_n) e^{i\frac{2\pi kn}{N}}\right|$$
$$\leq \sum_{n=-N/2}^{N/2-1} |X_n - c_n| \leq \sum_{|\nu|\geq N/2} |c_\nu|. \quad (11.14)$$

In the formulas (11.13) and (11.14) the numbers c_n are the Fourier coefficients of the signal $x(t)$, and $\{X_n\}_{n=-N/2}^{N/2-1}$ is the DFT of $\{x(t_k)\}_{k=-N/2}^{N/2-1}$ with $t_k = 2\pi k/N$.

Theorem 11.13. *If $x(t)$ is periodic and belongs to the Sobolev space $W_2^m(-\pi,\pi)$ for some $m = 1,2,\ldots$, then*

$$X_n = c_n + o\left(\frac{1}{N^{m-1/2}}\right), \quad N \to \infty \quad (11.15)$$

and

$$\mathscr{F}^{-1}(X_n)_k = \sum_{n=-N/2}^{N/2-1} c_n e^{i\frac{2\pi kn}{N}} + o\left(\frac{1}{N^{m-1/2}}\right), \quad N \to \infty \quad (11.16)$$

uniformly in n and k from the set $\{-N/2,\ldots,N/2-1\}$.

Proof. Using Hölder's inequality, we have

$$\sum_{|\nu|\geq N/2} |c_\nu| \leq \left(\sum_{|\nu|\geq N/2} |\nu|^{2m}|c_\nu|^2\right)^{1/2} \left(\sum_{|\nu|\geq N/2} |\nu|^{-2m}\right)^{1/2}.$$

The first sum on the right-hand side tends to zero as $N \to \infty$ due to Parseval's equality for a function from the Sobolev space $W_2^m(-\pi,\pi)$. The second sum can be estimated precisely. Namely, since for every $m = 1,2,\ldots$ we have

$$\left(\sum_{|\nu|\geq N/2} |\nu|^{-2m}\right)^{1/2} \asymp \left(\int_{N/2}^{\infty} t^{-2m} dt\right)^{1/2} \asymp N^{-m+1/2},$$

we conclude that (11.15) and (11.16) follow from the last estimate and (11.13) and (11.14), respectively. □

Corollary 11.14. *An unknown periodic function $x(t) \in W_2^m(-\pi,\pi), m = 1,2,\ldots$, can be recovered from its IDFT as*

$$x\left(\frac{2\pi k}{N}\right) = \mathscr{F}^{-1}(X_n)_k + o\left(\frac{1}{N^{m-1/2}}\right), \quad N \to \infty$$

uniformly in k from the set $\{-N/2,\ldots,N/2-1\}$.

Exercise 11.3. (1) Show that

$$\mathscr{F}\left(\{a^k\}_{k=-N/2}^{N/2-1}\right)_n = \begin{cases} 1, & a = e^{i\frac{2\pi n}{N}}, \\ \frac{1}{N}(-1)^n \frac{a^{-N/2}-a^{N/2}}{1-ae^{-i\frac{2\pi n}{N}}}, & a \neq e^{i\frac{2\pi n}{N}}. \end{cases}$$

(2) If a sequence $Y = \{Y_k\}_{k=-N/2}^{N/2-1}$ is real, then show that

$$\mathscr{F}(Y)_n = \overline{\mathscr{F}(Y)_{N-n}}.$$

Chapter 12
Connection Between the Discrete Fourier Transform and the Fourier Transform.

If a function $f(x)$ is integrable over the whole line, i.e.,

$$\int_{-\infty}^{\infty} |f(x)|\mathrm{d}x < \infty,$$

then its *Fourier transform* is defined as

$$\mathscr{F}f(\xi) = \widehat{f}(\xi) := \frac{1}{\sqrt{2\pi}}\int_{-\infty}^{\infty} f(x)\mathrm{e}^{-\mathrm{i}x\xi}\mathrm{d}x.$$

Similarly, the *inverse Fourier transform* of an integrable function $g(\xi)$ is defined as

$$\mathscr{F}^{-1}g(x) := \frac{1}{\sqrt{2\pi}}\int_{-\infty}^{\infty} g(\xi)\mathrm{e}^{\mathrm{i}x\xi}\mathrm{d}\xi.$$

Theorem 12.1 (Riemann–Lebesgue lemma). *For an integrable function $f(x)$, its Fourier transform $\mathscr{F}f(\xi)$ is continuous, and*

$$\lim_{\xi\to\pm\infty} \mathscr{F}f(\xi) = 0.$$

Proof. Since $\mathrm{e}^{\mathrm{i}\pi} = -1$, we have

$$\widehat{f}(\xi) = -\frac{1}{\sqrt{2\pi}}\int_{-\infty}^{\infty} f(x)\mathrm{e}^{-\mathrm{i}x\xi+\mathrm{i}\pi}\mathrm{d}x = -\frac{1}{\sqrt{2\pi}}\int_{-\infty}^{\infty} f(y+\pi/\xi)\mathrm{e}^{-\mathrm{i}\xi y}\mathrm{d}y.$$

This fact implies that

$$-2\widehat{f}(\xi) = \frac{1}{\sqrt{2\pi}}\int_{-\infty}^{\infty} (f(x+\pi/\xi) - f(x))\mathrm{e}^{-\mathrm{i}\xi x}\mathrm{d}x.$$

V. Serov, *Fourier Series, Fourier Transform and Their Applications to Mathematical Physics*, Applied Mathematical Sciences 197,
DOI 10.1007/978-3-319-65262-7_12

Hence

$$2|\widehat{f}(\xi)| \leq \frac{1}{\sqrt{2\pi}} \int_{-\infty}^{\infty} |f(x+\pi/\xi) - f(x)| \mathrm{d}x \to 0$$

as $|\xi| \to \infty$, since f is integrable on the whole line. This is a well known property of integrable functions. Continuity of $\widehat{f}(\xi)$ follows from the representation

$$\widehat{f}(\xi+h) - \widehat{f}(\xi) = \frac{1}{\sqrt{2\pi}} \int_{-\infty}^{\infty} f(x) \mathrm{e}^{-\mathrm{i}x\xi} (\mathrm{e}^{-\mathrm{i}xh} - 1) \mathrm{d}x$$

and its consequence

$$\begin{aligned} &|\widehat{f}(\xi+h) - \widehat{f}(\xi)| \\ &\quad \leq \frac{1}{\sqrt{2\pi}} \int_{|xh|<\delta} |f(x)||\mathrm{e}^{-\mathrm{i}xh} - 1| \mathrm{d}x + \frac{2}{\sqrt{2\pi}} \int_{|xh|>\delta} |f(x)| \mathrm{d}x =: I_1 + I_2. \end{aligned}$$

For the first term I_1 we have the estimate

$$I_1 \leq \frac{1}{\sqrt{2\pi}} \int_{|xh|<\delta} |f(x)||xh| \mathrm{d}x < \frac{\delta}{\sqrt{2\pi}} \int_{-\infty}^{\infty} |f(x)| \mathrm{d}x \to 0, \quad \delta \to 0.$$

For the second term I_2 we have

$$I_2 = \frac{2}{\sqrt{2\pi}} \int_{|x|>\delta/|h|} |f(x)| \mathrm{d}x \to 0$$

as $|h| \to 0$. If we choose $\delta = |h|^{1/2}$, then both I_1 and I_2 tend to zero as $|h| \to 0$. This completes the proof. □

If a function $f(x)$ has integrable derivatives $f^{(k)}(x)$ of order $k = 0, 1, 2, \ldots, m$, then we say that f belongs to the Sobolev space $W_1^m(\mathbb{R})$.

Exercise 12.1. Prove that if $f \in W_1^1(\mathbb{R})$, then $\lim_{x\to\pm\infty} f(x) = 0$.

Theorem 12.2 (Fourier inversion formula). *Suppose that f belongs to $W_1^1(\mathbb{R})$. Then*

$$\mathscr{F}^{-1}(\mathscr{F} f)(x) = f(x)$$

at every point $x \in \mathbb{R}$.

Proof. First we prove that

$$\int_{-\infty}^{\infty} f(x)\widehat{g}(x) \mathrm{d}x = \int_{-\infty}^{\infty} \widehat{f}(\xi) g(\xi) \mathrm{d}\xi$$

for every pair of integrable functions f and g. Indeed,

$$\int_{-\infty}^{\infty} f(x)\widehat{g}(x)\mathrm{d}x = \frac{1}{\sqrt{2\pi}}\int_{-\infty}^{\infty} f(x)\int_{-\infty}^{\infty} g(\xi)\mathrm{e}^{-\mathrm{i}x\xi}\mathrm{d}\xi\mathrm{d}x$$
$$= \frac{1}{\sqrt{2\pi}}\int_{-\infty}^{\infty} g(\xi)\int_{-\infty}^{\infty} f(x)\mathrm{e}^{-\mathrm{i}x\xi}\mathrm{d}x\mathrm{d}\xi = \int_{-\infty}^{\infty}\widehat{f}(\xi)g(\xi)\mathrm{d}\xi$$

by Fubini's theorem. Suppose now that $g(\xi)$ is given by

$$g(\xi) = \begin{cases} 1, & |\xi| < n, \\ 0, & |\xi| > n. \end{cases}$$

Its Fourier transform is equal to

$$\widehat{g}(x) = \frac{1}{\sqrt{2\pi}}\int_{-n}^{n}\mathrm{e}^{-\mathrm{i}x\xi}\mathrm{d}\xi = \frac{1}{\sqrt{2\pi}}\left(\frac{\mathrm{e}^{-\mathrm{i}xn}}{-\mathrm{i}x} - \frac{\mathrm{e}^{\mathrm{i}xn}}{-\mathrm{i}x}\right) = \sqrt{\frac{2}{\pi}}\frac{\sin(nx)}{x}.$$

Thus, we have the equality

$$\sqrt{\frac{2}{\pi}}\int_{-\infty}^{\infty} f(x)\frac{\sin(nx)}{x}\mathrm{d}x = \int_{-n}^{n}\widehat{f}(\xi)\mathrm{d}\xi,$$

where $f \in W_1^1(\mathbb{R})$. Letting $n \to \infty$, we obtain

$$\mathrm{p.v.}\int_{-\infty}^{\infty}\widehat{f}(\xi)\mathrm{d}\xi = \lim_{n\to\infty}\sqrt{\frac{2}{\pi}}\int_{-\infty}^{\infty} f(x)\frac{\sin(nx)}{x}\mathrm{d}x. \tag{12.1}$$

We will prove that the limit in (12.1) is actually equal to $\sqrt{2\pi}f(0)$. Since

$$\int_{-\infty}^{\infty}\frac{\sin(nx)}{x}\mathrm{d}x = \pi,$$

the limit in (12.1) can be rewritten as

$$\lim_{n\to\infty}\sqrt{\frac{2}{\pi}}\int_{-\infty}^{\infty} f(x)\frac{\sin(nx)}{x}\mathrm{d}x$$
$$= \sqrt{2\pi}f(0) + \lim_{n\to\infty}\sqrt{\frac{2}{\pi}}\int_{-\infty}^{\infty}(f(x) - f(0))\frac{\sin(nx)}{x}\mathrm{d}x$$
$$= \sqrt{2\pi}f(0) + \lim_{n\to\infty}\sqrt{\frac{2}{\pi}}\int_{-\infty}^{\infty}(f(t/n) - f(0))\frac{\sin(t)}{t}\mathrm{d}t.$$

It remains to show that the latter limit is equal to zero. In order to prove this fact we split the above integral as follows:

$$\int_{-\infty}^{\infty}(f(t/n)-f(0))\frac{\sin(t)}{t}\mathrm{d}t$$
$$=\int_{|t|<1}(f(t/n)-f(0))\frac{\sin(t)}{t}\mathrm{d}t+\int_{|t|>1}(f(t/n)-f(0))\frac{\sin(t)}{t}\mathrm{d}t=:I_1+I_2.$$

Since $f\in W_1^1(\mathbb{R})$, it is continuous with respect to the definition of Sobolev spaces in this chapter. Therefore,

$$|I_1|\le\sup_{|t|<1}|f(t/n)-f(0)|\int_{|t|<1}\left|\frac{\sin(t)}{t}\right|\mathrm{d}t\to 0,\quad n\to\infty.$$

For the second term, I_2, we first change variables to obtain

$$\begin{aligned}I_2&=\int_{|z|>1/n}(f(z)-f(0))\frac{\sin(nz)}{z}\mathrm{d}z\\&=\int_{\frac{1}{n}}^{\infty}(f(z)-f(0))\left(\int_0^z\frac{\sin(ny)}{y}\mathrm{d}y\right)'\mathrm{d}z\\&\quad+\int_{-\infty}^{-\frac{1}{n}}(f(z)-f(0))\left(\int_0^z\frac{\sin(ny)}{y}\mathrm{d}y\right)'\mathrm{d}z.\end{aligned}$$

Integration by parts in these integrals leads to

$$\begin{aligned}I_2&=(f(z)-f(0))\int_0^z\frac{\sin(ny)}{y}\mathrm{d}y\Big|_{1/n}^{\infty}-\int_{1/n}^{\infty}f'(z)\int_0^z\frac{\sin(ny)}{y}\mathrm{d}y\mathrm{d}z\\&\quad+(f(z)-f(0))\int_0^z\frac{\sin(ny)}{y}\mathrm{d}y\Big|_{-\infty}^{-1/n}-\int_{-\infty}^{-1/n}f'(z)\int_0^z\frac{\sin(ny)}{y}\mathrm{d}y\mathrm{d}z\\&=\left(\lim_{z\to\infty}f(z)-f(0)\right)\int_0^{\infty}\frac{\sin(ny)}{y}\mathrm{d}y-(f(1/n)-f(0))\int_0^{1/n}\frac{\sin(ny)}{y}\mathrm{d}y\\&\quad-\int_{1/n}^{\infty}f'(z)\int_0^{nz}\frac{\sin(t)}{t}\mathrm{d}t\mathrm{d}z+(f(-1/n)-f(0))\int_0^{-1/n}\frac{\sin(ny)}{y}\mathrm{d}y\\&\quad-\left(\lim_{z\to-\infty}f(z)-f(0)\right)\int_0^{-\infty}\frac{\sin(ny)}{y}\mathrm{d}y-\int_{-\infty}^{-1/n}f'(z)\int_0^{nz}\frac{\sin(t)}{t}\mathrm{d}t\mathrm{d}z.\end{aligned}$$

Since $\lim_{z\to\pm\infty}f(z)=0$ (see Exercise 12.1) and since f is continuous, we obtain (as $n\to\infty$)

$$\begin{aligned}I_2\to&-f(0)\frac{\pi}{2}-\lim_{n\to\infty}\int_{1/n}^{\infty}f'(z)\int_0^{nz}\frac{\sin(t)}{t}\mathrm{d}t\mathrm{d}z\\&-f(0)\frac{\pi}{2}-\lim_{n\to\infty}\int_{-\infty}^{-1/n}f'(z)\int_0^{nz}\frac{\sin(t)}{t}\mathrm{d}t\mathrm{d}z\end{aligned}$$

$$\begin{aligned}
&= -\pi f(0) - \lim_{n\to\infty}\int_{1/n}^{\infty} f'(z)\int_0^{nz}\frac{\sin(t)}{t}\mathrm{d}t\mathrm{d}z \\
&\quad - \lim_{n\to\infty}\int_{-\infty}^{-1/n} f'(z)\int_0^{nz}\frac{\sin(t)}{t}\mathrm{d}t\mathrm{d}z \\
&= -\pi f(0) - \int_0^{\infty} f'(z)\frac{\pi}{2}\mathrm{d}z + \int_{-\infty}^{0} f'(z)\frac{\pi}{2}\mathrm{d}z \\
&= -\pi f(0) + f(0)\frac{\pi}{2} + f(0)\frac{\pi}{2} = 0.
\end{aligned}$$

Here we used again the fact that $\lim_{z\to\pm\infty} f(z) = 0$ and Lebesgue's dominated convergence theorem. Thus, (12.1) transforms to

$$\text{p. v.}\int_{-\infty}^{\infty}\widehat{f}(\xi)\mathrm{d}\xi = \sqrt{2\pi}f(0),$$

or equivalently,

$$\mathscr{F}^{-1}(\mathscr{F}f)(0) = f(0).$$

In order to prove the Fourier inversion formula for every $x\in\mathbb{R}$, let us note that

$$\begin{aligned}
\widehat{f_x(y)}(\xi) = \widehat{f(x+y)}(\xi) &= \frac{1}{\sqrt{2\pi}}\int_{-\infty}^{\infty} f(x+y)\mathrm{e}^{-\mathrm{i}y\xi}\mathrm{d}y \\
&= \frac{1}{\sqrt{2\pi}}\int_{-\infty}^{\infty} f(z)\mathrm{e}^{-\mathrm{i}z\xi}\mathrm{e}^{\mathrm{i}x\xi}\mathrm{d}z = \mathrm{e}^{\mathrm{i}x\xi}\widehat{f}(\xi).
\end{aligned}$$

Since $f_x(0) = f(x)$, it follows that

$$\begin{aligned}
f(x) &= f_x(0) = \mathscr{F}^{-1}(\mathscr{F}f_x)(0) \\
&= \frac{1}{\sqrt{2\pi}}\int_{-\infty}^{\infty}(\mathscr{F}f_x)(\xi)\mathrm{d}\xi = \frac{1}{\sqrt{2\pi}}\int_{-\infty}^{\infty}\mathrm{e}^{\mathrm{i}x\xi}\widehat{f}(\xi)\mathrm{d}\xi = \mathscr{F}^{-1}(\mathscr{F}f)(x).
\end{aligned}$$

This completes the proof. □

Remark 12.3. As a by-product of the above proof we record the limit

$$\lim_{n\to\infty}\frac{1}{\pi}\int_{-\infty}^{\infty} f(x)\frac{\sin(nx)}{x}\mathrm{d}x = f(0)$$

for every $f\in W_1^1(\mathbb{R})$.

Lemma 12.4. *If f belongs to the Sobolev space $W_1^m(\mathbb{R})$ for some $m = 1,2,\ldots$, then*

$$\widehat{f}(\xi) = o\left(\frac{1}{|\xi|^m}\right) \tag{12.2}$$

as $|\xi|\to\infty$.

Proof. Since $f \in W_1^m(\mathbb{R})$, it follows that $f', f'', \dots, f^{(m-1)} \in W_1^1(\mathbb{R})$. By Exercise 12.1, we have

$$\lim_{x\to\pm\infty} f^{(k)}(x) = 0$$

for each $k = 0, 1, \dots, m-1$. This fact and repeated integration by parts give us

$$\int_{-\infty}^{\infty} f(x)\mathrm{e}^{-\mathrm{i}x\xi}\mathrm{d}x = \frac{\mathrm{e}^{-\mathrm{i}x\xi}}{-\mathrm{i}\xi}f(x)\Big|_{-\infty}^{\infty} + \frac{1}{\mathrm{i}\xi}\int_{-\infty}^{\infty} f'(x)\mathrm{e}^{-\mathrm{i}x\xi}\mathrm{d}x = \frac{1}{\mathrm{i}\xi}\int_{-\infty}^{\infty} f'(x)\mathrm{e}^{-\mathrm{i}x\xi}\mathrm{d}x$$

$$= -\frac{\mathrm{e}^{-\mathrm{i}x\xi}}{(\mathrm{i}\xi)^2}f'(x)\Big|_{-\infty}^{\infty} + \frac{1}{(\mathrm{i}\xi)^2}\int_{-\infty}^{\infty} f''(x)\mathrm{e}^{-\mathrm{i}x\xi}\mathrm{d}x$$

$$= \frac{1}{(\mathrm{i}\xi)^2}\int_{-\infty}^{\infty} f''(x)\mathrm{e}^{-\mathrm{i}x\xi}\mathrm{d}x = \cdots = \frac{1}{(\mathrm{i}\xi)^m}\int_{-\infty}^{\infty} f^{(m)}(x)\mathrm{e}^{-\mathrm{i}x\xi}\mathrm{d}x = o\left(\frac{1}{|\xi|^m}\right)$$

due to the Riemann–Lebesgue lemma. □

The equality (12.2) allows us to consider (with respect to the accuracy of calculations) the Fourier transform only on the interval $(-R,R)$ with $R > 0$ sufficiently large, i.e., we may neglect the values of $\mathcal{F}f(\xi)$ for $|\xi| > R$. This simplification justifies the following approximation of the inverse Fourier transform:

$$f^*(x) := \frac{1}{\sqrt{2\pi}}\int_{-R}^{R} \mathcal{F}f(\xi)\mathrm{e}^{\mathrm{i}x\xi}\mathrm{d}\xi.$$

At the same time and without loss of generality we may assume that the function $f(x)$ is equal to zero outside some finite interval. In that case it can be proved that $\mathcal{F}f(\xi)$ is a smooth function for which (12.2) holds.

Definition 12.5. We say that $f \in \mathring{W}_1^m(-R,R)$ if $f \in W_1^m(\mathbb{R})$ and $f \equiv 0$ for $x \notin (-R,R)$.

Theorem 12.6. *Suppose that $f \in \mathring{W}_1^m(-R,R)$ is supported in a fixed interval $[a,b] \subset (-R,R)$ with $R > 0$ sufficiently large for some $m = 2,3,\dots$. Then*

$$f(x) = \sqrt{\frac{2}{\pi}}\frac{1}{N^{m/(m+2)}}\sum_{n=-N/2}^{N/2-1}\mathcal{F}f\left(\frac{2n+1}{N^{m/(m+2)}}\right)\mathrm{e}^{\mathrm{i}\frac{x(2n+1)}{N^{m/(m+2)}}}$$
$$+ O\left(\frac{1}{N^{(2m-2)/(m+2)}}\right) \qquad (12.3)$$

uniformly in $x \in (-R,R)$ for $R = N^{2/(m+2)}$ and even $N \to \infty$.

Proof. Since $f(x) = 0$ for $x \notin (-R,R)$, using the Fourier inversion formula (see Theorem 12.2) we have

$$f(x) - f^*(x) = \frac{1}{\sqrt{2\pi}} \int_{-\infty}^{\infty} \mathscr{F}f(\xi)\mathrm{e}^{\mathrm{i}x\xi}\mathrm{d}\xi - \frac{1}{\sqrt{2\pi}} \int_{-R}^{R} \mathscr{F}f(\xi)\mathrm{e}^{\mathrm{i}x\xi}\mathrm{d}\xi$$
$$= \frac{1}{\sqrt{2\pi}} \int_{|\xi|>R} \mathscr{F}f(\xi)\mathrm{e}^{\mathrm{i}x\xi}\mathrm{d}\xi.$$

Lemma 12.4 implies then that

$$f(x) - f^*(x) = o\left(\frac{1}{R^{m-1}}\right). \tag{12.4}$$

Let us divide the interval $[-R,R]$ into $N+1$ subintervals $[\xi_n, \xi_{n+1}]$, where $n = -N/2, \ldots, N/2-1$ such that

$$-R = \xi_{-N/2} < \xi_{-N/2+1} < \cdots < \xi_{N/2} = R,$$

where

$$\xi_n = \frac{2Rn}{N}, \quad \xi_{n+1} - \xi_n = \frac{2R}{N}.$$

Let us also set

$$\xi_n^* = \frac{\xi_n + \xi_{n+1}}{2} = \frac{R}{N}(2n+1).$$

Then we obtain

$$f^*(x) = \frac{1}{\sqrt{2\pi}} \int_{-R}^{R} \mathscr{F}f(\xi)\mathrm{e}^{\mathrm{i}x\xi}\mathrm{d}\xi$$
$$= \frac{1}{\sqrt{2\pi}} \sum_{n=-N/2}^{N/2-1} \mathscr{F}f(\xi_n^*)\mathrm{e}^{\mathrm{i}x\xi_n^*} \frac{2R}{N}$$
$$+ \frac{1}{\sqrt{2\pi}} \sum_{n=-N/2}^{N/2-1} \int_{\xi_n}^{\xi_{n+1}} \left(\mathscr{F}f(\xi)\mathrm{e}^{\mathrm{i}x\xi} - \mathscr{F}f(\xi_n^*)\mathrm{e}^{\mathrm{i}x\xi_n^*}\right) \mathrm{d}\xi$$
$$= \sqrt{\frac{2}{\pi}} \frac{R}{N} \sum_{n=-N/2}^{N/2-1} \mathscr{F}f(R(2n+1)/N)\mathrm{e}^{\mathrm{i}xR(2n+1)/N} + O\left(\frac{R^3}{N^2}\right). \tag{12.5}$$

Exercise 12.2. Prove that

$$\frac{1}{\sqrt{2\pi}}\sum_{n=-N/2}^{N/2-1}\int_{\xi_n}^{\xi_{n+1}}\left(\mathscr{F}f(\xi)\mathrm{e}^{\mathrm{i}x\xi}-\mathscr{F}f(\xi_n^*)\mathrm{e}^{\mathrm{i}x\xi_n^*}\right)\mathrm{d}\xi = \begin{cases} O\left(\frac{R^3}{N^2}\right), & \operatorname{supp} f=[a,b] \\ O\left(\frac{R^5}{N^2}\right), & \operatorname{supp} f\subset(-R,R)\end{cases}$$

uniformly in $x \in \operatorname{supp} f$.

Hint. Use the Taylor expansion for the smooth function $\mathscr{F}f(\xi)\mathrm{e}^{\mathrm{i}x\xi}$ at the point ξ_n^*.

If we combine (12.5) with (12.4) and choose $R = N^{2/(m+2)}$, then we obtain (12.3). This completes the proof. □

Remark 12.7. The main part of (12.3) represents some kind of inverse discrete Fourier transform. In order to reconstruct f at a point $x \in [-R,R]$ we need to know only the Fourier transform of this unknown function at the points

$$\frac{2n+1}{N^{m/(m+2)}}, \quad n=-N/2,\ldots,N/2-1,$$

where m is the smoothness index of f. What is more, the formula (12.3) shows us that it is effective if

$$\frac{2m-2}{m+2} > \frac{m}{m+2}$$

or $m > 2$. This means that f must belong to the Sobolev space $\overset{\circ}{W}{}_1^m(-R,R)$ with some $m \geq 3$.

Chapter 13
Some Applications of the Discrete Fourier Transform.

First we prove the Poisson summation formula.

Definition 13.1. Let f be a function such that

$$\lim_{N\to\infty} \sum_{|n|\le N} f(x+2\pi n)$$

exists pointwise in $x \in \mathbb{R}$. Then

$$f_{\mathrm{p}}(x) := \sum_{n=-\infty}^{\infty} f(x+2\pi n) \tag{13.1}$$

is called the *periodization* of f.

Remark 13.2. It is clear that $f_{\mathrm{p}}(x)$ is periodic with period 2π. Hence we will consider it only on the interval $[-\pi, \pi]$.

Theorem 13.3 (Poisson summation formula). *Suppose that $f \in L^1(\mathbb{R})$. Then $f_{\mathrm{p}}(x)$ from* (13.1) *is finite almost everywhere, satisfies $f_{\mathrm{p}}(x+2\pi) = f_{\mathrm{p}}(x)$ almost everywhere, and is integrable on the interval $[-\pi, \pi]$. The Fourier coefficients of $f_{\mathrm{p}}(x)$ are given by*

$$\frac{1}{2\pi}\int_{-\pi}^{\pi} f_{\mathrm{p}}(x)\mathrm{e}^{-imx}\mathrm{d}x = \frac{1}{\sqrt{2\pi}}\mathscr{F}f(m).$$

If in addition

$$\sum_{m=-\infty}^{\infty} |\mathscr{F}f(m)| < \infty,$$

V. Serov, *Fourier Series, Fourier Transform and Their Applications to Mathematical Physics*, Applied Mathematical Sciences 197,
DOI 10.1007/978-3-319-65262-7_13

then

$$\sum_{n=-\infty}^{\infty} f(x+2\pi n) = \frac{1}{\sqrt{2\pi}} \sum_{m=-\infty}^{\infty} \mathscr{F}f(m)\mathrm{e}^{imx}. \tag{13.2}$$

In particular, $f_{\mathrm{p}}(x)$ *is continuous, and we have the* Poisson identity

$$\sum_{n=-\infty}^{\infty} f(2\pi n) = \frac{1}{\sqrt{2\pi}} \sum_{m=-\infty}^{\infty} \mathscr{F}f(m). \tag{13.3}$$

Proof. Since $f \in L^1(\mathbb{R})$, we have

$$\begin{aligned}\int_{-\pi}^{\pi} |f_{\mathrm{p}}(x)|\mathrm{d}x &\le \int_{-\pi}^{\pi} \sum_{n=-\infty}^{\infty} |f(x+2\pi n)|\mathrm{d}x = \sum_{n=-\infty}^{\infty} \int_{-\pi}^{\pi} |f(x+2\pi n)|\mathrm{d}x \\ &= \sum_{n=-\infty}^{\infty} \int_{-\pi+2\pi n}^{\pi+2\pi n} |f(t)|\mathrm{d}t = \int_{-\infty}^{\infty} |f(t)|\mathrm{d}t < \infty.\end{aligned}$$

This shows that f_{p} is finite almost everywhere and integrable on $[-\pi,\pi]$. Applying the same calculation to $f_{\mathrm{p}}(x)\mathrm{e}^{-imx}$ allows us to integrate term by term to obtain

$$\begin{aligned}c_m(f_{\mathrm{p}}) &= \frac{1}{2\pi}\int_{-\pi}^{\pi} f_{\mathrm{p}}(x)\mathrm{e}^{-imx}\mathrm{d}x = \sum_{n=-\infty}^{\infty} \frac{1}{2\pi}\int_{-\pi}^{\pi} f(x+2\pi n)\mathrm{e}^{-imx}\mathrm{d}x \\ &= \sum_{n=-\infty}^{\infty} \frac{1}{2\pi}\int_{-\pi+2\pi n}^{\pi+2\pi n} f(t)\mathrm{e}^{-im(t-2\pi n)}\mathrm{d}t \\ &= \frac{1}{\sqrt{2\pi}} \sum_{n=-\infty}^{\infty} \frac{1}{\sqrt{2\pi}}\int_{-\pi+2\pi n}^{\pi+2\pi n} f(t)\mathrm{e}^{-imt}\mathrm{d}t \\ &= \frac{1}{\sqrt{2\pi}}\frac{1}{\sqrt{2\pi}}\int_{-\infty}^{\infty} f(t)\mathrm{e}^{-imt}\mathrm{d}t = \frac{1}{\sqrt{2\pi}}\mathscr{F}f(m).\end{aligned}$$

Now, if the series

$$\sum_{m=-\infty}^{\infty} |\mathscr{F}f(m)|$$

converges, then that convergence is equivalent to the fact that

$$\sum_{m=-\infty}^{\infty} |c_m(f_{\mathrm{p}})| < \infty.$$

Thus, $f_{\mathrm{p}}(x)$ can be represented by its Fourier series at least almost everywhere (and we can redefine $f_{\mathrm{p}}(x)$ so that this representation holds pointwise), i.e.,

$$f_{\mathrm{p}}(x) = \sum_{m=-\infty}^{\infty} c_m(f_{\mathrm{p}})\mathrm{e}^{imx}.$$

This means that (see (13.1))

$$\sum_{n=-\infty}^{\infty} f(x+2\pi n) = \frac{1}{\sqrt{2\pi}} \sum_{m=-\infty}^{\infty} \mathscr{F} f(m) e^{imx}.$$

Finally, set $x = 0$ to obtain the Poisson identity (13.3). □

Example 13.4. If

$$f(x) = \frac{1}{\sqrt{4\pi t}} e^{-\frac{x^2}{4t}}, \quad x \in \mathbb{R},$$

where $t > 0$ is a parameter, then it is very well known (see, e.g., Example 16.7 and Exercise 16.4) that

$$\mathscr{F} f(\xi) = \frac{1}{\sqrt{2\pi}} e^{-t\xi^2}.$$

Formula (13.2) transforms in this case to

$$\frac{1}{\sqrt{4\pi t}} \sum_{n=-\infty}^{\infty} e^{-\frac{(x+2\pi n)^2}{4t}} = \frac{1}{2\pi} \sum_{m=-\infty}^{\infty} e^{-tm^2} e^{imx},$$

and the Poisson identity transforms to

$$\sqrt{\frac{\pi}{t}} \sum_{n=-\infty}^{\infty} e^{-\pi^2 n^2/t} = \sum_{m=-\infty}^{\infty} e^{-tm^2}.$$

As an application of the Poisson summation formula we consider the problem of reconstructing a band-limited signal from its values on the integers.

Definition 13.5. A signal $f(t)$ is called *band-limited* if it has a representation

$$f(t) = \frac{1}{\sqrt{2\pi}} \int_{-2\pi\lambda}^{2\pi\lambda} F(\xi) e^{it\xi} d\xi, \tag{13.4}$$

where λ is a positive parameter and F is some integrable function.

Remark 13.6. If we set $F(\xi) = 0$ for $|\xi| > 2\pi\lambda$, then (13.4) is the inverse Fourier transform of $F \in L^1(\mathbb{R})$. In that case f is bounded and continuous.

Theorem 13.7 (Whittaker, Shannon, Boas). *Suppose that $F \in L^1(\mathbb{R})$ and $F(\xi) = 0$ for $|\xi| > 2\pi\lambda$. If $\lambda \le 1/2$, then for every $t \in \mathbb{R}$ we have*

$$f(t) = \sum_{n=-\infty}^{\infty} f(n) \frac{\sin \pi(t-n)}{\pi(t-n)}, \tag{13.5}$$

where the fraction is equal to 1 *when* $t = n$. *If* $\lambda > 1/2$, *we have*

$$\left| f(t) - \sum_{n=-\infty}^{\infty} f(n) \frac{\sin \pi(t-n)}{\pi(t-n)} \right| \leq \sqrt{\frac{2}{\pi}} \int_{\pi < |\xi| < 2\pi\lambda} |F(\xi)| \mathrm{d}\xi.$$

Proof. Let

$$F_{\mathrm{p}}(\xi) = \sum_{n=-\infty}^{\infty} F(\xi + 2\pi n)$$

be the periodization of F. Formula (13.4) shows that

$$\mathscr{F}(F)(t) = f(-t),$$

where $\mathscr{F}(F)$ denotes the Fourier transform of $F(\xi)$. Hence, by the Poisson summation formula, we have (see (13.2))

$$\sum_{n=-\infty}^{\infty} F(\xi + 2\pi n) = \frac{1}{\sqrt{2\pi}} \sum_{m=-\infty}^{\infty} f(-m) \mathrm{e}^{im\xi} = \frac{1}{\sqrt{2\pi}} \sum_{m=-\infty}^{\infty} f(m) \mathrm{e}^{-im\xi}.$$

Since every trigonometric Fourier series can be integrated term by term, we obtain

$$\begin{aligned}
\int_{-\pi}^{\pi} F_{\mathrm{p}}(\xi) \mathrm{e}^{it\xi} \mathrm{d}\xi &= \int_{-\pi}^{\pi} \sum_{n=-\infty}^{\infty} F(\xi + 2\pi n) \mathrm{e}^{it\xi} \mathrm{d}\xi \\
&= \frac{1}{\sqrt{2\pi}} \sum_{m=-\infty}^{\infty} f(m) \int_{-\pi}^{\pi} \mathrm{e}^{\mathrm{i}(t-m)\xi} \mathrm{d}\xi \\
&= \frac{1}{\sqrt{2\pi}} \sum_{m=-\infty}^{\infty} f(m) \frac{\mathrm{e}^{\mathrm{i}(t-m)\pi} - \mathrm{e}^{-\mathrm{i}(t-m)\pi}}{\mathrm{i}(t-m)} \\
&= \frac{2\pi}{\sqrt{2\pi}} \sum_{m=-\infty}^{\infty} f(m) \frac{\sin \pi(t-m)}{\pi(t-m)}.
\end{aligned}$$

Now, if $\lambda \leq 1/2$, then $F(\xi)$ for $|\xi| \leq \pi$ is equal to its periodization $F_{\mathrm{p}}(\xi)$ (see Definition 13.1) and

$$\int_{-\pi}^{\pi} F_{\mathrm{p}}(\xi) \mathrm{e}^{it\xi} \mathrm{d}\xi = \sqrt{2\pi} f(t).$$

These equalities imply immediately that

$$f(t) = \sum_{n=-\infty}^{\infty} f(n) \frac{\sin \pi(t-n)}{\pi(t-n)},$$

so that (13.5) is proved. If $\lambda > 1/2$, then we cannot expect that $F(\xi) = F_{\mathrm{p}}(\xi)$ for $|\xi| > \pi$, but using Definition 13.1 we have

$$f(t) = \frac{1}{\sqrt{2\pi}} \int_{-\pi}^{\pi} F(\xi) e^{it\xi} d\xi + \sqrt{\frac{2}{\pi}} \int_{\pi<|\xi|<2\pi\lambda} F(\xi) e^{it\xi} d\xi$$
$$= \sum_{n=-\infty}^{\infty} \widetilde{f}(n) \frac{\sin \pi(t-n)}{\pi(t-n)} + \sqrt{\frac{2}{\pi}} \int_{\pi<|\xi|<2\pi\lambda} F(\xi) e^{it\xi} d\xi,$$

where

$$\widetilde{f}(n) = \frac{1}{\sqrt{2\pi}} \int_{-\pi}^{\pi} F(\xi) e^{in\xi} d\xi.$$

Therefore,

$$f(t) = \sum_{n=-\infty}^{\infty} f(n) \frac{\sin \pi(t-n)}{\pi(t-n)}$$
$$+ \sum_{n=-\infty}^{\infty} \left(\widetilde{f}(n) - f(n)\right) \frac{\sin \pi(t-n)}{\pi(t-n)} + \frac{1}{\sqrt{2\pi}} \int_{\pi<|\xi|<2\pi\lambda} F(\xi) e^{it\xi} d\xi.$$

Here the middle series is equal to

$$-\frac{1}{\sqrt{2\pi}} \sum_{n=-\infty}^{\infty} \left(\int_{\pi<|\xi|<2\pi\lambda} F(\xi) e^{in\xi} d\xi \right) \frac{\sin \pi(t-n)}{\pi(t-n)},$$

or

$$-\frac{1}{\sqrt{2\pi}} \int_{\pi<|\xi|<2\pi\lambda} F(\xi) \left(\sum_{n=-\infty}^{\infty} \frac{\sin \pi(t-n)}{\pi(t-n)} e^{in\xi} \right) d\xi.$$

Exercise 13.1. Prove that

$$\sum_{n=-\infty}^{\infty} \frac{\sin \pi(t-n)}{\pi(t-n)} e^{in\xi} = e^{it\xi}, \quad \xi \in [-\pi, \pi].$$

Hint. Show that

$$c_n \left(e^{it\xi} \right) = \frac{\sin \pi(t-n)}{\pi(t-n)}.$$

Using Exercise 13.1 and the periodicity of $e^{in\xi}$, we have

$$\left| f(t) - \sum_{n=-\infty}^{\infty} f(n) \frac{\sin \pi(t-n)}{\pi(t-n)} \right| \le \sqrt{\frac{2}{\pi}} \int_{\pi<|\xi|<2\pi\lambda} |F(\xi)| d\xi.$$

This proves the theorem. □

Theorem 13.7 shows that in order to reconstruct a band-limited signal $f(t)$, it is enough to know the values of the signal at integers n. In turn, to evaluate $f(n)$ it is enough to use the IDFT of $F(\xi)$; see (13.4). Indeed, let us assume without loss of generality that $\lambda = 1/2$. Then

$$f(n) = \frac{1}{\sqrt{2\pi}} \int_{-\pi}^{\pi} F(\xi) e^{in\xi} d\xi.$$

If F is smooth enough (say $F \in C^2[-\pi,\pi]$), then formula (12.5) gives ($R = \pi$ and $N \gg 1$)

$$f(n) = \frac{\sqrt{2\pi}}{N} \sum_{k=-N/2}^{N/2-1} F_k e^{i\frac{\pi n(2k+1)}{N}} + O\left(\frac{1}{N^2}\right)$$

$$= \frac{\sqrt{2\pi}}{N} e^{i\frac{\pi n}{N}} \sum_{k=-N/2}^{N/2-1} F_k e^{i\frac{2\pi kn}{N}} + O\left(\frac{1}{N^2}\right),$$

where F_k denotes the value of $F(\xi)$ at the point $\pi(2k+1)/N$. Therefore, up to the accuracy of calculations,

$$f(n) \approx \frac{\sqrt{2\pi}}{N} e^{i\frac{\pi n}{N}} \mathscr{F}^{-1}(F_k)_n,$$

i.e., for N a sufficiently large even integer,

$$f(t) \approx \sum_{n=-\infty}^{\infty} \frac{\sqrt{2\pi}}{N} e^{i\frac{\pi n}{N}} \mathscr{F}^{-1}(F_k)_n \frac{\sin \pi(t-n)}{\pi(t-n)}.$$

Chapter 14
Applications to Solving Some Model Equations

14.1 The One-Dimensional Heat Equation

Let us consider a heat conduction problem for a straight bar of uniform cross section and homogeneous material. Let $x = 0$ and $x = L$ denote the ends of the bar (the x-axis is chosen to lie along the axis of the bar). Suppose that no heat passes through the sides of the bar. We also assume that the cross-sectional dimensions are so small that the temperature u can be considered constant on any given cross section (Figure 14.1).

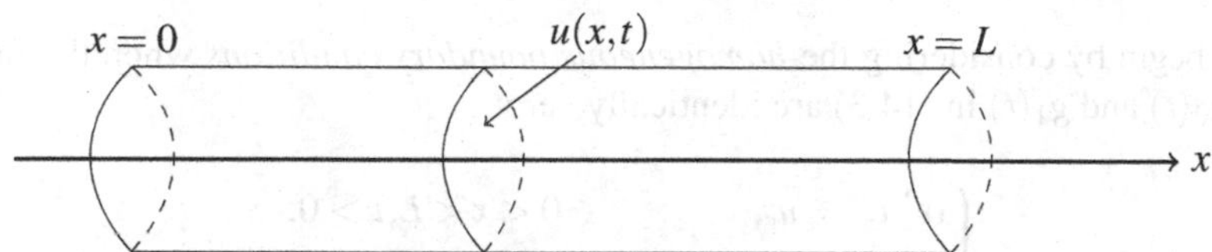

Fig. 14.1 Geometry of the heat conduction problem for a bar.

Then u is a function only of the coordinate x and the time t. The variation of temperature in the bar is governed by the partial differential equation

$$\alpha^2 u_{xx}(x,t) = u_t(x,t), \quad 0 < x < L, t > 0, \tag{14.1}$$

where α^2 is a constant known as the thermal diffusivity. This equation is called the *heat conduction equation* or *heat equation.*

In addition, we assume that the initial temperature distribution in the bar is given by

$$u(x,0) = f(x), \quad 0 \leq x \leq L, \tag{14.2}$$

where f is a given function. Finally, we assume that the temperature at each end of the bar is given by

V. Serov, *Fourier Series, Fourier Transform and Their Applications to Mathematical Physics*, Applied Mathematical Sciences 197,
DOI 10.1007/978-3-319-65262-7_14

$$u(0,t) = g_0(t), \quad u(L,t) = g_1(t), \quad t > 0, \tag{14.3}$$

where g_0 and g_1 are given functions. The problem (14.1), (14.2), (14.3) is an initial value problem in the time variable t. With respect to the space variable x it is a *boundary value problem*, and the conditions of (14.3) are called the *boundary conditions*. Alternatively, this problem can be considered a boundary value problem in the xt-plane (Figure 14.2):

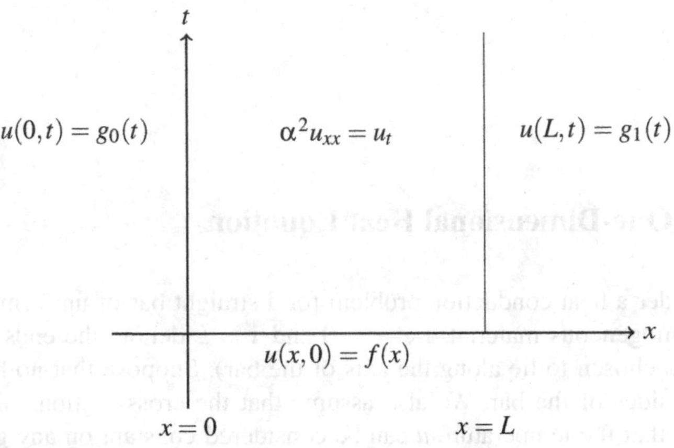

Fig. 14.2 Geometric illustration of the heat equation as a boundary value problem.

We begin by considering the *homogeneous boundary conditions* when the functions $g_0(t)$ and $g_1(t)$ in (14.3) are identically zero:

$$\begin{cases} \alpha^2 u_{xx} = u_t, & 0 < x < L, t > 0, \\ u(0,t) = u(L,t) = 0, & t > 0, \\ u(x,0) = f(x), & 0 \le x \le L. \end{cases} \tag{14.4}$$

We look for a solution to the problem (14.4) in the form

$$u(x,t) = X(x)T(t). \tag{14.5}$$

Such a method is called a *separation of variables* (or *Fourier's method*). Substituting (14.5) into (14.1) yields

$$\alpha^2 X''(x)T(t) = X(x)T'(t),$$

or

$$\frac{X''(x)}{X(x)} = \frac{1}{\alpha^2}\frac{T'(t)}{T(t)},$$

in which the variables are separated, that is, the left-hand side depends only on x, and the right-hand side only on t. This is possible only when both sides are equal to the same constant:

$$\frac{X''}{X} = \frac{1}{\alpha^2}\frac{T'}{T} = -\lambda.$$

Hence, we obtain two ordinary differential equations for $X(x)$ and $T(t)$:

$$X'' + \lambda X = 0,$$

$$T' + \alpha^2 \lambda T = 0. \tag{14.6}$$

The boundary condition for $u(x,t)$ at $x = 0$ leads to

$$u(0,t) = X(0)T(t) = 0.$$

It follows that

$$X(0) = 0,$$

since otherwise, $T \equiv 0$, and so $u \equiv 0$, which we do not accept. Similarly, the boundary condition at $x = L$ requires that

$$X(L) = 0.$$

So, for the function $X(x)$ we obtain the homogeneous boundary value problem

$$\begin{cases} X'' + \lambda X = 0, & 0 < x < L, \\ X(0) = X(L) = 0. \end{cases} \tag{14.7}$$

The values of λ for which nontrivial solutions of (14.7) exist are called *eigenvalues*, and the corresponding nontrivial solutions are called *eigenfunctions*. The problem (14.7) is called an *eigenvalue problem.*

Lemma 14.1. *The problem* (14.7) *has an infinite sequence of positive eigenvalues*

$$\lambda_n = \frac{n^2\pi^2}{L^2}, \quad n = 1, 2, \ldots,$$

with the corresponding eigenfunctions

$$X_n(x) = c \sin \frac{n\pi x}{L},$$

where c is an arbitrary nonzero constant.

Proof. Suppose first that $\lambda > 0$, i.e., $\lambda = \mu^2$. The characteristic equation for (14.7) is $r^2 + \mu^2 = 0$ with roots $r = \pm i\mu$, so the general solution is

$$X(x) = c_1 \cos \mu x + c_2 \sin \mu x.$$

Note that μ is nonzero, and there is no loss of generality if we assume that $\mu > 0$. The first boundary condition in (14.7) implies

$$X(0) = c_1 = 0,$$

and the second reduces to

$$c_2 \sin \mu L = 0,$$

or

$$\sin \mu L = 0,$$

since we do not allow $c_2 = 0$ either. It follows that

$$\mu L = n\pi, \quad n = 1, 2, \ldots,$$

or

$$\lambda_n = \frac{n^2 \pi^2}{L^2}, \quad n = 1, 2, \ldots.$$

Hence the corresponding eigenfunctions are

$$X_n(x) = c \sin \frac{n\pi x}{L}.$$

If $\lambda = -\mu^2 < 0$, $\mu > 0$, then the characteristic equation for (14.7) is $r^2 - \mu^2 = 0$ with roots $r = \pm\mu$. Hence the general solution is

$$X(x) = c_1 \cosh \mu x + c_2 \sinh \mu x.$$

Since

$$\cosh \mu x = \frac{e^{\mu x} + e^{-\mu x}}{2} \quad \text{and} \quad \sinh \mu x = \frac{e^{\mu x} - e^{-\mu x}}{2},$$

this is equivalent to

$$X(x) = c_1' e^{\mu x} + c_2' e^{-\mu x}.$$

The first boundary condition requires again that $c_1 = 0$, while the second gives

$$c_2 \sinh \mu L = 0.$$

Since $\mu \neq 0$ ($\mu > 0$), it follows that $\sinh \mu L \neq 0$, and therefore we must have $c_2 = 0$. Consequently, $X \equiv 0$, i.e., there are no nontrivial solutions for $\lambda < 0$.

If $\lambda = 0$, the general solution is

$$X(x) = c_1 x + c_2.$$

The boundary conditions can be satisfied only if $c_1 = c_2 = 0$, so there is only the trivial solution in this case as well. □

Turning now to (14.6) for $T(t)$ and substituting $\frac{n^2\pi^2}{L^2}$ for λ, we have

$$T(t) = c\mathrm{e}^{-\left(\frac{n\pi\alpha}{L}\right)^2 t}.$$

Hence the functions

$$u_n(x,t) = \mathrm{e}^{-\left(\frac{n\pi\alpha}{L}\right)^2 t} \sin\frac{n\pi x}{L} \tag{14.8}$$

satisfy (14.1) and the homogeneous boundary conditions from (14.4) for each $n = 1, 2, \ldots$. The *linear superposition principle* gives that every linear combination

$$u(x,t) = \sum_{n=1}^{N} c_n \mathrm{e}^{-\left(\frac{n\pi\alpha}{L}\right)^2 t} \sin\frac{n\pi x}{L}$$

is also a solution of the same problem. In order to take into account infinitely many functions (14.8), we assume that

$$u(x,t) = \sum_{n=1}^{\infty} c_n \mathrm{e}^{-\left(\frac{n\pi\alpha}{L}\right)^2 t} \sin\frac{n\pi x}{L}, \tag{14.9}$$

where the coefficients c_n are still undetermined, and the series converges in some sense. To satisfy the initial condition from (14.4) we must have

$$u(x,0) = \sum_{n=1}^{\infty} c_n \sin\frac{n\pi x}{L} = f(x), \quad 0 \le x \le L. \tag{14.10}$$

In other words, we need to choose the coefficients c_n so that the series (14.10) converges to the initial temperature distribution $f(x)$. We extend f from $[0,L]$ to $[-L,L]$ as an odd function and then obtain that

$$c_n = \frac{2}{L}\int_0^L f(x) \sin\frac{n\pi x}{L} \mathrm{d}x.$$

It is not difficult to prove that for $t > 0$, $0 < x < L$, the series (14.9) converges (with any derivative with respect to x and t) and solves (14.1) with boundary conditions (14.4). Only one question remains: can every function $f(x)$ be represented by a Fourier sine series (14.10)? Some sufficient conditions for such a representation are given in Chapter 10.

Remark 14.2. We can consider the boundary value problem for a linear differential equation

$$y'' + p(x)y' + q(x)y = g(x) \tag{14.11}$$

of order two on the interval (a, b) with the boundary conditions

$$y(a) = y_0, \quad y(b) = y_1, \tag{14.12}$$

where y_0 and y_1 are given constants. Let us assume that we have found a fundamental set of solutions $y_1(x)$ and $y_2(x)$ to the corresponding homogeneous equation

$$y'' + p(x)y' + q(x)y = 0.$$

Then the general solution to (14.11) is

$$y(x) = c_1 y_1(x) + c_2 y_2(x) + y_p(x),$$

where $y_p(x)$ is a particular solution to (14.11) and c_1 and c_2 are arbitrary constants.

To satisfy the boundary conditions (14.12) we have the linear inhomogeneous algebraic system

$$\begin{cases} c_1 y_1(a) + c_2 y_2(a) = y_0 - y_p(a), \\ c_1 y_1(b) + c_2 y_2(b) = y_1 - y_p(b). \end{cases} \tag{14.13}$$

If the determinant

$$\begin{vmatrix} y_1(a) & y_2(a) \\ y_1(b) & y_2(b) \end{vmatrix}$$

is nonzero, then the constants c_1 and c_2 can be determined uniquely, and therefore the boundary value problem (14.11)–(14.12) has a unique solution. If

$$\begin{vmatrix} y_1(a) & y_2(a) \\ y_1(b) & y_2(b) \end{vmatrix} = 0,$$

then (14.11)–(14.12) has either no solutions or infinitely many solutions.

Example 14.3. Let us consider the boundary value problem

$$\begin{cases} y'' + \mu^2 y = 1, & 0 < x < 1, \\ y(0) = y_0,\ y(1) = y_1, \end{cases}$$

where $\mu > 0$ is fixed. This differential equation has a particular solution $y_p(x) = \frac{1}{\mu^2}$. Hence, the system (14.13) becomes

$$\begin{cases} c_1 \sin 0 + c_2 \cos 0 = y_0 - \frac{1}{\mu^2}, \\ c_1 \sin \mu + c_2 \cos \mu = y_1 - \frac{1}{\mu^2}, \end{cases}$$

or

$$\begin{cases} c_2 = y_0 - \frac{1}{\mu^2}, \\ c_1 \sin \mu = y_1 - \frac{1}{\mu^2} - \left(y_0 - \frac{1}{\mu^2} \right) \cos \mu. \end{cases}$$

If

$$\begin{vmatrix} 0 & 1 \\ \sin \mu & \cos \mu \end{vmatrix} \neq 0,$$

i.e., $\sin \mu \neq 0$, then c_1 is uniquely determined, and the boundary value problem in question has a unique solution. If $\sin \mu = 0$, then the problem has solutions (in fact, infinitely many) if and only if

$$y_1 - \frac{1}{\mu^2} = \left(y_0 - \frac{1}{\mu^2} \right) \cos \mu.$$

If $\mu = 2\pi k$, then $\sin \mu = 0$ and $\cos \mu = 1$ and the following equation must hold:

$$y_1 - \frac{1}{\mu^2} = y_0 - \frac{1}{\mu^2},$$

i.e., $y_1 = y_0$. If $\mu = \pi + 2\pi k$, then $\sin \mu = 0$ and $\cos \mu = -1$, and we must have

$$y_1 + y_0 = \frac{2}{\mu^2}.$$

Suppose now that the ends of the bar are held at constant temperatures T_1 and T_2. The corresponding boundary value problem is then

$$\begin{cases} \alpha^2 u_{xx} = u_t, & 0 < x < L, t > 0, \\ u(0,t) = T_1, u(L,t) = T_2, & t > 0, \\ u(x,0) = f(x). \end{cases} \tag{14.14}$$

After a long time ($t \to \infty$) we anticipate that a steady temperature distribution $v(x)$ will be reached that is independent of time and the initial condition. Since the solution of (14.14) with $T_1 = T_2 = 0$ tends to zero as $t \to \infty$, see (14.9), we look for the solution to (14.14) in the form

$$u(x,t) = v(x) + w(x,t). \tag{14.15}$$

Substituting (14.15) into (14.14) leads to

$$\begin{cases} \alpha^2(v_{xx}+w_{xx}) = w_t, \\ v(0)+w(0,t) = T_1, v(L)+w(L,t) = T_2, \\ v(x)+w(x,0) = f(x). \end{cases}$$

Let us assume that $v(x)$ satisfies the steady-state problem

$$\begin{cases} v''(x) = 0, & 0 < x < L, \\ v(0) = T_1, & v(L) = T_2. \end{cases} \tag{14.16}$$

Then $w(x,t)$ satisfies the homogeneous boundary value problem for the heat equation:

$$\begin{cases} \alpha^2 w_{xx} = w_t, & 0 < x < L, t > 0, \\ w(0,t) = w(L,t) = 0, \\ w(x,0) = \tilde{f}(x), \end{cases} \tag{14.17}$$

where $\tilde{f}(x) = f(x) - v(x)$. Since the solution of (14.16) is

$$v(x) = \frac{T_2 - T_1}{L}x + T_1, \tag{14.18}$$

the solution of (14.17) is

$$w(x,t) = \sum_{n=1}^{\infty} c_n \mathrm{e}^{-\left(\frac{n\pi\alpha}{L}\right)^2 t} \sin\frac{n\pi x}{L}, \tag{14.19}$$

where the coefficients c_n are given by

$$c_n = \frac{2}{L}\int_0^L \left[f(x) - \frac{T_2 - T_1}{L}x - T_1\right] \sin\frac{n\pi x}{L}\mathrm{d}x.$$

Combining (14.18) and (14.19), we obtain

$$u(x,t) = \frac{T_2 - T_1}{L}x + T_1 + \sum_{n=1}^{\infty} c_n \mathrm{e}^{-\left(\frac{n\pi\alpha}{L}\right)^2 t} \sin\frac{n\pi x}{L}.$$

Let us slightly complicate the problem (14.14), namely assume that

$$\begin{cases} \alpha^2 u_{xx} = u_t + p(x), & 0 < x < L, t > 0, \\ u(0,t) = T_1, u(L,t) = T_2, & t > 0, \\ u(x,0) = f(x). \end{cases} \tag{14.20}$$

We begin by assuming that the solution to (14.20) consists of a steady-state solution $v(x)$ and a transient solution $w(x,t)$ that tends to zero as $t \to \infty$:

$$u(x,t) = v(x) + w(x,t).$$

Then for $v(x)$ we will have the problem

$$\begin{cases} v''(x) = \frac{1}{\alpha^2} p(x), \quad 0 < x < L, \\ v(0) = T_1, v(L) = T_2. \end{cases} \tag{14.21}$$

To solve this we integrate twice to get

$$v(x) = \frac{1}{\alpha^2} \int_0^x \mathrm{d}y \int_0^y p(s)\mathrm{d}s + c_1 x + c_2.$$

The boundary conditions yield $c_2 = T_1$ and

$$c_1 = \frac{1}{L}\left\{ T_2 - T_1 - \frac{1}{\alpha^2} \int_0^L \mathrm{d}y \int_0^y p(s)\mathrm{d}s \right\}.$$

Therefore, the solution of (14.21) has the form

$$v(x) = \frac{T_2 - T_1}{L} x - \frac{x}{L\alpha^2} \int_0^L \mathrm{d}y \int_0^y p(s)\mathrm{d}s + \frac{1}{\alpha^2} \int_0^x \mathrm{d}y \int_0^y p(s)\mathrm{d}s + T_1.$$

For $w(x,t)$ we will have the homogeneous problem

$$\begin{cases} \alpha^2 w_{xx} = w_t, \quad 0 < x < L, t > 0, \\ w(0,t) = w(L,t) = 0, t > 0, \\ w(x,0) = \tilde{f}(x) := f(x) - v(x). \end{cases}$$

A different problem occurs if the ends of the bar are insulated so that there is no passage of heat through them. Thus, in the case of no heat flow, the boundary value problem is

$$\begin{cases} \alpha^2 u_{xx} = u_t, \quad 0 < x < L, t > 0, \\ u_x(0,t) = u_x(L,t) = 0, t > 0, \\ u(x,0) = f(x). \end{cases} \tag{14.22}$$

This problem can also be solved by the method of separation of variables. If we let $u(x,t) = X(x)T(t)$, it follows that

$$X'' + \lambda X = 0, \quad T' + \alpha^2 \lambda T = 0. \tag{14.23}$$

The boundary conditions now yield

$$X'(0) = X'(L) = 0. \tag{14.24}$$

If $\lambda = -\mu^2 < 0$, $\mu > 0$, then (14.23) for $X(x)$ becomes $X'' - \mu^2 X = 0$ with general solution

$$X(x) = c_1 \sinh \mu x + c_2 \cosh \mu x.$$

Therefore, the conditions (14.24) give $c_1 = 0$ and $c_2 = 0$, which is unacceptable. Hence λ cannot be negative.

If $\lambda = 0$, then

$$X(x) = c_1 x + c_2.$$

Thus $X'(0) = c_1 = 0$ and $X'(L) = 0$ for every c_2, leaving c_2 undetermined. Therefore $\lambda = 0$ is an eigenvalue, corresponding to the eigenfunction $X_0(x) = 1$. It follows from (14.23) that $T(t)$ is also a constant. Hence, for $\lambda = 0$ we obtain the constant solution $u_0(x,t) = c_2$.

If $\lambda = \mu^2 > 0$, then $X'' + \mu^2 X = 0$ and consequently

$$X(x) = c_1 \sin \mu x + c_2 \cos \mu x.$$

The boundary conditions imply $c_1 = 0$ and $\mu = \frac{n\pi}{L}$, $n = 1,2,\ldots$, leaving c_2 arbitrary. Thus we have an infinite sequence of positive eigenvalues $\lambda_n = \frac{n^2\pi^2}{L^2}$ with the corresponding eigenfunctions

$$X_n(x) = \cos \frac{n\pi x}{L}, \quad n = 1,2,\ldots.$$

If we combine these eigenvalues and eigenfunctions with zero eigenvalue $\lambda_0 = 0$ and $X_0(x) = 1$, we may conclude that we have the infinite sequences

$$\lambda_n = \frac{n^2\pi^2}{L^2}, \quad X_n(x) = \cos \frac{n\pi x}{L}, \quad n = 0,1,2,\ldots,$$

and

$$u_n(x,t) = \cos \frac{n\pi x}{L} \mathrm{e}^{-\left(\frac{n\pi\alpha}{L}\right)^2 t}, \quad n = 0,1,2,\ldots.$$

Each of these functions satisfies the equation and boundary conditions from (14.22). It remains to satisfy the initial condition. In order to do so, we assume that $u(x,t)$ has the form

$$u(x,t) = \frac{c_0}{2} + \sum_{n=1}^{\infty} c_n \cos \frac{n\pi x}{L} \mathrm{e}^{-\left(\frac{n\pi\alpha}{L}\right)^2 t}, \tag{14.25}$$

where the coefficients c_n are determined by the requirement that

$$u(x,0) = \frac{c_0}{2} + \sum_{n=1}^{\infty} c_n \cos\frac{n\pi x}{L} = f(x), \quad 0 \le x \le L.$$

Thus the unknown coefficients in (14.25) must be the Fourier coefficients in the Fourier cosine series of period $2L$ for the even extension of f. Hence

$$c_n = \frac{2}{L}\int_0^L f(x)\cos\frac{n\pi x}{L}\mathrm{d}x, \quad n = 0,1,2,\ldots,$$

and the series (14.25) provides the solution to the heat conduction problem (14.22) for a bar with insulated ends. The physical interpretation of the term

$$\frac{c_0}{2} = \frac{1}{L}\int_0^L f(x)\mathrm{d}x$$

is that it is the mean value of the original temperature distribution and a steady-state solution in this case.

Exercise 14.1. Let $v(x)$ be a solution of the problem

$$\begin{cases} v''(x) = 0, & 0 < x < L, \\ v'(0) = T_1, v'(L) = T_2. \end{cases}$$

Show that the problem

$$\begin{cases} \alpha^2 u_{xx} = u_t, & 0 < x < L, t > 0, \\ u_x(0,t) = T_1, u_x(L,t) = T_2, t > 0, \\ u(x,0) = f(x), \end{cases}$$

has a solution of the form $u(x,t) = v(x) + w(x,t)$ if and only if $T_1 = T_2$.

Example 14.4.

$$\begin{cases} u_{xx} = u_t, & 0 < x < 1, t > 0, \\ u(0,t) = u(1,t) = 0, \\ u(x,0) = \sum_{n=1}^{\infty} \frac{1}{n^2}\sin(n\pi x) := f(x). \end{cases}$$

As we know, the solution of this problem is given by

$$u(x,t) = \sum_{n=1}^{\infty} c_n \sin(n\pi x)\mathrm{e}^{-(n\pi)^2 t}.$$

Since

$$u(x,0) = \sum_{n=1}^{\infty} c_n \sin(n\pi x) = \sum_{n=1}^{\infty} \frac{1}{n^2} \sin(n\pi x),$$

we conclude that necessarily $c_n = \frac{1}{n^2}$ (since the Fourier series is unique). Hence the solution is

$$u(x,t) = \sum_{n=1}^{\infty} \frac{1}{n^2} \sin(n\pi x) \mathrm{e}^{-(n\pi)^2 t}.$$

Exercise 14.2. Find a solution of the problem

$$\begin{cases} u_{xx} = u_t, & 0 < x < \pi, t > 0, \\ u_x(0,t) = u_x(\pi,t) = 0, t > 0, \\ u(x,0) = 1 - \sin x, \end{cases}$$

using the method of separation of variables.

Let us consider a bar with mixed boundary conditions at the ends. Assume that the temperature at $x = 0$ is zero, while the end $x = L$ is insulated so that no heat passes through it:

$$\begin{cases} \alpha^2 u_{xx} = u_t, & 0 < x < L, t > 0, \\ u(0,t) = u_x(L,t) = 0, t > 0, \\ u(x,0) = f(x). \end{cases}$$

Separation of variables leads to

$$\begin{cases} X'' + \lambda X = 0, & 0 < x < L, \\ X(0) = X'(L) = 0, \end{cases} \tag{14.26}$$

and

$$T' + \alpha^2 \lambda T = 0, \quad t > 0.$$

As above, one can show that (14.26) has nontrivial solutions only for $\lambda > 0$, namely

$$\lambda_m = \frac{(2m-1)^2 \pi^2}{4L^2}, \quad X_m(x) = \sin \frac{(2m-1)\pi x}{2L}, \quad m = 1, 2, 3, \ldots.$$

The solution to the mixed boundary value problem is

$$u(x,t) = \sum_{m=1}^{\infty} c_m \sin \frac{(2m-1)\pi x}{2L} \mathrm{e}^{-\left(\frac{(2m-1)\pi\alpha}{2L}\right)^2 t}$$

with arbitrary constants c_m. To satisfy the initial condition we have

$$f(x) = \sum_{m=1}^{\infty} c_m \sin\frac{(2m-1)\pi x}{2L}, \quad 0 \le x \le L.$$

This is a Fourier sine series but in some specific form. We show that the coefficients c_m can be calculated as

$$c_m = \frac{2}{L}\int_0^L f(x)\sin\frac{(2m-1)\pi x}{2L}\mathrm{d}x,$$

and such a representation is possible.

In order to prove this, let us first extend $f(x)$ to the interval $0 \le x \le 2L$ so that it is symmetric about $x = L$, i.e., $f(2L-x) = f(x)$ for $0 \le x \le L$. We then extend the resulting function to the interval $(-2L, 0)$ as an odd function and elsewhere as a periodic function $\widetilde{f}$ of period $4L$. In this procedure we need to define

$$\widetilde{f}(0) = \widetilde{f}(2L) = \widetilde{f}(-2L) = 0.$$

Then the Fourier series contains only sines:

$$\widetilde{f}(x) = \sum_{n=1}^{\infty} c_n \sin\frac{n\pi x}{2L},$$

with the Fourier coefficients

$$c_n = \frac{2}{2L}\int_0^{2L} \widetilde{f}(x)\sin\frac{n\pi x}{2L}\mathrm{d}x.$$

Let us show that $c_n = 0$ for even $n = 2m$. Indeed,

$$\begin{aligned}
c_{2m} &= \frac{1}{L}\int_0^{2L} \widetilde{f}(x)\sin\frac{m\pi x}{L}\mathrm{d}x \\
&= \frac{1}{L}\int_0^L f(x)\sin\frac{m\pi x}{L}\mathrm{d}x + \frac{1}{L}\int_L^{2L} f(2L-x)\sin\frac{m\pi x}{L}\mathrm{d}x \\
&= \frac{1}{L}\int_0^L f(x)\sin\frac{m\pi x}{L}\mathrm{d}x - \frac{1}{L}\int_L^0 f(y)\sin\frac{m\pi(2L-y)}{L}\mathrm{d}y \\
&= \frac{1}{L}\int_0^L f(x)\sin\frac{m\pi x}{L}\mathrm{d}x + \frac{1}{L}\int_L^0 f(y)\sin\frac{m\pi y}{L}\mathrm{d}y = 0,
\end{aligned}$$

which is why

$$\widetilde{f}(x) = \sum_{m=1}^{\infty} c_{2m-1}\sin\frac{(2m-1)\pi x}{2L},$$

where

$$\begin{aligned} c_{2m-1} &= \frac{1}{L}\int_0^{2L} \widetilde{f}(x)\sin\frac{(2m-1)\pi x}{2L}\mathrm{d}x \\ &= \frac{1}{L}\int_0^{L} f(x)\sin\frac{(2m-1)\pi x}{2L}\mathrm{d}x + \frac{1}{L}\int_L^{2L} f(2L-x)\sin\frac{(2m-1)\pi x}{2L}\mathrm{d}x \\ &= \frac{2}{L}\int_0^{L} f(x)\sin\frac{(2m-1)\pi x}{2L}\mathrm{d}x, \end{aligned}$$

as claimed. Let us remark that the series

$$\sum_{m=1}^{\infty} c_m \sin\frac{(2m-1)\pi x}{2L}$$

represents $f(x)$ on $(0,L]$.

Remark 14.5. For the boundary conditions

$$u_x(0,t) = u(L,t) = 0$$

the function $f(x)$ must be extended to the interval $0 \le x \le 2L$ as $f(x) = -f(2L-x)$ with $f(L) = 0$. Furthermore, $\widetilde{f}$ is an even extension to the interval $(-2L,0)$. Then the corresponding Fourier series represents $f(x)$ on the interval $[0,L)$.

Exercise 14.3. (1) Let $u(x,t)$ satisfy

$$\begin{cases} u_{xx} = u_t, & 0 < x < 1, t > 0, \\ u(0,t) = u(1,t) = 0, & t \ge 0, \\ u(x,0) = f(x), & 0 \le x \le 1, \end{cases}$$

where $f \in C[0,1]$. Show that for every $T \ge 0$ we have

$$\int_0^1 |u(x,T)|^2 \mathrm{d}x \le \int_0^1 |f(x)|^2 \mathrm{d}x.$$

Hint: Use the identity $2u(u_t - u_{xx}) = \partial_t u^2 - \partial_x(u \cdot u_x) + 2(u_x)^2$.

(2) Use Fourier's method to solve

$$\begin{cases} u_{xx} = u_t, & 0 < x < 1, t > 0, \\ u(0,t) = u(1,t), u_x(0,t) = u_x(1,t) & t \ge 0, \\ u(x,0) = f(x), & 0 \le x \le 1, \end{cases}$$

where $f \in C[0,1]$ with piecewise continuous derivative.

(3) Use Fourier's method to solve

$$\begin{cases} u_{xx} + u = u_t, & 0 < x < 1, t > 0, \\ u(0,t) = u_x(0,t) = 0 & t \geq 0, \\ u(x,0) = x(1-x), & 0 \leq x \leq 1. \end{cases}$$

(4) Show that

$$u(x,t) = \frac{1}{\sqrt{4\pi t}} \int_{-\infty}^{\infty} \varphi(\xi) \mathrm{e}^{-(x-\xi+t)^2/(4t)} \mathrm{d}\xi$$

solves the problem

$$\begin{cases} u_{xx} + u_x = u_t, & -\infty < x < \infty, t > 0, \\ u(x,0) = \varphi(x), & -\infty < x < \infty, \\ u(x,t) \text{ is bounded for } -\infty < x < \infty, t \geq 0. \end{cases}$$

14.2 The One-Dimensional Wave Equation

Another situation in which the separation of variables applies occurs in the study of a vibrating string. Suppose that an elastic string of length L is tightly stretched between two supports, so that the x-axis lies along the string. Let $u(x,t)$ denote the vertical displacement experienced by the string at the point x at time t. It turns out that if damping effects are neglected, and if the amplitude of the motion is not too large, then $u(x,t)$ satisfies the partial differential equation

$$a^2 u_{xx} = u_{tt}, \quad 0 < x < L, t > 0. \tag{14.27}$$

Equation (14.27) is known as the one-dimensional *wave equation*. The constant a^2 is equal to T/ρ, where T is the force in the string and ρ is the mass per unit length of the string material (Figure 14.3).

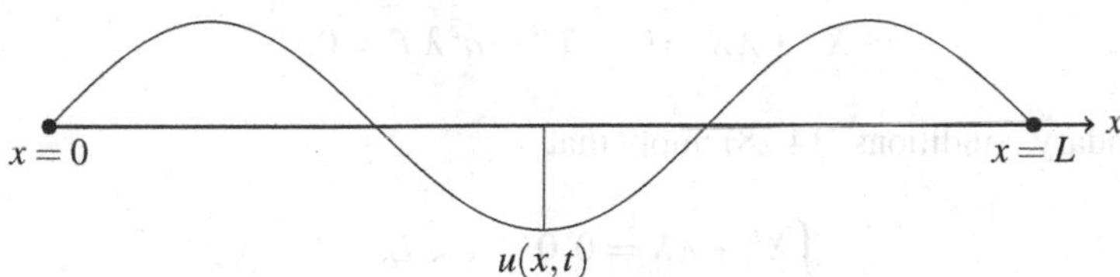

Fig. 14.3 Geometry of the one dimensional wave equation.

To describe the motion completely it is necessary also to specify suitable initial and boundary conditions for the displacement $u(x,t)$. The ends of the string are assumed to remain fixed:

$$u(0,t) = u(L,t) = 0, \quad t \geq 0. \tag{14.28}$$

The initial conditions are (since (14.27) is of second order with respect to t)

$$u(x,0) = f(x), \quad u_t(x,0) = g(x), \quad 0 \leq x \leq L, \tag{14.29}$$

where f and g are given functions. In order for (14.28) and (14.29) to be consistent, it is also necessary to require that

$$f(0) = f(L) = g(0) = g(L) = 0. \tag{14.30}$$

Equations (14.27)–(14.30) can be interpreted as the following boundary value problem for the wave equation (Figure 14.4):

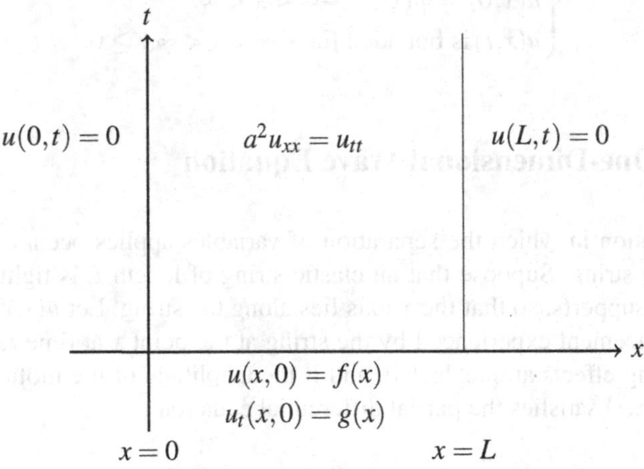

Fig. 14.4 Geometric illustration of the wave equation as a boundary value problem.

Let us apply the method of separation of variables to this homogeneous boundary value problem. Assuming that $u(x,t) = X(x)T(t)$, we obtain

$$X'' + \lambda X = 0, \quad T'' + a^2\lambda T = 0.$$

The boundary conditions (14.28) imply that

$$\begin{cases} X'' + \lambda X = 0, 0 < x < L, \\ X(0) = X(L) = 0. \end{cases}$$

This is the same boundary value problem that we have considered previously. Hence,

$$\lambda_n = \frac{n^2\pi^2}{L^2}; \quad X_n(x) = \sin\frac{n\pi x}{L}, \quad n = 1, 2, \ldots.$$

Taking $\lambda = \lambda_n$ in the equation for $T(t)$, we have

$$T''(t) + \left(\frac{n\pi a}{L}\right)^2 T(t) = 0.$$

The general solution to this equation is

$$T(t) = k_1 \cos\frac{n\pi at}{L} + k_2 \sin\frac{n\pi at}{L},$$

where k_1 and k_2 are arbitrary constants. Using the linear superposition principle, we consider the series

$$u(x,t) = \sum_{n=1}^{\infty} \sin\frac{n\pi x}{L}\left(a_n \cos\frac{n\pi at}{L} + b_n \sin\frac{n\pi at}{L}\right), \tag{14.31}$$

where the coefficients a_n and b_n are to be determined. It is clear that $u(x,t)$ from (14.31) satisfies (14.27) and (14.28) (at least formally). The initial conditions (14.29) imply

$$\begin{aligned} f(x) &= \sum_{n=1}^{\infty} a_n \sin\frac{n\pi x}{L}, \quad 0 \le x \le L, \\ g(x) &= \sum_{n=1}^{\infty} \frac{n\pi a}{L} b_n \sin\frac{n\pi x}{L}, \quad 0 \le x \le L. \end{aligned} \tag{14.32}$$

Since (14.30) are fulfilled it follows that equations (14.32) are the Fourier sine series for f and g, respectively. Therefore,

$$\begin{aligned} a_n &= \frac{2}{L}\int_0^L f(x) \sin\frac{n\pi x}{L} \mathrm{d}x, \\ b_n &= \frac{2}{n\pi a}\int_0^L g(x) \sin\frac{n\pi x}{L} \mathrm{d}x. \end{aligned} \tag{14.33}$$

Finally, we may conclude that the series (14.31) with the coefficients (14.33) solves (at least formally) the boundary value problem (14.27)–(14.30).

Each displacement pattern

$$u_n(x,t) = \sin\frac{n\pi x}{L}\left(a_n \cos\frac{n\pi at}{L} + b_n \sin\frac{n\pi at}{L}\right)$$

is called a *natural mode* of vibration and is periodic in both the space variable x and time variable t. The spatial period $\frac{2L}{n}$ in x is called the *wavelength*, while the numbers $\frac{n\pi a}{L}$ are called the *natural frequencies*.

Exercise 14.4. Find a solution of the problem

$$\begin{cases} u_{xx} = u_{tt}, 0 < x < 1, t > 0, \\ u(0,t) = u(1,t) = 0, t \geq 0, \\ u(x,0) = x(1-x), u_t(x,0) = \sin(7\pi x), \end{cases}$$

using the method of separation of variables.

If we compare the two series

$$u(x,t) = \sum_{n=1}^{\infty} \sin\frac{n\pi x}{L}\left(a_n \cos\frac{n\pi at}{L} + b_n \sin\frac{n\pi at}{L}\right),$$

$$u(x,t) = \sum_{n=1}^{\infty} c_n \sin\frac{n\pi x}{L} e^{-\left(\frac{n\pi a}{L}\right)^2 t}$$

for the wave and heat equations, we see that the second series has an exponential factor that decays rapidly with n for every $t > 0$. This guarantees convergence of the series as well as the smoothness of the sum. This is no longer true for the first series, because it contains only oscillatory terms that do not decay with increasing n. This means that the solution of the heat equation is a C^∞ function in the corresponding domain, but the solution of the wave equation is not necessarily smooth.

The boundary value problem for the wave equation with free ends of the string can be formulated as follows:

$$\begin{cases} a^2 u_{xx} = u_{tt}, 0 < x < L, t > 0, \\ u_x(0,t) = u_x(L,t) = 0, t \geq 0, \\ u(x,0) = f(x), u_t(x,0) = g(x), 0 \leq x \leq L. \end{cases}$$

Let us first note that the boundary conditions imply that $f(x)$ and $g(x)$ must satisfy

$$f'(0) = f'(L) = g'(0) = g'(L) = 0.$$

The method of separation of variables gives that the eigenvalues are

$$\lambda_n = \left(\frac{n\pi}{L}\right)^2, \quad n = 0, 1, 2, \ldots,$$

and the formal solution $u(x,t)$ is

$$u(x,t) = \frac{b_0 t + a_0}{2} + \sum_{n=1}^{\infty} \cos\frac{n\pi x}{L}\left(a_n \cos\frac{n\pi at}{L} + b_n \sin\frac{n\pi at}{L}\right).$$

The initial conditions are satisfied when

$$f(x) = \frac{a_0}{2} + \sum_{n=1}^{\infty} a_n \cos\frac{n\pi x}{L}$$

and

$$g(x) = \frac{b_0}{2} + \sum_{n=1}^{\infty} b_n \frac{n\pi a}{L} \cos\frac{n\pi x}{L},$$

where

$$a_n = \frac{2}{L}\int_0^L f(x)\cos\frac{n\pi x}{L}\mathrm{d}x, \quad n = 0,1,2,\ldots,$$
$$b_0 = \frac{2}{L}\int_0^L g(x)\mathrm{d}x,$$

and

$$b_n = \frac{2}{n\pi a}\int_0^L g(x)\cos\frac{n\pi x}{L}\mathrm{d}x, \quad n = 1,2,\ldots.$$

Exercise 14.5. (1) Show that there is no uniqueness for the problem

$$\begin{cases} u_{xx} = u_{tt}, & 0 < x < L, t > 0, \\ u(0,t) = u(L,t), u_x(0,t) = u_x(L,t), & t \geq 0, \\ u(x,0) = f(x), u_t(x,0) = g(x), & 0 \leq x \leq L, \end{cases}$$

i.e., this problem is ill posed.

(2) Find a solution of the problem

$$\begin{cases} u_{xx} = u_{tt}, & 0 < x < L, t > 0, \\ u_x(0,t) = u(L,t) = 0, & t \geq 0, \\ u(x,0) = f(x), u_t(x,0) = g(x), & 0 \leq x \leq L. \end{cases}$$

Let us consider the wave equation on the whole line. It corresponds, so to speak, to an infinite string. In that case we no longer have the boundary conditions, but we have the initial conditions

$$\begin{cases} a^2 u_{xx} = u_{tt}, -\infty < x < \infty, t > 0, \\ u(x,0) = f(x), u_t(x,0) = g(x). \end{cases} \tag{14.34}$$

Proposition 14.6. *The solution $u(x,t)$ of the wave equation is of the form*

$$u(x,t) = \varphi(x - at) + \psi(x + at),$$

where φ and ψ are two arbitrary C^2 functions of one variable.

Proof. By the chain rule,

$$\partial_{tt}u - a^2\partial_{xx}u = 0$$

if and only if

$$\partial_\xi \partial_\eta u = 0,$$

where $\xi = x + at$ and $\eta = x - at$ (and so $\partial_x = \partial_\xi + \partial_\eta, \frac{1}{a}\partial_t = \partial_\xi - \partial_\eta$). It follows that

$$\partial_\xi u = \Theta(\xi),$$

or

$$u = \psi(\xi) + \varphi(\eta),$$

where $\psi' = \Theta$. □

To satisfy the initial conditions, we have

$$f(x) = \varphi(x) + \psi(x), \quad g(x) = -a\varphi'(x) + a\psi'(x).$$

It follows that

$$\varphi'(x) = \frac{1}{2}f'(x) - \frac{1}{2a}g(x), \quad \psi'(x) = \frac{1}{2}f'(x) + \frac{1}{2a}g(x).$$

Integrating, we obtain

$$\varphi(x) = \frac{1}{2}f(x) - \frac{1}{2a}\int_0^x g(s)\mathrm{d}s + c_1, \quad \psi(x) = \frac{1}{2}f(x) + \frac{1}{2a}\int_0^x g(s)\mathrm{d}s + c_2,$$

where c_1 and c_2 are arbitrary constants. But $\varphi(x) + \psi(x) = f(x)$ implies $c_1 + c_2 = 0$. Therefore, the solution of the initial value problem is

$$u(x,t) = \frac{1}{2}(f(x-at) + f(x+at)) + \frac{1}{2a}\int_{x-at}^{x+at} g(s)\mathrm{d}s. \tag{14.35}$$

This formula is called *d'Alembert's formula.*

Exercise 14.6. Prove that if f is a C^2 function and g is a C^1 function, then u from (14.35) is a C^2 function that satisfies (14.34) in the classical sense (pointwise).

Exercise 14.7. Prove that if f and g are merely locally integrable, then u from (14.35) is a solution of (14.34) in the sense of integral equalities and the initial conditions are satisfied pointwise.

Example 14.7. The solution of

$$\begin{cases} u_{xx} = u_{tt}, -\infty < x < \infty, t > 0, \\ u(x,0) = f(x), u_t(x,0) = 0, \end{cases}$$

where

$$f(x) = \begin{cases} 1, & |x| \leq 1, \\ 0, & |x| > 1, \end{cases}$$

is given by d'Alembert's formula

$$u(x,t) = \frac{1}{2}\left(f(x-t) + f(x+t)\right).$$

Some solutions are graphed below (Figure 14.5).

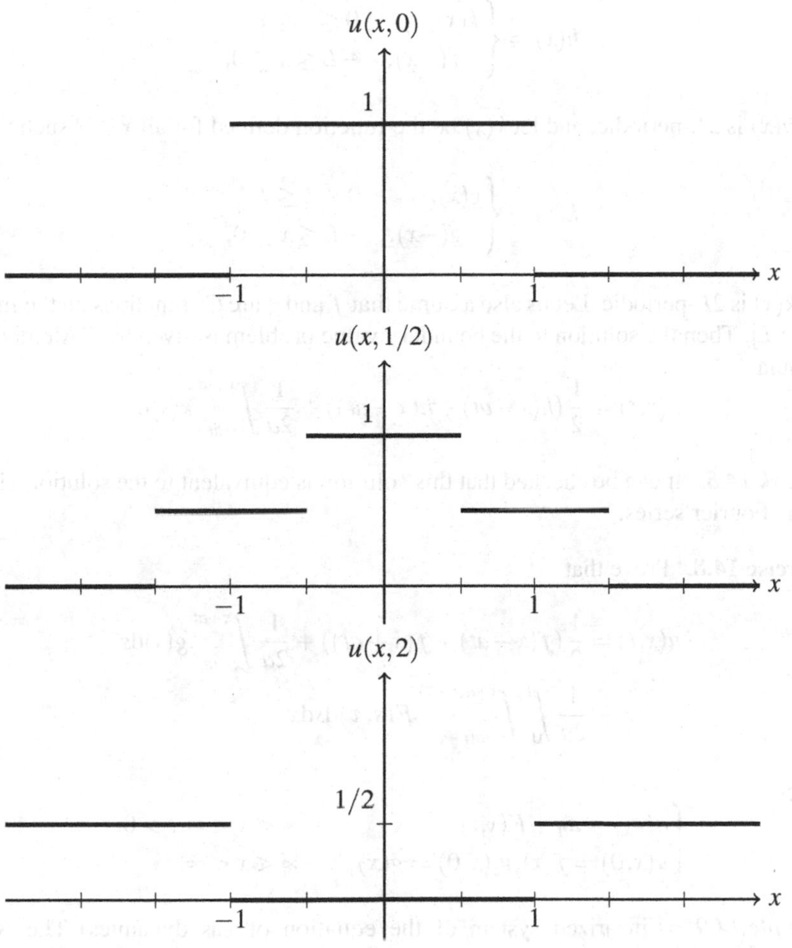

Fig. 14.5 Some solutions of Example 14.7.

We can also apply d'Alembert's formula to the finite string. Consider again the boundary value problem with homogeneous boundary conditions with fixed ends of the string:

$$\begin{cases} a^2 u_{xx} = u_{tt}, 0 < x < L, t > 0, \\ u(0,t) = u(L,t) = 0, t \geq 0, \\ u(x,0) = f(x), u_t(x,0) = g(x), 0 \leq x \leq L, \\ f(0) = f(L) = g(0) = g(L) = 0. \end{cases}$$

Let $h(x)$ be the function defined for all $x \in \mathbb{R}$ such that

$$h(x) = \begin{cases} f(x), & 0 \leq x \leq L, \\ -f(-x), & -L \leq x \leq 0, \end{cases}$$

and $h(x)$ is $2L$-periodic, and let $k(x)$ be the function defined for all $x \in \mathbb{R}$ such that

$$k(x) = \begin{cases} g(x), & 0 \leq x \leq L, \\ -g(-x), & -L \leq x \leq 0, \end{cases}$$

and $k(x)$ is $2L$-periodic. Let us also assume that f and g are C^2 functions on the interval $[0,L]$. Then the solution to the boundary value problem is given by d'Alembert's formula

$$u(x,t) = \frac{1}{2}(h(x-at) + h(x+at)) + \frac{1}{2a}\int_{x-at}^{x+at} k(s)\mathrm{d}s.$$

Remark 14.8. It can be checked that this solution is equivalent to the solution given by the Fourier series.

Exercise 14.8. Prove that

$$u(x,t) = \frac{1}{2}(f(x-at) + f(x+at)) + \frac{1}{2a}\int_{x-at}^{x+at} g(s)\mathrm{d}s$$
$$- \frac{1}{2a}\int_0^t \int_{x-a(t-\tau)}^{x+a(t-\tau)} F(s,\tau)\mathrm{d}s\mathrm{d}\tau$$

solves

$$\begin{cases} a^2 u_{xx} = u_{tt} + F(x,t), & -\infty < x < \infty, t > 0, \\ u(x,0) = f(x), u_t(x,0) = g(x), & -\infty < x < \infty. \end{cases}$$

Example 14.9. (Linearized system of the equation of gas dynamics) The isentropic (the entropy is assumed to be constant) flow of an inviscid gas in the one-dimensional case satisfies the nonlinear equations

$$\begin{cases} u_t + u \cdot u_x + \dfrac{c^2}{\rho} \rho_x = 0, \\ \rho_t + u \cdot \rho_x + \rho u_x = 0, \qquad -\infty < x < \infty, t > 0, \end{cases}$$

where $u(x,t)$ is the velocity of the gas at x and time t, $\rho(x,t)$ is the density, and $c = c(\rho)$ is the known local speed of sound.

As a first step in understanding the general nature of solutions to this system we assume that $u(x,t)$ and $\rho(x,t)$ are not very different from their values at time $t = 0$ and that these values and their derivatives are "small." Neglecting products of terms of "small" order, we arrive at the linear system (which is the linearization of the original system)

$$\begin{cases} u_t + \dfrac{c_0^2}{\rho_0} \rho_x = 0, \\ \rho_t + \rho_0 u_x = 0, \end{cases} \quad \text{so} \quad \begin{cases} c_0 \rho_{xx} = \rho_{tt}, \\ \dfrac{c_0^2}{\rho_0} u_{xx} = u_{tt}, \end{cases}$$

where ρ_0 is the density of the fluid at rest and $c_0 = c(\rho_0)$. This is just a wave equation for ρ and u. Thus we have that

$$u(x,t) = \frac{c_0}{\sqrt{\rho_0}} \left(f(x - c_0 t) + g(x + c_0 t)\right)$$
$$\rho(x,t) = \sqrt{\rho_0} \left(f(x - c_0 t) - g(x + c_0 t)\right),$$

where f and g are arbitrary C^2 functions, and they are the same here due to the linearized system. If in addition

$$u(x,0) = \varphi_1(x), \quad \rho(x,0) = \varphi_2(x),$$

then

$$u(x,t) = \frac{1}{2} \left(\varphi_1(x - c_0 t) + \varphi_1(x + c_0 t)\right) + \frac{c_0}{2\rho_0} \left(\varphi_2(x - c_0 t) - \varphi_2(x + c_0 t)\right),$$
$$\rho(x,t) = \frac{1}{2} \left(\varphi_2(x - c_0 t) + \varphi_2(x + c_0 t)\right) + \frac{\rho_0}{2c_0} \left(\varphi_1(x - c_0 t) - \varphi_1(x + c_0 t)\right).$$

14.3 The Laplace Equation in a Rectangle and in a Disk

One of the most important of all partial differential equations in applied mathematics is the *Laplace equation*:

$$\begin{aligned} u_{xx} + u_{yy} &= 0 \quad \text{2D equation,} \\ u_{xx} + u_{yy} + u_{zz} &= 0 \quad \text{3D equation.} \end{aligned} \tag{14.36}$$

The Laplace equation appears quite naturally in many applications. For example, a steady-state solution of the heat equation in two space dimensions,

$$\alpha^2(u_{xx}+u_{yy}) = u_t,$$

satisfies the 2D Laplace equation (14.36). When electrostatic fields are considered, the electric potential function must satisfy either the 2D or the 3D equation (14.36).

A typical boundary value problem for the Laplace equation is (in dimension two)

$$\begin{cases} u_{xx}+u_{yy}=0, & (x,y)\in\Omega\subset\mathbb{R}^2, \\ u(x,y)=f(x,y), & (x,y)\in\partial\Omega, \end{cases} \tag{14.37}$$

where f is a given function on the boundary $\partial\Omega$ of the domain Ω. The problem (14.37) is called the *Dirichlet problem* (Dirichlet boundary conditions). The problem

$$\begin{cases} u_{xx}+u_{yy}=0, & (x,y)\in\Omega, \\ \frac{\partial u}{\partial \nu}(x,y)=g(x,y), & (x,y)\in\partial\Omega, \end{cases}$$

where g is given and $\frac{\partial u}{\partial \nu}$ is the outward normal derivative, is called the *Neumann problem* (Neumann boundary conditions) (Figure 14.6).

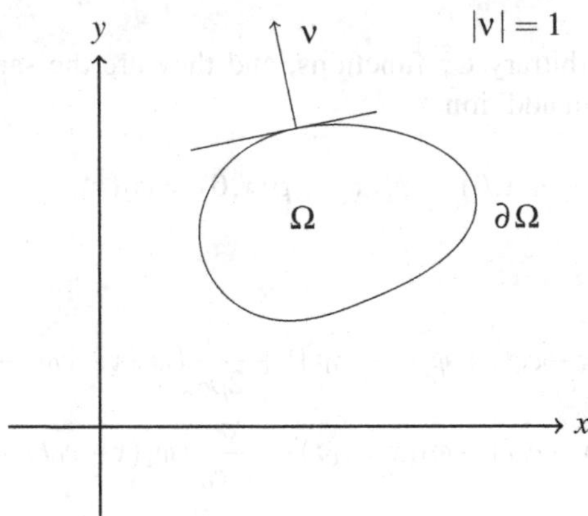

Fig. 14.6 Domain Ω and the outward unit normal vector ν on the boundary $\partial\Omega$.

Dirichlet problem for a rectangle

Consider the boundary value problem in most general form:

$$\begin{cases} w_{xx}+w_{yy}=0, & 0<x<a, 0<y<b, \\ w(x,0)=g_1(x), w(x,b)=f_1(x), & 0<x<a, \\ w(0,y)=g_2(y), w(a,y)=f_2(y), & 0\le y\le b, \end{cases}$$

for fixed $a > 0$ and $b > 0$. The solution of this problem can be reduced to the solutions of

$$\begin{cases} u_{xx} + u_{yy} = 0, & 0 < x < a, 0 < y < b, \\ u(x,0) = u(x,b) = 0, & 0 < x < a, \\ u(0,y) = g_2(y), u(a,y) = f_2(y), & 0 \le y \le b, \end{cases} \tag{14.38}$$

and

$$\begin{cases} u_{xx} + u_{yy} = 0, & 0 < x < a, 0 < y < b, \\ u(x,0) = g_1(x), u(x,b) = f_1(x), & 0 < x < a, \\ u(0,y) = 0, u(a,y) = 0, & 0 \le y \le b. \end{cases}$$

Due to symmetry in x and y, we consider (14.38) only (Figure 14.7).

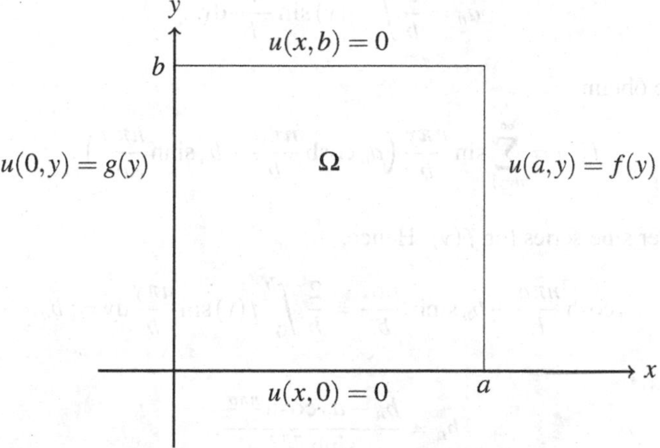

Fig. 14.7 Geometric illustration of the boundary value problem (14.38).

The method of separation of variables gives for $u(x,y) = X(x)Y(y)$,

$$\begin{cases} Y'' + \lambda Y = 0, & 0 < y < b, \\ Y(0) = Y(b) = 0, \end{cases} \tag{14.39}$$

and

$$X'' - \lambda X = 0, \quad 0 < x < a. \tag{14.40}$$

From (14.39) one obtains the eigenvalues and eigenfunctions

$$\lambda_n = \left(\frac{n\pi}{b}\right)^2, \quad Y_n(y) = \sin\frac{n\pi y}{b}, \quad n = 1,2,\ldots.$$

We substitute λ_n into (14.40) to get the general solution

$$X(x) = c_1 \cosh \frac{n\pi x}{b} + c_2 \sinh \frac{n\pi x}{b}.$$

As above, we represent the solution to (14.38) in the form

$$u(x,y) = \sum_{n=1}^{\infty} \sin \frac{n\pi y}{b} \left(a_n \cosh \frac{n\pi x}{b} + b_n \sinh \frac{n\pi x}{b} \right). \tag{14.41}$$

The boundary condition at $x = 0$ gives

$$g(y) = \sum_{n=1}^{\infty} a_n \sin \frac{n\pi y}{b},$$

with

$$a_n = \frac{2}{b} \int_0^b g(y) \sin \frac{n\pi y}{b} \mathrm{d}y.$$

At $x = a$ we obtain

$$f(y) = \sum_{n=1}^{\infty} \sin \frac{n\pi y}{b} \left(a_n \cosh \frac{n\pi a}{b} + b_n \sinh \frac{n\pi a}{b} \right).$$

It is a Fourier sine series for $f(y)$. Hence,

$$a_n \cosh \frac{n\pi a}{b} + b_n \sinh \frac{n\pi a}{b} = \frac{2}{b} \int_0^b f(y) \sin \frac{n\pi y}{b} \mathrm{d}y =: \widetilde{b}_n.$$

This implies

$$b_n = \frac{\widetilde{b}_n - a_n \cosh \frac{n\pi a}{b}}{\sinh \frac{n\pi a}{b}}. \tag{14.42}$$

Substituting (14.42) into (14.41) gives

$$\begin{aligned}
u(x,y) &= \sum_{n=1}^{\infty} \sin \frac{n\pi y}{b} \left(a_n \cosh \frac{n\pi x}{b} + \frac{\widetilde{b}_n - a_n \cosh \frac{n\pi a}{b}}{\sinh \frac{n\pi a}{b}} \sinh \frac{n\pi x}{b} \right) \\
&= \sum_{n=1}^{\infty} \sin \frac{n\pi y}{b} \widetilde{b}_n \frac{\sinh \frac{n\pi x}{b}}{\sinh \frac{n\pi a}{b}} \\
&\quad + \sum_{n=1}^{\infty} \sin \frac{n\pi y}{b} a_n \left(\frac{\cosh \frac{n\pi x}{b} \sinh \frac{n\pi a}{b} - \cosh \frac{n\pi a}{b} \sinh \frac{n\pi x}{b}}{\sinh \frac{n\pi a}{b}} \right) \\
&= \sum_{n=1}^{\infty} \sin \frac{n\pi y}{b} \widetilde{b}_n \frac{\sinh \frac{n\pi x}{b}}{\sinh \frac{n\pi a}{b}} + \sum_{n=1}^{\infty} \sin \frac{n\pi y}{b} a_n \frac{\sinh \frac{n\pi (a-x)}{b}}{\sinh \frac{n\pi a}{b}},
\end{aligned}$$

because $\cosh \alpha \sinh \beta - \sinh \alpha \cosh \beta = \sinh(\beta - \alpha)$. Using the properties of $\sinh \alpha$ and $\cosh \alpha$ for large α, we may conclude that inside of the rectangle, i.e., for

$0 < x < a, 0 < y < b$, we may differentiate this series term by term as often as we wish, and so u is a C^∞ function there.

Exercise 14.9. Find a solution of the problem

$$\begin{cases} u_{xx} + u_{yy} = 0, & 0 < x < 2, 0 < y < 1, \\ u(x,0) = u(x,1) = 0, & 0 < x < 2, \\ u(0,y) = 0, u(2,y) = y(1-y), & 0 \le y \le 1, \end{cases}$$

using the method of separation of variables.

Exercise 14.10. Find a solution of the problem

$$\begin{cases} u_{xx} + u_{yy} = 0, & 0 < x < a, 0 < y < b, \\ u_y(x,0) = u_y(x,b) = 0, & 0 < x < a, \\ u_x(0,y) = f(y), u_x(a,y) = g(y), & 0 < y < b, \end{cases}$$

using the method of separation of variables.

Dirichlet problem for a disk

Consider the problem of solving the Laplace equation in a disk $\{x \in \mathbb{R}^2 : |x| < a\}$ subject to the boundary condition

$$u(a,\theta) = f(\theta), \tag{14.43}$$

where f is a given function on $0 \le \theta \le 2\pi$. In polar coordinates $x = r\cos\theta$, $y = r\sin\theta$, the Laplace equation takes the form

$$u_{rr} + \frac{1}{r}u_r + \frac{1}{r^2}u_{\theta\theta} = 0. \tag{14.44}$$

We apply again the method of separation of variables and assume that

$$u(r,\theta) = R(r)T(\theta). \tag{14.45}$$

Substitution for u in (14.44) yields

$$R''T + \frac{1}{r}R'T + \frac{1}{r^2}RT'' = 0,$$

or

$$\begin{cases} r^2R'' + rR' - \lambda R = 0, \\ T'' + \lambda T = 0. \end{cases}$$

There are no homogeneous boundary conditions; however, we require $T(\theta)$ to be 2π-periodic and also bounded. This fact in particular leads to ($f(0) = f(2\pi)$ and

$f'(0) = f'(2\pi)$)

$$T(0) = T(2\pi), \quad T'(0) = T'(2\pi). \tag{14.46}$$

It is possible to show that equations (14.46) require λ to be real. In what follows we will consider the three possible cases.

If $\lambda = -\mu^2 < 0$, $\mu > 0$, then the equation for T becomes $T'' - \mu^2 T = 0$, and consequently

$$T(\theta) = c_1 e^{\mu\theta} + c_2 e^{-\mu\theta}.$$

It follows from (14.46) that

$$\begin{cases} c_1 + c_2 = c_1 e^{2\pi\mu} + c_2 e^{-2\pi\mu}, \\ c_1 - c_2 = c_1 e^{2\pi\mu} - c_2 e^{-2\pi\mu}, \end{cases}$$

so that $c_1 = c_2 = 0$.

If $\lambda = 0$, then $T'' = 0$ and $T(\theta) = c_1 + c_2\theta$. The first condition in (14.46) implies then that $c_2 = 0$ and therefore $T(\theta) \equiv$ constant.

If $\lambda = \mu^2 > 0$, $\mu > 0$, then

$$T(\theta) = c_1 \cos(\mu\theta) + c_2 \sin(\mu\theta).$$

Now the conditions (14.46) imply that

$$\begin{cases} c_1 = c_1 \cos(2\pi\mu) + c_2 \sin(2\pi\mu), \\ c_2 = -c_1 \sin(2\pi\mu) + c_2 \cos(2\pi\mu), \end{cases}$$

or

$$\begin{cases} c_1 \sin^2(\mu\pi) = c_2 \sin(\mu\pi)\cos(\mu\pi), \\ c_2 \sin^2(\mu\pi) = -c_1 \sin(\mu\pi)\cos(\mu\pi). \end{cases}$$

If $\sin(\mu\pi) \neq 0$, then

$$\begin{cases} c_1 = c_2 \cot(\mu\pi), \\ c_2 = -c_1 \cot(\mu\pi). \end{cases}$$

Hence $c_1^2 + c_2^2 = 0$, i.e., $c_1 = c_2 = 0$. Thus we must have $\sin(\mu\pi) = 0$, and so

$$\lambda_n = n^2, \quad T_n(\theta) = c_1 \cos(n\theta) + c_2 \sin(n\theta), \quad n = 0, 1, 2, \ldots. \tag{14.47}$$

Turning now to R, for $\lambda = 0$ we have $r^2 R'' + rR' = 0$, i.e., $R(r) = k_1 + k_2 \log r$. Since $\log r \to -\infty$ as $r \to 0$, we must choose $k_2 = 0$ in order for R (and u) to be bounded. That is why

$$R_0(r) \equiv \text{constant}. \tag{14.48}$$

For $\lambda = \mu^2 = n^2$ the equation for R becomes

$$r^2R'' + rR' - n^2R = 0.$$

Hence

$$R(r) = k_1r^n + k_2r^{-n}.$$

Again, we must choose $k_2 = 0$, and therefore

$$R_n(r) = k_1r^n, \quad n = 1,2,\dots. \tag{14.49}$$

Combining (14.45), (14.47), (14.48), and (14.49), we obtain

$$u(r,\theta) = \frac{a_0}{2} + \sum_{n=1}^{\infty} r^n(a_n\cos(n\theta) + b_n\sin(n\theta)). \tag{14.50}$$

The boundary condition (14.43) then requires

$$u(a,\theta) = \frac{a_0}{2} + \sum_{n=1}^{\infty} a^n(a_n\cos(n\theta) + b_n\sin(n\theta)) = f(\theta).$$

Hence the coefficients are given by

$$a_0 = \frac{1}{\pi}\int_0^{2\pi} f(\theta)\mathrm{d}\theta,$$

$$a_n = \frac{1}{\pi a^n}\int_0^{2\pi} f(\theta)\cos(n\theta)\mathrm{d}\theta,$$

and

$$b_n = \frac{1}{\pi a^n}\int_0^{2\pi} f(\theta)\sin(n\theta)\mathrm{d}\theta.$$

This procedure can be used also to study the Neumann problem, i.e., the problem in the disk with the boundary condition

$$\frac{\partial u}{\partial r}(a,\theta) = f(\theta). \tag{14.51}$$

Also in this case the solution $u(r,\theta)$ has the form (14.50). The boundary condition (14.51) implies that

$$\left.\frac{\partial u}{\partial r}(r,\theta)\right|_{r=a} = \sum_{n=1}^{\infty} na^{n-1}(a_n\cos(n\theta) + b_n\sin(n\theta)) = f(\theta).$$

Hence

$$a_n = \frac{1}{\pi n a^{n-1}} \int_0^{2\pi} f(\theta)\cos(n\theta)\mathrm{d}\theta$$

and

$$b_n = \frac{1}{\pi n a^{n-1}} \int_0^{2\pi} f(\theta)\sin(n\theta)\mathrm{d}\theta.$$

Remark 14.10. For the Neumann problem a solution is defined up to an arbitrary constant $\frac{a_0}{2}$. Moreover, f must satisfy the consistency condition

$$\int_0^{2\pi} f(\theta)\mathrm{d}\theta = 0,$$

since integrating

$$f(\theta) = \sum_{n=1}^{\infty} n a^{n-1}(a_n\cos(n\theta) + b_n\sin(n\theta))$$

termwise gives us zero.

Remark 14.11. The solution $u(r,\theta)$ of the Laplace equation in a disk $\{x \in \mathbb{R}^2 : |x| < a\}$ subject to the boundary condition $u(a,\theta) = f(\theta)$ without the assumption that $f'(0) = f'(2\pi)$ (but with the assumption $f(0) = f(2\pi)$) can be obtained as

$$u(r,\theta) = \frac{a_0}{2} + \sum_{n=1}^{\infty} r^n(a_n\cos(n\theta) + b_n\sin(n\theta)) + \sum_{n=1}^{\infty} r^{n-1/2}c_n\sin(n-1/2)\theta. \tag{14.52}$$

Exercise 14.11. Prove (14.52) and then show that we have no uniqueness in this boundary value problem. Hint: Use the fact that if $c_n = 0$, $n = 1,2,\ldots$, then we may uniquely determine a_0, a_n, b_n satisfying the boundary condition $u(a,\theta) = f(\theta)$, and if $a_n = b_n = 0$, $n = 1,2,\ldots$, we may uniquely define $c_n, n = 1,2,\ldots$ that depend parametrically on an arbitrary constant a_0.

Remark 14.12. The solution $u(r,\theta)$ of the Laplace equation in the disk $\{x \in \mathbb{R}^2 : |x| < a\}$ subject to the boundary condition $u_\theta(a,\theta) = f(\theta)$ with (possibly non-smooth) periodic function $f(\theta)$ can be obtained as

$$u(r,\theta) = \frac{a_0\theta + b_0}{2} + \sum_{n=1}^{\infty} r^n(a_n\cos(n\theta) + b_n\sin(n\theta)) + \sum_{n=1}^{\infty} r^{n-1/2}c_n\cos(n-1/2)\theta. \tag{14.53}$$

Exercise 14.12. Prove (14.53) and then show that we have no uniqueness in this boundary value problem (see the previous exercise).

Part II
Fourier Transform and Distributions

Chapter 15
Introduction

In this part we assume that the reader is familiar with the following concepts:

(1) Metric spaces and their completeness.
(2) The Lebesgue integral in a bounded domain $\Omega \subset \mathbb{R}^n$ and in $\mathbb{R}^n$.
(3) The Banach spaces (L^p, $1 \le p \le \infty$, C^k) and Hilbert spaces (L^2): If $1 \le p < \infty$, then we set

$$L^p(\Omega) := \{f : \Omega \to \mathbb{C}\,\text{measurable} : \|f\|_{L^p(\Omega)} := \left(\int_\Omega |f(x)|^p \mathrm{d}x\right)^{1/p} < \infty\},$$

while

$$L^\infty(\Omega) := \{f : \Omega \to \mathbb{C}\,\text{measurable} : \|f\|_{L^\infty(\Omega)} := \operatorname*{ess\,sup}_{x\in\Omega} |f(x)| < \infty\}.$$

Moreover,

$$C^k(\overline{\Omega}) := \{f : \overline{\Omega} \to \mathbb{C} : \|f\|_{C^k(\overline{\Omega})} := \max_{x\in\overline{\Omega}} \sum_{|\alpha|\le k} |\partial^\alpha f(x)| < \infty\},$$

where $\overline{\Omega}$ is the closure of Ω. We say that $f \in C^\infty(\Omega)$ if $f \in C^k(\overline{\Omega_1})$ for all $k \in \mathbb{N}$ and for all bounded subsets $\Omega_1 \subset \Omega$. The space $C^\infty(\Omega)$ is not a normed space. The inner product in $L^2(\Omega)$ is denoted by

$$(f,g)_{L^2(\Omega)} = \int_\Omega f(x)\overline{g(x)}\mathrm{d}x.$$

Also in $L^2(\Omega)$, the duality pairing is given by

$$\langle f,g\rangle_{L^2(\Omega)} = \int_\Omega f(x)g(x)\mathrm{d}x.$$

V. Serov, *Fourier Series, Fourier Transform and Their Applications to Mathematical Physics*, Applied Mathematical Sciences 197,
DOI 10.1007/978-3-319-65262-7_15

(4) Hölder's inequality: Let $1 \le p \le \infty$, $u \in L^p$, and $v \in L^{p'}$ with

$$\frac{1}{p} + \frac{1}{p'} = 1.$$

Then $uv \in L^1$ and

$$\int_\Omega |u(x)v(x)|\mathrm{d}x \le \left(\int_\Omega |u(x)|^p \mathrm{d}x\right)^{\frac{1}{p}} \left(\int_\Omega |v(x)|^{p'} \mathrm{d}x\right)^{\frac{1}{p'}},$$

where the Hölder conjugate exponent p' of p is obtained via

$$p' = \frac{p}{p-1}$$

with the understanding that $p' = \infty$ if $p = 1$ and $p' = 1$ if $p = \infty$.

(5) Lebesgue's dominated convergence theorem:
Let $A \subset \mathbb{R}^n$ be measurable and let $\{f_k\}_{k=1}^\infty$ be a sequence of measurable functions converging to $f(x)$ pointwise in A. If there exists a function $g \in L^1(A)$ such that $|f_k(x)| \le g(x)$ in A, then $f \in L^1(A)$ and

$$\lim_{k\to\infty} \int_A f_k(x)\mathrm{d}x = \int_A f(x)\mathrm{d}x.$$

(6) Fubini's theorem on the interchange of the order of integration:

$$\int_{X\times Y} f(x,y)\mathrm{d}x\mathrm{d}y = \int_X \mathrm{d}x \left(\int_Y f(x,y)\mathrm{d}y\right) = \int_Y \mathrm{d}y \left(\int_X f(x,y)\mathrm{d}x\right)$$

if $f \in L^1(X \times Y)$.

(7) The divergence theorem: Let $\Omega \subset \mathbb{R}^n$ be a bounded domain with C^1 boundary $\partial\Omega$, and let $\vec{F}$ be a C^1 vector-valued function on $\overline{\Omega}$. Then

$$\int_\Omega \operatorname{div} \vec{F}(x)\mathrm{d}x = \int_{\partial\Omega} \vec{\nu} \cdot \vec{F} \mathrm{d}\sigma(x),$$

where $\vec{\nu}$ is the outward normal vector to $\partial\Omega$.

Chapter 16
The Fourier Transform in Schwartz Space

Consider the Euclidean space $\mathbb{R}^n$, $n \geq 1$, with $x = (x_1, \ldots, x_n) \in \mathbb{R}^n$ and with $|x| = \sqrt{x_1^2 + \cdots + x_n^2}$ and scalar product $(x, y) = \sum_{j=1}^n x_j y_j$. The open ball of radius $\delta > 0$ centered at $x \in \mathbb{R}^n$ is denoted by

$$U_\delta(x) := \{y \in \mathbb{R}^n : |x - y| < \delta\}.$$

Recall the Cauchy–Bunyakovsky–Schwarz inequality

$$|(x, y)| \leq |x||y|.$$

Following Laurent Schwartz, we call an n-tuple $\alpha = (\alpha_1, \ldots, \alpha_n)$, $\alpha_j \in \mathbb{N} \cup \{0\} \equiv \mathbb{N}_0$ an n-dimensional multi-index. Define

$$|\alpha| = \alpha_1 + \cdots + \alpha_n, \quad \alpha! = \alpha_1! \cdots \alpha_n!$$

and

$$x^\alpha = x_1^{\alpha_1} \cdots x_n^{\alpha_n}, \quad 0^0 = 1, \quad 0! = 1.$$

Moreover, multi-indices α and β can be ordered according to

$$\alpha \leq \beta$$

if $\alpha_j \leq \beta_j$ for all $j = 1, 2, \ldots, n$. Let us also introduce a shorthand notation

$$\partial^\alpha = \partial_1^{\alpha_1} \cdots \partial_n^{\alpha_n}, \quad \partial_j = \frac{\partial}{\partial x_j}.$$

V. Serov, *Fourier Series, Fourier Transform and Their Applications to Mathematical Physics*, Applied Mathematical Sciences 197,
DOI 10.1007/978-3-319-65262-7_16

Definition 16.1. The *Schwartz space* $S(\mathbb{R}^n)$ of rapidly decaying functions is defined as

$$S(\mathbb{R}^n) = \{f \in C^\infty(\mathbb{R}^n) : |f|_{\alpha,\beta} := \sup_{x\in\mathbb{R}^n} \left|x^\alpha \partial^\beta f(x)\right| < \infty \text{ for any } \alpha,\beta \in \mathbb{N}_0^n\}.$$

The following properties of $S = S(\mathbb{R}^n)$ are readily verified.

(1) S is a linear space.
(2) $\partial^\alpha : S \to S$ for every $\alpha \geq 0$.
(3) $x^\beta \cdot : S \to S$ for every $\beta \geq 0$.
(4) If $f \in S(\mathbb{R}^n)$, then $|f(x)| \leq c_m(1+|x|)^{-m}$ for every $m \in \mathbb{N}$. The converse is not true (see part (3) of Example 16.2).
(5) It follows from part (4) that $S(\mathbb{R}^n) \subset L^p(\mathbb{R}^n)$ for every $1 \leq p \leq \infty$.

Example 16.2.

(1) $f(x) = \mathrm{e}^{-a|x|^2} \in S$ for every $a > 0$.
(2) $f(x) = \mathrm{e}^{-a(1+|x|^2)^a} \in S$ for every $a > 0$.
(3) $f(x) = \mathrm{e}^{-|x|} \notin S$.
(4) $C_0^\infty(\mathbb{R}^n) \subset S(\mathbb{R}^n)$, where

$$C_0^\infty(\mathbb{R}^n) = \{f \in C^\infty(\mathbb{R}^n) : \operatorname{supp} f \text{ compact in } \mathbb{R}^n\}$$

and $\operatorname{supp} f = \overline{\{x \in \mathbb{R}^n : f(x) \neq 0\}}$.

The space $S(\mathbb{R}^n)$ is generated by a countable family of seminorms because $|f|_{\alpha,\beta}$ is only a seminorm for $\alpha \geq 0$ and $\beta > 0$, i.e., the condition

$$|f|_{\alpha,\beta} = 0 \quad \text{if and only if} \quad f = 0$$

fails to hold for, e.g., a constant function f. The space (S,ρ) is not normable but it is a metric space if the metric ρ is defined by

$$\rho(f,g) = \sum_{\alpha,\beta\geq 0} 2^{-|\alpha|-|\beta|} \cdot \frac{|f-g|_{\alpha,\beta}}{1+|f-g|_{\alpha,\beta}}.$$

Exercise 16.1. Prove that ρ is a metric, that is,

(1) $\rho(f,g) \geq 0$ and $\rho(f,g) = 0$ if and only if $f = g$.
(2) $\rho(f,g) = \rho(g,f)$.
(3) $\rho(g,h) \leq \rho(g,f) + \rho(f,h)$.

Prove also that $|\rho(f,h) - \rho(g,h)| \leq \rho(f,g)$.

Theorem 16.3 (Completeness). *The space (S,ρ) is a complete metric space, i.e., every Cauchy sequence converges.*

Proof. Let $\{f_k\}_{k=1}^{\infty}$, $f_k \in S$, be a Cauchy sequence, that is, for every $\varepsilon > 0$ there exists $n_0(\varepsilon) \in \mathbb{N}$ such that

$$\rho(f_k, f_m) < \varepsilon, \quad k, m \geq n_0(\varepsilon).$$

It follows that

$$\sup_{x \in K} \left| \partial^\beta (f_k - f_m) \right| < \varepsilon$$

for every $\beta \geq 0$ and for every compact set K in $\mathbb{R}^n$. This means that $\{f_k\}_{k=1}^{\infty}$ is a Cauchy sequence in the Banach space $C^{|\beta|}(K)$. Hence there exists a function $f \in C^{|\beta|}(K)$ such that

$$\lim_{k \to \infty} f_k \overset{C^{|\beta|}(K)}{=} f.$$

Thus we may conclude that our function f is in $C^\infty(\mathbb{R}^n)$. It remains only to prove that $f \in S$. It is clear that

$$\begin{aligned} \sup_{x \in K} |x^\alpha \partial^\beta f| &\leq \sup_{x \in K} |x^\alpha \partial^\beta (f_k - f)| + \sup_{x \in K} |x^\alpha \partial^\beta f_k| \\ &\leq C_\alpha(K) \sup_{x \in K} |\partial^\beta (f_k - f)| + \sup_{x \in K} |x^\alpha \partial^\beta f_k|. \end{aligned}$$

Taking $k \to \infty$, we obtain

$$\sup_{x \in K} |x^\alpha \partial^\beta f| \leq \limsup_{k \to \infty} |f_k|_{\alpha,\beta} < \infty.$$

The last inequality is valid, since $\{f_k\}_{k=1}^{\infty}$ is a Cauchy sequence, so that $|f_k|_{\alpha,\beta}$ is bounded. The last inequality doesn't depend on K, and we may conclude that $|f|_{\alpha,\beta} < \infty$, or $f \in S$. □

Definition 16.4. We say that $f_k \xrightarrow{S} f$ as $k \to \infty$ if

$$|f_k - f|_{\alpha,\beta} \to 0, \quad k \to \infty$$

for all $\alpha, \beta \geq 0$.

Exercise 16.2. Prove that $\overline{C_0^\infty(\mathbb{R}^n)} = S$, that is, for every $f \in S$ there exists $\{f_k\}_{k=1}^{\infty}$, $f_k \in C_0^\infty(\mathbb{R}^n)$, such that $f_k \xrightarrow{S} f, k \to \infty$.

Now we are in position to define the Fourier transform in $S(\mathbb{R}^n)$.

Definition 16.5. The *Fourier transform* $\mathscr{F} f(\xi)$ or $\widehat{f}(\xi)$ of the function $f(x) \in S$ is defined by

$$\mathscr{F} f(\xi) \equiv \widehat{f}(\xi) := (2\pi)^{-n/2} \int_{\mathbb{R}^n} \mathrm{e}^{-\mathrm{i}(x,\xi)} f(x) \mathrm{d}x, \quad \xi \in \mathbb{R}^n.$$

Remark 16.6. This integral is well defined, since

$$\left|\widehat{f}(\xi)\right| \le c_m(2\pi)^{-n/2}\int_{\mathbb{R}^n}(1+|x|)^{-m}\mathrm{d}x < \infty,$$

for $m > n$.

Next we prove the following properties of the Fourier transform:

(1) $\mathscr{F}$ is a linear continuous map from S into S.
(2) $\xi^\alpha \partial_\xi^\beta \widehat{f}(\xi) = (-\mathrm{i})^{|\alpha|+|\beta|}\partial_x^\alpha \widehat{(x^\beta f(x))}$.

Indeed, we have

$$\partial_\xi^\beta \widehat{f}(\xi) = (2\pi)^{-n/2}\int_{\mathbb{R}^n}(-\mathrm{i}x)^\beta \mathrm{e}^{-\mathrm{i}(x,\xi)}f(x)\mathrm{d}x$$

and hence

$$\left\|\partial_\xi^\beta \widehat{f}(\xi)\right\|_{L^\infty(\mathbb{R}^n)} \le c_m(2\pi)^{-n/2}\int_{\mathbb{R}^n}\frac{|x|^{|\beta|}}{(1+|x|)^m}\mathrm{d}x < \infty$$

if we choose $m > n + |\beta|$. At the same time we have obtained the formula

$$\partial_\xi^\beta \widehat{f}(\xi) = \widehat{(-\mathrm{i}x)^\beta f}(x). \tag{16.1}$$

Further, integration by parts gives us

$$\xi^\alpha \widehat{f}(\xi) = (-\mathrm{i})^{|\alpha|}(2\pi)^{-n/2}\int_{\mathbb{R}^n}\mathrm{e}^{-\mathrm{i}(x,\xi)}\partial_x^\alpha f(x)\mathrm{d}x,$$

from which we have the estimate

$$\left\|\xi^\alpha \widehat{f}\right\|_{L^\infty(\mathbb{R}^n)} \le c\int_{\mathbb{R}^n}|\partial_x^\alpha f(x)|\,\mathrm{d}x < \infty,$$

since $\partial_x^\alpha f(x) \in S$ for every $\alpha \ge 0$ if $f(x) \in S$. And also we have the formula

$$\xi^\alpha \widehat{f} = \widehat{(-\mathrm{i})^{|\alpha|}\partial_x^\alpha f}. \tag{16.2}$$

If we combine these last two estimates, we may conclude that $\mathscr{F} : S \to S$ and $\mathscr{F}$ is a continuous map (in the sense of the metric space (S,ρ)), since $\mathscr{F}$ maps every bounded set from S again to a bounded set from S.

The formulas (16.1) and (16.2) show us that it is more convenient to use the following notation:

$$D_j = -\mathrm{i}\partial_j = -\mathrm{i}\frac{\partial}{\partial x_j}, \quad D^\alpha = D_1^{\alpha_1}\cdots D_n^{\alpha_n}.$$

For this new derivative the formulas (16.1) and (16.2) can be rewritten as

$$D_\xi^\alpha \widehat{f} = (-1)^{|\alpha|}\widehat{x^\alpha f}, \quad \xi^\alpha \widehat{f} = \widehat{D^\alpha f}.$$

Example 16.7. It is true that

$$\mathscr{F}(\mathrm{e}^{-\frac{1}{2}|x|^2})(\xi) = \mathrm{e}^{-\frac{1}{2}|\xi|^2}.$$

Proof. The definition gives us directly

$$\begin{aligned}\mathscr{F}(\mathrm{e}^{-\frac{1}{2}|x|^2})(\xi) &= (2\pi)^{-n/2}\int_{\mathbb{R}^n} \mathrm{e}^{-\mathrm{i}(x,\xi)-\frac{1}{2}|x|^2}\mathrm{d}x \\ &= (2\pi)^{-n/2}\mathrm{e}^{-\frac{1}{2}|\xi|^2}\int_{\mathbb{R}^n} \mathrm{e}^{-\frac{1}{2}(|x|^2+2\mathrm{i}(x,\xi)-|\xi|^2)}\mathrm{d}x \\ &= (2\pi)^{-n/2}\mathrm{e}^{-\frac{1}{2}|\xi|^2}\prod_{j=1}^n\int_{-\infty}^{\infty} \mathrm{e}^{-\frac{1}{2}(t+\mathrm{i}\xi_j)^2}\mathrm{d}t.\end{aligned}$$

In order to calculate the last integral, we consider the function $f(z) = \mathrm{e}^{-\frac{z^2}{2}}$ of the complex variable z and the domain D_R depicted in Figure 16.1.

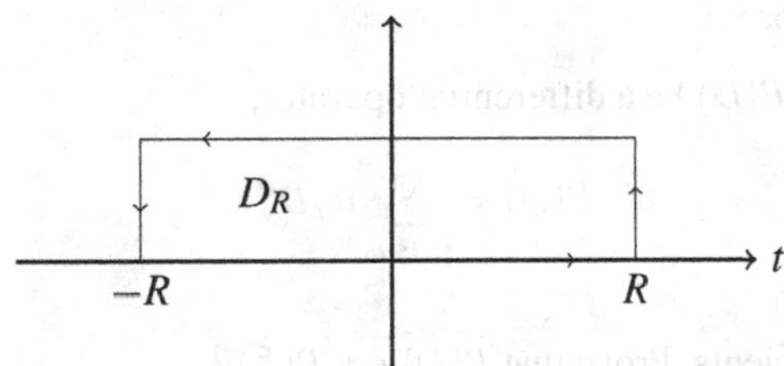

Fig. 16.1 Domain D_R.

We consider the positive direction of going around the boundary ∂D_R. It is clear that $f(z)$ is a holomorphic function in this domain, and by Cauchy's theorem we have

$$\oint_{\partial D_R} \mathrm{e}^{-\frac{z^2}{2}}\mathrm{d}z = 0.$$

But

$$\begin{aligned}\oint_{\partial D_R} \mathrm{e}^{-\frac{z^2}{2}}\mathrm{d}z &= \int_{-R}^{R} \mathrm{e}^{-\frac{t^2}{2}}\mathrm{d}t + \mathrm{i}\int_0^{\xi_j} \mathrm{e}^{-\frac{1}{2}(R+\mathrm{i}\tau)^2}\mathrm{d}\tau \\ &+ \int_R^{-R} \mathrm{e}^{-\frac{1}{2}(t+\mathrm{i}\xi_j)^2}\mathrm{d}t + i\int_{\xi_j}^{0} \mathrm{e}^{-\frac{1}{2}(-R+\mathrm{i}\tau)^2}\mathrm{d}\tau.\end{aligned}$$

If $R \to \infty$, then

$$\int_0^{\xi_j} e^{-\frac{1}{2}(\pm R+i\tau)^2} d\tau \to 0.$$

Hence

$$\int_{-\infty}^{\infty} e^{-\frac{1}{2}(t+i\xi_j)^2} dt = \int_{-\infty}^{\infty} e^{-\frac{t^2}{2}} dt, \quad j = 1, \ldots, n.$$

Using Fubini's theorem and polar coordinates, we can evaluate the last integral as

$$\left(\int_{-\infty}^{\infty} e^{-\frac{t^2}{2}} dt\right)^2 = \int_{\mathbb{R}^2} e^{-\frac{1}{2}(t^2+s^2)} dt ds = \int_0^{2\pi} d\theta \int_0^{\infty} e^{-\frac{r^2}{2}} r dr$$
$$= 2\pi \int_0^{\infty} e^{-m} dm = 2\pi.$$

Thus

$$\int_{-\infty}^{\infty} e^{-\frac{1}{2}(t+i\xi_j)^2} dt = \sqrt{2\pi}$$

and

$$F(e^{-\frac{|x|^2}{2}})(\xi) = (2\pi)^{-\frac{n}{2}} e^{-\frac{1}{2}|\xi|^2} \prod_{j=1}^{n} \sqrt{2\pi} = e^{-\frac{1}{2}|\xi|^2}.$$

This completes the proof. □

Exercise 16.3. Let $P(D)$ be a differential operator,

$$P(D) = \sum_{|\alpha| \le m} a_\alpha D^\alpha,$$

with constant coefficients. Prove that $\widehat{P(D)u} = P(\xi)\widehat{u}$.

Definition 16.8. We adopt the following notation for *translation* and *dilation* of a function:

$$(\tau_h f)(x) := f(x-h), \quad (\sigma_\lambda f)(x) := f(\lambda x), \quad \lambda \neq 0.$$

Exercise 16.4. Let $f \in S(\mathbb{R}^n)$, $h \in \mathbb{R}^n$, and $\lambda \in \mathbb{R}$, $\lambda \neq 0$. Prove that

(1) $\widehat{\sigma_\lambda f}(\xi) = |\lambda|^{-n} \sigma_{\frac{1}{\lambda}} \widehat{f}(\xi)$ and $\sigma_\lambda \widehat{f}(\xi) = |\lambda|^{-n} \widehat{\sigma_{\frac{1}{\lambda}} f}(\xi)$;

(2) $\widehat{\tau_h f}(\xi) = e^{-i(h,\xi)} \widehat{f}(\xi)$ and $\tau_h \widehat{f}(\xi) = \widehat{e^{i(h,\cdot)} f}(\xi)$.

Exercise 16.5. Let A be a real-valued $n \times n$ matrix such that A^{-1} exists. Define $f_A(x) := f(A^{-1}x)$. Prove that

$$\widehat{f_A}(\xi) = (\widehat{f})_A(\xi)$$

if and only if A is an orthogonal matrix (a rotation), that is, $A^T = A^{-1}$.

Let us now consider f and g from $S(\mathbb{R}^n)$. Then

$$(\mathscr{F}f,g)_{L^2} = \int_{\mathbb{R}^n} \widehat{f}(\xi)\overline{g(\xi)}\mathrm{d}\xi = (2\pi)^{-n/2}\int_{\mathbb{R}^n} \overline{g(\xi)}\left(\int_{\mathbb{R}^n} \mathrm{e}^{-\mathrm{i}(x,\xi)}f(x)\mathrm{d}x\right)\mathrm{d}\xi$$
$$= (2\pi)^{-n/2}\int_{\mathbb{R}^n} f(x)\left(\int_{\mathbb{R}^n} \overline{\mathrm{e}^{\mathrm{i}(x,\xi)}g(\xi)}\mathrm{d}\xi\right)\mathrm{d}x = (f,\mathscr{F}^*g)_{L^2},$$

where $\mathscr{F}^*g(x) := \mathscr{F}g(-x)$.

Remark 16.9. Here $\mathscr{F}^*$ is the adjoint operator (in the sense of L^2), which maps S into S since $\mathscr{F} : S \to S$. The inverse Fourier transform $\mathscr{F}^{-1}$ is defined as $\mathscr{F}^{-1} := \mathscr{F}^*$.

In order to justify this definition we will prove the following theorem.

Theorem 16.10 (Fourier inversion formula). *Let f be a function from $S(\mathbb{R}^n)$. Then*

$$\mathscr{F}^*\mathscr{F}f = f.$$

To this end we will prove first the following (somewhat technical) lemma.

Lemma 16.11. *Let $f_0(x)$ be a function from $L^1(\mathbb{R}^n)$ with $\int_{\mathbb{R}^n} f_0(x)\mathrm{d}x = 1$ and let $f(x)$ be a function from $L^\infty(\mathbb{R}^n)$ that is continuous at $\{0\}$. Then*

$$\lim_{\varepsilon\to 0+}\int_{\mathbb{R}^n} \varepsilon^{-n} f_0\left(\frac{x}{\varepsilon}\right) f(x)\mathrm{d}x = f(0).$$

Proof. Since

$$\int_{\mathbb{R}^n} \varepsilon^{-n} f_0\left(\frac{x}{\varepsilon}\right) f(x)\mathrm{d}x - f(0) = \int_{\mathbb{R}^n} \varepsilon^{-n} f_0\left(\frac{x}{\varepsilon}\right)(f(x)-f(0))\mathrm{d}x,$$

we may assume without loss of generality that $f(0) = 0$. Since f is continuous at $\{0\}$, there exists $\delta > 0$ for every $\eta > 0$ such that

$$|f(x)| < \frac{\eta}{\|f_0\|_{L^1}}$$

whenever $|x| < \delta$. Note that

$$\left|\int_{\mathbb{R}^n} f_0(x)\mathrm{d}x\right| \le \|f_0\|_{L^1}.$$

We may therefore conclude that

$$\begin{aligned}\left|\int_{\mathbb{R}^n}\varepsilon^{-n}f_0\left(\frac{x}{\varepsilon}\right)f(x)\mathrm{d}x\right| &\leq \frac{\eta}{\|f_0\|_{L^1}}\cdot\varepsilon^{-n}\int_{|x|<\delta}\left|f_0\left(\frac{x}{\varepsilon}\right)\right|\mathrm{d}x\\ &\quad+\|f\|_{L^\infty}\varepsilon^{-n}\int_{|x|>\delta}\left|f_0\left(\frac{x}{\varepsilon}\right)\right|\mathrm{d}x\\ &\leq\frac{\eta}{\|f_0\|_{L^1}}\cdot\|f_0\|_{L^1}+\|f\|_{L^\infty}\int_{|y|>\frac{\delta}{\varepsilon}}|f_0(y)|\,\mathrm{d}y\\ &=\eta+\|f\|_{L^\infty}I_\varepsilon.\end{aligned}$$

But $I_\varepsilon \to 0$ as $\varepsilon \to 0+$. This proves the lemma.

Proof (*Proof of theorem* 16.10). Let us consider $v(x) = \mathrm{e}^{-\frac{|x|^2}{2}}$. We know from Example 16.7 that $\int_{\mathbb{R}^n} v(x)\mathrm{d}x = (2\pi)^{n/2}$ and $Fv = v$. If we apply Lemma 16.11 with $f_0 = (2\pi)^{-n/2}v(x)$ and $f \in S(\mathbb{R}^n)$, then by Exercise 16.4,

$$\begin{aligned}(2\pi)^{\frac{n}{2}}f(0) &= \lim_{\varepsilon\to0+}\int_{\mathbb{R}^n}\varepsilon^{-n}v\left(\frac{x}{\varepsilon}\right)f(x)\mathrm{d}x = \lim_{\varepsilon\to0+}\langle f,\varepsilon^{-n}\sigma_{\frac{1}{\varepsilon}}v\rangle_{L^2}\\ &= \lim_{\varepsilon\to0+}\langle f,\varepsilon^{-n}\sigma_{\frac{1}{\varepsilon}}\mathscr{F}v\rangle_{L^2} = \lim_{\varepsilon\to0+}\langle f,\mathscr{F}(\sigma_\varepsilon v)\rangle_{L^2}\\ &= \lim_{\varepsilon\to0+}\langle\mathscr{F}f,\sigma_\varepsilon v\rangle_{L^2} = \langle\mathscr{F}f,1\rangle = \int_{\mathbb{R}^n}\mathscr{F}f(\xi)\mathrm{e}^{\mathrm{i}(0,\xi)}\mathrm{d}\xi,\end{aligned}$$

where we have used Lebesgue's dominated convergence theorem in the last step. This proves that

$$f(0) = (2\pi)^{-n/2}\int_{\mathbb{R}^n}\mathscr{F}f(\xi)\mathrm{e}^{\mathrm{i}(0,\xi)}\mathrm{d}\xi = (\mathscr{F}^*\mathscr{F}f)(0).$$

The proof is now finished by

$$\begin{aligned}f(x) = (\tau_{-x}f)(0) = (\mathscr{F}^*\mathscr{F}(\tau_{-x}f))(0) &= (2\pi)^{-n/2}\int_{\mathbb{R}^n}\mathscr{F}(\tau_{-x}f)(\xi)\mathrm{e}^{\mathrm{i}(0,\xi)}\mathrm{d}\xi\\ &= (2\pi)^{-n/2}\int_{\mathbb{R}^n}\mathrm{e}^{\mathrm{i}(x,\xi)}\mathscr{F}f(\xi)\mathrm{d}\xi = \mathscr{F}^*\mathscr{F}f(x),\end{aligned}$$

where we have used Exercise 16.4. □

Corollary 16.12. *The Fourier transform is an isometry (in the sense of L^2).*

Proof. The fact that the Fourier transform preserves the norm of $f \in S$ follows from

$$\|\mathscr{F}f\|_{L^2}^2 = (\mathscr{F}f,\mathscr{F}f)_{L^2} = (f,\mathscr{F}^*\mathscr{F}f)_{L^2} = (f,f)_{L^2} = \|f\|_{L^2}^2.$$

This is called *Parseval's equality*. □

Note that

$$(\mathscr{F}f,g)_{L^2} = (f,\mathscr{F}^*g)_{L^2}$$

means that

$$\int_{\mathbb{R}^n} \widehat{f}(\xi)\overline{g(\xi)}\mathrm{d}\xi = \int_{\mathbb{R}^n} f(x)\overline{\mathscr{F}^* g(x)}\mathrm{d}x = \int_{\mathbb{R}^n} f(x)\mathscr{F}(\overline{g})(x)\mathrm{d}x.$$

This implies that

$$\int_{\mathbb{R}^n} \widehat{f}(\xi)g(\xi)\mathrm{d}\xi = \int_{\mathbb{R}^n} f(x)\widehat{g}(x)\mathrm{d}x,$$

or

$$\langle \mathscr{F} f, g\rangle_{L^2(\mathbb{R}^n)} = \langle f, \mathscr{F} g\rangle_{L^2(\mathbb{R}^n)}.$$

Chapter 17
The Fourier Transform in $L^p(\mathbb{R}^n)$, $1 \le p \le 2$

Let us begin with a preliminary proposition.

Proposition 17.1. *Let X be a linear normed space and $E \subset X$ a subspace of X such that $\overline{E} = X$; that is, the closure $\overline{E}$ of E in the sense of the norm in X is equal to X. Let Y be a Banach space. If $T : E \to Y$ is a continuous linear map, i.e., there exists $M > 0$ such that*

$$\|Tu\|_Y \le M \|u\|_X, \quad u \in E,$$

then there exists a unique linear continuous map $T_{\text{ex}} : X \to Y$ such that $T_{\text{ex}}|_E = T$ and

$$\|T_{\text{ex}}u\|_Y \le M \|u\|_X, \quad u \in X.$$

Exercise 17.1. Prove the previous proposition.

Lemma 17.2. *Let $1 \le p < \infty$. Then*

$$\overline{C_0^\infty(\mathbb{R}^n)}^{L^p} = L^p(\mathbb{R}^n),$$

that is, $C_0^\infty(\mathbb{R}^n)$ is dense in $L^p(\mathbb{R}^n)$ in the sense of L^p-norm.

Proof. We will use the fact that the set of finite linear combinations of characteristic functions of bounded measurable sets in $\mathbb{R}^n$ is dense in $L^p(\mathbb{R}^n)$, $1 \le p < \infty$. This is a well known fact from functional analysis.

Let now $A \subset \mathbb{R}^n$ be a bounded measurable set and let $\varepsilon > 0$. Then there exist a closed set F and an open set Q such that $F \subset A \subset Q$ and $\mu(Q \setminus F) < \varepsilon^p$ (or only $\mu(Q) < \varepsilon^p$ if there is no closed set $F \subset A$). Here μ is the Lebesgue measure in $\mathbb{R}^n$. Let now φ be a function from $C_0^\infty(\mathbb{R}^n)$ such that $\operatorname{supp}\varphi \subset Q$, $\varphi|_F \equiv 1$ and $0 \le \varphi \le 1$; see [19]. Then

$$\|\varphi - \chi_A\|_{L^p(\mathbb{R}^n)}^p = \int_{\mathbb{R}^n} |\varphi(x) - \chi_A(x)|^p \,\mathrm{d}x \le \int_{Q \setminus F} 1 \mathrm{d}x = \mu(Q \setminus F) < \varepsilon^p,$$

V. Serov, *Fourier Series, Fourier Transform and Their Applications to Mathematical Physics*, Applied Mathematical Sciences 197,
DOI 10.1007/978-3-319-65262-7_17

or

$$\|\varphi - \chi_A\|_{L^p(\mathbb{R}^n)} < \varepsilon,$$

where χ_A denotes the characteristic function of A, i.e.,

$$\chi_A(x) = \begin{cases} 1, & x \in A, \\ 0, & x \notin A. \end{cases}$$

Thus, we may conclude that $\overline{C_0^\infty(\mathbb{R}^n)} = L^p(\mathbb{R}^n)$ for $1 \le p < \infty$. □

Remark 17.3. Lemma 17.2 does not hold for $p = \infty$. Indeed, for a function $f \equiv c_0 \neq 0$ and for every function $\varphi \in C_0^\infty(\mathbb{R}^n)$, we have that

$$\|f - \varphi\|_{L^\infty(\mathbb{R}^n)} \ge |c_0| > 0.$$

Hence we cannot approximate a function from $L^\infty(\mathbb{R}^n)$ by functions from $C_0^\infty(\mathbb{R}^n)$. This means that

$$\overline{C_0^\infty(\mathbb{R}^n)}^{L^\infty} \neq L^\infty(\mathbb{R}^n).$$

But the following result holds:

Exercise 17.2. Prove that $\overline{S(\mathbb{R}^n)} \stackrel{L^\infty}{=} \dot{C}(\mathbb{R}^n)$, where

$$\dot{C}(\mathbb{R}^n) := \{f \in C(\mathbb{R}^n) : \lim_{|x|\to\infty} f(x) = 0\}.$$

Now we are in a position to extend F from $S \subset L^1$ to L^1.

Theorem 17.4 (Riemann–Lebesgue lemma). *Let $\mathscr{F} : S \to S$ be the Fourier transform in Schwartz space $S(\mathbb{R}^n)$. Then there exists a unique extension $\mathscr{F}_{\text{ex}}$ as a map from $L^1(\mathbb{R}^n)$ to $\dot{C}(\mathbb{R}^n)$ with norm $\|\mathscr{F}_{\text{ex}}\|_{L^1\to L^\infty} = (2\pi)^{-n/2}$.*

Proof. We know that $\|\mathscr{F} f\|_{L^\infty} \le (2\pi)^{-n/2}\|f\|_{L^1}$ for $f \in S$. Now we apply the preliminary proposition to $E = S$, $X = L^1$, and $Y = L^\infty$. Since $\overline{S} \stackrel{L^1}{=} L^1$ (which follows from $C_0^\infty \subset S$ and $\overline{C_0^\infty} \stackrel{L^1}{=} L^1$) for every $f \in L^1(\mathbb{R}^n)$, there exists $\{f_k\} \subset S$ such that $\|f_k - f\|_{L^1} \to 0$ as $k \to \infty$. In that case, we can define

$$\mathscr{F}_{\text{ex}} f \stackrel{L^\infty}{:=} \lim_{k\to\infty} \mathscr{F} f_k.$$

Since $\overline{S} \stackrel{L^\infty}{=} \dot{C}$ (see Exercise 17.2), it follows that $\mathscr{F}_{\text{ex}} f \in \dot{C}$ and $\|\mathscr{F}_{\text{ex}}\|_{L^1\to L^\infty} \le (2\pi)^{-n/2}$. On the other hand,

$$\|\mathscr{F} f\|_{L^\infty} \ge |\widehat{f}(0)| = (2\pi)^{-n/2}\|f\|_{L^1}$$

for $f \in L^1$ and $f \ge 0$. Hence $\|\mathscr{F}_{\text{ex}}\|_{L^1 \to L^\infty} = (2\pi)^{-n/2}$. □

Proof (Alternative proof). If $f \in L^1(\mathbb{R}^n)$, then we can define the Fourier transform $\mathscr{F}f(\xi)$ directly by

$$\mathscr{F}f(\xi) := (2\pi)^{-n/2} \int_{\mathbb{R}^n} \mathrm{e}^{-\mathrm{i}(x,\xi)} f(x)\mathrm{d}x,$$

since

$$\left| \int_{\mathbb{R}^n} \mathrm{e}^{-\mathrm{i}(x,\xi)} f(x)\mathrm{d}x \right| \le \int_{\mathbb{R}^n} |f(x)|\mathrm{d}x = \|f\|_{L^1}.$$

Also we have

$$\begin{aligned}(2\pi)^{n/2} \left\| \widehat{f}(\xi+h) - \widehat{f}(\xi) \right\|_{L^\infty(\mathbb{R}^n)} &= \sup_{\xi \in \mathbb{R}^n} \left| \int_{\mathbb{R}^n} \mathrm{e}^{-\mathrm{i}(\xi,x)} (\mathrm{e}^{-\mathrm{i}(h,x)} - 1) f(x)\mathrm{d}x \right| \\ &\le 2 \int_{|x| > \frac{\varepsilon}{|h|}} |f(x)|\mathrm{d}x + \varepsilon \int_{|x||h| \le \varepsilon} |f(x)|\mathrm{d}x =: I_1 + I_2.\end{aligned}$$

Here we have used the fact that $|\mathrm{e}^{\mathrm{i}y} - 1| \le |y|$ for $y \in \mathbb{R}$ with $|y| \le 1$. It is easily seen that $I_1 \to 0$ for $|h| \to 0$ and $I_2 \to 0$ for $\varepsilon \to 0$, since $f \in L^1(\mathbb{R}^n)$.

This means that the Fourier transform $\widehat{f}(\xi)$ is continuous (even uniformly continuous) on $\mathbb{R}^n$. Moreover, we have

$$2\widehat{f}(\xi) = (2\pi)^{-n/2} \int_{\mathbb{R}^n} \mathrm{e}^{-\mathrm{i}(x,\xi)} \left(f(x) - f\left(x + \frac{\xi\pi}{|\xi|^2}\right)\right) \mathrm{d}x.$$

This equality follows from

$$\begin{aligned}\widehat{f}(\xi) = -(2\pi)^{-n/2} \int_{\mathbb{R}^n} \mathrm{e}^{\mathrm{i}\pi} \mathrm{e}^{-\mathrm{i}(x,\xi)} f(x)\mathrm{d}x \\ = -(2\pi)^{-n/2} \int_{\mathbb{R}^n} \mathrm{e}^{\mathrm{i}\left(\xi, \frac{\xi}{|\xi|^2}\right)\pi} \mathrm{e}^{-\mathrm{i}(x,\xi)} f(x)\mathrm{d}x\end{aligned}$$

and Exercise 16.4. Thus,

$$2|\widehat{f}(\xi)| \le (2\pi)^{-n/2} \left\| f(x) - f\left(x + \frac{\xi\pi}{|\xi|^2}\right)\right\|_{L^1(\mathbb{R}^n)} \to 0$$

for $|\xi| \to \infty$. □

Theorem 17.5 (Plancherel). *Let $\mathscr{F} : S \to S$ be the Fourier transform in S with $\|\mathscr{F}f\|_{L^2} = \|f\|_{L^2}$. Then there exists a unique extension $\mathscr{F}_{\text{ex}}$ of $\mathscr{F}$ to L^2-space such that $\mathscr{F}_{\text{ex}} : L^2 \overset{\text{onto}}{\to} L^2$ and $\|\mathscr{F}_{\text{ex}}\|_{L^2 \to L^2} = 1$. Also Parseval's equality remains valid.*

Proof. We know that $\overline{S}\overset{L^2}{=}L^2$, since $\overline{C_0^\infty}\overset{L^2}{=}L^2$. Thus for every $f\in L^2(\mathbb{R}^n)$ there exists $\{f_k\}_{k=1}^\infty\subset S(\mathbb{R}^n)$ such that $\|f_k-f\|_{L^2(\mathbb{R}^n)}\to 0$ as $k\to\infty$. By Parseval's equality in S we get

$$\|\mathscr{F}f_k-\mathscr{F}f_l\|_{L^2}=\|f_k-f_l\|_{L^2}\to 0,\quad k,l\to\infty.$$

Thus $\{\mathscr{F}f_k\}_{k=1}^\infty$ is a Cauchy sequence in $L^2(\mathbb{R}^n)$, and therefore, $\mathscr{F}f_k\overset{L^2}{\to}g$, where $g\in L^2$. Therefore, we may put $\mathscr{F}_{\text{ex}}f:=g$. Also we have Parseval's equality

$$\|\mathscr{F}_{\text{ex}}f\|_{L^2}=\lim_{k\to\infty}\|\mathscr{F}f_k\|_{L^2}=\lim_{k\to\infty}\|f_k\|_{L^2}=\|f\|_{L^2},$$

which proves the statement about the operator norm. □

Remark 17.6. In L^2-space we also have the Fourier inversion formula $\mathscr{F}_{\text{ex}}^*\mathscr{F}_{\text{ex}}f=f$, or $\mathscr{F}_{\text{ex}}^{-1}\mathscr{F}_{\text{ex}}f=f$.

Exercise 17.3. Prove that if $f\in L^2(\mathbb{R}^n)$, then

(1) $\mathscr{F}_{\text{ex}}f(\xi)\overset{L^2}{=}\lim_{R\to+\infty}\mathscr{F}f_R(\xi)$, where $f_R(x)=\chi_{\{x:|x|\le R\}}(x)f(x)$.
(2) $\mathscr{F}_{\text{ex}}f(\xi)\overset{L^2}{=}\lim_{\varepsilon\to 0+}\mathscr{F}(\mathrm{e}^{-\varepsilon|x|}f)$.

Exercise 17.4. Let us assume that $f\in L^1(\mathbb{R}^n)$ and $\mathscr{F}f(\xi)\in L^1(\mathbb{R}^n)$. Prove that

$$f(x)=(2\pi)^{-n/2}\int_{\mathbb{R}^n}\mathrm{e}^{\mathrm{i}(x,\xi)}\mathscr{F}f(\xi)\mathrm{d}\xi=\mathscr{F}^{-1}\mathscr{F}f(x).$$

This means that the Fourier inversion formula is valid.

Exercise 17.5. Let f_1 and f_2 belong to $L^2(\mathbb{R}^n)$. Prove that

$$(f_1,f_2)_{L^2}=(\mathscr{F}f_1,\mathscr{F}f_2)_{L^2}.$$

Theorem 17.7 (Riesz–Thorin interpolation theorem). *Let T be a linear continuous map from $L^{p_1}(\mathbb{R}^n)$ to $L^{q_1}(\mathbb{R}^n)$ with norm estimate M_1 and from $L^{p_2}(\mathbb{R}^n)$ to $L^{q_2}(\mathbb{R}^n)$ with norm estimate M_2. Then T is a linear continuous map from $L^p(\mathbb{R}^n)$ to $L^q(\mathbb{R}^n)$ with p and q such that*

$$\frac{1}{p}=\frac{\theta}{p_1}+\frac{1-\theta}{p_2},\quad \frac{1}{q}=\frac{\theta}{q_1}+\frac{1-\theta}{q_2},\quad 0\le\theta\le 1,$$

with norm estimate $M_1^\theta M_2^{1-\theta}$.

Proof. Let F and G be two functions with the following properties:

(1) $F,G\ge 0$,
(2) $\|F\|_{L^1}=\|G\|_{L^1}=1$.

Let us consider now the function $\Phi(z)$ of a complex variable $z \in \mathbb{C}$ given by

$$\Phi(z) := M_1^{-z} M_2^{-1+z} \int_{\mathbb{R}^n} T(f_0 F^{\frac{z}{p_1}+\frac{1-z}{p_2}})(x) g_0 G^{\frac{z}{q_1'}+\frac{1-z}{q_2'}}(x) \mathrm{d}x,$$

where $\frac{1}{q_1}+\frac{1}{q_1'}=1$, $\frac{1}{q_2}+\frac{1}{q_2'}=1$, $|f_0| \le 1$, and $|g_0| \le 1$. The two functions f_0 and g_0 will be selected later. We assume also that $0 \le \mathrm{Re}(z) \le 1$.

Our aim is to prove the inequality

$$|\langle Tf, g\rangle_{L^2}| \le M_1^{\theta} M_2^{1-\theta} \|f\|_{L^p} \|g\|_{L^{q'}},$$

where

$$\frac{1}{p} = \frac{\theta}{p_1} + \frac{1-\theta}{p_2}, \quad \frac{1}{q} = \frac{\theta}{q_1} + \frac{1-\theta}{q_2}, \quad \frac{1}{q} + \frac{1}{q'} = 1.$$

Since T is a linear continuous map and $F^{\frac{z}{p_1}+\frac{1-z}{p_2}}$, $G^{\frac{z}{q_1'}+\frac{1-z}{q_2'}}$ are holomorphic functions with respect to z (consider $a^z = \mathrm{e}^{z\log a}$, $a > 0$), we may conclude that $\Phi(z)$ is a holomorphic function also.

(1) Let us assume now that $\mathrm{Re}(z) = 0$, i.e., $z = \mathrm{i}y$. Then we have

$$\Phi(\mathrm{i}y) = M_1^{-\mathrm{i}y} M_2^{-1+\mathrm{i}y} \langle T(f_0 F^{\frac{\mathrm{i}y}{p_1}+\frac{1-\mathrm{i}y}{p_2}}), g_0 G^{\frac{\mathrm{i}y}{q_1'}+\frac{1-\mathrm{i}y}{q_2'}} \rangle_{L^2}.$$

Since $|a^{\mathrm{i}x}| = 1$ for $a, x \in \mathbb{R}$, $a > 0$, it follows from Hölder's inequality and the assumptions on T that

$$\begin{aligned} |\Phi(\mathrm{i}y)| &\le M_2^{-1} M_2 \left\| f_0 F^{\frac{\mathrm{i}y}{p_1}+\frac{1-\mathrm{i}y}{p_2}} \right\|_{L^{p_2}} \left\| g_0 G^{\frac{\mathrm{i}y}{q_1'}+\frac{1-\mathrm{i}y}{q_2'}} \right\|_{L^{q_2'}} \\ &= \left\| |f_0| F^{\frac{1}{p_2}} \right\|_{L^{p_2}} \left\| |g_0| G^{\frac{1}{q_2'}} \right\|_{L^{q_2'}} \le \|F\|_{L^1}^{\frac{1}{p_2}} \|G\|_{L^1}^{\frac{1}{q_2'}} = 1. \end{aligned}$$

(2) Let us assume now that $\mathrm{Re}(z) = 1$, i.e., $z = 1 + \mathrm{i}y$. Then we have similarly that

$$\begin{aligned} |\Phi(1+\mathrm{i}y)| &\le M_1^{-1} M_1 \left\| f_0 F^{\frac{1+\mathrm{i}y}{p_1}+\frac{-\mathrm{i}y}{p_2}} \right\|_{L^{p_1}} \left\| g_0 G^{\frac{1+\mathrm{i}y}{q_1'}+\frac{-\mathrm{i}y}{q_2'}} \right\|_{L^{q_1'}} \\ &= \left\| |f_0| F^{\frac{1}{p_1}} \right\|_{L^{p_1}} \left\| |g_0| G^{\frac{1}{q_1'}} \right\|_{L^{q_1'}} \le \|F\|_{L^1}^{\frac{1}{p_1}} \|G\|_{L^1}^{\frac{1}{q_1'}} = 1. \end{aligned}$$

If we apply now the *Phragmén–Lindelöf theorem* for the domain $0 < \mathrm{Re}(z) < 1$, we obtain that $|\Phi(z)| \le 1$ for every z such that $0 < \mathrm{Re}(z) < 1$. Then $|\Phi(\theta)| \le 1$ also for $0 < \theta < 1$. But this is equivalent to the estimate

$$|\langle T(f_0F^{\frac{1}{p}}), g_0G^{\frac{1}{q'}}\rangle_{L^2}| \le M_1^\theta M_2^{1-\theta}, \tag{17.1}$$

where $\frac{1}{p} = \frac{\theta}{p_1} + \frac{1-\theta}{p_2}, \frac{1}{q} = \frac{\theta}{q_1} + \frac{1-\theta}{q_2}$ and $\frac{1}{q} + \frac{1}{q'} = 1$. In order to finish the proof of this theorem let us choose (for arbitrary functions $f \in L^p$ and $g \in L^{q'}$ with p and q' as above) the functions F, G, f_0 and g_0 as follows:

$$F = |f_1|^p, \quad G = |g_1|^{q'}, \quad f_0 = \operatorname{sgn} f_1, \quad g_0 = \operatorname{sgn} g_1,$$

where $f_1 = \frac{f}{\|f\|_{L^p}}, g_1 = \frac{g}{\|g\|_{L^{q'}}}$, and

$$\operatorname{sgn} f = \begin{cases} 1, & f > 0, \\ 0, & f = 0, \\ -1, & f < 0. \end{cases}$$

In that case, $f_1 = f_0F^{\frac{1}{p}}$ and $g_1 = g_0G^{\frac{1}{q'}}$. Applying the estimate (17.1), we obtain

$$\left|\left\langle T\left(\frac{f}{\|f\|_{L^p}}\right), \frac{g}{\|g\|_{L^{q'}}}\right\rangle_{L^2}\right| \le M_1^\theta M_2^{1-\theta},$$

which is equivalent to

$$|\langle Tf, g\rangle_{L^2}| \le M_1^\theta M_2^{1-\theta} \|f\|_{L^p} \|g\|_{L^{q'}}.$$

This implies the desired final estimate

$$\|Tf\|_{L^q} \le M_1^\theta M_2^{1-\theta} \|f\|_{L^p}$$

and finishes the proof. □

Theorem 17.8 (Hausdorff–Young). *Let $\mathscr{F} : S \to S$ be the Fourier transform in Schwartz space. Then there exists a unique extension $\mathscr{F}_{\text{ex}}$ as a linear continuous map*

$$\mathscr{F}_{\text{ex}} : L^p(\mathbb{R}^n) \to L^{p'}(\mathbb{R}^n),$$

where $1 \le p \le 2$ and $\frac{1}{p} + \frac{1}{p'} = 1$. What is more, we have the norm estimate

$$\|\mathscr{F}_{\text{ex}}\|_{L^p \to L^{p'}} \le (2\pi)^{-n\left(\frac{1}{p} - \frac{1}{2}\right)}.$$

This is called the Hausdorff–Young inequality.

Proof. We know from Theorems 17.4 and 17.5 that there exists a unique extension $\mathscr{F}_{\text{ex}}$ of the Fourier transform from S to S for spaces:

(1) $\mathscr{F}_{\text{ex}} : L^1(\mathbb{R}^n) \to L^\infty(\mathbb{R}^n)$ with norm estimate $M_1 = (2\pi)^{-\frac{n}{2}}$;
(2) $\mathscr{F}_{\text{ex}} : L^2(\mathbb{R}^n) \to L^2(\mathbb{R}^n)$ with norm estimate $M_2 = 1$.

Applying now Theorem 17.7, we obtain that $\mathscr{F}_{\text{ex}} : L^p \to L^q$, where

$$\frac{1}{p} = \frac{\theta}{1} + \frac{1-\theta}{2} = \frac{1}{2} + \frac{\theta}{2}, \quad \frac{1}{q} = \frac{\theta}{\infty} + \frac{1-\theta}{2} = \frac{1}{2} - \frac{\theta}{2}.$$

It follows that

$$\frac{1}{p} + \frac{1}{q} = 1,$$

i.e., $q = p'$ and $\theta = \frac{2}{p} - 1$. For θ to satisfy the condition $0 \le \theta \le 1$ we get $1 \le p \le 2$. We may also conclude that

$$\|\mathscr{F}_{\text{ex}}\|_{L^p \to L^{p'}} \le ((2\pi)^{-\frac{n}{2}})^\theta 1^{1-\theta} = (2\pi)^{-n\left(\frac{1}{p}-\frac{1}{2}\right)}$$

to obtain the desired norm estimate. □

Remark 17.9. In order to obtain $\mathscr{F}_{\text{ex}}$ in $L^p(\mathbb{R}^n)$, $1 \le p \le 2$, constructively we can apply the following procedure. Let us assume that $f \in L^p(\mathbb{R}^n)$, $1 \le p \le 2$, and $\{f_k\}_{k=1}^\infty \subset S(\mathbb{R}^n)$ such that

$$\|f_k - f\|_{L^p(\mathbb{R}^n)} \to 0, \quad k \to \infty.$$

It follows from the Hausdorff–Young inequality that

$$\|\mathscr{F} f_k - \mathscr{F} f_l\|_{L^{p'}(\mathbb{R}^n)} \le C_n \|f_k - f_l\|_{L^p(\mathbb{R}^n)}.$$

This means that $\{\mathscr{F} f_k\}_{k=1}^\infty$ is a Cauchy sequence in $L^{p'}(\mathbb{R}^n)$. We can therefore define

$$\mathscr{F}_{\text{ex}} f \overset{L^{p'}}{:=} \lim_{k\to\infty} \mathscr{F} f_k.$$

And we also have the Hausdorff–Young inequality

$$\|\mathscr{F}_{\text{ex}} f\|_{L^{p'}} = \lim_{k\to\infty} \|\mathscr{F} f_k\|_{L^{p'}} \le \lim_{k\to\infty} C_n \|f_k\|_{L^p} = C_n \|f\|_{L^p}.$$

Example 17.10 (Fourier transform on the line). Let $f_2(x) = \frac{1}{(x-\mathrm{i}\varepsilon)^2}$, where $\varepsilon > 0$ is fixed. It is clear that $f_2 \in L^1(\mathbb{R})$ and

$$\widehat{f_2}(\xi) = \frac{1}{\sqrt{2\pi}} \int_{-\infty}^{+\infty} \frac{\mathrm{e}^{-\mathrm{i}x\xi}\,\mathrm{d}x}{(x-\mathrm{i}\varepsilon)^2}.$$

In order to calculate this integral we consider the function $F(z) := \frac{e^{-iz\xi}}{(z-i\varepsilon)^2}$ of a complex variable $z \in \mathbb{C}$. It is easily seen that $z = i\varepsilon$, $\varepsilon > 0$, is a pole of order 2. We consider the cases $\xi > 0$ and $\xi < 0$ separately; see Figure 17.1.

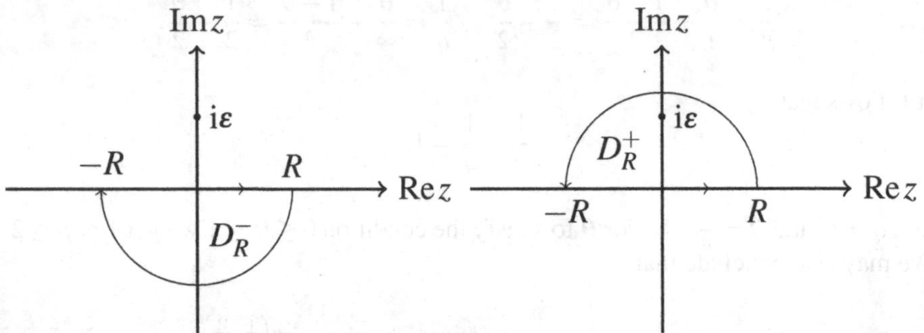

Fig. 17.1 Domains D_R^- and D_R^+ of integration.

(1) Let $\xi > 0$. By Cauchy's theorem we have

$$\oint_{\partial D_R^-} F(z)\mathrm{d}z = 0 = \int_{-R}^{R} F(z)\mathrm{d}z + \int_{\substack{|z|=R \\ \mathrm{Im}z<0}} F(z)\mathrm{d}z =: I_1 + I_2.$$

It follows that

$$I_1 \to \int_{-\infty}^{+\infty} \frac{e^{-ix\xi}\mathrm{d}x}{(x-i\varepsilon)^2}, \quad R \to \infty,$$

and

$$I_2 \to 0, \quad R \to \infty$$

due to Jordan's lemma, since $\xi \,\mathrm{Im} z < 0$. We therefore may conclude that

$$\int_{-\infty}^{+\infty} \frac{e^{-ix\xi}\mathrm{d}x}{(x-i\varepsilon)^2} = 0$$

for $\xi > 0$.

(2) Let $\xi < 0$. In this case again $\xi \,\mathrm{Im} z < 0$. So we may apply Jordan's lemma again and obtain

$$\oint_{\partial D_R^+} F(z)\mathrm{d}z = \int_{-R}^{R} \frac{e^{-ix\xi}\mathrm{d}x}{(x-i\varepsilon)^2} + \int_{\substack{|z|=R \\ \mathrm{Im}z>0}} F(z)\mathrm{d}z = 2\pi i \operatorname*{Res}_{z=i\varepsilon} F(z).$$

Hence

$$\int_{-\infty}^{\infty} \frac{e^{-ix\xi}dx}{(x-i\varepsilon)^2} = 2\pi i((z-i\varepsilon)^2 F(z))' |_{z=i\varepsilon} = 2\pi\xi e^{\varepsilon\xi}.$$

If we combine these two cases, we obtain

$$\widehat{\frac{1}{(x-i\varepsilon)^2}}(\xi) = \sqrt{2\pi}\xi H(-\xi)e^{\varepsilon\xi},$$

where

$$H(t) = \begin{cases} 1, & t \ge 0, \\ 0, & t < 0, \end{cases}$$

is the Heaviside function. Similarly we obtain

$$\widehat{\frac{1}{(x+i\varepsilon)^2}}(\xi) = -\sqrt{2\pi}\xi H(\xi)e^{-\varepsilon\xi}.$$

Example 17.11. Let $f_1(x) = \frac{1}{x-i\varepsilon}$, where $\varepsilon > 0$ is fixed. It is clear that $f_1 \notin L^1(\mathbb{R})$, but $f_1 \in L^p(\mathbb{R})$, $1 < p \le 2$. Analogously to Example 17.10 we obtain

$$\widehat{\frac{1}{x+i\varepsilon}}(\xi) = \begin{cases} -i\sqrt{2\pi}H(\xi)e^{-\varepsilon\xi}, & \xi \ne 0, \\ -i\sqrt{\frac{\pi}{2}}, & \xi = 0, \end{cases}$$

and

$$\widehat{\frac{1}{x-i\varepsilon}}(\xi) = \begin{cases} i\sqrt{2\pi}H(-\xi)e^{\varepsilon\xi}, & \xi \ne 0, \\ i\sqrt{\frac{\pi}{2}}, & \xi = 0. \end{cases}$$

Exercise 17.6. Find the Fourier transforms of the following functions on the line.

(1) $f(x) = \begin{cases} e^{-x}, & x > 0, \\ 0, & x \le 0, \end{cases}$

(2) $f(x) = e^{-|x|}$ and $f(x) = \frac{1}{1+x^2}$,

(3) $f_3(x) = \frac{1}{(x \pm i\varepsilon)^3}$, $\varepsilon > 0$.

Exercise 17.7. Define the Laplace transform by

$$L(p) := \int_0^{\infty} f(x)e^{-px}dx,$$

where $|f(x)| \le Me^{ax}$, $x > 0$, $f(x) = 0$, $x < 0$, and $p = p_1 + ip_2$, $p_1 > a$. Prove that

(1) $L(p) = \sqrt{2\pi}\widehat{f(x)\mathrm{e}^{-p_1 x}}(p_2)$.
(2) Apply the Fourier inversion formula to prove the Mellin formula

$$f(x) = \frac{1}{2\pi \mathrm{i}} \int_{p_1 - \mathrm{i}\infty}^{p_1 + \mathrm{i}\infty} L(p)\mathrm{e}^{px}\mathrm{d}p, \quad p_1 > a.$$

Chapter 18
Tempered Distributions

In this chapter we will consider two types of distributions: Schwartz distributions and tempered distributions. To that end we consider the space $\mathscr{D} := C_0^\infty(\mathbb{R}^n)$ of test functions. It is clear that $\mathscr{D}$ is a linear space and $\mathscr{D} \subset S$. A notion of convergence is given in the following definition.

Definition 18.1. A sequence $\{\varphi_k\}_{k=1}^\infty$ is a *null sequence* in $\mathscr{D}$ if

(1) there exists a compact set $K \subset \mathbb{R}^n$ such that $\operatorname{supp}\varphi_k \subset K$ for every k and
(2) for every $\alpha \geq 0$ we have

$$\sup_{x\in K} |D^\alpha \varphi_k(x)| \to 0, \quad k \to \infty.$$

We denote this fact by $\varphi_k \xrightarrow{\mathscr{D}} 0$. As usual, $\varphi_k \xrightarrow{\mathscr{D}} \varphi \in \mathscr{D}$ means that $\varphi_k - \varphi \xrightarrow{\mathscr{D}} 0$.

Now we are in a position to define the *Schwartz distribution space*.

Definition 18.2. A functional $T : \mathscr{D} \to \mathbb{C}$ is a Schwartz distribution if it is linear and continuous, that is,

(1) $T(\alpha_1\varphi_1 + \alpha_2\varphi_2) = \alpha_1 T(\varphi_1) + \alpha_2 T(\varphi_2)$ for every $\varphi_1, \varphi_2 \in \mathscr{D}$ and $\alpha_1, \alpha_2 \in \mathbb{C}$
(2) for every null sequence φ_k in $\mathscr{D}$, one has $T(\varphi_k) \to 0$ in $\mathbb{C}$ as $k \to \infty$.

The linear space of Schwartz distributions is denoted by $\mathscr{D}'$. The action of T on φ is denoted by $T(\varphi) = \langle T, \varphi\rangle$.

Example 18.3. Every locally integrable function f, that is, $f \in L^1_{\text{loc}}(\mathbb{R}^n)$, defines a Schwartz distribution by the formula

$$\langle T_f, \varphi\rangle := \int_{\mathbb{R}^n} f(x)\varphi(x)\mathrm{d}x.$$

V. Serov, *Fourier Series, Fourier Transform and Their Applications to Mathematical Physics*, Applied Mathematical Sciences 197,
DOI 10.1007/978-3-319-65262-7_18

It is clear that T_f is a linear map. It remains only to prove that T_f is a continuous map on $\mathscr{D}$. Let $\{\varphi_k\}_{k=1}^{\infty}$ be a null sequence in $\mathscr{D}$. Then

$$|\langle T_f, \varphi_k\rangle| \le \sup_{x\in K}|\varphi_k(x)| \int_K |f(x)|\mathrm{d}x \to 0, \quad k\to\infty$$

by the definition of null sequence.

Example 18.4. If $\langle T,\varphi\rangle := \varphi(0)$, then $T\in\mathscr{D}'$. Indeed, T is linear, and if $\varphi_k \xrightarrow{\mathscr{D}} 0$, then $\langle T,\varphi_k\rangle = \varphi_k(0)\to 0$ for $k\to\infty$. This distribution is called the δ-function and is denoted by δ, i.e.,

$$\langle \delta,\varphi\rangle = \varphi(0), \quad \varphi\in\mathscr{D}.$$

Remark 18.5. A distribution T is *regular* if it can be written in the form

$$\langle T,\varphi\rangle = \int_{\mathbb{R}^n} f(x)\varphi(x)\mathrm{d}x$$

for some locally integrable function f. All other distributions are *singular*.

Exercise 18.1. Prove that δ is a singular distribution.

Definition 18.6. The functional T defined by

$$\langle T,\varphi\rangle := \lim_{\varepsilon\to 0+}\int_{|x|>\varepsilon}\frac{\varphi(x)}{x}\mathrm{d}x \equiv \mathrm{p.\,v.}\int_{-\infty}^{\infty}\frac{\varphi(x)}{x}\mathrm{d}x$$

on $\mathscr{D}(\mathbb{R})$ is called the *principal value* of $\frac{1}{x}$. We denote it by $T = \mathrm{p.\,v.}\frac{1}{x}$.

Remark 18.7. Note that $\frac{1}{x}\notin L^1_{\mathrm{loc}}(\mathbb{R})$, but we have the following result.

Exercise 18.2. Prove that

$$\langle \mathrm{p.\,v.}\frac{1}{x},\varphi\rangle = \int_0^{\infty}\frac{\varphi(x)-\varphi(-x)}{x}\mathrm{d}x = \mathrm{p.\,v.}\int_{-\infty}^{\infty}\frac{\varphi(x)-\varphi(0)}{x}\mathrm{d}x.$$

Example 18.8. Let σ be a hypersurface of dimension $n-1$ in $\mathbb{R}^n$ and let $\mathrm{d}\sigma$ stand for an element of surface area on σ. Consider the functional

$$\langle T,\varphi\rangle = \int_{\sigma} a(x)\varphi(x)\mathrm{d}\sigma$$

on $\mathscr{D}$, where $a(x)$ is a locally integrable function on σ. We can interpret T in terms of surface source. Indeed,

$$\langle \int_{\sigma} a(\xi)\delta(x-\xi)\mathrm{d}\sigma_{\xi},\varphi\rangle := \int_{\sigma} a(\xi)\langle\delta(x-\xi),\varphi(x)\rangle\mathrm{d}\sigma_{\xi} = \int_{\sigma} a(\xi)\varphi(\xi)\mathrm{d}\sigma_{\xi}.$$

It is easy to see that T is a singular distribution. This distribution is known as the *simple layer*.

Definition 18.9. If $T \in \mathscr{D}'$ and $g \in C^\infty(\mathbb{R}^n)$, then we may define the product gT by

$$\langle gT, \varphi \rangle := \langle T, g\varphi \rangle, \quad \varphi \in \mathscr{D}.$$

This product is well defined, because $g\varphi \in \mathscr{D}$.

If f is a locally integrable function whose derivative $\frac{\partial f}{\partial x_j}$ is also locally integrable, then

$$\left\langle \frac{\partial f}{\partial x_j}, \varphi \right\rangle = \int_{\mathbb{R}^n} \frac{\partial f}{\partial x_j} \varphi(x) \mathrm{d}x = -\int_{\mathbb{R}^n} f \frac{\partial \varphi}{\partial x_j} \mathrm{d}x = -\left\langle f, \frac{\partial \varphi}{\partial x_j} \right\rangle, \quad \varphi \in \mathscr{D}$$

by integration by parts. This property is used to define the derivative of any distribution.

Definition 18.10. Let T be a distribution from $\mathscr{D}'$. For a multi-index α we define the derivative $\partial^\alpha T$ by

$$\langle \partial^\alpha T, \varphi \rangle := \langle T, (-1)^{|\alpha|} \partial^\alpha \varphi \rangle, \quad \varphi \in \mathscr{D}.$$

It is easily seen that $\partial^\alpha T \in \mathscr{D}'$.

Example 18.11. Consider the Heaviside function $H(x)$. Since $H \in L^1_{\mathrm{loc}}(\mathbb{R})$, it follows that

$$\langle H', \varphi \rangle = -\langle H, \varphi' \rangle = -\int_0^\infty \varphi'(x) \mathrm{d}x = \varphi(0) = \langle \delta, \varphi \rangle.$$

Hence $H' = \delta$.

Example 18.12. Let us prove that $(\log|x|)' = \mathrm{p.v.}\frac{1}{x}$ in the sense of Schwartz distributions. Indeed,

$$\begin{aligned}
&\langle (\log|x|)', \varphi \rangle = -\langle \log|x|, \varphi' \rangle = -\int_{-\infty}^{\infty} \log(|x|) \varphi'(x) \mathrm{d}x \\
&= -\int_0^\infty \log(x) \varphi'(x) \mathrm{d}x - \int_{-\infty}^0 \log(-x) \varphi'(x) \mathrm{d}x \\
&= -\int_0^\infty \log(x)(\varphi'(x) + \varphi'(-x)) \mathrm{d}x = -\int_0^\infty \log(x)(\varphi(x) - \varphi(-x))' \mathrm{d}x \\
&= -\log(x)\left[\varphi(x) - \varphi(-x)\right]_0^\infty + \int_0^\infty \frac{\varphi(x) - \varphi(-x)}{x} \mathrm{d}x = \langle \mathrm{p.v.}\frac{1}{x}, \varphi \rangle
\end{aligned}$$

by integration by parts and Exercise 18.2.

Exercise 18.3. Prove that

$$\left(\text{p.v.}\frac{1}{x}\right)' = -\text{p.v.}\frac{1}{x^2},$$

where $\text{p.v.}\frac{1}{x^2}$ is defined as

$$\langle \text{p.v.}\frac{1}{x^2}, \varphi\rangle = \lim_{\varepsilon\to 0+}\int_{|x|>\varepsilon} \frac{\varphi(x)-\varphi(0)}{x^2}\mathrm{d}x.$$

The following characterization of $\mathscr{D}'$ is given without proof: $T \in \mathscr{D}'$ if and only if for every compact $K \subset \mathbb{R}^n$ there exists $n_0(K) \in \mathbb{N}_0$ such that

$$|\langle T, \varphi\rangle| \le C_0 \sum_{|\alpha|\le n_0} \sup_{x\in K} |D^\alpha \varphi|$$

for every $\varphi \in \mathscr{D}$ with $\operatorname{supp}\varphi \subset K$.

Definition 18.13. A functional $T : S \to \mathbb{C}$ is a *tempered distribution* if

(1) T is linear, i.e., $\langle T, \alpha\varphi + \beta\psi\rangle = \alpha\langle T, \varphi\rangle + \beta\langle T, \psi\rangle$ for all $\alpha, \beta \in \mathbb{C}$ and $\varphi, \psi \in S$.
(2) T is continuous on S, i.e., there exist $n_0 \in \mathbb{N}_0$ and a constant $c_0 > 0$ such that

$$|\langle T, \varphi\rangle| \le c_0 \sum_{|\alpha|,|\beta|\le n_0} |\varphi|_{\alpha,\beta}$$

for every $\varphi \in S$.

The space of tempered distributions is denoted by S'. In addition, for $T_k, T \in S'$ the convergence $T_k \xrightarrow{S'} T$ means that $\langle T_k, \varphi\rangle \xrightarrow{\mathbb{C}} \langle T, \varphi\rangle$ for all $\varphi \in S$.

Remark 18.14. Since $\mathscr{D} \subset S$, the space of tempered distributions is narrower than the space of Schwartz distributions, $S' \subset \mathscr{D}'$. Later we will consider the even narrower distribution space $\mathscr{E}'$, which consists of continuous linear functionals on the (widest test function) space $\mathscr{E} := C^\infty(\mathbb{R}^n)$. In short, $\mathscr{D} \subset S \subset \mathscr{E}$ implies that

$$\mathscr{E}' \subset S' \subset \mathscr{D}'.$$

It turns out that members of $\mathscr{E}'$ have compact support, and they are therefore called *distributions with compact support*. But more on that later.

Example 18.15. Let us consider $\mathbb{R}^1$.

(1) It is clear that $f(x) = \mathrm{e}^{|x|^2}$ is a Schwartz distribution but not a tempered distribution, because part (2) of the previous definition is not satisfied.
(2) If $f(x) = \sum_{k=0}^{m} a_k x^k$ is a polynomial, then $f(x) \in S'$, since

$$\begin{aligned}|\langle T_f,\varphi\rangle| &= \left|\int_{\mathbb{R}}\sum_{k=0}^{m} a_k x^k \varphi(x)\mathrm{d}x\right| \\ &\le \sum_{k=0}^{m}|a_k|\int_{\mathbb{R}}(1+|x|)^{-1-\delta}(1+|x|)^{1+\delta}|x|^k|\varphi(x)|\mathrm{d}x \\ &\le C\sum_{k=0}^{m}|a_k||\varphi|_{0,k+1+\delta}\int_{\mathbb{R}}(1+|x|)^{-1-\delta}\mathrm{d}x,\end{aligned}$$

so condition (2) is satisfied, e.g., for $\delta = 1$, $n_0 = m+2$. This polynomial is a regular distribution, since $\langle T_f,\varphi\rangle = \int_{\mathbb{R}} f(x)\varphi(x)\mathrm{d}x$ is well defined.

Definition 18.16. Let T be a distribution from $\mathscr{D}'$. Then the support of T is defined by

$$\operatorname{supp} T := \mathbb{R}^n \setminus A,$$

where $A = \{x \in \mathbb{R}^n : \langle T,\varphi\rangle = 0 \text{ for all } \varphi \in C^\infty \text{ with } \operatorname{supp}\varphi \subset U_\delta(x)\}$.

Exercise 18.4. Prove that

(1) if f is continuous, then $\operatorname{supp} T_f = \operatorname{supp} f$;
(2) $\operatorname{supp}(\partial^\alpha T) \subset \operatorname{supp} T$;
(3) $\operatorname{supp}\delta = \{0\}$.

Example 18.17. (1) The weighted Lebesgue spaces are defined as

$$L^p_\sigma(\mathbb{R}^n) := \{f \in L^p_{\mathrm{loc}}(\mathbb{R}^n) : \|f\|_{L^p_\sigma} := \left(\int_{\mathbb{R}^n}(1+|x|)^{\sigma p}|f(x)|^p\mathrm{d}x\right)^{\frac{1}{p}} < \infty\}$$

for $1 \le p < \infty$ and

$$L^\infty_\sigma(\mathbb{R}^n) := \{f \in L^\infty_{\mathrm{loc}}(\mathbb{R}^n) : \|f\|_{L^\infty_\sigma} := \operatorname*{ess\,sup}_{x\in\mathbb{R}^n}(1+|x|)^\sigma|f(x)| < \infty\}.$$

If $f \in L^1_{-\delta}(\mathbb{R}^n)$ for some $\delta > 0$, then $T_f \in S'$. In fact,

$$|\langle T_f,\varphi\rangle| = \left|\int_{\mathbb{R}^n} f\varphi\mathrm{d}x\right| \le \|f\|_{L^1_{-\delta}}\|\varphi\|_{L^\infty_\delta}.$$

This means that $\int_{\mathbb{R}} f\varphi\mathrm{d}x$ is well defined in this case and

$$\langle T_f,\varphi\rangle := \int_{\mathbb{R}^n} f\varphi\mathrm{d}x.$$

(2) If $f \in L^p$, $1 \le p \le \infty$, then $f \in S'$. Indeed,

$$L^p(\mathbb{R}^n) \subset L^1_{-\delta}(\mathbb{R}^n) \quad \text{for} \quad \delta > \frac{n}{p'},$$

where $\frac{1}{p}+\frac{1}{p'}=1$. This follows from Hölder's inequality

$$\int_{\mathbb{R}}(1+|x|)^{-\delta}|f(x)|\mathrm{d}x \le \left(\int_{\mathbb{R}}(1+|x|)^{-\delta p'}\mathrm{d}x\right)^{\frac{1}{p'}} \|f\|_{L^p}.$$

(3) Let $T\in S'$ and $\varphi_0(x)\in C_0^\infty(\mathbb{R}^n)$ with $\varphi_0(0)=1$. The product $\varphi_0\left(\frac{x}{k}\right)T$ is well defined in S' by

$$\left\langle \varphi_0\left(\frac{x}{k}\right)T,\varphi\right\rangle := \left\langle T,\varphi_0\left(\frac{x}{k}\right)\varphi\right\rangle.$$

If we consider the sequence $T_k:=\varphi_0\left(\frac{x}{k}\right)T$, then

(a) $\langle T_k,\varphi\rangle \equiv \langle T,\varphi_0(\frac{x}{k})\varphi\rangle \overset{k\to\infty}{\to} \langle T,\varphi\rangle$ (since $\varphi_0(\frac{x}{k})\varphi \overset{S}{\to}\varphi$), so that $T_k\overset{S'}{\to}T$.

(b) T_k has compact support as a tempered distribution. This fact follows from the compactness of support of $\varphi_k=\varphi_0(\frac{x}{k})$.

Now we are ready to prove a more serious and more useful fact.

Theorem 18.18. *Let $T\in S'$. Then there exists $T_k\in S$ such that*

$$\langle T_k,\varphi\rangle = \int_{\mathbb{R}^n} T_k(x)\varphi(x)\mathrm{d}x \to \langle T,\varphi\rangle,\quad k\to\infty,$$

where $\varphi\in S$. In short, $\overline{S}\overset{S'}{=}S'$.

Proof. Let $j(x)$ be a function from $\mathscr{D}\equiv C_0^\infty(\mathbb{R}^n)$ with $\int_{\mathbb{R}^n} j(x)\mathrm{d}x=1$ and $j(-x)=j(x)$. Let $j_k(x):=k^n j(kx)$. By Lemma 16.11 we have

$$\lim_{k\to\infty}\langle j_k,\varphi\rangle = \lim_{k\to\infty}\int_{\mathbb{R}^n} j_k(x)\varphi(x)\mathrm{d}x=\varphi(0)$$

for every $\varphi\in S$. That is, $j_k(x)\overset{S'}{\to}\delta(x)$.

The *convolution* of two integrable functions g and φ is defined by

$$(g*\varphi)(x):=\int_{\mathbb{R}^n} g(x-y)\varphi(y)\mathrm{d}y.$$

If h and g are integrable functions and $\varphi\in S$, then it follows from Fubini's theorem that

$$\begin{aligned}\langle h*g,\varphi\rangle &= \int_{\mathbb{R}^n}\varphi(x)\mathrm{d}x\int_{\mathbb{R}^n} h(x-y)g(y)\mathrm{d}y = \int_{\mathbb{R}^n} g(y)\mathrm{d}y\int_{\mathbb{R}^n} h(x-y)\varphi(x)\mathrm{d}x\\ &= \int_{\mathbb{R}^n} g(y)\mathrm{d}y\int_{\mathbb{R}^n} Rh(y-x)\varphi(x)\mathrm{d}x = \langle g, Rh*\varphi\rangle,\end{aligned}$$

where $Rh(z):=h(-z)$ is the reflection of h.

Let now $\varphi_0(x) \in \mathscr{D}$ with $\varphi_0(0) = 1$. For every $T \in S'$ let us put $T_k := j_k * \widetilde{T_k}$, where $\widetilde{T_k} = \varphi_0\left(\frac{x}{k}\right)T$. From the above considerations we know that $\langle j_k * \widetilde{T_k}, \varphi\rangle = \langle \widetilde{T_k}, Rj_k * \varphi\rangle$.

Let us prove that this T_k meets the requirements of the theorem. First of all,

$$\langle T_k, \varphi\rangle \equiv \langle j_k * \widetilde{T_k}, \varphi\rangle = \langle \widetilde{T_k}, Rj_k * \varphi\rangle = \langle \varphi_0\left(\frac{x}{k}\right)T, j_k * \varphi\rangle$$
$$= \langle T, \varphi_0\left(\frac{x}{k}\right)(j_k * \varphi)\rangle \to \langle T, \varphi\rangle, \quad k \to \infty,$$

because

(a) $\varphi_0\left(\frac{x}{k}\right) \to 1$ pointwise for $k \to \infty$, since $\varphi_0(0) = 1$ and $\varphi_0(\frac{x}{k})\varphi \xrightarrow{S} \varphi$,
(b) $j_k * \varphi \xrightarrow{S} \varphi$ for $k \to \infty$ by Lemma 16.11:

$$\int_{\mathbb{R}^n} j_k(x-y)\varphi(y)\mathrm{d}y = \int_{\mathbb{R}^n} j_k(z)\varphi(x-z)\mathrm{d}z \to \varphi(x).$$

Finally, $j_k(x) \in C_0^\infty(\mathbb{R}^n)$ implies that $T_k \in C_0^\infty(\mathbb{R}^n) \subset S$ also. □

Definition 18.19. Let us assume that $L : S \to S$ is a continuous linear map. The adjoint map $L' : S' \to S'$ is defined by

$$\langle L'T, \varphi\rangle := \langle T, L\varphi\rangle, \quad T \in S'.$$

Clearly, L' is also a continuous linear map.

Corollary 18.20. *Every continuous linear map (operator) $L : S \to S$ admits a continuous linear extension $\widetilde{L} : S' \to S'$.*

Proof. If $T \in S'$, then by Theorem 18.18 there exists $T_k \in S$ such that $T_k \xrightarrow{S'} T$. Then

$$\langle LT_k, \varphi\rangle = \langle T_k, L'\varphi\rangle \to \langle T, L'\varphi\rangle =: \langle \widetilde{L}T, \varphi\rangle$$

as $k \to \infty$. □

Now we are in a position to formulate the following theorem.

Theorem 18.21 (Properties of tempered distributions). The following linear continuous operators from S into S admit unique continuous linear extensions as maps from S' into S':

(1) $\langle uT, \varphi\rangle := \langle T, u\varphi\rangle, \quad u \in S;$
(2) $\langle \partial^\alpha T, \varphi\rangle := \langle T, (-1)^{|\alpha|}\partial^\alpha \varphi\rangle;$
(3) $\langle \tau_h T, \varphi\rangle := \langle T, \tau_{-h}\varphi\rangle;$

(4) $\langle \sigma_\lambda T, \varphi \rangle := \langle T, |\lambda|^{-n} \sigma_{\frac{1}{\lambda}} \varphi \rangle, \quad \lambda \neq 0;$
(5) $\langle \mathscr{F} T, \varphi \rangle := \langle T, \mathscr{F} \varphi \rangle.$

Proof. See Theorem 18.18, Definition 18.19, and Corollary 18.20. □

Remark 18.22. Since $\langle \mathscr{F}^{-1} \mathscr{F} T, \varphi \rangle = \langle \mathscr{F} T, \mathscr{F}^{-1} \varphi \rangle = \langle T, \mathscr{F} \mathscr{F}^{-1} \varphi \rangle = \langle T, \varphi \rangle$, we have that $\mathscr{F}^{-1} \mathscr{F} = \mathscr{F} \mathscr{F}^{-1} = I$ in S'.

Example 18.23. (1) Since

$$\langle \mathscr{F} 1, \varphi \rangle \equiv \langle 1, \mathscr{F} \varphi \rangle = \int_{\mathbb{R}^n} (\mathscr{F} \varphi)(\xi) \mathrm{d}\xi = (2\pi)^{\frac{n}{2}} (2\pi)^{-\frac{n}{2}} \int_{\mathbb{R}^n} \mathrm{e}^{\mathrm{i}(0,\xi)} \mathscr{F} \varphi \mathrm{d}\xi$$
$$= (2\pi)^{\frac{n}{2}} \mathscr{F}^{-1} \mathscr{F} \varphi(0) = (2\pi)^{\frac{n}{2}} \varphi(0) = (2\pi)^{\frac{n}{2}} \langle \delta, \varphi \rangle$$

for every $\varphi \in S$, we have that

$$\widehat{1} = (2\pi)^{\frac{n}{2}} \delta$$

in S'.
(2) $\widehat{\delta} = (2\pi)^{-\frac{n}{2}} \cdot 1$, since for $\varphi \in S$ we have

$$\langle \widehat{\delta}, \varphi \rangle = \langle \delta, \mathscr{F} \varphi \rangle = \mathscr{F} \varphi(0) = (2\pi)^{-\frac{n}{2}} \int_{\mathbb{R}^n} \mathrm{e}^{-\mathrm{i}(0,x)} \varphi(x) \mathrm{d}x = (2\pi)^{-\frac{n}{2}} \langle 1, \varphi \rangle.$$

Moreover, $\mathscr{F}^{-1} \delta = (2\pi)^{-\frac{n}{2}} \cdot 1$ in S'.
(3) $\widehat{\mathrm{e}^{-a\frac{x^2}{2}}} = a^{-\frac{n}{2}} \mathrm{e}^{-\frac{\xi^2}{2a}}$, $\operatorname{Re} a \geq 0$, $a \neq 0$. Indeed, for $a > 0$ we know that

$$\mathscr{F}(\mathrm{e}^{-a\frac{x^2}{2}}) = \mathscr{F}(\mathrm{e}^{-\frac{(\sqrt{a}x)^2}{2}}) = a^{-\frac{n}{2}} \mathrm{e}^{-\frac{\xi^2}{2a}}.$$

If a is such that $\operatorname{Re} a \geq 0$, $a \neq 0$, then we can use analytic continuation of these formulas.
(4) Consider $(1 - \Delta)u = f$, where $\Delta = \frac{\partial^2}{\partial x_1^2} + \cdots + \frac{\partial^2}{\partial x_n^2}$ is the Laplacian in $\mathbb{R}^n$ and and $u, f \in S'$. This equation can be solved in S' using the Fourier transform. Indeed, we get

$$(1 + |\xi|^2) \widehat{u} = \widehat{f},$$

or

$$\widehat{u} = (1 + |\xi|^2)^{-1} \widehat{f},$$

or

$$u = \mathscr{F}^{-1}((1 + |\xi|^2)^{-1} \mathscr{F} f).$$

If $f \in S$, then $\mathscr{F} f \in S$ and $(1+|\xi|^2)^{-1}\mathscr{F} f \in S$ also, and then $u \in S$ exists. If $f \in S'$, then by Theorem 18.18 there exists $f_k \in S$ such that $f_k \xrightarrow{S'} f$. We conclude that

$$u \stackrel{S'}{=} \lim_{k\to\infty} u_k,$$

where $u_k = \mathscr{F}^{-1}((1+|\xi|^2)^{-1}\mathscr{F} f_k)$.

Exercise 18.5. Let $P(D)$ be an elliptic partial differential operator

$$P(D) = \sum_{|\alpha|\le m} a_\alpha D^\alpha$$

with constant coefficients and $P(\xi) \neq 0$ for $\xi \neq 0$. Prove that if $u \in S'$ and $Pu = 0$, then u is a polynomial.

Corollary 18.24. *If $\Delta u = 0$ in S' and $|u|$ is less than or equal to some constant, then u is constant.*

Exercise 18.6. Prove that

(1) $\mathscr{F}(\text{p.v.}\frac{1}{x}) = -\mathrm{i}\sqrt{\frac{\pi}{2}}\operatorname{sgn}\xi$;
(2) $\mathscr{F}(\text{p.v.}\frac{1}{x^2}) = -\sqrt{\frac{\pi}{2}}|\xi|$.

Definition 18.25. Let us introduce the tempered distributions

$$\frac{1}{x\pm \mathrm{i}0} := \lim_{\varepsilon\to 0+} \frac{1}{x\pm \mathrm{i}\varepsilon}$$

(if they exist), i.e.,

$$\left\langle \frac{1}{x\pm \mathrm{i}0}, \varphi \right\rangle = \lim_{\varepsilon\to 0+} \left\langle \frac{1}{x\pm \mathrm{i}\varepsilon}, \varphi \right\rangle, \quad \varphi \in S.$$

In a similar fashion,

$$\frac{1}{(x\pm \mathrm{i}0)^2} := \lim_{\varepsilon\to 0+} \frac{1}{(x\pm \mathrm{i}\varepsilon)^2}$$

in S' (if they exist).

Example 18.26. We know from Example 17.11 that

$$\widehat{\frac{1}{x+\mathrm{i}\varepsilon}}(\xi) = \begin{cases} -\mathrm{i}\sqrt{2\pi}H(\xi)e^{-\varepsilon\xi}, & \xi \neq 0, \\ -\mathrm{i}\sqrt{\frac{\pi}{2}}, & \xi = 0, \end{cases}$$

and
$$\widehat{\frac{1}{x-\mathrm{i}\varepsilon}}(\xi)=\begin{cases}\mathrm{i}\sqrt{2\pi}H(-\xi)\mathrm{e}^{\varepsilon\xi}, & \xi\neq 0,\\ \mathrm{i}\sqrt{\frac{\pi}{2}}, & \xi=0.\end{cases}$$

Hence
$$\widehat{\frac{1}{x+\mathrm{i}0}}=\lim_{\varepsilon\to 0+}\widehat{\frac{1}{x+\mathrm{i}\varepsilon}}=\begin{cases}-\mathrm{i}\sqrt{2\pi}H(\xi), & \xi\neq 0,\\ -\mathrm{i}\sqrt{\frac{\pi}{2}}, & \xi=0,\end{cases}$$

and
$$\widehat{\frac{1}{x-\mathrm{i}0}}=\lim_{\varepsilon\to 0+}\widehat{\frac{1}{x-\mathrm{i}\varepsilon}}=\begin{cases}\mathrm{i}\sqrt{2\pi}H(-\xi), & \xi\neq 0,\\ \mathrm{i}\sqrt{\frac{\pi}{2}}, & \xi=0.\end{cases}$$

It follows from Exercise 18.6 that
$$\widehat{\frac{1}{x+\mathrm{i}0}}+\widehat{\frac{1}{x-\mathrm{i}0}}=-\mathrm{i}\sqrt{2\pi}\operatorname{sgn}\xi=2\left(-\mathrm{i}\sqrt{\frac{\pi}{2}}\operatorname{sgn}\xi\right)=2\widehat{\mathrm{p.v.}\frac{1}{x}}$$

and thus
$$\frac{1}{x+\mathrm{i}0}+\frac{1}{x-\mathrm{i}0}=2\,\mathrm{p.v.}\frac{1}{x}.$$

In a similar fashion,
$$\widehat{\frac{1}{x-\mathrm{i}0}}-\widehat{\frac{1}{x+\mathrm{i}0}}=\mathrm{i}\sqrt{2\pi}\cdot 1=\mathrm{i}\sqrt{2\pi}\sqrt{2\pi}\widehat{\delta}=2\pi\mathrm{i}\widehat{\delta},$$

and so
$$\frac{1}{x-\mathrm{i}0}-\frac{1}{x+\mathrm{i}0}=2\pi\mathrm{i}\delta.$$

We add and subtract to get finally
$$\frac{1}{x+\mathrm{i}0}=\mathrm{p.v.}\frac{1}{x}-\mathrm{i}\pi\delta \quad\text{and}\quad \frac{1}{x-\mathrm{i}0}=\mathrm{p.v.}\frac{1}{x}+\mathrm{i}\pi\delta.$$

Exercise 18.7. Prove that

(1)
$$\widehat{\frac{1}{(x+\mathrm{i}0)^2}}=-\sqrt{2\pi}\xi H(\xi)\quad\text{and}\quad\widehat{\frac{1}{(x-\mathrm{i}0)^2}}=\sqrt{2\pi}\xi H(-\xi);$$

(2)
$$\frac{1}{(x+\mathrm{i}0)^2}+\frac{1}{(x-\mathrm{i}0)^2}=2\,\mathrm{p.v.}\frac{1}{x^2}\quad\text{and}\quad\frac{1}{(x-\mathrm{i}0)^2}-\frac{1}{(x+\mathrm{i}0)^2}=-2\pi\mathrm{i}\delta';$$

(3)
$$\frac{1}{(x+i0)^2} = \text{p. v.}\frac{1}{x^2} + \pi i\delta' \quad \text{and} \quad \frac{1}{(x-i0)^2} = \text{p. v.}\frac{1}{x^2} - \pi i\delta';$$

(4)
$$\widehat{\log|x|} = -\sqrt{\frac{\pi}{2}}\,\text{p. v.}\frac{1}{|\xi|};$$

(5)
$$\widehat{x^\beta} = (2\pi)^{n/2} i^{|\beta|} \partial^\beta \delta.$$

Exercise 18.8. Prove that

(1)
$$\widehat{H}(\xi) = -\frac{i}{\sqrt{2\pi}} \cdot \frac{1}{\xi - i0};$$

(2)
$$\widehat{\text{sgn}}(\xi) = -\frac{i}{\sqrt{\frac{\pi}{2}}}\,\text{p. v.}\frac{1}{\xi}.$$

Example 18.27. Since

$$\begin{aligned}\langle \widehat{\partial^\alpha \delta}, \varphi\rangle &= \langle \partial^\alpha \delta, \widehat{\varphi}\rangle = (-1)^{|\alpha|}\langle \delta, \partial^\alpha \widehat{\varphi}\rangle = \langle \delta, i^{|\alpha|}\widehat{\xi^\alpha \varphi}\rangle \\ &= \langle \widehat{\delta}, (i\xi)^\alpha \varphi\rangle = \langle (2\pi)^{-\frac{n}{2}}, (i\xi)^\alpha \varphi\rangle = \langle (2\pi)^{-\frac{n}{2}}(i\xi)^\alpha, \varphi\rangle,\end{aligned}$$

we get
$$\widehat{\partial^\alpha \delta} = (2\pi)^{-\frac{n}{2}}(i\xi)^\alpha.$$

In particular, in dimension one,

$$\widehat{\delta^{(k)}} = \frac{1}{\sqrt{2\pi}} i^k \xi^k, \quad \widehat{x^k} = \sqrt{2\pi} i^k \delta^{(k)}.$$

Let us consider the *Cauchy–Riemann operator*

$$\overline{\partial} := \frac{1}{2}\left(\frac{\partial}{\partial x} + i\frac{\partial}{\partial y}\right)$$

and
$$\partial := \frac{1}{2}\left(\frac{\partial}{\partial x} - i\frac{\partial}{\partial y}\right)$$

in $\mathbb{R}^2$.

Let us prove the following facts about these operators:

(1)
$$\partial \cdot \bar{\partial} = \bar{\partial} \cdot \partial = \frac{1}{4}\Delta,$$

(2)
$$\frac{1}{\pi}\bar{\partial}\left(\frac{1}{z}\right) = \delta \quad \text{in } \mathbb{R}^2.$$

The last fact means that
$$\frac{1}{\pi}\frac{1}{x+\mathrm{i}y}$$

is the fundamental solution (see Chapter 22) of $\bar{\partial}$. Taking the Fourier transform of (2) gives us
$$\widehat{\bar{\partial}\frac{1}{z}}(\xi) = \pi\widehat{\delta}(\xi),$$

which is equivalent to
$$\frac{1}{2}(\mathrm{i}\xi_1 - \xi_2)\cdot\widehat{\frac{1}{z}}(\xi) = \pi\cdot(2\pi)^{-1}\cdot 1 = \frac{1}{2},$$

or
$$\widehat{\frac{1}{z}}(\xi) = \frac{1}{\mathrm{i}\xi_1 - \xi_2} = -\mathrm{i}\frac{1}{\xi_1 + \mathrm{i}\xi_2}, \quad \xi \neq 0.$$

Let us check that this is indeed the case. We have, by Example 17.11,
$$\begin{aligned}
\widehat{\frac{1}{z}}(\xi) &= \frac{1}{2\pi}\int_{\mathbb{R}^2}\frac{\mathrm{e}^{-\mathrm{i}(\xi_1 x+\xi_2 y)}}{x+\mathrm{i}y}\mathrm{d}x\mathrm{d}y = \frac{1}{2\pi}\int_{-\infty}^{\infty}\mathrm{e}^{-\mathrm{i}\xi_2 y}\mathrm{d}y\int_{-\infty}^{\infty}\frac{\mathrm{e}^{-\mathrm{i}\xi_1 x}}{x+\mathrm{i}y}\mathrm{d}x\\
&= \frac{1}{2\pi}\int_0^{\infty}\mathrm{e}^{-\mathrm{i}\xi_2 y}\mathrm{d}y\int_{-\infty}^{\infty}\frac{\mathrm{e}^{-\mathrm{i}\xi_1 x}}{x+\mathrm{i}y}\mathrm{d}x + \frac{1}{2\pi}\int_{-\infty}^{0}\mathrm{e}^{-\mathrm{i}\xi_2 y}\mathrm{d}y\int_{-\infty}^{\infty}\frac{\mathrm{e}^{-\mathrm{i}\xi_1 x}}{x+\mathrm{i}y}\mathrm{d}x\\
&= \frac{1}{2\pi}\int_0^{\infty}\mathrm{e}^{-\mathrm{i}\xi_2 y}\sqrt{2\pi}(-\mathrm{i}\sqrt{2\pi}H(\xi_1)\mathrm{e}^{-y\xi_1})\mathrm{d}y\\
&+ \frac{1}{2\pi}\int_{-\infty}^{0}\mathrm{e}^{-\mathrm{i}\xi_2 y}\sqrt{2\pi}(\mathrm{i}\sqrt{2\pi}H(-\xi_1)\mathrm{e}^{-y\xi_1})\mathrm{d}y\\
&= -\mathrm{i}\left(H(\xi_1)\int_0^{\infty}\mathrm{e}^{-y(\xi_1+\mathrm{i}\xi_2)}\mathrm{d}y - H(-\xi_1)\int_{-\infty}^{0}\mathrm{e}^{-y(\xi_1+\mathrm{i}\xi_2)}\mathrm{d}y\right).
\end{aligned}$$

For $\xi_1 > 0$ we have
$$-\mathrm{i}\int_0^{\infty}\mathrm{e}^{-y(\xi_1+\mathrm{i}\xi_2)}\mathrm{d}y = \mathrm{i}\left.\frac{\mathrm{e}^{-y(\xi_1+\mathrm{i}\xi_2)}}{\xi_1+\mathrm{i}\xi_2}\right|_0^{\infty} = -\mathrm{i}\frac{1}{\xi_1+\mathrm{i}\xi_2}.$$

For $\xi_1 < 0$ we have

$$\mathrm{i}\int_{-\infty}^{0} \mathrm{e}^{-y(\xi_1+\mathrm{i}\xi_2)}\mathrm{d}y = -\mathrm{i}\,\frac{\mathrm{e}^{-y(\xi_1+\mathrm{i}\xi_2)}}{\xi_1+\mathrm{i}\xi_2}\bigg|_{-\infty}^{0} = -\mathrm{i}\frac{1}{\xi_1+\mathrm{i}\xi_2}.$$

Hence

$$\widehat{\frac{1}{z}}(\xi) = -\frac{\mathrm{i}}{\xi_1+\mathrm{i}\xi_2},$$

which proves (2). Part (1) is established with a simple calculation:

$$\partial\cdot\overline{\partial} = \frac{1}{4}\left(\frac{\partial}{\partial x}-\mathrm{i}\frac{\partial}{\partial y}\right)\left(\frac{\partial}{\partial x}+\mathrm{i}\frac{\partial}{\partial y}\right) = \frac{1}{4}\left(\left(\frac{\partial}{\partial x}\right)^2+\left(\frac{\partial}{\partial y}\right)^2\right) = \frac{1}{4}\Delta = \overline{\partial}\cdot\partial.$$

Chapter 19
Convolutions in S and S'

Let us consider first the direct product of distributions. Let us assume that $T_1, \dots, T_n$ are one-dimensional tempered distributions, $T_j \in S'(\mathbb{R})$, $j = 1, 2, \dots, n$. The product $T_1(x_1) \cdots T_n(x_n)$ can be formally defined by

$$\begin{aligned}\langle T_1(x_1) \cdots T_n(x_n), \varphi(x_1, \dots, x_n) \rangle &= \langle T_1(x_1) \cdots T_{n-1}(x_{n-1}), \varphi_1(x_1, \dots, x_{n-1}) \rangle \\ &= \langle T_1(x_1) \cdots T_{n-2}(x_{n-2}), \varphi_2(x_1, \dots, x_{n-2}) \rangle \\ &= \cdots = \langle T_1(x_1), \varphi_{n-1}(x_1) \rangle,\end{aligned}$$

where

$$\begin{aligned}\varphi_1(x_1, \dots, x_{n-1}) &:= \langle T_n(x_n), \varphi(x_1, \dots, x_n) \rangle \in S(\mathbb{R}^{n-1}), \\ \varphi_j(x_1, \dots, x_{n-j}) &:= \langle T_{n-j+1}, \varphi_{j-1}(x_1, \dots, x_{n-j+1}) \rangle \in S(\mathbb{R}^{n-j}).\end{aligned}$$

In this sense, it is clear that

$$\delta(x_1, \dots, x_n) = \delta(x_1) \cdots \delta(x_n).$$

But the product $T_1(x)T_2(x)$, where the x are the same, in general case does not exist, that is, it is impossible to define such a product. We remedy this by recalling the following definition.

Definition 19.1. The *convolution* $\varphi * \psi$ of the functions $\varphi \in S$ and $\psi \in S$ is defined as

$$(\varphi * \psi)(x) := \int_{\mathbb{R}^n} \varphi(x-y)\psi(y)\mathrm{d}y.$$

We can observe the following immediately.

V. Serov, *Fourier Series, Fourier Transform and Their Applications to Mathematical Physics*, Applied Mathematical Sciences 197,
DOI 10.1007/978-3-319-65262-7_19

(1) The convolution is commutative for every $n \geq 1$. If $n \geq 2$, then

$$(\varphi * \psi)(x) = \int_{\mathbb{R}^n} \varphi(x-y)\psi(y)\mathrm{d}y = \int_{\mathbb{R}^n} \varphi(z)\psi(x-z)\mathrm{d}z = (\psi * \varphi)(x).$$

If $n = 1$, then

$$\begin{aligned}(\varphi * \psi)(x) &= \int_{-\infty}^{\infty} \varphi(x-y)\psi(y)\mathrm{d}y = -\int_{\infty}^{-\infty} \varphi(z)\psi(x-z)\mathrm{d}z \\ &= \int_{-\infty}^{\infty} \psi(x-z)\varphi(z)\mathrm{d}z = (\psi * \varphi)(x).\end{aligned}$$

(2) It is also clear that the convolution is well defined for φ and ψ from S, and moreover, for every $\alpha \geq 0$,

$$\begin{aligned}\partial_x^\alpha(\varphi * \psi)(x) &= (\partial^\alpha \varphi * \psi)(x) = \int_{\mathbb{R}^n} \partial_x^\alpha \varphi(x-y)\psi(y)\mathrm{d}y \\ &= \int_{\mathbb{R}^n} (-1)^{|\alpha|}\partial_y^\alpha \varphi(x-y)\psi(y)\mathrm{d}y \\ &= (-1)^{2|\alpha|}\int_{\mathbb{R}^n} \varphi(x-y)\partial_y^\alpha \psi(y)\mathrm{d}y = (\varphi * \partial^\alpha \psi)(x),\end{aligned}$$

where we integrated by parts and used the fact that $\partial_{x_j}\varphi(x-y) = -\partial_{y_j}\varphi(x-y)$.

We would like to prove that for φ and ψ from S it follows that $\varphi * \psi$ from S also. In fact,

(1) $\varphi * \psi \in C^\infty(\mathbb{R}^n)$ since $\partial^\alpha(\varphi * \psi) = \varphi * \partial^\alpha \psi$ and $\partial^\alpha : S \to S$.
(2) $\varphi * \psi$ decreases at infinity faster than any inverse power:

$$\begin{aligned}\left|\int_{\mathbb{R}^n} \varphi(x-y)\psi(y)\mathrm{d}y\right| &\leq c_1 \int_{|y|\leq \frac{|x|}{2}} \frac{1}{|x-y|^m}|\psi(y)|\mathrm{d}y + c_2 \int_{|y|>\frac{|x|}{2}} |\psi(y)|\mathrm{d}y \\ &\leq \frac{c_1'}{|x|^m}\int_{|y|\leq \frac{|x|}{2}} |\psi(y)|\mathrm{d}y + c_2 \int_{|y|>\frac{|x|}{2}} |y|^{-m}|y|^m|\psi(y)|\mathrm{d}y \\ &\leq \frac{c_1''}{|x|^m} + \frac{c_2''}{|x|^m} = c|x|^{-m}, \quad m \in \mathbb{N}.\end{aligned}$$

Next we collect some important inequalities involving the convolution.

(1) Hölder's inequality implies that

$$\|\varphi * \psi\|_{L^\infty(\mathbb{R}^n)} \leq \|\varphi\|_{L^p(\mathbb{R}^n)} \cdot \|\psi\|_{L^{p'}(\mathbb{R}^n)}, \tag{19.1}$$

where $\frac{1}{p} + \frac{1}{p'} = 1, 1 \leq p \leq \infty$. This means that the convolution is well defined even for $\varphi \in L^p(\mathbb{R}^n)$ and $\psi \in L^{p'}(\mathbb{R}^n)$. In particular,

$$\|\varphi * \psi\|_{L^\infty(\mathbb{R}^n)} \le \|\varphi\|_{L^1(\mathbb{R}^n)} \cdot \|\psi\|_{L^\infty(\mathbb{R}^n)} . \tag{19.2}$$

(2) It follows from Fubini's theorem that

$$\begin{aligned} \|\varphi * \psi\|_{L^1} &\le \int_{\mathbb{R}^n} \mathrm{d}x \int_{\mathbb{R}^n} |\varphi(x-y)||\psi(y)|\mathrm{d}y \\ &= \int_{\mathbb{R}^n} |\psi(y)|\mathrm{d}y \int_{\mathbb{R}^n} |\varphi(x-y)|\mathrm{d}x = \|\varphi\|_{L^1} \|\psi\|_{L^1} . \end{aligned} \tag{19.3}$$

(3) Interpolating (19.2) and (19.3) leads us to (see the Riesz–Thorin theorem, Theorem 17.7)

$$\|\varphi * \psi\|_{L^p} \le \|\psi\|_{L^1} \cdot \|\varphi\|_{L^p} . \tag{19.4}$$

(4) Interpolating (19.1) and (19.4) leads us to (again by the Riesz–Thorin theorem)

$$\|\varphi * \psi\|_{L^s} \le \|\psi\|_{L^r} \cdot \|\varphi\|_{L^p} ,$$

where

$$1 + \frac{1}{s} = \frac{1}{r} + \frac{1}{p}.$$

Indeed, the linear operator $T\psi = \varphi * \psi$ with $\varphi \in L^p(\mathbb{R}^n)$ maps as

$$T : L^{p'}(\mathbb{R}^n) \to L^\infty(\mathbb{R}^n), \quad \frac{1}{p'} + \frac{1}{p} = 1$$

and

$$T : L^1(\mathbb{R}^n) \to L^p(\mathbb{R}^n);$$

see (19.1) and (19.4), respectively. Thus

$$T : L^r(\mathbb{R}^n) \to L^s(\mathbb{R}^n),$$

where

$$\frac{1}{r} = \frac{\theta}{p'} + \frac{1-\theta}{1} = 1 - \frac{\theta}{p}$$

and

$$\frac{1}{s} = \frac{\theta}{\infty} + \frac{1-\theta}{p} = \frac{1}{p} - \frac{\theta}{p}.$$

This gives

$$\frac{1}{r} - \frac{1}{s} = 1 - \frac{1}{p}.$$

Now we are in a position to consider the Fourier transform of a convolution.

(1) Let $\varphi, \psi \in S$. Then $\varphi * \psi \in S$ and $\mathscr{F}(\varphi * \psi) \in S$. Moreover,

$$\begin{aligned}\mathscr{F}(\varphi * \psi) &= (2\pi)^{-\frac{n}{2}} \int_{\mathbb{R}^n} \mathrm{e}^{-\mathrm{i}(x,\xi)}\mathrm{d}x \int_{\mathbb{R}^n} \varphi(x-y)\psi(y)\mathrm{d}y \\ &= (2\pi)^{-\frac{n}{2}} \int_{\mathbb{R}^n} \psi(y)\mathrm{d}y \int_{\mathbb{R}^n} \mathrm{e}^{-\mathrm{i}(x,\xi)}\varphi(x-y)\mathrm{d}x \\ &= (2\pi)^{-\frac{n}{2}} \int_{\mathbb{R}^n} \psi(y)\mathrm{e}^{-\mathrm{i}(y,\xi)}\mathrm{d}y \int_{\mathbb{R}^n} \varphi(z)\mathrm{e}^{-\mathrm{i}(z,\xi)}\mathrm{d}z = (2\pi)^{\frac{n}{2}} \mathscr{F}\varphi \cdot \mathscr{F}\psi,\end{aligned}$$

i.e.,

$$\widehat{\varphi * \psi} = (2\pi)^{\frac{n}{2}} \widehat{\varphi} \cdot \widehat{\psi}.$$

Similarly,

$$\mathscr{F}^{-1}(\varphi * \psi) = (2\pi)^{\frac{n}{2}} \mathscr{F}^{-1}\varphi \cdot \mathscr{F}^{-1}\psi.$$

Hence

$$\varphi * \psi = (2\pi)^{\frac{n}{2}} \mathscr{F}(\mathscr{F}^{-1}\varphi \cdot \mathscr{F}^{-1}\psi),$$

which implies that

$$\mathscr{F}\varphi_1 * \mathscr{F}\psi_1 = (2\pi)^{\frac{n}{2}} \mathscr{F}(\varphi_1 \cdot \psi_1),$$

or

$$\widehat{\varphi \cdot \psi} = (2\pi)^{-\frac{n}{2}} \widehat{\varphi} * \widehat{\psi}.$$

(2) Let us assume that $\varphi \in L^1$ and $\psi \in L^p$, $1 \le p \le 2$. Then (19.4) implies that $\varphi * \psi \in L^p$, $1 \le p \le 2$. Further, $\mathscr{F}(\varphi * \psi)$ belongs to $L^{p'}$ by the Hausdorff–Young inequality. Thus,

$$\widehat{\varphi * \psi} = (2\pi)^{\frac{n}{2}} \widehat{\varphi} \cdot \widehat{\psi} \in L^{p'}.$$

Lemma 19.2. *Let $\varphi(x)$ be a function from $L^1(\mathbb{R}^n)$ with $\int_{\mathbb{R}^n} \varphi(x)\mathrm{d}x = 1$ and let $\psi(x)$ be a function from $L^2(\mathbb{R}^n)$. Let us set $\varphi_\varepsilon(x) := \varepsilon^{-n}\varphi\left(\frac{x}{\varepsilon}\right)$, $\varepsilon > 0$. Then*

$$\lim_{\varepsilon \to 0+} \varphi_\varepsilon * \psi \overset{L^2(\mathbb{R}^n)}{=} \psi.$$

Proof. By (19.4) we have that $\varphi_\varepsilon * \psi \in L^2(\mathbb{R}^n)$. Then

$$\widehat{\varphi_\varepsilon * \psi} = (2\pi)^{\frac{n}{2}} \widehat{\varphi_\varepsilon} \cdot \widehat{\psi}$$

in L^2. But

$$\widehat{\varphi_\varepsilon} = \varepsilon^{-n} \widehat{\varphi\left(\frac{x}{\varepsilon}\right)} = \varepsilon^{-n} \widehat{\sigma_{\frac{1}{\varepsilon}}\varphi}(\xi) = \varepsilon^{-n} \left(\frac{1}{\varepsilon}\right)^{-n} \widehat{\varphi}(\varepsilon\xi) = \widehat{\varphi}(\varepsilon\xi) \overset{L^\infty}{\to} \widehat{\varphi}(0)$$

as $\varepsilon \to 0+$. Note also that

$$\widehat{\varphi}(0) = (2\pi)^{-\frac{n}{2}} \int_{\mathbb{R}^n} \mathrm{e}^{-\mathrm{i}(0,x)} \varphi(x)\mathrm{d}x = (2\pi)^{-\frac{n}{2}}.$$

Hence

$$\widehat{\varphi_\varepsilon * \psi} = (2\pi)^{\frac{n}{2}} \widehat{\varphi}(\varepsilon\xi) \cdot \widehat{\psi}(\xi) \xrightarrow{L^2} \widehat{\psi}(\xi), \quad \varepsilon \to 0+.$$

By the Fourier inversion formula it follows that

$$\varphi_\varepsilon * \psi \xrightarrow{L^2} \psi$$

as $\varepsilon \to 0+$. □

Theorem 19.3. *For every fixed function φ from $S(\mathbb{R}^n)$ the map $\varphi * T$ has, as a continuous linear map from S to S (with respect to T), a unique continuous linear extension as a map from S' to S' (with respect to T) as follows:*

$$\langle \varphi * T, \psi \rangle := \langle T, R\varphi * \psi \rangle,$$

where $R\varphi(x) := \varphi(-x)$. Moreover, this extension has the properties

(1) $\widehat{\varphi * T} = (2\pi)^{\frac{n}{2}} \widehat{\varphi} \cdot \widehat{T}$,
(2) $\partial^\alpha(\varphi * T) = \partial^\alpha \varphi * T = \varphi * \partial^\alpha T$.

Proof. Let us assume that φ, ψ, and T belong to the Schwartz space $S(\mathbb{R}^n)$. Then we have checked already the properties (1) and (2) above. But we can easily check that for such functions the definition is also true. In fact,

$$\begin{aligned}
\langle \varphi * T, \psi \rangle &= \int_{\mathbb{R}^n} (\varphi * T)(x)\psi(x)\mathrm{d}x = \int_{\mathbb{R}^n}\int_{\mathbb{R}^n} \varphi(x-y)T(y)\mathrm{d}y\psi(x)\mathrm{d}x \\
&= \int_{\mathbb{R}^n} T(y) \int_{\mathbb{R}^n} \varphi(x-y)\psi(x)\mathrm{d}x\mathrm{d}y \\
&= \int_{\mathbb{R}^n} T(y)\mathrm{d}y \int_{\mathbb{R}^n} R\varphi(y-x)\psi(x)\mathrm{d}x = \langle T, R\varphi * \psi \rangle.
\end{aligned}$$

For the case $T \in S'$ the statement of this theorem follows from the fact that $\overline{S} \overset{S'}{=} S'$ (see Theorem 18.18). □

Corollary 19.4. *Since $\varphi * T = T * \varphi$ for φ and T from S, we may define $T * \varphi$ as follows (for $T \in S'$):*

$$\langle T * \varphi, \psi \rangle := \langle T, R\varphi * \psi \rangle.$$

Example 19.5.

(1) It is true that $\delta * \varphi = \varphi$. Indeed,

$$\langle \delta * \varphi, \psi \rangle = \langle \delta, R\varphi * \psi \rangle = (R\varphi * \psi)(0) = \int_{\mathbb{R}^n} \varphi(y)\psi(y)\mathrm{d}y = \langle \varphi, \psi \rangle.$$

Alternatively, we note that

$$\widehat{\delta * \varphi} = (2\pi)^{\frac{n}{2}} \widehat{\delta} \cdot \widehat{\varphi} = 1 \cdot \widehat{\varphi} = \widehat{\varphi}$$

is equivalent to

$$\delta * \varphi = \varphi$$

in S'.

(2) Property (2) of Theorem 19.3 and part (1) of this example imply that

$$\partial^\alpha (\delta * \varphi) = \delta * \partial^\alpha \varphi = \partial^\alpha \varphi.$$

(3) Let us consider again the equation $(1 - \Delta)u = f$ for u and $f \in L^2$ (or even from S'). Then $(1 + |\xi|^2)\widehat{u} = \widehat{f}$ is still valid in L^2 and $\widehat{u} = (1 + |\xi|^2)^{-1}\widehat{f}$ or

$$\begin{aligned} u(x) &= \mathscr{F}^{-1}\left(\frac{1}{1+|\xi|^2}\widehat{f}\right) \\ &= (2\pi)^{-\frac{n}{2}} \mathscr{F}^{-1}\left(\frac{1}{1+|\xi|^2}\right) * f = \int_{\mathbb{R}^n} K(x-y) f(y)\mathrm{d}y, \end{aligned}$$

where

$$K(x-y) := \frac{1}{(2\pi)^n} \int_{\mathbb{R}^n} \frac{\mathrm{e}^{\mathrm{i}(x-y,\xi)}}{1+|\xi|^2} \mathrm{d}\xi.$$

This is the inverse Fourier transform of a locally integrable function. This function K is the *free space Green's function* of the operator $1 - \Delta$ in $\mathbb{R}^n$. We will calculate this integral precisely in Chapter 22.

Lemma 19.6. *Let $j(x)$ be a function from $L^1(\mathbb{R}^n)$ with $\int_{\mathbb{R}^n} j(x)\mathrm{d}x = 1$. Set $j_\varepsilon(x) = \varepsilon^{-n} j\left(\frac{x}{\varepsilon}\right)$, $\varepsilon > 0$. Then*

$$\|j_\varepsilon * f - f\|_{L^p} \to 0, \quad \varepsilon \to 0+,$$

for every function $f \in L^p(\mathbb{R}^n)$, $1 \le p < \infty$. In the case $p = \infty$ we can state only the fact

$$\int_{\mathbb{R}^n} (j_\varepsilon * f)\overline{g}\mathrm{d}x \to \int_{\mathbb{R}^n} f \cdot \overline{g}\mathrm{d}x, \quad \varepsilon \to 0+$$

for every $g \in L^1(\mathbb{R}^n)$.

Exercise 19.1. Prove Lemma 19.6 and find a counterexample showing that the first part fails for $p = \infty$.

Remark 19.7. If $j \in C_0^\infty(\mathbb{R}^n)$ or $S(\mathbb{R}^n)$, then $j_\varepsilon * f \in C_0^\infty(\mathbb{R}^n)$ or $S(\mathbb{R}^n)$ also for every $f \in L^p(\mathbb{R}^n)$, $1 \le p < \infty$.

Chapter 20
Sobolev Spaces

Lemma 20.1. *For every function $f \in L^2(\mathbb{R}^n)$ the following statements are equivalent:*

(1) $\frac{\partial f}{\partial x_j}(x) \in L^2(\mathbb{R}^n)$,

(2) $\xi_j \widehat{f}(\xi) \in L^2(\mathbb{R}^n)$,

(3) $\lim\limits_{t\to 0} \frac{\Delta_j^t f(x)}{t}$ *exists in* $L^2(\mathbb{R}^n)$. *Here* $\Delta_j^t f(x) := f(x+te_j) - f(x)$ *with* $t \in \mathbb{R}$ *and* $e_j = (0,\dots,1,0,\dots,0)$.

(4) There exists $\{f_k\}_{k=1}^{\infty}$, $f_k \in S$, *such that* $f_k \xrightarrow{L^2} f$ *and* $\frac{\partial f_k}{\partial x_j}$ *has a limit in* $L^2(\mathbb{R}^n)$.

Proof. (1) $\Leftrightarrow$ (2): Since

$$\widehat{D_j f} = \xi_j \widehat{f},$$

we have

$$\left\| \xi_j \widehat{f} \right\|_{L^2} = \left\| D_j f \right\|_{L^2}$$

by Parseval's equality.

(2) $\Rightarrow$ (3): Let $\xi_j \widehat{f}$ be a function from $L^2(\mathbb{R}^n)$. Then the equality

$$\widehat{\frac{1}{t}\Delta_j^t f}(\xi) = \frac{1}{t}(\mathrm{e}^{\mathrm{i}t\xi_j} - 1)\widehat{f}(\xi) = \frac{\mathrm{e}^{\mathrm{i}t\xi_j} - 1}{t\xi_j} \cdot \xi_j \widehat{f}(\xi)$$

holds. But

$$\frac{\mathrm{e}^{\mathrm{i}t\xi_j} - 1}{t\xi_j} \to \mathrm{i}$$

V. Serov, *Fourier Series, Fourier Transform and Their Applications to Mathematical Physics*, Applied Mathematical Sciences 197,
DOI 10.1007/978-3-319-65262-7_20

pointwise as $t \to 0$. Hence

$$\widehat{\frac{1}{t}\Delta_j^t f} \xrightarrow{L^2} \mathrm{i}\xi_j \widehat{f}, \quad t \to 0$$

i.e. (again due to Parseval's equality),

$$\frac{1}{t}\Delta_j^t f \xrightarrow{L^2} \frac{\partial f}{\partial x_j}, \quad t \to 0.$$

The same arguments lead us to the statement that (3) $\Rightarrow$ (1).

(4) $\Rightarrow$ (1): Let f_k be a sequence from S such that $f_k \xrightarrow{L^2} f$. Then $f_k \xrightarrow{S'} f$ and $\frac{\partial f_k}{\partial x_j} \xrightarrow{S'} \frac{\partial f}{\partial x_j}$ also. By condition (4) we have that the limit $\lim_{k\to\infty} \frac{\partial f_k}{\partial x_j} \overset{L^2}{=} g$ exists. We may conclude that $\frac{\partial f_k}{\partial x_j} \xrightarrow{S'} g$. This means that $g = \frac{\partial f}{\partial x_j}$ in S'.

(2) $\Rightarrow$ (4): Let us write $\widehat{f}(\xi)$ as the sum of two functions $\widehat{f}(\xi) = g(\xi) + h(\xi)$, where

$$g(\xi) = \widehat{f}(\xi)\chi_{\{|\xi_j|<1\}}, \quad h(\xi) = \widehat{f}(\xi)\chi_{\{|\xi_j|>1\}}.$$

Let $\{g_k\}$ be a sequence in S such that $g_k \xrightarrow{L^2} g$ and $\operatorname{supp} g_k \subset \{|\xi_j| < 2\}$. Let $\{h_k\}$ be a sequence in S such that $h_k \xrightarrow{L^2} \xi_j h$ and $\operatorname{supp} h_k \subset \{|\xi_j| > \frac{1}{2}\}$. If we define the sequence $f_k(x) = \mathscr{F}^{-1}\left(g_k + \frac{h_k}{\xi_j}\right)(x)$, then

$$\widehat{f_k}(\xi) = g_k + \frac{h_k}{\xi_j} \xrightarrow{L^2} g(\xi) + h(\xi) = \widehat{f}(\xi).$$

But

$$\widehat{\frac{\partial f_k}{\partial x_j}} = \mathrm{i}\xi_j g_k + \mathrm{i}h_k \xrightarrow{L^2} \mathrm{i}\xi_j(g+h) = \mathrm{i}\xi_j \widehat{f}.$$

This means that (by the Fourier inversion formula or Parseval's equality)

$$\frac{\partial f_k}{\partial x_j} \xrightarrow{L^2} \mathscr{F}^{-1}(\mathrm{i}\xi_j \widehat{f}) = \frac{\partial f}{\partial x_j}.$$

This completes the proof. □

We have also the following generalization of Lemma 20.1 to a multi-index α.

Lemma 20.2. *Let f be a function from $L^2(\mathbb{R}^n)$ and let $s \in \mathbb{N}$. Then the following statements are equivalent:*

(1) $D^\alpha f \in L^2(\mathbb{R}^n), |\alpha| \le s$;

(2) $\xi^\alpha \widehat{f} \in L^2(\mathbb{R}^n), |\alpha| \le s$;

(3) $\lim\limits_{h\to 0} \frac{\Delta_h^\alpha f}{h^\alpha}$ *exists in* $L^2(\mathbb{R}^n)$, $|\alpha| \le s$. *Here* $\Delta_h^\alpha f := (\Delta_{h_1}^{\alpha_1} \cdots \Delta_{h_n}^{\alpha_n}) f$ *and* $h \in \mathbb{R}^n$ *with* $h_j \neq 0$ *for all* $j = 1, 2, \ldots, n$.

(4) There exists $f_k \in S$ *such that* $f_k \xrightarrow{L^2} f$ *and* $D^\alpha f_k$ *has a limit in* $L^2(\mathbb{R}^n)$ *for* $|\alpha| \le s$.

Proof. The result follows from Lemma 20.1 by induction on $|\alpha|$. □

Definition 20.3. Let $s > 0$ be an integer. Then

$$H^s(\mathbb{R}^n) := \{f \in L^2(\mathbb{R}^n) : \sum_{|\alpha|\le s} \|D^\alpha f\|_{L^2} < \infty\}$$

is called the (L^2-based) *Sobolev space* of order s with norm

$$\|f\|_{H^s(\mathbb{R}^n)} := \left(\sum_{|\alpha|\le s} \|D^\alpha f\|^2_{L^2(\mathbb{R}^n)} \right)^{1/2}.$$

Remark 20.4. It is easy to check that $H^s(\mathbb{R}^n)$, $s \in \mathbb{N}$, can be characterized by

$$H^s(\mathbb{R}^n) = \{f \in L^2(\mathbb{R}^n) : \int_{\mathbb{R}^n} (1 + |\xi|^2)^s |\widehat{f}(\xi)|^2 d\xi < \infty\}.$$

Proof. It follows from Parseval's equality that

$$\begin{aligned}\sum_{|\alpha|\le s} \|D^\alpha f\|^2_{L^2} &= \sum_{|\alpha|\le s} \left\|\widehat{D^\alpha f}\right\|^2_{L^2} = \sum_{|\alpha|\le s} \left\|\xi^\alpha \widehat{f}\right\|^2_{L^2} \\ &= \sum_{|\alpha|\le s} \int_{\mathbb{R}^n} |\xi^\alpha|^2 |\widehat{f}(\xi)|^2 d\xi = \int_{\mathbb{R}^n} \sum_{|\alpha|\le s} |\xi^\alpha|^2 |\widehat{f}(\xi)|^2 d\xi.\end{aligned}$$

But it is easily seen that there are positive constants c_1 and c_2 such that

$$c_1(1 + |\xi|^2)^s \le \sum_{|\alpha|\le s} |\xi^\alpha|^2 \le c_2(1 + |\xi|^2)^s,$$

or

$$\sum_{|\alpha|\le s} |\xi^\alpha|^2 \asymp (1 + |\xi|^2)^s.$$

Therefore we may conclude that

$$\sum_{|\alpha|\le s}\|D^\alpha f\|_{L^2}<\infty \Leftrightarrow \sum_{|\alpha|\le s}\|D^\alpha f\|_{L^2}^2<\infty \Leftrightarrow \int_{\mathbb{R}^n}(1+|\xi|^2)^s|\widehat{f}(\xi)|^2\mathrm{d}\xi<\infty.$$

This establishes the characterization. □

This property of an integer s justifies the following definition.

Definition 20.5. Let s be a real number. Then we define

$$H^s(\mathbb{R}^n):=\{f\in S' : (1+|\xi|^2)^{\frac{s}{2}}\widehat{f}\in L^2(\mathbb{R}^n)\}$$

with the norm

$$\|f\|_{H^s(\mathbb{R}^n)}:=\left(\int_{\mathbb{R}^n}(1+|\xi|^2)^s|\widehat{f}(\xi)|^2\mathrm{d}\xi\right)^{\frac{1}{2}}.$$

Definition 20.6. Let $s>0$ be an integer and $1\le p\le\infty$. Then

$$W_p^s(\mathbb{R}^n):=\{f\in L^p(\mathbb{R}^n) : \sum_{|\alpha|\le s}\|D^\alpha f\|_{L^p(\mathbb{R}^n)}<\infty\}$$

is called the Sobolev space with norm

$$\|f\|_{W_p^s(\mathbb{R}^n)}:=\left(\sum_{|\alpha|\le s}\|D^\alpha f\|_{L^p(\mathbb{R}^n)}^p\right)^{1/p}.$$

Exercise 20.1. Let $s>0$ be an even integer and $1\le p\le\infty$. Prove that

$$|f|_{W_p^s(\mathbb{R}^n)}:=\left(\int_{\mathbb{R}^n}|\mathscr{F}^{-1}((1+|\xi|^2)^{\frac{s}{2}}\widehat{f})|^p\mathrm{d}x\right)^{\frac{1}{p}}$$

is an equivalent norm in $W_p^s(\mathbb{R}^n)$.

Definition 20.7. Let $s>0$ be a real number and $1\le p\le\infty$. Then

$$W_p^s(\mathbb{R}^n):=\{f\in S' : \left(\int_{\mathbb{R}^n}|\mathscr{F}^{-1}((1+|\xi|^2)^{\frac{s}{2}}\widehat{f})|^p\mathrm{d}x\right)^{\frac{1}{p}}<\infty\}$$

with the norm

$$\|f\|_{W_p^s(\mathbb{R}^n)}:=\left(\int_{\mathbb{R}^n}|\mathscr{F}^{-1}((1+|\xi|^2)^{\frac{s}{2}}\widehat{f})|^p\mathrm{d}x\right)^{\frac{1}{p}}.$$

Exercise 20.2. Let $s \in \mathbb{R}$. Prove that

$$f \in H^s(\mathbb{R}^n)$$

if and only if

$$\widehat{f} \in L^2_s(\mathbb{R}^n).$$

Proposition 20.8. *Let us assume that* $0 < s < 1$. *Then*

$$\int_{\mathbb{R}^n} (1+|\xi|^{2s})|\widehat{f}(\xi)|^2 \mathrm{d}\xi = \int_{\mathbb{R}^n} |f(x)|^2 \mathrm{d}x + A_s \int_{\mathbb{R}^n}\int_{\mathbb{R}^n} \frac{|f(x)-f(y)|^2}{|x-y|^{n+2s}} \mathrm{d}x\mathrm{d}y, \quad (20.1)$$

where A_s *is a positive constant depending on* s *and* n.

Remark 20.9. Since $1+|\xi|^{2s} \asymp (1+|\xi|^2)^s$, $0 < s < 1$, the right-hand side of (20.1) is an equivalent norm in $H^s(\mathbb{R}^n)$.

Proof. Denote by I the double integral appearing on the right-hand side of (20.1). Then

$$\begin{aligned} I &= \int_{\mathbb{R}^n}\int_{\mathbb{R}^n} |f(y+z)-f(y)|^2 |z|^{-n-2s} \mathrm{d}y\mathrm{d}z \\ &= \int_{\mathbb{R}^n} |z|^{-n-2s} \mathrm{d}z \int_{\mathbb{R}^n} |\mathrm{e}^{\mathrm{i}(z,\xi)}-1|^2 |\widehat{f}(\xi)|^2 \mathrm{d}\xi = \int_{\mathbb{R}^n} |\widehat{f}(\xi)|^2 \mathrm{d}\xi \int_{\mathbb{R}^n} \frac{|\mathrm{e}^{\mathrm{i}(z,\xi)}-1|^2}{|z|^{n+2s}} \mathrm{d}z \end{aligned}$$

by Parseval's equality and Exercise 16.4. We claim that

$$\int_{\mathbb{R}^n} \frac{|\mathrm{e}^{\mathrm{i}(z,\xi)}-1|^2}{|z|^{n+2s}} \mathrm{d}z = |\xi|^{2s} A_s^{-1}.$$

Indeed, if we consider the Householder reflection matrix

$$A := I - \frac{2vv^T}{|v|^2}, \quad v = \xi - |\xi| e_1, \quad \xi \in \mathbb{R}^n,$$

then $A^T = A^{-1} = A$ and $A\xi = |\xi| e_1 = (|\xi|, 0, \ldots, 0)$. It follows that

$$\begin{aligned} |\xi|^{-2s} \int_{\mathbb{R}^n} \frac{|\mathrm{e}^{\mathrm{i}(z,\xi)}-1|^2}{|z|^{n+2s}} \mathrm{d}z &= |\xi|^{-2s} \int_{\mathbb{R}^n} \frac{|\mathrm{e}^{\mathrm{i}(Az,A\xi)}-1|^2}{|z|^{n+2s}} \mathrm{d}z \\ &= |\xi|^{-2s} \int_{\mathbb{R}^n} \frac{|\mathrm{e}^{\mathrm{i}(y,A\xi)}-1|^2}{|y|^{n+2s}} \mathrm{d}y = |\xi|^{-2s} \int_{\mathbb{R}^n} \frac{|\mathrm{e}^{\mathrm{i}y_1|\xi|}-1|^2}{|y|^{n+2s}} \mathrm{d}y \\ &= \int_{\mathbb{R}^n} \frac{|\mathrm{e}^{\mathrm{i}z_1}-1|^2}{|z|^{n+2s}} \mathrm{d}z =: A_s^{-1}. \end{aligned}$$

Therefore,

$$\int_{\mathbb{R}^n} |f(x)|^2 dx + A_s \int_{\mathbb{R}^n}\int_{\mathbb{R}^n} |f(x)-f(y)|^2 |x-y|^{-n-2s} dxdy \\ = \int_{\mathbb{R}^n} |\widehat{f}(\xi)|^2 d\xi + \int_{\mathbb{R}^n} |\xi|^{2s} |\widehat{f}(\xi)|^2 d\xi.$$

This completes the proof. □

Remark 20.10. Note that A_s exists only for $0 < s < 1$.

Exercise 20.3. Prove that

$$\|f\|^2_{H^{k+s}(\mathbb{R}^n)} \asymp \int_{\mathbb{R}^n} (1+|\xi|^{2k+2s}) |\widehat{f}(\xi)|^2 d\xi \asymp \left\|\widehat{f}\right\|^2_{L^2} + \sum_{|\alpha|=k} \int_{\mathbb{R}^n} |\xi|^{2s} |\widehat{D^\alpha f}|^2 d\xi \\ = \|f\|^2_{L^2} + A_s \sum_{|\alpha|=k} \int_{\mathbb{R}^n}\int_{\mathbb{R}^n} |D^\alpha f(x) - D^\alpha f(y)|^2 |x-y|^{-n-2s} dxdy.$$

Example 20.11. $1 \notin H^s(\mathbb{R}^n)$ for all s. Indeed, assume that $1 \in H^{s_0}(\mathbb{R}^n)$ for some s_0 (it is clear that $s_0 > 0$). This means that $(1+|\xi|^2)^{\frac{s_0}{2}} \widehat{1} \in L^2(\mathbb{R}^n)$. It follows from this fact that $\widehat{1} \in L^2_{\text{loc}}(\mathbb{R}^n)$, and further, $\widehat{1} \in L^1_{\text{loc}}(\mathbb{R}^n)$. But $\widehat{1} = (2\pi)^{\frac{n}{2}} \delta$, and we know that δ is not a regular distribution.

Next we list some properties of $H^s(\mathbb{R}^n)$.

(1) Since $f \in H^s(\mathbb{R}^n)$ if and only if $\widehat{f}(\xi) \in L^2_s(\mathbb{R}^n)$ and $L^2_s(\mathbb{R}^n)$ is a separable Hilbert space with the scalar product

$$(f_1, g_1)_{L^2_s(\mathbb{R}^n)} = \int_{\mathbb{R}^n} (1+|\xi|^2)^s f_1 \cdot \overline{g_1} d\xi,$$

it follows that $H^s(\mathbb{R}^n)$ is also a separable Hilbert space, and the scalar product can be defined by

$$(f, g)_{H^s(\mathbb{R}^n)} = \int_{\mathbb{R}^n} (1+|\xi|^2)^s \widehat{f} \cdot \overline{\widehat{g}} d\xi.$$

We may prove the following property:

$$H^{-s}(\mathbb{R}^n) = (H^s(\mathbb{R}^n))^*, \quad s \in \mathbb{R}$$

in the sense that

$$\|f\|_{H^{-s}(\mathbb{R}^n)} := \sup_{0 \neq g \in H^s(\mathbb{R}^n)} \frac{|(f,g)_{L^2(\mathbb{R}^n)}|}{\|g\|_{H^s(\mathbb{R}^n)}}.$$

This means that $H^{-s}(\mathbb{R}^n)$ is the dual space to $H^s(\mathbb{R}^n)$ with respect to the Hilbert space $L^2(\mathbb{R}^n)$. Indeed, since by Parseval's equality

$$(f,g)_{L^2(\mathbb{R}^n)} = (\widehat{f},\widehat{g})_{L^2(\mathbb{R}^n)} = ((1+|\xi|^2)^{-s/2}\widehat{f},(1+|\xi|^2)^{s/2}\widehat{g})_{L^2(\mathbb{R}^n)},$$

it follows that

$$\sup_{0\neq g\in H^s(\mathbb{R}^n)} \frac{|(f,g)_{L^2(\mathbb{R}^n)}|}{\|g\|_{H^s(\mathbb{R}^n)}} = \sup_{0\neq \widehat{g}\in L^2_s(\mathbb{R}^n)} \frac{|(\widehat{f},\widehat{g})_{L^2(\mathbb{R}^n)}|}{\|\widehat{g}\|_{L^2_s(\mathbb{R}^n)}} = \left\|\widehat{f}\right\|_{L^2_{-s}(\mathbb{R}^n)} = \|f\|_{H^{-s}(\mathbb{R}^n)}.$$

We have used here the fact that the space $L^2_{-s}(\mathbb{R}^n)$ is dual to $L^2_s(\mathbb{R}^n)$ for every $s\in\mathbb{R}$.

(2) For $-\infty < s < t < \infty$, it follows that $S \subset H^t(\mathbb{R}^n) \subset H^s(\mathbb{R}^n) \subset S'$.

Example 20.12. $\delta \in H^s(\mathbb{R}^n)$ if and only if $s < -\frac{n}{2}$. Indeed, if we define

$$\langle\xi\rangle := (1+|\xi|^2)^{\frac{1}{2}},$$

then $\delta \in H^s(\mathbb{R}^n)$ is equivalent to $(2\pi)^{-\frac{n}{2}}\langle\xi\rangle^s \in L^2(\mathbb{R}^n)$, which in turn is equivalent to $s < -\frac{n}{2}$.

(3) Let φ be a function from $H^s(\mathbb{R}^n)$, and ψ a function from $H^{-s}(\mathbb{R}^n)$. Then $\widehat{\varphi} \in L^2_s(\mathbb{R}^n)$ and $\widehat{\psi} \in L^2_{-s}(\mathbb{R}^n)$, so that $\widehat{\varphi}\cdot\widehat{\psi} \in L^1(\mathbb{R}^n)$ by Hölder's inequality. We may therefore define (temporarily, and with slight abuse of notation)

$$\langle\varphi,\psi\rangle_{L^2(\mathbb{R}^n)} := \int_{\mathbb{R}^n} \widehat{\varphi}\cdot\widehat{\psi}\,\mathrm{d}\xi$$

and obtain

$$|\langle\varphi,\psi\rangle_{L^2(\mathbb{R}^n)}| \le \|\varphi\|_{H^s(\mathbb{R}^n)}\cdot\|\psi\|_{H^{-s}(\mathbb{R}^n)}.$$

For example, if φ is a function from $H^{\frac{n}{2}+1+\varepsilon}(\mathbb{R}^n)$, $\varepsilon > 0$, and $\psi = \frac{\partial\delta}{\partial x_j}$, then

$$\left\langle \frac{\partial\delta}{\partial x_j},\varphi\right\rangle_{L^2(\mathbb{R}^n)} = \int_{\mathbb{R}^n} \widehat{\frac{\partial\delta}{\partial x_j}}\cdot\widehat{\varphi}\,\mathrm{d}\xi = \mathrm{i}(2\pi)^{-\frac{n}{2}}\int_{\mathbb{R}^n}\xi_j\widehat{\varphi}(\xi)\,\mathrm{d}\xi$$

is well defined, since $\widehat{\varphi} \in L^2_{\frac{n}{2}+1+\varepsilon}(\mathbb{R}^n)$ and $\xi_j \in L^2_{-\frac{n}{2}-1-\varepsilon}(\mathbb{R}^n)$.

(4) Consider a differential operator with constant coefficients

$$P(D) = \sum_{|\alpha|\le m} a_\alpha D^\alpha.$$

Then $P(D): H^s(\mathbb{R}^n) \to H^{s-m}(\mathbb{R}^n)$ for all real s.

Proof. By the properties of the Fourier transform we have

$$\begin{aligned}\|P(D)f\|^2_{H^{s-m}(\mathbb{R}^n)} &= \int_{\mathbb{R}^n}(1+|\xi|^2)^{s-m}|\widehat{P(D)f}|^2 \mathrm{d}\xi \\ &= \int_{\mathbb{R}^n}(1+|\xi|^2)^{s-m}|P(\xi)|^2\cdot|\widehat{f}(\xi)|^2 \mathrm{d}\xi \\ &\le c\int_{\mathbb{R}^n}(1+|\xi|^2)^{s-m}(1+|\xi|^2)^{m}|\widehat{f}(\xi)|^2 \mathrm{d}\xi = c\|f\|^2_{H^s(\mathbb{R}^n)}\end{aligned}$$

for all real s. □

There is a generalization of this result. Let $P(x,D)$ be a differential operator

$$P(x,D) = \sum_{|\alpha|\le m} a_\alpha(x)D^\alpha$$

with variable coefficients such that $|a_\alpha(x)| \le c_0$ for all $x\in\mathbb{R}^n$ and $|\alpha|\le m$. Then

$$P(x,D): H^m(\mathbb{R}^n)\to L^2(\mathbb{R}^n).$$

Indeed,

$$\begin{aligned}\|P(x,D)f\|_{L^2} &\le c_0\sum_{|\alpha|\le m}\|D^\alpha f\|_{L^2} = c_0\sum_{|\alpha|\le m}\left\|\xi^\alpha\widehat{f}\right\|_{L^2} \\ &\le c_0'\left\|(1+|\xi|^2)^{\frac{m}{2}}\widehat{f}\right\|_{L^2} = c_0'\|f\|_{H^m}.\end{aligned}$$

(5) We have the following lemma.

Lemma 20.13. *Let φ be a function from S, and f a function from $H^s(\mathbb{R}^n)$ for $s\in\mathbb{R}$. Then $\varphi\cdot f\in H^s(\mathbb{R}^n)$ and*

$$\|\varphi f\|_{H^s} \le c\left\|(1+|\xi|^2)^{\frac{|s|}{2}}\widehat{\varphi}\right\|_{L^1}\cdot\|f\|_{H^s}.$$

Proof. We know that

$$\widehat{\varphi\cdot f}(\xi) = (2\pi)^{-\frac{n}{2}}\int_{\mathbb{R}^n}\widehat{\varphi}(\xi-\eta)\widehat{f}(\eta)\mathrm{d}\eta.$$

Hence

$$\langle\xi\rangle^s\widehat{\varphi\cdot f} = (2\pi)^{-\frac{n}{2}}\int_{\mathbb{R}^n}\frac{\langle\xi\rangle^s}{\langle\eta\rangle^s}\widehat{\varphi}(\xi-\eta)\langle\eta\rangle^s\widehat{f}(\eta)\mathrm{d}\eta.$$

Let us prove that

$$\langle\xi\rangle^s \le 2^{\frac{|s|}{2}}\langle\eta\rangle^s\cdot\langle\xi-\eta\rangle^{|s|}$$

for all $s \in \mathbb{R}$. Indeed,

$$\langle \xi \rangle = (1+|\xi|^2)^{\frac{1}{2}} \leq (1+|\eta|^2)^{\frac{1}{2}} + |\xi - \eta| = \langle \eta \rangle + |\xi - \eta| \leq \langle \eta \rangle (1+|\xi - \eta|).$$

Since $1+|\xi-\eta| \leq \sqrt{2}\langle \xi - \eta \rangle$, we have

$$\langle \xi \rangle^s \leq 2^{\frac{s}{2}} \langle \eta \rangle^s \cdot \langle \xi - \eta \rangle^s$$

for $s \geq 0$. Moreover, for $s < 0$ we have

$$\frac{\langle \xi \rangle^s}{\langle \eta \rangle^s} = \frac{\langle \eta \rangle^{|s|}}{\langle \xi \rangle^{|s|}} \leq 2^{\frac{|s|}{2}} \langle \eta - \xi \rangle^{|s|}.$$

It now follows from (19.4) that

$$\|\varphi f\|_{H^s} = \left\|\langle \xi \rangle^s \widehat{\varphi f}\right\|_{L^2} \leq c \left\| |\langle \xi \rangle^{|s|} \widehat{\varphi}| * |\langle \eta \rangle^s \widehat{f}| \right\|_{L^2} \leq c \left\| \langle \xi \rangle^{|s|} \widehat{\varphi} \right\|_{L^1} \left\| \langle \eta \rangle^s \widehat{f} \right\|_{L^2}$$

for all $s \in \mathbb{R}$. □

Exercise 20.4. Suppose $s > n/2$. Show that if $u, v \in L^2(\mathbb{R}^n)$ and

$$w(\xi) = \int_{\mathbb{R}^n} \frac{u(\xi - \eta) v(\eta) d\eta}{(1+|\eta|^2)^{s/2}},$$

then $w \in L^2(\mathbb{R}^n)$ and $\|w\|_{L^2} \leq C \|u\|_{L^2} \|v\|_{L^2}$.

(6) Let us now consider distributions with compact support in greater detail than what we saw in Chapter 18.

Definition 20.14. Set $\mathscr{E} = C^\infty(\mathbb{R}^n)$. We say that $T \in \mathscr{E}'$ if T is a linear functional on $\mathscr{E}$ that is also continuous, i.e., $\varphi_k \to 0$ in $\mathscr{E}$ implies that $\langle T, \varphi_k \rangle \to 0$ in $\mathbb{C}$. Here $\varphi_k \to 0$ in $\mathscr{E}$ means that

$$\sup_K |\partial^\alpha \varphi_k| \to 0, \quad k \to \infty$$

for every compact subset $K \subset \mathbb{R}^n$ and multi-index α.

It can be proved that $T \in \mathscr{E}'$ if and only if there exist $c_0 > 0$, $R_0 > 0$, and $n_0 \in \mathbb{N}_0$ such that

$$|\langle T, \varphi \rangle| \leq c_0 \sum_{|\alpha| \leq n_0} \sup_{|x| \leq R_0} |D^\alpha \varphi(x)|$$

for all $\varphi \in C^\infty(\mathbb{R}^n)$. Moreover, members of $\mathscr{E}'$ have compact support.

Assume that $T \in \mathscr{E}'$. Since $\varphi(x) = \mathrm{e}^{-\mathrm{i}(x,\xi)} \in C^\infty(\mathbb{R}^n)$, it follows that $\langle T, \mathrm{e}^{-\mathrm{i}(x,\xi)} \rangle$ is well defined and that there exist $c_0 > 0$, $R_0 > 0$, and $n_0 \in \mathbb{N}_0$ such that

$$|\langle T, \mathrm{e}^{-\mathrm{i}(x,\xi)} \rangle| \le c_0 \sum_{|\alpha| \le n_0} \sup_{|x| \le R_0} |D_x^\alpha \mathrm{e}^{-\mathrm{i}(x,\xi)}| \le c_0 \sum_{|\alpha| \le n_0} |\xi^\alpha| \asymp (1+|\xi|^2)^{\frac{n_0}{2}}.$$

If we now set

$$\widehat{T}(\xi) := (2\pi)^{-n/2} \langle T, \mathrm{e}^{-\mathrm{i}(x,\xi)} \rangle,$$

then $\widehat{T}$ is a usual function of ξ. The same is true for

$$\partial^\alpha \widehat{T}(\xi) = (2\pi)^{-n/2} (-1)^{|\alpha|} \langle T, \partial^\alpha \mathrm{e}^{-\mathrm{i}(x,\xi)} \rangle$$

and hence $\widehat{T} \in C^\infty(\mathbb{R}^n)$. On the other hand, $|\langle T, \mathrm{e}^{-\mathrm{i}(x,\xi)} \rangle| \le c_0 \langle \xi \rangle^{n_0}$ implies that $|\widehat{T}(\xi)| \le c_0' \langle \xi \rangle^{n_0}$ and hence $\widehat{T} \in L_s^2(\mathbb{R}^n)$ for $s < -n_0 - \frac{n}{2}$. So, by Exercise 20.2, we may conclude that every $T \in \mathscr{E}'$ belongs to $H^s(\mathbb{R}^n)$ for $s < -n_0 - \frac{n}{2}$.

(7) We have the following lemma.

Lemma 20.15. *The closure of $C_0^\infty(\mathbb{R}^n)$ in the norm of $H^s(\mathbb{R}^n)$ is $H^s(\mathbb{R}^n)$ for all $s \in \mathbb{R}$. In short, $\overline{C_0^\infty(\mathbb{R}^n)} \overset{H^s}{=} H^s(\mathbb{R}^n)$.*

Proof. Let f be an arbitrary function from $H^s(\mathbb{R}^n)$ and let f_R be a new function such that

$$\widehat{f_R}(\xi) = \chi_R(\xi) \widehat{f}(\xi) = \begin{cases} \widehat{f}(\xi), & |\xi| < R, \\ 0, & |\xi| > R. \end{cases}$$

Then $f_R(x) = \mathscr{F}^{-1}(\chi_R \widehat{f})(x) = (2\pi)^{-\frac{n}{2}} (\mathscr{F}^{-1}\chi_R * f)(x)$. It follows from the above considerations that $\mathscr{F}^{-1}\chi_R \in C^\infty(\mathbb{R}^n)$ as the inverse Fourier transform of a compactly supported function (but $\notin C_0^\infty(\mathbb{R}^n)$) and

$$\|f - f_R\|_{H^s}^2 = \int_{\mathbb{R}^n} |\widehat{f}(\xi) - \widehat{f_R}(\xi)|^2 \langle \xi \rangle^{2s} \mathrm{d}\xi = \int_{|\xi| > R} |\widehat{f}(\xi)|^2 \langle \xi \rangle^{2s} \mathrm{d}\xi \to 0$$

as $R \to \infty$, since $f \in H^s(\mathbb{R}^n)$. This completes the first step.
The second step is as follows. Let $j(\xi) \in C_0^\infty(|\xi| < 1)$ with $\int_{\mathbb{R}^n} j(\xi) \mathrm{d}\xi = 1$. Let us set $j_k(\xi) := k^n j(k\xi)$. We recall from Lemma 19.6 that $j_k * g \overset{L^p}{\to} g$, $1 \le p < \infty$. Define the sequence $v_k := \mathscr{F}^{-1}(j_k * \widehat{f_R})$. Since $\widehat{v_k} = j_k * \widehat{f_R}$, it follows that $\operatorname{supp} \widehat{v_k} \subset U_{R+1}(0)$, and so $\widehat{v_k} \in C_0^\infty(\mathbb{R}^n)$. Hence $v_k \in S$. Therefore, $v_k \in H^s(\mathbb{R}^n)$ and

$$\begin{aligned} \|v_k - f_R\|_{H^s(\mathbb{R}^n)}^2 &= \int_{|\xi| < R+1} \langle \xi \rangle^{2s} |j_k * \widehat{f_R} - \widehat{f_R}|^2 \mathrm{d}\xi \\ &\le C_R \int_{|\xi| < R+1} |j_k * \widehat{f_R} - \widehat{f_R}|^2 \mathrm{d}\xi \to 0, \quad k \to \infty. \end{aligned}$$

Since $v_k \notin C_0^\infty(\mathbb{R}^n)$, we take a function $\kappa \in C_0^\infty(\mathbb{R}^n)$ with $\kappa(0) = 1$. Then

$$\kappa\left(\frac{x}{A}\right) v_k \xrightarrow{S} v_k, \quad A \to \infty.$$

This fact implies that $\kappa\left(\frac{x}{A}\right) v_k \xrightarrow{H^s} v_k$ as $A \to \infty$. Setting $f_k(x) := \kappa\left(\frac{x}{A}\right) v_k(x) \in C_0^\infty(\mathbb{R}^n)$, we get finally

$$\|f - f_k\|_{H^s} \le \|f - f_R\|_{H^s} + \|f_R - v_k\|_{H^s} + \left\|v_k - \kappa\left(\frac{x}{A}\right) v_k\right\|_{H^s} \to 0$$

if A, k, and R are sufficiently large. □

Now we are in a position to formulate the main result concerning $H^s(\mathbb{R}^n)$.

Theorem 20.16 (Sobolev embedding theorem). *Let f be a function from $H^s(\mathbb{R}^n)$ for $s > k + \frac{n}{2}$, where $k \in \mathbb{N}_0$. Then $D^\alpha f \in \dot{C}(\mathbb{R}^n)$ for all α such that $|\alpha| \le k$. In short,*

$$H^s \subset \dot{C}^k(\mathbb{R}^n), \quad s > k + \frac{n}{2}.$$

Proof. Let $f \in H^s(\mathbb{R}^n) \subset S'$. Then

$$D^\alpha f = \mathscr{F}^{-1}\mathscr{F}(D^\alpha f) = \mathscr{F}^{-1}(\xi^\alpha \widehat{f}(\xi)).$$

What is more,

$$\int_{\mathbb{R}^n} |\xi^\alpha \widehat{f}(\xi)| \mathrm{d}\xi \le c \int_{\mathbb{R}^n} |\xi|^{|\alpha|} |\widehat{f}(\xi)| \mathrm{d}\xi = c \int_{\mathbb{R}^n} \frac{|\xi|^{|\alpha|}}{\langle\xi\rangle^s} \langle\xi\rangle^s |\widehat{f}(\xi)| \mathrm{d}\xi$$
$$\le c \left(\int_{\mathbb{R}^n} \frac{|\xi|^{2|\alpha|}}{\langle\xi\rangle^{2s}} \mathrm{d}\xi\right)^{1/2} \left(\int_{\mathbb{R}^n} \langle\xi\rangle^{2s} |\widehat{f}(\xi)|^2 \mathrm{d}\xi\right)^{1/2} \le c' \|f\|_{H^s(\mathbb{R}^n)}$$

if and only if $2s - 2|\alpha| > n$, or $s > |\alpha| + n/2$.

This means that for such s and α the function $D^\alpha f$ is the Fourier transform of some function from $L^1(\mathbb{R}^n)$. By the Riemann–Lebesgue lemma we have that $D^\alpha f$ from $\dot{C}(\mathbb{R}^n)$. □

Lemma 20.17. *$L_s^2(\mathbb{R}^n) \subset L^q(\mathbb{R}^n)$ if and only if $q = 2$ and $s \ge 0$ or $1 \le q < 2$ and $s > n\left(\frac{1}{q} - \frac{1}{2}\right)$.*

Exercise 20.5. Prove Lemma 20.17.

Lemma 20.18 (Hörmander).

(1) $\mathscr{F} : H^s(\mathbb{R}^n) \to L^q(\mathbb{R}^n)$ *for* $1 \le q < 2$ *and* $s > n\left(\frac{1}{q} - \frac{1}{2}\right)$.
(2) $\mathscr{F} : L^p(\mathbb{R}^n) \to H^{-s}(\mathbb{R}^n)$ *for* $2 < p \le \infty$ *and* $s > n\left(\frac{1}{2} - \frac{1}{p}\right)$.
(3) $\mathscr{F} : L^2(\mathbb{R}^n) \to L^2(\mathbb{R}^n)$.

Proof. (1) See Lemma 20.17.
(2) Let f be a function from $L^p(\mathbb{R}^n)$ for $2 < p \le \infty$. Then $f \in S'$ and $|\langle \widehat{f}, \varphi\rangle_{L^2(\mathbb{R}^n)}| = |\langle f, \widehat{\varphi}\rangle_{L^2(\mathbb{R}^n)}| \le \|f\|_p \cdot \|\widehat{\varphi}\|_{p'}$, where $1 \le p' < 2$. But if $\varphi \in H^s(\mathbb{R}^n)$ for $s > n\left(\frac{1}{p'} - \frac{1}{2}\right)$, then $\|\widehat{\varphi}\|_{L^{p'}} \le c\|\varphi\|_{H^s}$. So

$$|\langle \widehat{f}, \varphi\rangle_{L^2(\mathbb{R}^n)}| \le c\|f\|_p \cdot \|\varphi\|_{H^s}.$$

Therefore (by duality),

$$\left\|\widehat{f}\right\|_{H^{-s}} \le c\|f\|_{L^p}$$

for $s > n\left(\frac{1}{p'} - \frac{1}{2}\right) = n\left(\frac{1}{2} - \frac{1}{p}\right)$.
(3) This is simply Parseval's equality $\left\|\widehat{f}\right\|_{L^2} = \|f\|_{L^2}$.
This completes the proof. □

Exercise 20.6. Prove that

(1) $\chi_{[0,1]} \in H^s(\mathbb{R})$ if and only if $s < 1/2$.
(2) $\chi_{[0,1]\times[0,1]} \in H^s(\mathbb{R}^2)$ if and only if $s < 1/2$.
(3) $K(x) := \mathscr{F}^{-1}\left(\frac{1}{1+|\xi|^2}\right) \in H^s(\mathbb{R}^n)$ if and only if $s < 2 - n/2$.
(4) Let $f(x) = \chi(x)\log\log|x|^{-1}$ in $\mathbb{R}^2$, where $\chi(x) \in C_0^\infty(|x| < 1/3)$. Prove that $f \in H^1(\mathbb{R}^2)$ but $f \notin L^\infty(\mathbb{R}^2)$.

Remark 20.19. This counterexample shows us that the Sobolev embedding theorem is sharp.

Lemma 20.20. *Let us assume that* φ *and* f *from* $H^s(\mathbb{R}^n)$ *for* $s > \frac{n}{2}$. *Then* $\mathscr{F}(\varphi f) \in L^1(\mathbb{R}^n)$.

Proof. Since $f, \varphi \in H^s(\mathbb{R}^n)$, it follows that $\widehat{f}, \widehat{\varphi} \in L^2_s(\mathbb{R}^n)$ for $s > \frac{n}{2}$. But this implies (see Lemma 20.17) that $\widehat{f}$ and $\widehat{\varphi} \in L^1(\mathbb{R}^n)$ and

$$\mathscr{F}(\varphi f) = (2\pi)^{-\frac{n}{2}}\widehat{\varphi} * \widehat{f}$$

also belongs to $L^1(\mathbb{R}^n)$. □

Remark 20.21. It is possible to prove that if $\varphi, f \in H^s(\mathbb{R}^n)$ for $s > \frac{n}{2}$, then $\varphi f \in H^s(\mathbb{R}^n)$ with the same s.

Exercise 20.7. Prove that $W_p^1(\mathbb{R}^n) \cdot W_p^1(\mathbb{R}^n) \subset W_p^1(\mathbb{R}^n)$ if $p > n$.

Next, we consider the trace map τ, defined initially on $S(\mathbb{R}^n)$ by $\tau u = f$, where $f(x') = u(0, x')$ if $x = (x_1, \ldots, x_n)$ and $x' = (x_2, \ldots, x_n)$.

Proposition 20.22. *The map τ extends uniquely to a continuous linear map*

$$\tau : H^s(\mathbb{R}^n) \to H^{s-1/2}(\mathbb{R}^{n-1})$$

for all $s > 1/2$, and this map is surjective.

Proof. If $f = \tau u$, then for all $u \in S(\mathbb{R}^n)$ we may define

$$\widehat{f}(\xi') = \int_{-\infty}^{\infty} \widehat{u}(\xi) d\xi_1.$$

Hence, using the Cauchy-Bunyakovsky-Schwarz inequality, we have

$$|\widehat{f}(\xi')|^2 \le \int_{-\infty}^{\infty} |\widehat{u}(\xi)|^2 (1 + |\xi|^2)^s d\xi_1 \int_{-\infty}^{\infty} (1 + |\xi|^2)^{-s} d\xi_1.$$

Since

$$\int_{-\infty}^{\infty} \frac{1}{(1 + |\xi'|^2 + \xi_1^2)^s} d\xi_1 = \frac{1}{(1 + |\xi'|^2)^s} \int_{-\infty}^{\infty} \frac{1}{\left(1 + \left(\frac{\xi_1}{(1+|\xi'|^2)^{1/2}}\right)^2\right)^s} d\xi_1$$
$$= (1 + |\xi'|^2)^{-s+1/2} \int_{-\infty}^{\infty} \frac{d\rho}{(1 + \rho^2)^s}, \quad s > 1/2,$$

we have

$$(1 + |\xi'|^2)^{-s+1/2} |\widehat{f}(\xi')|^2 \le C_s \int_{-\infty}^{\infty} |\widehat{u}(\xi)|^2 (1 + |\xi|^2)^s d\xi_1,$$

where C_s denotes the latter convergent integral with respect to ρ. Integrating with respect to ξ' in the latter inequality leads to

$$\|f\|^2_{H^{s-1/2}(\mathbb{R}^{n-1})} \le C_s \|u\|^2_{H^s(\mathbb{R}^n)}.$$

This means that the first part of this proposition is proved, since $S(\mathbb{R}^n)$ is dense in $H^s(\mathbb{R}^n)$. Surjectivity follows also. If $g \in H^{s-1/2}(\mathbb{R}^{n-1})$, $s > 1/2$, we can define

$$\widehat{u}(\xi) := \widehat{g}(\xi')\frac{(1+|\xi'|^2)^{s/2-1/4}}{(1+|\xi|^2)^{s/2}}.$$

Then $u := \mathscr{F}^{-1}(\widehat{u}(\xi))$ defines $u \in H^s(\mathbb{R}^n)$ and $u(0,x') = Cg(x')$ with some nonzero constant C. □

20.1 Sobolev Spaces on Bounded Domains

Let Ω be a bounded domain in $\mathbb{R}^n$ with smooth boundary $\partial\Omega$. We define for integers $k \geq 0$ the *Sobolev space* $H^k(\Omega) = W_2^k(\Omega)$ as the set of all $f \in L^2(\Omega)$ for which there exist (by analogy with Lemma 20.1 if we extend f by zero outside of Ω) the generalized derivatives $\partial^\alpha f$ in $L^2(\Omega)$ for all $|\alpha| \leq k$. The norm in this space is defined then by

$$\|f\|_{H^k(\Omega)} = \left(\sum_{|\alpha|\leq k}\int_\Omega |\partial^\alpha f|^2 \mathrm{d}x\right)^{1/2}. \tag{20.2}$$

We define the space $H_0^k(\Omega)$ as the completion of $C_0^\infty(\Omega)$ with respect to the norm of $H^k(\Omega)$.

Theorem 20.23 (Poincaré's inequality). *Suppose $f \in H_0^k(\Omega)$, $k \geq 1$. Then there is a constant $M > 0$ such that*

$$\|f\|_{H^{k-1}(\Omega)} \leq M \sum_{|\beta|=k} \left\|\partial^\beta f\right\|_{L^2(\Omega)}. \tag{20.3}$$

Proof. We apply induction with respect to k. Since Ω is bounded, it can be enclosed in a cube

$$Q_n := \{x \in \mathbb{R}^n : |x_j| \leq A, j = 1,\dots,n\},$$

and $f \in C_0^\infty(\Omega)$ will continue to be identically zero outside of Ω. Then for all $x \in Q_n$ we have

$$f(x) = \int_{-A}^{x_1} \partial_{x_1} f(\xi_1,x')\mathrm{d}\xi_1, \quad x = (x_1,x'). \tag{20.4}$$

We have used here the fact $f = 0$ on $\partial\Omega$. Using the Cauchy-Bunyakovsky-Schwarz inequality, we obtain that

$$\int_{-A}^{A} |f(x)|^2 \mathrm{d}x_1 \leq 2A\int_{-A}^{A}\mathrm{d}x_1\int_{-A}^{x_1}|\partial_{x_1} f(\xi_1,x')|^2\mathrm{d}\xi_1 \leq 4A^2\int_{-A}^{A}|\partial_{x_1} f(\xi_1,x')|^2\mathrm{d}\xi_1.$$

Integrating now with respect to x', we obtain

$$\int_{Q_n} |f(x)|^2 \mathrm{d}x \le 4A^2 \int_{Q_n} |\partial_{x_1} f|^2 \mathrm{d}x \le 4A^2 \sum_{|\beta|=1} \int_{Q_n} |\partial^\beta f|^2 \mathrm{d}x.$$

Since $C_0^\infty(\Omega)$ is dense in $H_0^1(\Omega)$, the case $k = 1$ is established. Let us assume that for all $f \in H_0^k(\Omega)$, $k \ge 1$, we have

$$\|f\|_{H^{k-1}(\Omega)} \le M \sum_{|\beta|=k} \left\|\partial^\beta f\right\|_{L^2(\Omega)}.$$

Then for all $f \in H^{k+1}(\Omega)$ we have that $\partial_j f \in H_0^k(\Omega)$, $j = 1, \ldots, n$, and by induction we have

$$\left\|\partial_j f\right\|_{H^{k-1}(\Omega)} \le M \sum_{|\beta|=k} \left\|\partial^\beta (\partial_j f)\right\|_{L^2(\Omega)} \le M \sum_{|\gamma|=k+1} \|\partial^\gamma f\|_{L^2(\Omega)}.$$

Thus

$$\sum_{j=1}^n \left\|\partial_j f\right\|_{H^{k-1}(\Omega)} \le M' \sum_{|\gamma|=k+1} \|\partial^\gamma f\|_{L^2(\Omega)},$$

or

$$\|f\|_{H^k(\Omega)} \le M' \sum_{|\gamma|=k+1} \|\partial^\gamma f\|_{L^2(\Omega)}.$$

Hence the theorem is completely proved. □

For all real $s > 0$ the space $H^s(\Omega)$ with fractional s can be obtained as the interpolation space between $L^2(\Omega)$ and $H^k(\Omega)$ with some integer $k \ge 1$ (see, e.g., [39, p. 286] for details).

Corollary 20.24. *Suppose $f \in H_0^k(\Omega)$, $k \ge 1$. Then $\partial^\alpha f|_{\partial\Omega} = 0$ for all $|\alpha| \le k-1$.*

Proof. This fact can be proved also by induction on $k \ge 1$ using (20.4). Since $C_0^\infty(\Omega)$ is complete in $H_0^k(\Omega)$, it follows that for $k = 1$ and $x = (x_1, x') \in \partial\Omega$ we have from (20.4) that

$$f(x) = \int_{-A}^{x_1} \partial_{x_1} f(\xi_1, x') \mathrm{d}\xi_1 = 0.$$

Using induction on k and the representation

$$\partial^\alpha f(x) = \int_{-A}^{x_1} \partial_{x_1} (\partial^u f(\xi_1, x')) \mathrm{d}\xi_1$$

for $x \in \partial\Omega$ and $|\alpha| \le k-1$, we obtain the desired result. □

Also, the converse of this corollary holds: if $f \in H^k(\Omega)$ and $\partial^\alpha f|_{\partial\Omega} = 0$ for $|\alpha| \le k-1$, then $f \in H_0^k(\Omega)$. For details, see [11].

Exercise 20.8. Prove Rellich's theorem: if Ω is bounded and $s > t \ge 0$, then the inclusion map $H_0^s(\Omega) \hookrightarrow H_0^t(\Omega)$ is compact. Hint: Use the Ascoli–Arzelà theorem (Theorem 34.7).

We define for every integer $k \ge 0$ the space

$$H^k(\mathbb{R}_+^n) = \{f \in L^2(\mathbb{R}_+^n) : \partial^\alpha f \in L^2(\mathbb{R}_+^n), |\alpha| \le k\}$$

with the norm

$$\|f\|_{H^k(\mathbb{R}_+^n)} = \left(\sum_{|\alpha|\le k} \int_{\mathbb{R}^{n-1}} \mathrm{d}x' \int_0^\infty |\partial^\alpha f(x_1,x')|^2 \mathrm{d}x_1\right)^{1/2}. \tag{20.5}$$

We may say that an element $f \in H^k(\mathbb{R}_+^n)$ is the restriction on $\mathbb{R}_+^n$ of some element from $H^k(\mathbb{R}^n)$. More precisely, the following proposition holds.

Proposition 20.25. *For every integer $k \ge 0$ there is an extension linear operator E defined on $H^k(\mathbb{R}_+^n)$ with image in $H^k(\mathbb{R}^n)$ such that*

$$\|Ef\|_{H^k(\mathbb{R}^n)} \le C\|f\|_{H^k(\mathbb{R}_+^n)}, \tag{20.6}$$

where the constant $C > 0$ is independent of f.

Proof. Since $S(\mathbb{R}_+^n)$ is dense in $H^k(\mathbb{R}_+^n)$ (see, for example, Lemma 20.15), it follows that for every $f \in S(\mathbb{R}_+^n)$ and integer $k \ge 0$ we may define E as

$$Ef(x) = \begin{cases} f(x), & x_1 \ge 0, \\ \sum_{j=1}^{k+1} a_j f(-jx_1,x'), & x_1 < 0, \end{cases}$$

where a_j, $j = 1,\dots,k+1$, are to be determined. Let us prove that this operator E satisfies this proposition. It is clear that E is linear and that for $k = 0$ ($a_1 = 1$),

$$Ef(x) = \begin{cases} f(x), & x_1 \ge 0, \\ f(-x_1,x'), & x_1 < 0, \end{cases}$$

belongs to $L^2(\mathbb{R}^n)$ and (20.6) holds. If $k \ge 1$ is an integer, then for $0 \le l \le k$ we have formally

$$\partial_{x_1}^l (Ef(x)) = \begin{cases} \partial_{x_1}^l f(x), & x_1 > 0, \\ \sum_{j=1}^{k+1} a_j (-j)^l \partial_{x_1}^l f(-jx_1,x'), & x_1 < 0. \end{cases}$$

Let us choose a_j as the solution of the linear algebraic system

$$\sum_{j=1}^{k+1} a_j(-j)^l = 1, \quad l = 0, 1, \ldots, k.$$

The determinant of this system is the well known Vandermonde polynomial, and it is not equal to zero. Hence, this system has a unique solution with respect to the coefficients a_j, $j = 1, \ldots, k-1$. After these coefficients have been determined, the inequality (20.6) follows immediately. □

Remark 20.26. For arbitrary $s \geq 0$ the result of Proposition 20.25 can be obtained by interpolation of Sobolev spaces H^s; see [39, p. 285].

Let $\Omega \subset \mathbb{R}^n$ be a bounded domain with a C^∞ boundary $\partial\Omega$. Since $\partial\Omega$ is a compact set, it can be covered by finitely many open sets U_j, $j = 1, \ldots, m$, such that

$$\partial\Omega \subset \bigcup_{j=1}^m U_j$$

and there is a partition of unity $\{\varphi_j\}_{j=1}^m$ with $\varphi_j \in C_0^\infty(U_j)$ and $\sum_{j=1}^m \varphi_j \equiv 1$ on $\partial\Omega$. Thus for every function $u(x)$ defined on $\partial\Omega$ we have $u = \sum_{j=1}^m \varphi_j u$. The functions $\varphi_j u$ can be written in local coordinates y' as $\varphi_j u = (\varphi_j u)(x(y'))$, $y' \in \mathbb{R}^{n-1}$, and the U_j are mapped into the unit ball $\{y \in \mathbb{R}^n : |y| < 1\}$ so that (see Figure 20.1)

$$U_j \cap \Omega \to \{y \in \mathbb{R}^n : |y| < 1, y_1 > 0\}$$

and

$$U_j \cap \partial\Omega \to \{y \in \mathbb{R}^n : |y| < 1, y_1 = 0\}.$$

For these purposes we may use the extension operator E from Proposition 20.25 such that ($s \geq 0$)

$$E : H^s(\Omega) \to H^s(U), \quad U = \bigcup_{j=1}^m U_j.$$

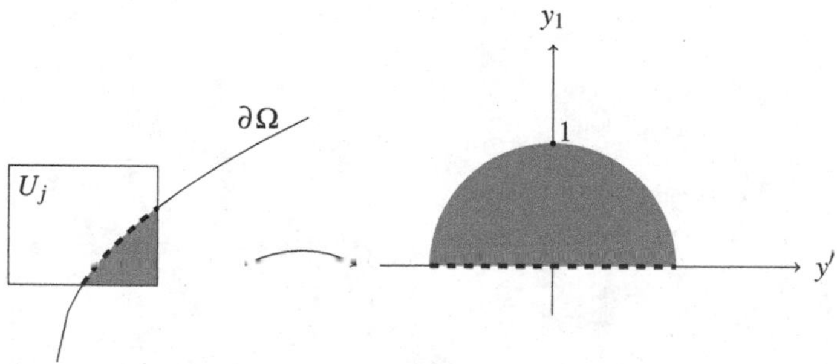

Fig. 20.1 Representation of the boundary in local coordinates.

Definition 20.27. A function u is said to belong to $H^s(\partial\Omega)$ if $\varphi_j u$, $j = 1,2,\ldots,m$, belong to $H^s(\mathbb{R}^{n-1})$ as a function of y'. The norm in $H^s(\partial\Omega)$, $s \geq 0$, is defined as

$$\|u\|_{H^s(\partial\Omega)} := \sum_{j=1}^{m} \|\varphi_j u\|_{H^s(\mathbb{R}^{n-1})}.$$

Remark 20.28. It can be shown that $H^s(\partial\Omega)$ is independent of the partition of unity and that the norms are equivalent with respect to the different partitions of unity.

Let τ be a trace map (linear):

$$\tau u := u|_{\partial\Omega}$$

for all $u \in C^\infty(\overline{\Omega})$ defined on Ω.

Proposition 20.29. *For $s > 1/2$ the map τ extends uniquely to a continuous linear map*

$$\tau : H^s(\Omega) \to H^{s-1/2}(\partial\Omega)$$

with the norm estimate

$$\|\tau u\|_{H^{s-1/2}(\partial\Omega)} \leq C \|u\|_{H^s(\Omega)}.$$

Moreover, this map is surjective.

Proof. The proof is based on Proposition 20.22, the definition of $H^s(\partial\Omega)$, and the following diagram:

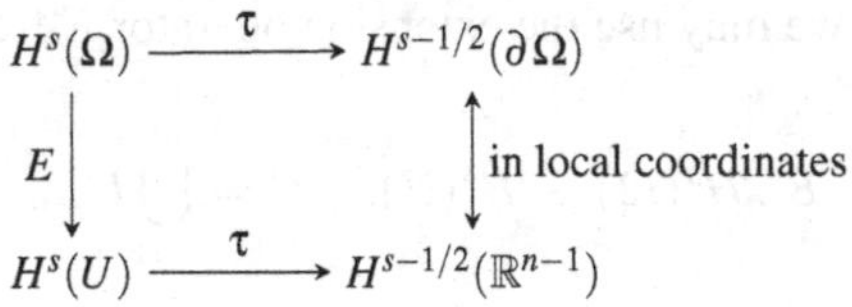

$$\begin{array}{ccc} H^s(\Omega) & \xrightarrow{\ \tau\ } & H^{s-1/2}(\partial\Omega) \\ {\scriptstyle E}\downarrow & & \updownarrow {\scriptstyle \text{in local coordinates}} \\ H^s(U) & \xrightarrow{\ \tau\ } & H^{s-1/2}(\mathbb{R}^{n-1}) \end{array}$$

See [39, p. 287] for details. □

Chapter 21
Homogeneous Distributions

We begin this chapter with the Fourier transform of a radially symmetric function.

Lemma 21.1. *Let $f(x)$ be a radially symmetric function in $\mathbb{R}^n$, i.e., $f(x) = f_1(|x|)$. Let us assume also that $f(x) \in L^1(\mathbb{R}^n)$. Then the Fourier transform $\widehat{f}(\xi)$ is also radial and*

$$\widehat{f}(\xi) = |\xi|^{1-\frac{n}{2}} \int_0^\infty f_1(r) r^{\frac{n}{2}} J_{\frac{n-2}{2}}(r|\xi|)\mathrm{d}r,$$

where $J_\nu(\cdot)$ is the Bessel function of order ν.

Proof. Let us take the Fourier transform

$$\begin{aligned}\widehat{f}(\xi) &= (2\pi)^{-\frac{n}{2}} \int_{\mathbb{R}^n} \mathrm{e}^{-\mathrm{i}(x,\xi)} f_1(|x|)\mathrm{d}x \\ &= (2\pi)^{-\frac{n}{2}} \int_0^\infty f_1(r) r^{n-1}\mathrm{d}r \int_{\mathbb{S}^{n-1}} \mathrm{e}^{-\mathrm{i}|\xi|r(\varphi,\theta)}\mathrm{d}\theta,\end{aligned}$$

where $x = r\theta$, $\xi = |\xi|\varphi$, and $\theta, \varphi \in \mathbb{S}^{n-1} := \{x \in \mathbb{R}^n : |x| = 1\}$. It is known [43] that

$$\int_{\mathbb{S}^{n-1}} \mathrm{e}^{-\mathrm{i}|\xi|r(\varphi,\theta)}\mathrm{d}\theta = \frac{2\pi^{\frac{n-1}{2}}}{\Gamma(\frac{n-1}{2})} \int_0^\pi \mathrm{e}^{-\mathrm{i}|\xi|r\cos\psi} (\sin\psi)^{n-2}\mathrm{d}\psi,$$

where Γ is the gamma function. This fact implies that $\widehat{f}(\xi)$ is a radial function, since the last integral depends only on $|\xi|$. A property of Bessel functions [23] is that

$$\int_0^\pi \mathrm{e}^{-\mathrm{i}|\xi|r\cos\psi} (\sin\psi)^{n-2}\mathrm{d}\psi = 2^{\frac{n}{2}-1}\sqrt{\pi}\Gamma\left(\frac{n-1}{2}\right) \frac{J_{\frac{n-2}{2}}(r|\xi|)}{(r|\xi|)^{\frac{n-2}{2}}}. \tag{21.1}$$

V. Serov, *Fourier Series, Fourier Transform and Their Applications to Mathematical Physics*, Applied Mathematical Sciences 197,
DOI 10.1007/978-3-319-65262-7_21

Collecting these things, we obtain

$$\widehat{f}(\xi)=|\xi|^{\frac{2-n}{2}}\int_0^\infty f_1(r)r^{\frac{n}{2}}J_{\frac{n-2}{2}}(r|\xi|)\mathrm{d}r,$$

and the claim follows. □

Remark 21.2. If we put the variable $u=\cos\psi$ in the integral I appearing in (21.1), then we obtain

$$I=\int_0^\pi \mathrm{e}^{-\mathrm{i}|\xi|r\cos\psi}(\sin\psi)^{n-2}\mathrm{d}\psi=\int_{-1}^1 \mathrm{e}^{-\mathrm{i}|\xi|ru}(\sqrt{1-u^2})^{n-3}\mathrm{d}u.$$

In particular, if $n=3$, then (21.1) implies that

$$I=\int_{-1}^1 \mathrm{e}^{-\mathrm{i}|\xi|ru}\mathrm{d}u=2\frac{\sin(|\xi|r)}{|\xi|r}=\sqrt{2\pi}\frac{J_{\frac{1}{2}}(r|\xi|)}{(r|\xi|)^{\frac{1}{2}}},$$

i.e.,

$$J_{\frac{1}{2}}(r|\xi|)=\sqrt{\frac{2}{\pi}}\frac{\sin(|\xi|r)}{(|\xi|r)^{\frac{1}{2}}}.$$

If $n=2$, then

$$I=\int_{-1}^1\frac{\mathrm{e}^{-\mathrm{i}|\xi|ru}}{\sqrt{1-u^2}}\mathrm{d}u=\pi J_0(r|\xi|),$$

i.e.,

$$J_0(r|\xi|)=\frac{1}{\pi}\int_{-1}^1\frac{\mathrm{e}^{-\mathrm{i}|\xi|ru}}{\sqrt{1-u^2}}\mathrm{d}u.$$

Remark 21.3. For later considerations we state the small- and large-argument asymptotics of J_ν for $\nu>-1$ as

$$J_\nu(|x|)\approx\begin{cases}c_\nu|x|^\nu, & |x|\to 0+,\\ c'_\nu\frac{1}{\sqrt{|x|}}\cos(A_\nu|x|+B_\nu), & |x|\to+\infty\end{cases}$$

(see [23]).

Exercise 21.1. Prove that $\widehat{f}(A\xi)=\widehat{f}(\xi)$ if A is a linear transformation in $\mathbb{R}^n$ with rotation $A'=A^{-1}$ and f is radially symmetric.

Let us return again to the distribution (cf. Example 19.5)

$$K_1(x):=\frac{1}{(2\pi)^{\frac{n}{2}}}\mathscr{F}^{-1}\left(\frac{1}{1+|\xi|^2}\right)(x).$$

Let us assume now that $n = 1, 2, 3, 4$. Then the last integral can be understood in the classical sense. It follows from Lemma 21.1 that

$$\begin{aligned} K_1(x) = \widetilde{K_1}(|x|) &= (2\pi)^{-\frac{n}{2}} |x|^{1-\frac{n}{2}} \int_0^\infty \frac{r^{\frac{n}{2}} J_{\frac{n-2}{2}}(r|x|)dr}{1+r^2} \\ &= (2\pi)^{-\frac{n}{2}} |x|^{2-n} \int_0^\infty \frac{\rho^{\frac{n}{2}} J_{\frac{n-2}{2}}(\rho)d\rho}{\rho^2 + |x|^2}. \end{aligned}$$

It is not too difficult to prove that for $|x| < 1$ we have

$$|K_1(x)| \le c \begin{cases} 1, & n = 1, \\ \log \frac{1}{|x|}, & n = 2, \\ |x|^{2-n}, & n = 3, 4. \end{cases}$$

Exercise 21.2. Prove this fact.

Remark 21.4. A little later we will prove estimates for $K_1(x)$ for every dimension and for all $x \in \mathbb{R}^n$.

There is one more important example. If we have the equation $(-1 - \Delta)u = f$ in $L^2(\mathbb{R}^n)$ (or even in S), then formally $u = (2\pi)^{-\frac{n}{2}} \mathscr{F}^{-1}\left(\frac{1}{|\xi|^2 - 1}\right) * f = K_{-1} * f$, where

$$K_{-1}(|x|) = (2\pi)^{-\frac{n}{2}} |x|^{2-n} \int_0^\infty \frac{\rho^{\frac{n}{2}} J_{\frac{n-2}{2}}(\rho)d\rho}{\rho^2 - |x|^2}.$$

But there is a problem with the convergence of this integral near $\rho = |x|$. Therefore, this integral must be regularized as

$$\lim_{\varepsilon \to 0+} \int_0^\infty \frac{\rho^{\frac{n}{2}} J_{\frac{n-2}{2}}(\rho)d\rho}{\rho^2 - |x|^2 - i\varepsilon}.$$

Recall that

(1) $\sigma_\lambda f(x) := f(\lambda x), \quad \lambda \neq 0$ and
(2) $\langle \sigma_\lambda T, \varphi \rangle := \lambda^{-n} \langle T, \sigma_{\frac{1}{\lambda}} \varphi \rangle, \quad \lambda > 0.$

Definition 21.5. A tempered distribution T is said to be a *homogeneous distribution* of degree $m \in \mathbb{C}$ if

$$\sigma_\lambda T = \lambda^m T$$

for every $\lambda > 0$. In other words,

$$\langle \sigma_\lambda T, \varphi \rangle = \lambda^m \langle T, \varphi \rangle,$$

or

$$\langle T, \varphi \rangle = \lambda^{-n-m} \langle T, \sigma_{\frac{1}{\lambda}} \varphi \rangle,$$

for $\varphi \in S$. The space of all such distributions is denoted by $H_m(\mathbb{R}^n)$.

Lemma 21.6. $\mathscr{F} : H_m(\mathbb{R}^n) \to H_{-m-n}(\mathbb{R}^n)$.

Proof. Let $T \in H_m(\mathbb{R}^n)$. Then

$$\begin{aligned}\langle \sigma_\lambda \widehat{T}, \varphi \rangle &= \lambda^{-n} \langle \widehat{T}, \sigma_{\frac{1}{\lambda}} \varphi \rangle = \lambda^{-n} \langle T, \widehat{\sigma_{\frac{1}{\lambda}} \varphi} \rangle = \lambda^{-n} \langle T, \lambda^n \sigma_\lambda \widehat{\varphi} \rangle \\ &= \langle T, \sigma_\lambda \widehat{\varphi} \rangle = \lambda^{-n} \langle \sigma_{\frac{1}{\lambda}} T, \widehat{\varphi} \rangle = \lambda^{-n} \lambda^{-m} \langle T, \widehat{\varphi} \rangle = \lambda^{-n-m} \langle \widehat{T}, \varphi \rangle\end{aligned}$$

for all $\varphi \in S$. □

Definition 21.7. We set $H_m^*(\mathbb{R}^n) := \{T \in H_m(\mathbb{R}^n) : T \in C^\infty(\mathbb{R}^n \setminus \{0\})\}$.

Exercise 21.3. Prove that

(1) if $T \in H_m^*$, then $D^\alpha T \in H_{m-|\alpha|}^*$ and $x^\alpha T \in H_{m+|\alpha|}^*$;
(2) $\mathscr{F} : H_m^* \to H_{-m-n}^*$.

Exercise 21.4. Let $\rho(x)$ be a function from $C^\infty(\mathbb{R}^n)$ with $|D^\alpha \rho(x)| \le c \langle x \rangle^{m-|\alpha|}$ for all $\alpha \ge 0$ and $m \in \mathbb{R}$. Prove that $\widehat{\rho}(\xi) \in C^\infty(\mathbb{R}^n \setminus \{0\})$ and $(1 - \varphi)\widehat{\rho} \in S$, where $\varphi \in C_0^\infty(\mathbb{R}^n)$ and $\varphi \equiv 1$ in $U_\delta(0)$.

Example 21.8. (1) $\delta \in H_{-n}^*(\mathbb{R}^n)$. Indeed,

$$\langle \sigma_\lambda \delta, \varphi \rangle = \lambda^{-n} \langle \delta, \sigma_{\frac{1}{\lambda}} \varphi \rangle = \lambda^{-n} \sigma_{\frac{1}{\lambda}} \varphi(0) = \lambda^{-n} \varphi(0) = \lambda^{-n} \langle \delta, \varphi \rangle.$$

But $\operatorname{supp} \delta = \{0\}$. This means that $\delta \in C^\infty(\mathbb{R}^n \setminus \{0\})$. Alternatively, one could note that

$$\widehat{\delta} = (2\pi)^{-\frac{n}{2}} \cdot 1 \in H_0^*(\mathbb{R}^n)$$

and use Exercise 21.3 to conclude that

$$\delta = \mathscr{F}^{-1}((2\pi)^{-\frac{n}{2}} \cdot 1) \in H_{-n}^*(\mathbb{R}^n).$$

(2) Let us assume that $\omega \in C^\infty(\mathbb{S}^{n-1})$ and $m > -n$. Set $T_m(x) := |x|^m \omega\left(\frac{x}{|x|}\right)$ for $x \in \mathbb{R}^n \setminus \{0\}$. Then $T_m(x) \in L^1_{\text{loc}}(\mathbb{R}^n)$ and $T_m \in H_m^*(\mathbb{R}^n)$. Indeed,

$$\langle \sigma_\lambda T_m, \varphi \rangle = \int_{\mathbb{R}^n} \sigma_\lambda T_m(x) \varphi(x) \mathrm{d}x = \int_{\mathbb{R}^n} |\lambda x|^m \omega\left(\frac{\lambda x}{|\lambda x|}\right) \varphi(x) \mathrm{d}x = \lambda^m \langle T_m, \varphi \rangle.$$

Since $|x|^m$ and $\omega\left(\frac{x}{|x|}\right)$ are from $C^\infty(\mathbb{R}^n \setminus \{0\})$, we have $T_m \in H_m^*(\mathbb{R}^n)$. Moreover, $D^\alpha T_m \in H_{m-|\alpha|}^*(\mathbb{R}^n)$ and $x^\alpha T_m \in H_{m+|\alpha|}^*(\mathbb{R}^n)$ by Exercise 21.3.

(3) Let now $m = -n$ in part (2) and in addition assume that $\int_{\mathbb{S}^{n-1}} \omega(\theta)\mathrm{d}\theta = 0$. Note that $T_{-n}(x) \notin L^1_{\mathrm{loc}}(\mathbb{R}^n)$. But we can define T_{-n} as a distribution from S' by

$$\langle \mathrm{p.v.}\, T_{-n}, \varphi\rangle := \int_{\mathbb{R}^n} T_{-n}(x)[\varphi(x) - \varphi(0)\psi(|x|)]\mathrm{d}x,$$

where $\varphi \in S(\mathbb{R}^n)$ and $\psi \in S(\mathbb{R})$ with $\psi(0) = 1$. We assume that ψ is fixed. But it is clear that this definition does not depend on ψ, because $\int_{\mathbb{S}^{n-1}} \omega(\theta)\mathrm{d}\theta = 0$.

Exercise 21.5. Prove that

$$\langle \mathrm{p.v.}\, T_{-n}, \varphi\rangle = \lim_{\varepsilon\to 0+} \int_{|x|\geq\varepsilon} T_{-n}(x)\varphi(x)\mathrm{d}x,$$

where $T_{-n} = |x|^{-n}\omega\left(\frac{x}{|x|}\right)$, $\int_{\mathbb{S}^{n-1}} \omega(\theta)\mathrm{d}\theta = 0$.

Let us prove the following:

(1) $\mathrm{p.v.}\, T_{-n} \in H^*_{-n}$;
(2) $\widehat{\mathrm{p.v.}\, T_{-n}} \in H^*_0(\mathbb{R}^n)$, and moreover, it is bounded.
(3) $\mathrm{p.v.}\, T_{-n}* : L^2(\mathbb{R}^n) \to L^2(\mathbb{R}^n)$.

Proof. Part (1) is clear. Part (2) follows from

$$\begin{aligned}
|\langle\widehat{\mathrm{p.v.}\, T_{-n}}, \varphi\rangle| = |\langle \mathrm{p.v.}\, T_{-n}, \widehat{\varphi}\rangle| &= \left|\int_{\mathbb{R}^n} T_{-n}(x)[\widehat{\varphi}(x) - \widehat{\varphi}(0)\psi(|x|)]\mathrm{d}x\right| \\
&\leq (2\pi)^{-\frac{n}{2}} \int_{\mathbb{R}^n} |\varphi(\xi)|\mathrm{d}\xi \left|\int_{\mathbb{R}^n} T_{-n}(x)[\mathrm{e}^{-\mathrm{i}(x,\xi)} - \psi(|x|)]\mathrm{d}x\right| \\
&= (2\pi)^{-\frac{n}{2}} \int_{\mathbb{R}^n} |\varphi(\xi)|\mathrm{d}\xi \left|\int_{\mathbb{S}^{n-1}} \omega(\theta)\mathrm{d}\theta \int_0^\infty \frac{1}{r}[\mathrm{e}^{-\mathrm{i}r(\theta,\xi)} - \psi(r)]\mathrm{d}r\right| \\
&\leq c\,\|\varphi\|_{L^1(\mathbb{R}^n)}.
\end{aligned}$$

Hence $\widehat{\mathrm{p.v.}\, T_{-n}} \in L^\infty(\mathbb{R}^n)$ by duality.

Finally, if $f \in L^2(\mathbb{R}^n)$, then

$$\mathscr{F}(\mathrm{p.v.}\, T_{-n} * f) = (2\pi)^{\frac{n}{2}}\widehat{\mathrm{p.v.}\, T_{-n}} \cdot \widehat{f},$$

which implies that

$$\|\mathrm{p.v.}\, T_{-n} * f\|_{L^2(\mathbb{R}^n)} < (2\pi)^{\frac{n}{2}} \left\|\widehat{\mathrm{p.v.}\, T_{-n}}\right\|_{L^\infty} \cdot \|f\|_{L^2}.$$

This proves part (3). □

Remark 21.9. In fact, it follows from Calderón–Zigmund theory that

$$\mathrm{p.v.}\, T_{-n}* : L^p(\mathbb{R}^n) \to L^p(\mathbb{R}^n), \quad 1 < p < \infty.$$

Next we want to consider a more difficult case than the previous one. Define

$$\langle \mathrm{p.v.} \frac{1}{|x|^n}, \varphi \rangle := \int_{\mathbb{R}^n} |x|^{-n} [\varphi(x) - \varphi(0)\psi(|x|)] \mathrm{d}x, \tag{21.2}$$

where $\varphi \in S$ and $\psi \in S$ with $\psi(0) = 1$. But now we don't have the condition $\int_{\mathbb{S}^{n-1}} \omega(\theta) \mathrm{d}\theta = 0$ as above. Therefore, (21.2) must depend on the function $\psi(|x|)$. We will try to choose an appropriate function ψ. Applying the operator σ_λ, we get

$$\begin{aligned}
\langle \sigma_\lambda \left(\mathrm{p.v.} \frac{1}{|x|^n} \right), \varphi \rangle &= \langle \mathrm{p.v.} \frac{1}{|x|^n}, \lambda^{-n} \sigma_{\frac{1}{\lambda}} \varphi \rangle \\
&= \int_{\mathbb{R}^n} |x|^{-n} \lambda^{-n} \left[\varphi\left(\frac{x}{\lambda}\right) - \varphi(0)\psi(|x|) \right] \mathrm{d}x \\
&= \lambda^{-n} \int_{\mathbb{R}^n} |y|^{-n} [\varphi(y) - \varphi(0)\psi(\lambda |y|)] \mathrm{d}y \\
&= \lambda^{-n} \int_{\mathbb{R}^n} |y|^{-n} [\varphi(y) - \varphi(0)\psi(|y|)] \mathrm{d}y \\
&\quad - \lambda^{-n} \int_{\mathbb{R}^n} |y|^{-n} \varphi(0) [\psi(\lambda |y|) - \psi(|y|)] \mathrm{d}y \\
&= \langle \lambda^{-n} \mathrm{p.v.} \frac{1}{|x|^n}, \varphi \rangle + \mathrm{Rest},
\end{aligned}$$

where

$$\begin{aligned}
\mathrm{Rest} &= -\lambda^{-n} \varphi(0) \int_{\mathbb{R}^n} |y|^{-n} [\psi(\lambda |y|) - \psi(|y|)] \mathrm{d}y \\
&= -\lambda^{-n} \langle \delta, \varphi \rangle \int_0^\infty \frac{\psi(\lambda r) - \psi(r)}{r} \mathrm{d}r \int_{\mathbb{S}^{n-1}} \mathrm{d}\theta \\
&= -\omega_n \lambda^{-n} \langle \delta, \varphi \rangle \int_0^\infty \frac{\psi(\lambda r) - \psi(r)}{r} \mathrm{d}r,
\end{aligned}$$

and $\omega_n = \frac{2\pi^{\frac{n}{2}}}{\Gamma(\frac{n}{2})}$ is the area of the unit sphere $\mathbb{S}^{n-1}$. Let us denote the last integral by $G(\lambda), \lambda > 0$. Then

$$G'(\lambda) = \int_0^\infty \psi'(\lambda r) \mathrm{d}r = \frac{1}{\lambda} \int_0^\infty \psi'(t) \mathrm{d}t = -\frac{1}{\lambda} \psi(0) = -\frac{1}{\lambda}.$$

We also have that $G(1) = 0$. We may therefore conclude that $G(\lambda) = -\log \lambda$, which implies that

$$\mathrm{Rest} = \omega_n \lambda^{-n} \log \lambda \langle \delta, \varphi \rangle,$$

and so

$$\sigma_\lambda\left(\mathrm{p.v.}\frac{1}{|x|^n}\right)=\lambda^{-n}\,\mathrm{p.v.}\frac{1}{|x|^n}+\omega_n\lambda^{-n}\log\lambda\cdot\delta(x).$$

Taking the Fourier transform, we get

$$\mathscr{F}\left(\sigma_\lambda\left(\mathrm{p.v.}\frac{1}{|x|^n}\right)\right)=\lambda^{-n}\mathscr{F}\left(\mathrm{p.v.}\frac{1}{|x|^n}\right)+(2\pi)^{-\frac{n}{2}}\omega_n\lambda^{-n}\log\lambda,$$

or

$$\lambda^{-n}\sigma_{\frac{1}{\lambda}}\mathscr{F}\left(\mathrm{p.v.}\frac{1}{|x|^n}\right)=\lambda^{-n}\mathscr{F}\left(\mathrm{p.v.}\frac{1}{|x|^n}\right)+(2\pi)^{-\frac{n}{2}}\omega_n\lambda^{-n}\log\lambda,$$

or

$$\mathscr{F}\left(\mathrm{p.v.}\frac{1}{|x|^n}\right)\left(\frac{\xi}{\lambda}\right)=\mathscr{F}\left(\mathrm{p.v.}\frac{1}{|x|^n}\right)(\xi)+(2\pi)^{-\frac{n}{2}}\omega_n\log\lambda.$$

Let us put now $\lambda=|\xi|$. Then

$$\mathscr{F}\left(\mathrm{p.v.}\frac{1}{|x|^n}\right)(\xi)=-(2\pi)^{-\frac{n}{2}}\omega_n\log|\xi|+\mathscr{F}\left(\mathrm{p.v.}\frac{1}{|x|^n}\right)\left(\frac{\xi}{|\xi|}\right).$$

Since $\mathrm{p.v.}\frac{1}{|x|^n}$ for such ψ is a radial homogeneous distribution, we must have that

$$\mathscr{F}\left(\mathrm{p.v.}\frac{1}{|x|^n}\right)$$

is also a radial homogeneous distribution. Therefore, $\mathscr{F}\left(\mathrm{p.v.}\frac{1}{|x|^n}\right)\left(\frac{\xi}{|\xi|}\right)$ depends only on $\left|\frac{\xi}{|\xi|}\right|=1$. So this term is a constant that depends on the choice of ψ. We will choose our function $\psi(|x|)$ so that this constant is zero. Then finally,

$$\mathscr{F}\left(\mathrm{p.v.}\frac{1}{|x|^n}\right)(\xi)=-(2\pi)^{-\frac{n}{2}}\omega_n\log|\xi|.$$

Now let us consider $T_{-m}=|x|^{-m}$, $0<m<n$. It is clear that $|x|^{-m}\in L^1_{\mathrm{loc}}(\mathbb{R}^n)$. Thus the situation here is simpler. We have

$$\langle\widehat{|x|^{-m}},\varphi\rangle=\langle|x|^{-m},\widehat{\varphi}\rangle=\int_{\mathbb{R}^n}|x|^{-m}\widehat{\varphi}(x)\mathrm{d}x.$$

Lemma 21.1 implies that

$$\widehat{|x|^{-m}}=|\xi|^{1-\frac{n}{2}}\int_0^\infty\frac{r^{\frac{n}{2}}J_{\frac{n-2}{2}}(r|\xi|)}{r^m}\mathrm{d}r=|\xi|^{-n+m}\int_0^\infty\rho^{\frac{n}{2}-m}J_{\frac{n-2}{2}}(\rho)\mathrm{d}\rho.$$

The last integral converges if $\frac{n-1}{2} < m < n$. We may therefore write that

$$\widehat{|x|^{-m}} = C_{n,m}|\xi|^{m-n}, \quad \frac{n-1}{2} < m < n.$$

In fact, this is true even for m such that $0 < \mathrm{Re}(m) < n$, which follows by analytic continuation on m. In order to calculate the constant $C_{n,m}$, let us apply this distribution to $\varphi = \mathrm{e}^{-\frac{|x|^2}{2}}$. Since $\widehat{\varphi} = \varphi$, we get

$$\langle |x|^{-m}, \mathrm{e}^{-\frac{|x|^2}{2}} \rangle = \langle C_{n,m}|\xi|^{m-n}, \mathrm{e}^{-\frac{|\xi|^2}{2}} \rangle.$$

The left-hand side is

$$\begin{aligned}\int_{\mathbb{R}^n} |x|^{-m}\mathrm{e}^{-\frac{|x|^2}{2}}\,\mathrm{d}x &= \omega_n \int_0^\infty r^{n-m-1}\mathrm{e}^{-\frac{r^2}{2}}\,\mathrm{d}r \\ &= 2^{\frac{n-m-2}{2}}\omega_n \int_0^\infty t^{\frac{n-m}{2}-1}\mathrm{e}^{-t}\,\mathrm{d}t = 2^{\frac{n-m-2}{2}}\omega_n \Gamma\left(\frac{n-m}{2}\right).\end{aligned}$$

Using this, the right-hand side becomes

$$C_{n,m}\langle |\xi|^{m-n}, \mathrm{e}^{-\frac{|\xi|^2}{2}} \rangle = C_{n,m} 2^{\frac{n-(n-m)-2}{2}} \omega_n \Gamma\left(\frac{m}{2}\right).$$

Therefore,

$$C_{n,m} 2^{\frac{m-2}{2}} \omega_n \Gamma\left(\frac{m}{2}\right) = 2^{\frac{n-m-2}{2}} \omega_n \Gamma\left(\frac{n-m}{2}\right),$$

which gives us

$$C_{n,m} = 2^{\frac{n}{2}-m}\frac{\Gamma\left(\frac{n-m}{2}\right)}{\Gamma\left(\frac{m}{2}\right)}.$$

Finally, we have

$$\widehat{|x|^{-m}} = 2^{\frac{n}{2}-m}\frac{\Gamma\left(\frac{n-m}{2}\right)}{\Gamma\left(\frac{m}{2}\right)} \cdot |\xi|^{m-n}. \tag{21.3}$$

Definition 21.10. The *Hilbert transform* Hf of $f \in S$ is defined by

$$Hf := \frac{1}{\pi}\left(\mathrm{p.v.}\frac{1}{x} * f\right),$$

i.e.,

$$Hf(x) = \frac{1}{\pi}\lim_{\varepsilon\to 0+}\int_{|x-t|\geq\varepsilon} \frac{f(t)\mathrm{d}t}{x-t}, \quad x \in \mathbb{R}.$$

Exercise 21.6. Prove that

(1) $\|Hf\|_{L^2(\mathbb{R})} = \|f\|_{L^2(\mathbb{R})}$.
(2) Hilbert transform has an extension to functions from $L^2(\mathbb{R})$.
(3) $H^2 = -I$, i.e., $H^{-1} = -H$.
(4) $(Hf_1, Hf_2)_{L^2} = (f_1, f_2)_{L^2}$ for $f_1 \in L^p$ and $f_2 \in L^{p'}$, where $\frac{1}{p} + \frac{1}{p'} = 1, 1 < p < \infty$.
(5) $H : L^p(\mathbb{R}) \to L^p(\mathbb{R})$, $1 < p < \infty$, i.e.,

$$\left\| \frac{1}{\pi} \int_{|x-t| \geq \varepsilon} \frac{f(t)\mathrm{d}t}{x-t} \right\|_{L^p} \leq c \|f\|_{L^p}$$

for ell $\varepsilon > 0$, where c does not depend on ε.

The multidimensional analogue of the Hilbert transform is developed in the following definition.

Definition 21.11. The functions

$$R_j(x) := \frac{x_j}{|x|^{n+1}}, \quad x \neq 0, \quad j = 1, 2, \ldots, n,$$

are called the *Riesz kernels*.

Remark 21.12. We can rewrite $R_j(x)$ in the form $R_j(x) = |x|^{-n}\omega_j(x)$, where $\omega_j(x) = \frac{x_j}{|x|}$ and conclude that

(1) $\int_{\mathbb{S}^{n-1}} \omega_j(\theta)\mathrm{d}\theta = 0$;
(2) $R_j(\lambda x) = \lambda^{-n} R_j(x), \quad \lambda > 0$.

These properties imply that we may define the *Riesz transform* by

$$R_j * f = \mathrm{p.v.} R_j * f,$$

because in our previous notation, $R_j(x) = T_{-n} \in H^*_{-n}(\mathbb{R}^n)$ is a homogeneous distribution. Let us calculate the Fourier transform of the Riesz kernels. By homogeneity, it suffices to consider $|\xi| = 1$. We have

$$\begin{aligned}\widehat{R_j}(\xi) = \widehat{\mathrm{p.v.} R_j}(\xi) &= (2\pi)^{-\frac{n}{2}} \int_{\mathbb{R}^n} \frac{\mathrm{e}^{-\mathrm{i}(x,\xi)} x_j}{|x|^{n+1}} \mathrm{d}x \\ &= \lim_{\substack{\varepsilon \to 0+ \\ \mu \to +\infty}} (2\pi)^{-\frac{n}{2}} \int_{\varepsilon < |x| < \mu} \frac{\mathrm{e}^{-\mathrm{i}(x,\xi)} x_j}{|x|^{n+1}} \mathrm{d}x.\end{aligned}$$

We split

$$\int_{\varepsilon < |x| < \mu} \frac{\mathrm{e}^{-\mathrm{i}(x,\xi)} x_j}{|x|^{n+1}} \mathrm{d}x = \int_{\varepsilon < |x| < 1} \frac{\mathrm{e}^{-\mathrm{i}(x,\xi)} x_j}{|x|^{n+1}} \mathrm{d}x + \int_{1 < |x| < \mu} \frac{\mathrm{e}^{-\mathrm{i}(x,\xi)} x_j}{|x|^{n+1}} \mathrm{d}x =: I_1 + I_2.$$

For I_1 we will use integration by parts:

$$\begin{aligned} I_1 &= \frac{1}{1-n}\int_{\varepsilon<|x|<1} \mathrm{e}^{-\mathrm{i}(x,\xi)} \frac{\partial}{\partial x_j}(|x|^{1-n})\mathrm{d}x \\ &= c_n \mathrm{i}\xi_j \int_{\varepsilon<|x|<1} \frac{\mathrm{e}^{-\mathrm{i}(x,\xi)}}{|x|^{n-1}}\mathrm{d}x + \int_{|x|=1} \frac{\mathrm{e}^{-\mathrm{i}(x,\xi)}x_j}{|x|^n}\mathrm{d}\sigma - \int_{|x|=\varepsilon} \frac{\mathrm{e}^{-\mathrm{i}(x,\xi)}x_j}{|x|^n}\mathrm{d}\sigma \\ &\to c_n \mathrm{i}\xi_j \int_{|x|<1} \frac{\mathrm{e}^{-\mathrm{i}(x,\xi)}}{|x|^{n-1}}\mathrm{d}x + \int_{|x|=1} \mathrm{e}^{-\mathrm{i}(x,\xi)} x_j \mathrm{d}\sigma - 0, \quad \varepsilon \to 0+. \end{aligned}$$

But

$$\begin{aligned} \int_{|x|=1} x_j \mathrm{e}^{-\mathrm{i}(x,\xi)}\mathrm{d}\sigma &= \mathrm{i}\frac{\partial}{\partial \xi_j}\int_{|x|=1} \mathrm{e}^{-\mathrm{i}(x,\xi)}\mathrm{d}\sigma = \mathrm{i}\frac{\partial}{\partial \xi_j}\int_{|x|=1} \cos(|\xi|\cdot x_1)\mathrm{d}\sigma \\ &= -\frac{\mathrm{i}\xi_j}{|\xi|}\int_{|x|=1} x_1 \cdot \sin(|\xi|\cdot x_1)\mathrm{d}\sigma = -\mathrm{i}\xi_j \cdot C_1, \quad |\xi| = 1, \end{aligned}$$

where we have used the fact that a rotation maps ξ to $(|\xi|, 0, \ldots, 0)$. Similarly, we may conclude that

$$\int_{|x|<1} \frac{\mathrm{e}^{-\mathrm{i}(x,\xi)}}{|x|^{n-1}}\mathrm{d}x = \int_{|x|<1} \cos(|\xi|x_1)|x|^{1-n}\mathrm{d}x = C_2, \quad |\xi| = 1.$$

If we collect all of these things, we obtain

$$I_1 \to C_n \mathrm{i}\xi_j, \quad |\xi| = 1$$

as $\varepsilon \to 0+$. For I_2 we will use the following technique:

$$\begin{aligned} I_2 &\to \int_{|x|>1} \frac{\mathrm{e}^{-\mathrm{i}(x,\xi)}x_j}{|x|^{n+1}}\mathrm{d}x = \mathrm{i}\frac{\partial}{\partial \xi_j}\int_{|x|>1} \frac{\mathrm{e}^{-\mathrm{i}(x,\xi)}}{|x|^{n+1}}\mathrm{d}x \\ &= \mathrm{i}\frac{\partial}{\partial \xi_j}\int_{|x|>1} |x|^{-n-1}\cos(|\xi|\cdot x_1)\mathrm{d}x \\ &= -\frac{\mathrm{i}\xi_j}{|\xi|}\int_{|x|>1} \frac{x_1 \sin(|\xi|x_1)}{|x|^{n+1}}\mathrm{d}x = -\mathrm{i}\xi_j \cdot \mathrm{const}, \quad |\xi| = 1, \quad \mu \to +\infty. \end{aligned}$$

Exercise 21.7. Prove the convergence of the last integral.

Collecting these integrals, we obtain that

$$\widehat{\mathrm{p.v.}R_j} = \mathrm{i}\xi_j \cdot C_n$$

for $|\xi| = 1$. But we know from Exercise 21.3 that $\widehat{\text{p. v.} R_j} \in H_0^*(\mathbb{R}^n)$. We conclude that $\widehat{\text{p. v.} R_j}(\xi) = \mathrm{i}C_n \frac{\xi_j}{|\xi|}$. Moreover, we have

$$\widehat{R_j * f} = (2\pi)^{\frac{n}{2}} \widehat{R_j} \cdot \widehat{f} = \mathrm{i}C_n' \frac{\xi_j}{|\xi|} \widehat{f},$$

or

$$R_j * f = \mathrm{i}C_n' \mathscr{F}^{-1} \left(\frac{\xi_j}{|\xi|} \widehat{f} \right).$$

It is easy to see that

$$C_n' = -\frac{\sqrt{\pi}}{2^{n/2} \Gamma((n+1)/2)}.$$

Corollary 21.13. *It is true that*

$$\sum_{j=1}^{n} R_j * R_j * = -\widetilde{C_n} \delta * .$$

Proof. By the above results we have

$$\sum_{j=1}^{n} \mathscr{F}(R_j * R_j * f) = \sum_{j=1}^{n} \mathrm{i}C_n' \frac{\xi_j}{|\xi|} \cdot \mathrm{i}C_n' \frac{\xi_j}{|\xi|} \widehat{f}(\xi) = -(C_n')^2 \widehat{f}(\xi).$$

Taking the inverse Fourier transform, we obtain the claim. □

Remark 21.14. By Parseval's equality we have

$$\|R_j * f\|_{L^2} = \left\| \widehat{R_j * f} \right\|_{L^2} = C \left\| \frac{\xi_j}{|\xi|} \widehat{f} \right\|_{L^2} \le C \left\| \widehat{f} \right\|_{L^2} = C \|f\|_{L^2},$$

i.e.,

$$R_j * : L^2(\mathbb{R}^n) \to L^2(\mathbb{R}^n),$$

and it follows from Calderón–Zigmund theory that

$$R_j * : L^p(\mathbb{R}^n) \to L^p(\mathbb{R}^n), \quad 1 < p < \infty.$$

Let us now introduce the *Riesz potential* by

$$I^{-1} f := \mathscr{F}^{-1} \left(\frac{1}{|\xi|} \widehat{f}(\xi) \right) = (2\pi)^{-\frac{n}{2}} \mathscr{F}^{-1} \left(\frac{1}{|\xi|} \right) * f = I_1 * f,$$

where by (21.3),

$$I_1(x) = c_n \frac{1}{|x|^{n-1}}.$$

Therefore, we have

$$I^{-1}f(x) = c_n \int_{\mathbb{R}^n} \frac{f(y)\mathrm{d}y}{|x-y|^{n-1}},$$

where

$$c_n = \frac{1}{2}\frac{\Gamma((n-1)/2)}{\pi^{(n+1)/2}}.$$

It is straightforward to verify that $\frac{\partial}{\partial x_j} I_1 = c_n' R_j$ and hence

$$\frac{\partial}{\partial x_j} I^{-1} f = c_n' R_j * f.$$

We would like to prove that

$$I^{-1} : L^s_\sigma(\mathbb{R}^n) \to W^1_s(\mathbb{R}^n)$$

for some s and σ. Since R_j* is a bounded map from $L^2(\mathbb{R}^n)$ to $L^2(\mathbb{R}^n)$, we may conclude that

$$\frac{\partial}{\partial x_j} I^{-1} : L^2(\mathbb{R}^n) \to L^2(\mathbb{R}^n). \tag{21.4}$$

Now let us assume for simplicity that $n \geq 3$. Let us try to prove that

$$I^{-1} : L^2_\sigma(\mathbb{R}^n) \to L^2(\mathbb{R}^n). \tag{21.5}$$

Indeed, for $f \in L^2(\mathbb{R}^n)$,

$$I^{-1} f \in L^2(\mathbb{R}^n)$$

if and only if

$$\frac{1}{|\xi|}\widehat{f} \in L^2(\mathbb{R}^n).$$

Let us assume now that $\sigma > 1$.

Lemma 20.17 implies that $L^2_\sigma(\mathbb{R}^n) \subset L^r(\mathbb{R}^n)$ for all $1 \leq r < 2$ and $\sigma > n\left(\frac{1}{r} - \frac{1}{2}\right)$. But for $\sigma > 1$ we may find appropriate r such that $r < \frac{2n}{n+2}$. We conclude that for a function $f \in L^2_\sigma(\mathbb{R}^n)$ with $\sigma > 1$ it follows from the Hausdorff–Young inequality that $\widehat{f} \in L^{r'}(\mathbb{R}^n)$ for some $r' > \frac{2n}{n-2}$ or $|\widehat{f}|^2 \in L^{\frac{r'}{2}}(\mathbb{R}^n)$. This fact implies that for $|\xi| < 1$ we have

$$|\xi|^{-1}\widehat{f}(\xi) \in L^2_{\mathrm{loc}}.$$

Indeed,

$$\int_{|\xi|<1} |\xi|^{-2}|\widehat{f}(\xi)|^2 \mathrm{d}\xi \leq \left(\int_{|\xi|<1} |\widehat{f}(\xi)|^{r'} \mathrm{d}\xi\right)^{2/r'} \left(\int_{|\xi|<1} |\xi|^{-2\left(\frac{r'}{2}\right)'} \mathrm{d}\xi\right)^{\frac{1}{\left(\frac{r'}{2}\right)'}} < \infty,$$

since $\frac{r'}{2} > \frac{n}{n-2}$ and $\left(\frac{r'}{2}\right)' < \frac{n}{2}$. For $|\xi| > 1$ the function $|\xi|^{-1}\widehat{f}(\xi)$ belongs to $L^2(\mathbb{R}^n)$. This fact follows from the inequality $|\xi|^{-1}|\widehat{f}(\xi)| < |\widehat{f}(\xi)|$ and from the positivity of σ (see Lemma 20.17). This proves (21.5) for $\sigma > 1$.

If we combine (21.4) and (21.5), we obtain that

$$I^{-1} : L^2_\sigma(\mathbb{R}^n) \to W^1_2(\mathbb{R}^n), \quad \sigma > 1.$$

Let us consider now $L^\infty_\sigma(\mathbb{R}^n)$ for $\sigma > 1$. If $f \in L^\infty_\sigma(\mathbb{R}^n)$, then $|f(x)| \le C(1+|x|)^{-\sigma}$ and thus

$$|I^{-1}f(x)| \le C\int_{\mathbb{R}^n} \frac{(1+|y|)^{-\sigma}dy}{|x-y|^{n-1}} < \infty.$$

This means that

$$I^{-1} : L^\infty_\sigma(\mathbb{R}^n) \to L^\infty(\mathbb{R}^n).$$

Interpolating this with (21.5), we can obtain the following result:

$$I^{-1} : L^s_\sigma(\mathbb{R}^n) \to L^s(\mathbb{R}^n), \quad 2 \le s \le \infty, \quad \sigma > 1.$$

If we recall the fact that $R_j* : L^s(\mathbb{R}^n) \to L^s(\mathbb{R}^n)$ for all $1 < s < \infty$, then we have

$$I^{-1} : L^s_\sigma(\mathbb{R}^n) \to W^1_s(\mathbb{R}^n), \qquad 2 \le s < \infty, \quad \sigma > 1.$$

Chapter 22
Fundamental Solution of the Helmholtz Operator

Let us consider a linear partial differential operator of order m in the form

$$L(x,D) = \sum_{|\alpha|\le m} a_\alpha(x)D^\alpha, \quad x\in\mathbb{R}^n,$$

where $\alpha = (\alpha_1,\ldots,\alpha_n)$ is a multi-index, $D^\alpha = D_1^{\alpha_1}\cdots D_n^{\alpha_n}$, and $D_j = \frac{1}{\mathrm{i}}\frac{\partial}{\partial x_j}$.

In this chapter, Ω is a bounded domain in $\mathbb{R}^n$, or $\Omega = \mathbb{R}^n$.

Definition 22.1. A *fundamental solution* for L in Ω is a distribution E in x that satisfies

$$L_xE(x|y) = \delta(x-y)$$

in $\mathscr{D}'(\Omega)$ with parameter $y\in\Omega$, i.e., $\langle L_xE,\varphi\rangle = \varphi(y)$ for $\varphi\in C_0^\infty(\Omega)$.

We understand that $\langle LE,\varphi\rangle$ is defined in distributional form

$$\langle LE,\varphi\rangle = \langle E,L'\varphi\rangle,$$

where L' is the formal adjoint operator of L given by

$$L'f = \sum_{|\alpha|\le m}(-1)^{|\alpha|}D^\alpha(a_\alpha(x)f(x)).$$

Here, $L'\varphi$ must be in $\mathscr{D}(\Omega)$ for φ from $\mathscr{D}(\Omega)$. This will be the case, for example, for $a_\alpha(x)\in C^\infty(\Omega)$.

Two fundamental solutions for L with the same parameter y differ by a solution of the homogeneous equation $Lu = 0$. Unless boundary conditions are imposed, the homogeneous equation will have many solutions, and the fundamental solution will not be uniquely determined. In most problems there are grounds of symmetry or causality for selecting the particular fundamental solution for the appropriate physical behavior.

V. Serov, *Fourier Series, Fourier Transform and Their Applications to Mathematical Physics*, Applied Mathematical Sciences 197,
DOI 10.1007/978-3-319-65262-7_22

We also observe that if L has constant coefficients, we can find the fundamental solution in the form $E(x|y) = E(x-y|0) := E(x-y)$. This fact follows from the properties of the Fourier transform:

$$\widehat{L_x E(x-y)} = \sum_{|\alpha|\le m} a_\alpha \xi^\alpha \widehat{E(x-y)} = \sum_{|\alpha|\le m} a_\alpha \xi^\alpha \mathrm{e}^{-\mathrm{i}(\xi,y)} \widehat{E(x)}$$
$$= \mathrm{e}^{-\mathrm{i}(\xi,y)} \widehat{\delta(x)} = \widehat{\delta(x-y)},$$

i.e.,

$$L_x E(x-y) = \delta(x-y).$$

Exercise 22.1. Let L be a differential operator with constant coefficients. Prove that $u = q * E = E * q$ solves the inhomogeneous equation

$$Lu = q$$

in D'.

Remark 22.2. In many cases the fundamental solution is a function. We can therefore write u as an integral

$$u(x) = \int_\Omega E(x-y)q(y)\mathrm{d}y.$$

Remark 22.3. In order for the convolution product $E * q$ (or $q * E$) to be well defined, we have to assume that, for example, q vanishes outside a finite sphere.

Remark 22.4. If L does not have constant coefficients, we can no longer appeal to convolution products; instead, one can often show that

$$u(x) = \int_\Omega E(x|y)q(y)\mathrm{d}y.$$

Definition 22.5. We denote by $a_0(x,\xi)$ the *main* (or *principal*) *symbol* of $L(x,D)$

$$a_0(x,\xi) = \sum_{|\alpha|=m} a_\alpha(x)\xi^\alpha, \quad \xi \in \mathbb{R}^n.$$

Assume that the $a_\alpha(x)$ are "smooth." An operator $L(x,D)$ is said to be *elliptic* in Ω if for every $x \in \Omega$ and $\xi \in \mathbb{R}^n \setminus \{0\}$ it follows that

$$a_0(x,\xi) \neq 0.$$

Exercise 22.2. Let $a_\alpha(x)$ be real for $|\alpha| = m$. Prove that the previous definition is equivalent to

(1) m is even,
(2) $a_0(x,\xi) \geq C_K|\xi|^m$ (or $-a_0(x,\xi) \geq C_K|\xi|^m$), $C_K > 0$, for every compact set $K \subset \Omega$ and for all $\xi \in \mathbb{R}^n$ and $x \in K$.

Let us consider the heat equation

$$\begin{cases} \dfrac{\partial u}{\partial t} = \Delta u, & t > 0, x \in \mathbb{R}^n, \\ u(x,0) = f(x), & x \in \mathbb{R}^n \end{cases}$$

in $S'(\mathbb{R}^n)$. Take the Fourier transform with respect to x to obtain

$$\begin{cases} \dfrac{\partial}{\partial t}\widehat{u}(\xi,t) = -\xi^2 \widehat{u}(\xi,t), & t > 0, \\ \widehat{u}(\xi,0) = \widehat{f}(\xi). \end{cases}$$

This initial value problem for an ordinary differential equation has the solution

$$\widehat{u}(\xi,t) = \mathrm{e}^{-t|\xi|^2}\widehat{f}(\xi).$$

Hence

$$u(x,t) = \mathscr{F}^{-1}(\mathrm{e}^{-t|\xi|^2}\widehat{f}(\xi)) = (2\pi)^{-\frac{n}{2}}\mathscr{F}^{-1}(\mathrm{e}^{-t|\xi|^2}) * f = P(\cdot,t) * f,$$

where

$$P(x,t) = (2\pi)^{-n}\int_{\mathbb{R}^n} \mathrm{e}^{-t|\xi|^2}\mathrm{e}^{\mathrm{i}(x,\xi)}\mathrm{d}\xi = \frac{1}{(4\pi t)^{\frac{n}{2}}}\mathrm{e}^{-\frac{|x|^2}{4t}}.$$

This formula implies that

$$u(x,t) = \frac{1}{(4\pi t)^{\frac{n}{2}}}\int_{\mathbb{R}^n} \mathrm{e}^{-\frac{|x-y|^2}{4t}} f(y)\mathrm{d}y.$$

Definition 22.6. The function $P(x,t)$ is the fundamental solution of the heat equation and satisfies

$$\begin{cases} \left(\dfrac{\partial}{\partial t} - \Delta\right) P(x,t) = 0, & t > 0, \\ \lim\limits_{t\to 0+} P(x,t) \overset{S'}{=} \delta(x). \end{cases}$$

It is also called the *Gaussian kernel* or *heat kernel*.

We can generalize this situation as follows. Let us consider an elliptic differential operator

$$L(D) = \sum_{|\alpha|\leq m} a_\alpha D^\alpha$$

with constant coefficients. Assume that $L(\xi)=\sum_{|\alpha|\le m}a_\alpha\xi^\alpha>0$ for all $\xi\in\mathbb{R}^n\setminus\{0\}$. If we consider $P_L(x,t)$ as a solution of

$$\begin{cases}\left(\frac{\partial}{\partial t}+L(D)\right)P_L(x,t)=0, & t>0,\\ \lim_{t\to 0+}P_L(x,t)\overset{S'}{=}\delta(x),\end{cases}$$

then $P_L(x,t)$ is the fundamental solution of $\frac{\partial}{\partial t}+L(D)$ and can be calculated by

$$P_L(x,t)=(2\pi)^{-n}\int_{\mathbb{R}^n}\mathrm{e}^{-tL(\xi)}\mathrm{e}^{\mathrm{i}(x,\xi)}\mathrm{d}\xi.$$

Lemma 22.7. *Let $P_L(x,t)$ be as above. Then the function*

$$F(x,\lambda)\overset{S'}{:=}\lim_{\varepsilon\to 0+}\int_\varepsilon^\infty \mathrm{e}^{-\lambda t}P_L(x,t)\mathrm{d}t$$

is a fundamental solution of the operator $L(D)+\lambda I$, $\lambda>0$.

Proof. By the definitions of F and P_L we have

$$\langle F(x,\lambda),\varphi\rangle=\lim_{\varepsilon\to 0+}\langle\int_\varepsilon^\infty \mathrm{e}^{-\lambda t}P_L(x,t)\mathrm{d}t,\varphi\rangle=\lim_{\varepsilon\to 0+}\int_\varepsilon^\infty \mathrm{e}^{-\lambda t}\langle P_L,\varphi\rangle\mathrm{d}t.$$

Therefore,

$$\begin{aligned}\langle (L(D)+\lambda)F,\varphi\rangle &=\lim_{\varepsilon\to 0+}\int_\varepsilon^\infty \mathrm{e}^{-\lambda t}\langle (L(D)+\lambda)P_L,\varphi\rangle\mathrm{d}t\\ &=\lim_{\varepsilon\to 0+}\int_\varepsilon^\infty \mathrm{e}^{-\lambda t}\langle L(D)P_L,\varphi\rangle\mathrm{d}t+\lambda\int_0^\infty \mathrm{e}^{-\lambda t}\langle P_L,\varphi\rangle\mathrm{d}t\\ &=\lim_{\varepsilon\to 0+}\int_\varepsilon^\infty \mathrm{e}^{-\lambda t}\langle -\frac{\partial}{\partial t}P_L,\varphi\rangle\mathrm{d}t+\lambda\langle F,\varphi\rangle\\ &=\lim_{\varepsilon\to 0+}\left[-\mathrm{e}^{-\lambda t}\langle P_L,\varphi\rangle|_\varepsilon^\infty-\lambda\int_\varepsilon^\infty \mathrm{e}^{-\lambda t}\langle P_L,\varphi\rangle\mathrm{d}t\right]+\lambda\langle F,\varphi\rangle\\ &=\lim_{\varepsilon\to 0+}\mathrm{e}^{-\lambda\varepsilon}\langle P_L(\cdot,\varepsilon),\varphi\rangle=\langle\delta,\varphi\rangle\end{aligned}$$

for all $\varphi\in S$. □

Exercise 22.3. Let us define a fundamental solution $\Gamma(x,t)$ of $\frac{\partial}{\partial t}+L(D)$ as a solution of

$$\begin{cases}(\frac{\partial}{\partial t}+L)\Gamma(x,t)=\delta(x)\delta(t),\\ \Gamma(x,0)=0.\end{cases}$$

Prove that

$$F(x,\lambda) := \int_0^\infty \mathrm{e}^{-\lambda t}\Gamma(x,t)\mathrm{d}t$$

is a fundamental solution of the operator $L(D) + \lambda I, \lambda > 0$.

Example 22.8. Let us consider $L(D) = \sum_{j=1}^n \left(\frac{1}{\mathrm{i}}\frac{\partial}{\partial x_j}\right)^2 = -\Delta$. Then $L(\xi) = |\xi|^2$, and the fundamental solution $F(x,\lambda)$ of the operator $L(D)+\lambda = -\Delta+\lambda$ has the form

$$\begin{aligned} F(x,\lambda) &= \int_0^\infty \frac{1}{(4\pi t)^{\frac{n}{2}}}\mathrm{e}^{-\lambda t}\cdot \mathrm{e}^{-\frac{x^2}{4t}}\mathrm{d}t = \frac{1}{(4\pi)^{\frac{n}{2}}}\int_0^\infty \mathrm{e}^{-\lambda t-\frac{x^2}{4t}}\cdot t^{-\frac{n}{2}}\mathrm{d}t \\ &= \frac{1}{(4\pi)^{\frac{n}{2}}}\lambda^{\frac{n}{2}-1}\int_0^\infty \mathrm{e}^{-\tau-\frac{(\sqrt{\lambda}|x|)^2}{4\tau}}\tau^{-\frac{n}{2}}\mathrm{d}\tau = \frac{1}{(4\pi)^{\frac{n}{2}}}\lambda^{\frac{n}{2}-1}\int_0^\infty \mathrm{e}^{-\tau-\frac{r^2}{4\tau}}\tau^{-\frac{n}{2}}\mathrm{d}\tau, \end{aligned}$$

where $r = \sqrt{\lambda}|x|$. From our previous considerations we know that

$$F(x,\lambda) = (2\pi)^{-n/2}\mathscr{F}^{-1}\left(\frac{1}{|\xi|^2+\lambda}\right)(x),$$

where $\mathscr{F}^{-1}$ is the inverse Fourier transform. The function

$$K_\nu(r) = \frac{1}{2}\left(\frac{r}{2}\right)^\nu \int_0^\infty \mathrm{e}^{-t-\frac{r^2}{4t}}\cdot t^{-1-\nu}\mathrm{d}t$$

is called the Macdonald function of order ν. So we have

$$F(x,\lambda) = (2\pi)^{-\frac{n}{2}}\left(\frac{|x|}{\sqrt{\lambda}}\right)^{1-\frac{n}{2}} K_{\frac{n}{2}-1}(\sqrt{\lambda}|x|).$$

It is known that

$$K_\nu(r) = \frac{\pi \mathrm{i}}{2}\mathrm{e}^{\mathrm{i}\pi\frac{\nu}{2}}H_\nu^{(1)}(\mathrm{i}r), \quad r>0,$$

where $H_\nu^{(1)}$ is the Hankel function of first kind of order ν.

Next we want to obtain estimates for $F(x,\lambda)$ for $x \in \mathbb{R}^n$, $\lambda > 0$, and $n \geq 1$. Let us consider the integral $\int_0^\infty \mathrm{e}^{-\tau-\frac{r^2}{4\tau}}\tau^{-\frac{n}{2}}\mathrm{d}\tau$ in two parts $I_1 + I_2 = \int_0^1 + \int_1^\infty$.

(1) If $0 < r < 1$, then

$$I_1 = \int_0^1 \mathrm{e}^{-y-\frac{r^2}{4y}}y^{-\frac{n}{2}}\mathrm{d}y \leq \int_0^1 \mathrm{e}^{-\frac{r^2}{4y}}y^{-\frac{n}{2}}\mathrm{d}y = c_n r^{2-n}\int_{\frac{r^2}{4}}^\infty \mathrm{e}^{-z}z^{\frac{n}{2}-2}\mathrm{d}z = c_n r^{2-n}I_1'.$$

Since

$$I_1' \sim c\begin{cases} r^{-1}, & n=1,\\ \log\frac{1}{r}, & n=2,\\ 1, & n\geq 3,\end{cases}$$

as $r\to 0+$, we have

$$|I_1| \leq c_n\begin{cases} 1, & n=1,\\ \log\frac{1}{r}, & n=2,\\ r^{2-n}, & n\geq 3.\end{cases}$$

For I_2 we can simply argue that

$$I_2 = \int_1^\infty \mathrm{e}^{-y-\frac{r^2}{4y}} y^{-\frac{n}{2}}\mathrm{d}y \leq \mathrm{e}^{-\frac{r^2}{4}}\int_1^\infty \mathrm{e}^{-y}\mathrm{d}y \leq \mathrm{e}^{-\frac{r^2}{4}} \leq 1, \quad r\to 0+.$$

(2) If $r>1$, then

$$I_1 \leq \int_0^1 \mathrm{e}^{-\frac{r^2}{4y}} y^{-\frac{n}{2}}\mathrm{d}y = c_n r^{2-n}\int_{\frac{r^2}{4}}^\infty \mathrm{e}^{-z} z^{\frac{n}{2}-2}\mathrm{d}z \leq c_n\begin{cases} r^{-2}\mathrm{e}^{-\frac{r^2}{4}}, & n=1,2,3,4\\ r^{2-n}\mathrm{e}^{-\delta r^2}, & n\geq 5,\end{cases}$$

where $0<\delta<\frac{1}{4}$. The last inequality follows from the fact that $z^{\frac{n}{2}-2} \leq c_\varepsilon \mathrm{e}^{\varepsilon z}$ for $\frac{n}{2}-2>0$ and all $\varepsilon>0$ $(z>1)$.

Since

$$I_2 \leq \int_1^\infty \mathrm{e}^{-y-\frac{r^2}{4y}}\mathrm{d}y,$$

we perform the change of variable $z := y+\frac{r^2}{4y}$. Then $z\geq r$ and $z\to+\infty$. Thus

$$\begin{aligned}\int_1^\infty \mathrm{e}^{-y-\frac{r^2}{4y}}\mathrm{d}y &= c\int_r^\infty \mathrm{e}^{-z}\left(1+\frac{z}{\sqrt{z^2-r^2}}\right)\mathrm{d}z\\ &= c\int_r^\infty \mathrm{e}^{-z}\mathrm{d}z + c\int_r^\infty \mathrm{e}^{-z}\frac{z\mathrm{d}z}{\sqrt{z^2-r^2}}\\ &= c\mathrm{e}^{-r} + c\left(\mathrm{e}^{-z}\sqrt{z^2-r^2}\Big|_r^\infty + \int_r^\infty \mathrm{e}^{-z}\sqrt{z^2-r^2}\mathrm{d}z\right)\\ &= c\left(\mathrm{e}^{-r} + \int_r^\infty \mathrm{e}^{-z}\sqrt{z^2-r^2}\mathrm{d}z\right) \leq c\mathrm{e}^{-\delta r}\end{aligned}$$

for all $0<\delta<1$.

If we collect all these estimates, we obtain the following:

(1) If $\sqrt{\lambda}|x| < 1$, then

$$|F(x,\lambda)| \le c_n \lambda^{\frac{n}{2}-1} \begin{cases} 1, & n=1, \\ \log \frac{1}{\sqrt{\lambda}|x|}, & n=2, \\ (\sqrt{\lambda}|x|)^{2-n}, & n \ge 3, \end{cases}$$

$$\le c_n' \lambda^{\frac{n}{2}-1} \mathrm{e}^{-\delta\sqrt{\lambda}|x|} \begin{cases} 1, & n=1, \\ \log \frac{1}{\sqrt{\lambda}|x|}, & n=2, \\ (\sqrt{\lambda}|x|)^{2-n}, & n \ge 3. \end{cases}$$

(2) If $\sqrt{\lambda}|x| > 1$, then

$$|F(x,\lambda)| \le c_n \mathrm{e}^{-\delta\sqrt{\lambda}|x|}, \quad n \ge 1.$$

We will rewrite these estimates in a more appropriate form for all $\lambda > 0$ and $x \in \mathbb{R}^n$ as

$$|F(x,\lambda)| \le c_n \mathrm{e}^{-\delta\sqrt{\lambda}|x|} \begin{cases} \frac{1}{\sqrt{\lambda}}, & n=1, \\ 1 + |\log \frac{1}{\sqrt{\lambda}|x|}|, & n=2, \\ |x|^{2-n}, & n \ge 3. \end{cases}$$

Remark 22.9. It is not too difficult to observe that $F(x,\lambda)$ is positive.

Example 22.10. Recall from Chapter 21 that the solution of the equation $(-1-\Delta)u = f$ can be written in the form

$$u(x) = K_{-1} * f = \mathscr{F}^{-1}\left(\frac{1}{|\xi|^2 - 1}\right) * f,$$

where

$$K_{-1}(|x|) = c_n |x|^{2-n} \lim_{\varepsilon \to 0+} \int_0^\infty \frac{\rho^{\frac{n}{2}} J_{\frac{n-2}{2}}(\rho)\mathrm{d}\rho}{\rho^2 - |x|^2 - \mathrm{i}\varepsilon}.$$

In fact, K_{-1} is a fundamental solution of the operator $-1-\Delta$. Let us consider the more general operator $-\Delta - \lambda$ for $\lambda > 0$ or even for $\lambda \in \mathbb{C}$. The operator $-\Delta - \lambda$ is called the *Helmholtz operator*. Its fundamental solution $E_n(x,\lambda)$ satisfies

$$-\Delta E_n - \lambda E_n = \delta(x).$$

We define $\sqrt{\lambda}$ with nonnegative imaginary part, i.e., $\sqrt{\lambda} = \alpha + \mathrm{i}\beta$, where $\beta \ge 0$ and $\beta = 0$ if and only if $\lambda \in [0,+\infty)$. We require that E_n is radially symmetric. Then for $x \ne 0$, E_n must solve the equation

$$(r^{n-1}u')' + \lambda r^{n-1}u = 0.$$

This equation can be reduced to one of Bessel type by making the substitution $u = wr^{1-\frac{n}{2}}$. A straightforward calculation shows that

$$(rw')' - \left(1 - \frac{n}{2}\right)^2 \frac{w}{r} + \lambda rw = 0,$$

or

$$w'' + \frac{w'}{r} + \left(\lambda - \left(1 - \frac{n}{2}\right)^2 \frac{1}{r^2}\right) w = 0,$$

or

$$v''(r\sqrt{\lambda}) + \frac{v'(r\sqrt{\lambda})}{r\sqrt{\lambda}} + \left(1 - \left(1 - \frac{n}{2}\right)^2 \frac{1}{\lambda r^2}\right) v(r\sqrt{\lambda}) = 0, \quad w(r) = v(r\sqrt{\lambda}).$$

This is the Bessel equation of order $\frac{n}{2} - 1$. Its two linearly independent solutions are the Bessel functions $J_{\frac{n}{2}-1}$ and $Y_{\frac{n}{2}-1}$ of the first and second kinds, respectively. Therefore the general solution is of the form

$$w(r) = c_0' J_{\frac{n}{2}-1}(\sqrt{\lambda}r) + c_1' Y_{\frac{n}{2}-1}(\sqrt{\lambda}r).$$

For us it is convenient to write it in terms of Hankel functions of the first and second kinds as

$$w(r) = c_0 H^{(1)}_{\frac{n}{2}-1}(\sqrt{\lambda}r) + c_1 H^{(2)}_{\frac{n}{2}-1}(\sqrt{\lambda}r),$$

where

$$H^{(1)}_{\nu}(z) = J_\nu(z) + \mathrm{i}Y_\nu(z), \quad H^{(2)}_{\nu}(z) = J_\nu(z) - \mathrm{i}Y_\nu(z).$$

The corresponding general solution u is

$$u(r) = r^{1-\frac{n}{2}} \left[c_0 H^{(1)}_{\frac{n}{2}-1}(\sqrt{\lambda}r) + c_1 H^{(2)}_{\frac{n}{2}-1}(\sqrt{\lambda}r) \right].$$

If $\lambda \notin [0, +\infty)$, then $\sqrt{\lambda}$ has positive imaginary part, and the solution $H^{(2)}_{\frac{n}{2}-1}(\sqrt{\lambda}r)$ is exponentially large at $z = +\infty$, whereas $H^{(1)}_{\frac{n}{2}-1}(\sqrt{\lambda}r)$ is exponentially small. Hence we take

$$E_n(x, \lambda) = c_0 r^{1-\frac{n}{2}} H^{(1)}_{\frac{n}{2}-1}(\sqrt{\lambda}r).$$

Exercise 22.4. Prove that

$$\lim_{\varepsilon \to 0+} \int_{|x|=\varepsilon} \frac{\partial E_n}{\partial r} \mathrm{d}\sigma(x) = 1$$

or

$$\lim_{r\to 0} r^{n-1}\omega_n \frac{\partial E_n}{\partial r} = 1,$$

where $\omega_n = |\mathbb{S}^{n-1}|$ is the area (measure) of the unit sphere $\mathbb{S}^{n-1}$.

For small values of r, we have the asymptotic expansions [23]

$$H^{(1)}_{\frac{n-2}{2}}(r) \sim -\frac{\mathrm{i}2^{\frac{n-2}{2}}\Gamma(\frac{n-2}{2})}{\pi} r^{-\frac{n-2}{2}}, \quad n \neq 2$$

and

$$H^{(1)}_0(r) \sim \frac{2\mathrm{i}}{\pi}\log r.$$

It can be proved using Exercise 22.4 that

$$c_0 = \frac{\mathrm{i}}{4}\left(\frac{\sqrt{\lambda}}{2\pi}\right)^{\frac{n-2}{2}}.$$

Thus for $n \geq 2$ and $\lambda \notin [0,+\infty)$ we obtain

$$E_n(x,\lambda) = \frac{\mathrm{i}}{4}\left(\frac{\sqrt{\lambda}}{2\pi|x|}\right)^{\frac{n-2}{2}} H^{(1)}_{\frac{n-2}{2}}(\sqrt{\lambda}|x|). \tag{22.1}$$

A direct calculation shows that for $n = 1$ we have

$$E_1(x,\lambda) = \frac{\mathrm{i}}{2\sqrt{\lambda}}\mathrm{e}^{i\sqrt{\lambda}|x|}$$

for all $\lambda \neq 0$. The formula (22.1) is valid also for $\lambda \in (0,+\infty)$. This fact follows from the definition:

$$E_n(x,\lambda) = \lim_{\varepsilon\to 0+} E_n(x,\lambda+\mathrm{i}\varepsilon) = \frac{\mathrm{i}}{4}\lim_{\varepsilon\to 0+}\left(\frac{\sqrt{\lambda+\mathrm{i}\varepsilon}}{2\pi|x|}\right)^{\frac{n-2}{2}} H^{(1)}_{\frac{n-2}{2}}(\sqrt{\lambda+\mathrm{i}\varepsilon}|x|)$$

$$= \frac{\mathrm{i}}{4}\left(\frac{\sqrt{\lambda}}{2\pi|x|}\right)^{\frac{n-2}{2}} H^{(1)}_{\frac{n-2}{2}}(\sqrt{\lambda}|x|).$$

Remark 22.11. We conclude that

$$(2\pi)^{-n/2}\mathscr{F}^{-1}\left(\frac{1}{|\xi|^2-\lambda-\mathrm{i}0}\right)=\frac{\mathrm{i}}{4}\left(\frac{\sqrt{\lambda}}{2\pi|x|}\right)^{\frac{n-2}{2}}H^{(1)}_{\frac{n-2}{2}}(\sqrt{\lambda}|x|),$$

for $\lambda>0$. A direct calculation shows that

$$E_n(x,0)=\begin{cases}-\frac{|x|}{2}, & n=1,\\ \frac{1}{2\pi}\log\frac{1}{|x|}, & n=2,\\ \frac{|x|^{2-n}}{(n-2)\omega_n}, & n\geq 3.\end{cases}$$

Chapter 23
Estimates for the Laplacian and Hamiltonian

Let us recall Agmon's $(2,2)$-estimate for the Laplacian [2]:

$$\|(-\Delta - k^2 - \mathrm{i}0)^{-1}\|_{L^2_\delta \to L^2_{-\delta}} \leq \frac{c}{|k|}, \tag{23.1}$$

where $(-\Delta - k^2 - \mathrm{i}0)^{-1}$ is an integral operator with kernel $E_n(x,k)$ from the previous chapter and $\delta > \frac{1}{2}$. In fact, this estimate allows us to consider the Hamiltonian with L^∞_{loc}-potentials only (if we want to preserve $(2,2)$-estimates). But we would like to consider the Hamiltonian with L^p_{loc}-potentials. We therefore need to prove (p,q)-estimates.

We proved in Example 18.26 that the limit $\lim\limits_{\varepsilon\to 0+} \frac{1}{x-\mathrm{i}\varepsilon} := \frac{1}{x-\mathrm{i}0}$ exists in the sense of tempered distributions and

$$\frac{1}{x-\mathrm{i}0} = \mathrm{p.\,v.}\frac{1}{x} + \mathrm{i}\pi\delta(x),$$

i.e.,

$$\langle \frac{1}{x-\mathrm{i}0}, \varphi\rangle = \lim_{\delta\to 0+} \int_{|x|>\delta} \frac{\varphi(x)}{x} \mathrm{d}x + \mathrm{i}\pi\varphi(0).$$

In Example 18.8 we considered the simple layer

$$\langle T, \varphi\rangle := \int_\sigma a(\xi)\varphi(\xi)\mathrm{d}\sigma_\xi,$$

where σ is a hypersurface of dimension $n-1$ in $\mathbb{R}^n$ and $a(\xi)$ is a density. These examples can be extended as follows. If $H : \mathbb{R}^n \to \mathbb{R}$ and $|\nabla H| \neq 0$ at every point where $H(\xi) = 0$, then we can define the distribution

$$(H(\xi) - \mathrm{i}0)^{-1} := \lim_{\varepsilon\to 0+} \frac{1}{H(\xi) - \mathrm{i}\varepsilon}$$

V. Serov, *Fourier Series, Fourier Transform and Their Applications to Mathematical Physics*, Applied Mathematical Sciences 197,
DOI 10.1007/978-3-319-65262-7_23

in $S'(\mathbb{R})$, and we can also prove that

$$(H(\xi)-\mathrm{i}0)^{-1}=\text{p. v.}\,\frac{1}{H(\xi)}+\mathrm{i}\pi\delta(H(\xi)=0),$$

where $\delta(H(\xi)=0)$ is defined as follows:

$$\langle\delta(H),\varphi\rangle=\int_{H(\xi)=0}\varphi(\xi)\mathrm{d}\sigma_\xi,\quad \varphi\in S(\mathbb{R}^n).$$

The equality $H(\xi)=0$ defines an $(n-1)$-dimensional hypersurface, and σ_ξ is any $(n-1)$-form such that $d\sigma_\xi\wedge\frac{dH}{|\nabla H|}=\mathrm{d}\xi$ (in local coordinates).

Exercise 23.1. Prove that

$$\delta(\alpha H)=\frac{1}{\alpha}\delta(H)$$

for every positive differentiable function α.

Due to Exercise 23.1 we may conclude that $\delta(H)=\frac{1}{|\nabla H|}\delta\left(\frac{H}{|\nabla H|}\right)$ if $|\nabla H|\neq 0$ for $H=0$.

Let us consider now $H(\xi):=-|\xi|^2+k^2, k>0$. Then $H(\xi)=0$ or $|\xi|=k$ is a sphere and $\nabla H(\xi)=-2\xi$ and $|\nabla H(\xi)|=2k$ at every point on this sphere. If we change variables, we then obtain

$$\langle\delta(H),\varphi\rangle=\int_{H(\xi)=0}\varphi(\xi)\mathrm{d}\sigma_\xi=\frac{1}{2k}\int_{\mathbb{S}^{n-1}}\varphi(k\theta)\mathrm{d}\theta.$$

We know that $(-\Delta-k^2-\mathrm{i}0)^{-1}f$ can be represented as

$$(-\Delta-k^2-\mathrm{i}0)^{-1}f=\int_{\mathbb{R}^n}G_k^+(|x-y|)f(y)\mathrm{d}y,$$

where $G_k^+(|x|)=\frac{\mathrm{i}}{4}\left(\frac{|k|}{2\pi|x|}\right)^{\frac{n-2}{2}}H^{(1)}_{\frac{n-2}{2}}(|k||x|)$. On the other hand, we can write

$$\begin{aligned}(-\Delta-k^2-\mathrm{i}0)^{-1}f&=\mathscr{F}^{-1}(\mathscr{F}[(-\Delta-k^2-\mathrm{i}0)^{-1}f])=(2\pi)^{-\frac{n}{2}}\int_{\mathbb{R}^n}\frac{\widehat{f}(\xi)\mathrm{e}^{\mathrm{i}(x,\xi)}\mathrm{d}\xi}{|\xi|^2-k^2-\mathrm{i}0}\\
&=(2\pi)^{-\frac{n}{2}}\int_{\mathbb{R}^n}\text{p. v.}\,\frac{1}{|\xi|^2-k^2}\widehat{f}(\xi)\mathrm{e}^{\mathrm{i}(x,\xi)}\mathrm{d}\xi+\frac{\mathrm{i}\pi(2\pi)^{-\frac{n}{2}}}{2k}\int_{\mathbb{R}^n}\delta(H)\widehat{f}(\xi)\mathrm{e}^{\mathrm{i}(x,\xi)}\mathrm{d}\xi\\
&=(2\pi)^{-\frac{n}{2}}\,\text{p. v.}\int_{\mathbb{R}^n}\frac{\widehat{f}(\xi)\mathrm{e}^{\mathrm{i}(x,\xi)}\mathrm{d}\xi}{|\xi|^2-k^2}+\frac{\mathrm{i}\pi}{2k(2\pi)^{\frac{n}{2}}}\int_{\mathbb{S}^{n-1}}\widehat{f}(k\theta)\mathrm{e}^{\mathrm{i}k(x,\theta)}\mathrm{d}\theta\\
&=(2\pi)^{-\frac{n}{2}}\,\text{p. v.}\int_{\mathbb{R}^n}\frac{\widehat{f}(\xi)\mathrm{e}^{\mathrm{i}(x,\xi)}\mathrm{d}\xi}{|\xi|^2-k^2}+\frac{\mathrm{i}\pi}{2k(2\pi)^n}\int_{\mathbb{R}^n}f(y)\mathrm{d}y\int_{\mathbb{S}^{n-1}}\mathrm{e}^{\mathrm{i}k(\theta,x-y)}\mathrm{d}\theta.\end{aligned}$$

Our aim is to prove the following result.

Theorem 23.1. *Let $k>0$ and $\frac{2}{n}\geq\frac{1}{p}-\frac{1}{p'}\geq\frac{2}{n+1}$ for $n\geq 3$ and $1>\frac{1}{p}-\frac{1}{p'}\geq\frac{2}{3}$ for $n=2$, where $\frac{1}{p}+\frac{1}{p'}=1$. Then there exists a constant C independent of k and f such that*

$$\|(-\Delta-k^2-\mathrm{i}0)^{-1}f\|_{L^{p'}(\mathbb{R}^n)}\leq Ck^{n\left(\frac{1}{p}-\frac{1}{p'}\right)-2}\|f\|_{L^p(\mathbb{R}^n)}.$$

Remark 23.2. In what follows we will use the notation $\widehat{G}_k$ instead of $(-\Delta-k^2-\mathrm{i}0)^{-1}$.

Proof. First we prove that if the claim holds for $k=1$, then it holds for every $k>0$. So let us assume that

$$\|\widehat{G}_1 f\|_{L^{p'}(\mathbb{R}^n)}\leq C\|f\|_{L^p(\mathbb{R}^n)}.$$

Set $T_\delta f:=f(\delta x),\delta>0$. It is clear that $\|T_\delta f\|_{L^p(\mathbb{R}^n)}=\delta^{-\frac{n}{p}}\|f\|_{L^p(\mathbb{R}^n)}$. It is not difficult to show that $\widehat{G}_k=k^{-2}T_k\widehat{G}_1T_{\frac{1}{k}}$. Indeed, since

$$\widehat{G}_k f=(2\pi)^{-n}\int_{\mathbb{R}^n}\int_{\mathbb{R}^n}\frac{\mathrm{e}^{\mathrm{i}(y,\xi)}f(x-y)\mathrm{d}\xi\mathrm{d}y}{|\xi|^2-k^2-\mathrm{i}0},$$

we get

$$\widehat{G}_1T_{\frac{1}{k}}f=(2\pi)^{-n}\int_{\mathbb{R}^n}\int_{\mathbb{R}^n}\frac{\mathrm{e}^{\mathrm{i}(y,\xi)}f(\frac{x-y}{k})\mathrm{d}\xi\mathrm{d}y}{|\xi|^2-1-\mathrm{i}0}.$$

It follows that

$$\begin{aligned}T_k\widehat{G}_1T_{\frac{1}{k}}f&=(2\pi)^{-n}\int_{\mathbb{R}^n}\int_{\mathbb{R}^n}\frac{\mathrm{e}^{\mathrm{i}(y,\xi)}f(x-\frac{y}{k})\mathrm{d}\xi\mathrm{d}y}{|\xi|^2-1-\mathrm{i}0}\\&=(2\pi)^{-n}\int_{\mathbb{R}^n}\int_{\mathbb{R}^n}\frac{\mathrm{e}^{\mathrm{i}(z,\eta)}k^{-n}f(x-z)k^n\mathrm{d}z\mathrm{d}\eta}{\frac{|\eta|^2}{k^2}-1-\mathrm{i}0}\\&=(2\pi)^{-n}k^2\int_{\mathbb{R}^n}\int_{\mathbb{R}^n}\frac{\mathrm{e}^{\mathrm{i}(z,\eta)}f(x-z)\mathrm{d}z\mathrm{d}\eta}{|\eta|^2-k^2-\mathrm{i}0}.\end{aligned}$$

This proves that

$$k^2\widehat{G}_kf\equiv T_k\widehat{G}_1T_{\frac{1}{k}}f,$$

which we use to get

$$\begin{aligned}\|\widehat{G}_kf\|_{L^{p'}}&=k^{-2}\|T_k\widehat{G}_1T_{\frac{1}{k}}f\|_{L^{p'}}=k^{-2}k^{-\frac{n}{p'}}\|\widehat{G}_1T_{\frac{1}{k}}f\|_{L^{p'}}\\&\leq Ck^{-2-\frac{n}{p'}}\|T_{\frac{1}{k}}f\|_{L^p}=Ck^{-2-\frac{n}{p'}}\left(\frac{1}{k}\right)^{-\frac{n}{p}}\|f\|_{L^p}=Ck^{n\left(\frac{1}{p}-\frac{1}{p'}\right)-2}\|f\|_{L^p}.\end{aligned}$$

It therefore suffices to prove this theorem for $k = 1$.

The rest of the proof makes use of the following lemmas.

Lemma 23.3. *Let $\omega(x) \in S(\mathbb{R}^n)$, $0 < \varepsilon < 1$, and $\sigma_\varepsilon \omega(\xi) = \varepsilon^{-n} \omega\left(\frac{\xi}{\varepsilon}\right)$. Let us set*

$$P_\varepsilon(\xi) := \text{p.v.}\left(\frac{1}{|\eta|^2 - 1} * \sigma_\varepsilon \omega\right)(\xi).$$

Then

$$|P_\varepsilon(\xi)| \le \frac{c}{\varepsilon}.$$

Proof. For P_ε we have the following representation:

$$P_\varepsilon = \text{p.v.}\left(\int_{1-\varepsilon \le |\eta| \le 1+\varepsilon} + \int_{|\eta| < 1-\varepsilon} + \int_{|\eta| > 1+\varepsilon}\right) \frac{\sigma_\varepsilon \omega(\xi - \eta)}{|\eta|^2 - 1} \mathrm{d}\eta =: I_1 + I_2 + I_3.$$

The integrals I_2 and I_3 can be easily bounded by $\varepsilon^{-1}\|\omega\|_{L^1}$, because $|\eta| < 1 - \varepsilon$ implies that $\left|\frac{1}{|\eta|^2-1}\right| = \frac{1}{1-|\eta|^2} < \frac{1}{\varepsilon}$ and $|\eta| > 1 + \varepsilon$ implies that $\left|\frac{1}{|\eta|^2-1}\right| = \frac{1}{|\eta|^2-1} < \frac{1}{\varepsilon}$. By the definition of p. v. we have

$$\begin{aligned} I_1 &= \lim_{\delta \to 0+} \int_{\delta < |1-|\eta|| < \varepsilon} \frac{\sigma_\varepsilon \omega(\xi - \eta)}{|\eta|^2 - 1} \mathrm{d}\eta \\ &= \lim_{\delta \to 0+} \left(\int_{1-\varepsilon}^{1-\delta} + \int_{1+\delta}^{1+\varepsilon}\right) \int_{\mathbb{S}^{n-1}} \sigma_\varepsilon \omega(\xi - r\theta) \frac{r^{n-1}}{r^2 - 1} \mathrm{d}\theta \mathrm{d}r. \end{aligned}$$

Replacing r with $2 - r$ in the latter integral, we obtain

$$I_1 = \lim_{\delta \to 0+} \int_{1-\varepsilon}^{1-\delta} \frac{F(r,\xi)}{r - 1} \mathrm{d}r,$$

where

$$F(r,\xi) = \int_{\mathbb{S}^{n-1}} \left[\sigma_\varepsilon \omega(\xi - r\theta) \frac{r^{n-1}}{r+1} - \sigma_\varepsilon \omega(\xi - (2-r)\theta) \frac{(2-r)^{n-1}}{3-r}\right] \mathrm{d}\theta.$$

If we observe that $F(1,\xi) = 0$, then we get by the mean value theorem (Lagrange formulas) that

$$\begin{aligned} \left|\int_{1-\varepsilon}^{1-\delta} \frac{F(r,\xi)}{r-1} \mathrm{d}r\right| &= \left|\int_{1-\varepsilon}^{1-\delta} \frac{F(r,\xi) - F(1,\xi)}{r-1} \mathrm{d}r\right| \le (\varepsilon - \delta) \sup_{1-\varepsilon < r < 1} \left|\frac{\partial F}{\partial r}(r,\xi)\right| \\ &\le \varepsilon \sup_{1-\varepsilon < r < 1} \left|\frac{\partial F}{\partial r}\right|. \end{aligned}$$

But

$$\frac{\partial F}{\partial r} = \left(\frac{r^{n-1}}{r+1}\right)' \int_{\mathbb{S}^{n-1}} \sigma_\varepsilon \omega(\xi - r\theta)\mathrm{d}\theta - \frac{r^{n-1}}{r+1}\int_{\mathbb{S}^{n-1}} \theta\cdot\nabla(\sigma_\varepsilon\omega(\xi - r\theta))\mathrm{d}\theta$$
$$-\left(\frac{(2-r)^{n-1}}{3-r}\right)'\int_{\mathbb{S}^{n-1}} \sigma_\varepsilon\omega(\xi-(2-r)\theta)\mathrm{d}\theta$$
$$-\frac{(2-r)^{n-1}}{3-r}\int_{\mathbb{S}^{n-1}} \theta\cdot\nabla(\sigma_\varepsilon\omega(\xi-(2-r)\theta))\mathrm{d}\theta =: \theta_1+\theta_2+\theta_3+\theta_4.$$

By the proof of Lemma 23.4 below we get $|\theta_1| \le c_1\varepsilon^{-1}$ and $|\theta_3| \le c_3\varepsilon^{-1}$, where the constants c_1 and c_3 depend on ω. The second integral, θ_2, can be estimated as (see Lemma 23.4)

$$\varepsilon^{-1}\sum_{j=1}^{n}\frac{r^{n-1}}{r+1}\int_{\mathbb{S}^{n-1}} \theta_j\sigma_\varepsilon\left(\frac{\partial}{\partial x_j}\omega\right)(\xi - r\theta)\mathrm{d}\theta \le c_2\varepsilon^{-2}.$$

The same estimate holds for θ_4. Thus, Lemma 23.3 is proved. □

Lemma 23.4. *Let us assume that $f \in L^\infty(\mathbb{S}^{n-1})$ and $\omega \in S(\mathbb{R}^n)$. Then*

$$\left\|\int_{\mathbb{S}^{n-1}} \sigma_\varepsilon\omega(\xi-\theta)f(\theta)\mathrm{d}\theta\right\|_{L^\infty(\mathbb{R}^n)} \le C\varepsilon^{-1}.$$

Proof. We can reduce the proof to compactly supported ω, since $\overline{C_0^\infty} \overset{S}{=} S$. Let us consider a C_0^∞ partition of unity in $\mathbb{R}^n$ such that $\sum_{j=0}^\infty \psi_j(\xi) = 1$ or even $\sum_{j=0}^\infty \psi_j\left(\frac{1}{\xi}\right) = 1$, where ψ_0 is supported in $|\xi| < 1$ and $\psi_j = \psi(2^{-j}\xi)$ for $j = 1,2,3,\dots$ with ψ supported in the annulus $1/2 < |\xi| < 2$. We may therefore write

$$\int_{\mathbb{S}^{n-1}} \sigma_\varepsilon\omega(\xi-\theta)f(\theta)\mathrm{d}\theta = \sum_{j=0}^\infty\int_{\mathbb{S}^{n-1}} \varepsilon^{-n}\psi_j\left(\frac{\xi-\theta}{\varepsilon}\right)\omega\left(\frac{\xi-\theta}{\varepsilon}\right)f(\theta)\mathrm{d}\theta.$$

For $j = 1,2,3,\dots$, the function $\psi_j\left(\frac{\xi-\theta}{\varepsilon}\right)\omega\left(\frac{\xi-\theta}{\varepsilon}\right)$ is supported in the annulus $2^{j-1} \le |\cdot| \le 2^{j+1}$. Since ω is rapidly decreasing, we have that in this annulus,

$$\left|\omega\left(\frac{\xi-\theta}{\varepsilon}\right)\right| \le \frac{C_M}{(1+2^j)^M},$$

for all $M \in \mathbb{N}$. Hence

$$\left|\int_{\mathbb{S}^{n-1}} \varepsilon^{-n}(\psi_j\omega)\left(\frac{\xi-\theta}{\varepsilon}\right)f(\theta)\mathrm{d}\theta\right| \le C_M\frac{(2^{j+1}\varepsilon)^{n-1}}{(1+2^j)^M}\varepsilon^{-n} \le C_M'\varepsilon^{-1}\frac{(2^j)^{n-1}}{(1+2^j)^M}.$$

Taking M large enough, we see that the sum in j converges to $C\varepsilon^{-1}$. To end the proof of Lemma 23.4, notice that the term for $j=0$ satisfies this inequality trivially. □

Exercise 23.2. Prove that $(-\Delta)^{-1}: L^2_{\delta}(\mathbb{R}^3) \to L^2_{-\delta}(\mathbb{R}^3)$ for $\delta > 1$.

Let us return to the proof of Theorem 23.1. We can rewrite $\widehat{G}_1 f$ in the form

$$\widehat{G}_1 f = C\,\mathrm{p.v.}\int_{\mathbb{R}^n} \frac{\widehat{f}(\xi)\mathrm{e}^{\mathrm{i}(x,\xi)}\mathrm{d}\xi}{|\xi|^2-1} + I_1 f,$$

where

$$I_1 f = C\int_{\mathbb{S}^{n-1}} \widehat{f}(\theta)\mathrm{e}^{\mathrm{i}(\theta,x)}\mathrm{d}\theta.$$

Let us take a partition of unity $\sum_{j=0}^{\infty}\psi_j(x)=1$ such that $\operatorname{supp}\psi_0\subset\{|x|<1\}$ and $\operatorname{supp}\psi_j\subset\{2^{j-1}<|x|<2^{j+1}\}$, where $\psi_j=\psi(2^{-j}x)$ with a fixed function $\psi\in S$. We set $\Psi_j := \psi_j G_1^+$ and $K_j f := \Psi_j * f$, where G_1^+ is the kernel of the integral operator $\widehat{G}_1$. Using the estimates of the Hankel function $H^{(1)}_{\frac{n-2}{2}}(|x|)$ for $|x|<2$, we obtain

$$|\Psi_0| \le C|x|^{2-n}, \quad n\ge 3,$$

and

$$|\Psi_0| \le C(|\log|x|| + 1), \quad n=2.$$

Exercise 23.3 (Sobolev inequality). Let $0<\alpha<n$, $1<p<q<\infty$, and $\frac{1}{q}=\frac{1}{p}-\frac{\alpha}{n}$. Prove that

$$\left\| \int_{\mathbb{R}^n} \frac{f(y)\mathrm{d}y}{|x-y|^{n-\alpha}} \right\|_{L^q} \le C\|f\|_{L^p}.$$

Hint: For $K := |x|^{-n+\alpha}$ use the representation $K = K_1 + K_2$, where

$$K_1 = \begin{cases} K, & |x|<\mu, \\ 0, & |x|>\mu, \end{cases} \quad \text{and} \quad K_2 = \begin{cases} 0, & |x|<\mu, \\ K, & |x|>\mu. \end{cases}$$

From Sobolev's inequality for $\alpha=2$ we may conclude that the operator K_0 is bounded from $L^p(\mathbb{R}^n)\to L^{p'}(\mathbb{R}^n)$ for the range $\frac{2}{n}\ge\frac{1}{p}-\frac{1}{p'}\ge 0$ if $n\ge 3$, and for the range $1>\frac{1}{p}-\frac{1}{p'}\ge 0$ if $n=2$. From Lemmas 23.3 and 23.4 with $\varepsilon=\frac{1}{2^j}$ we can obtain that

$$\|\mathscr{F}(\Psi_j)\|_\infty = \|(|\xi|^2-1-\mathrm{i}0)^{-1} * \psi_j\|_\infty \le C\cdot 2^j.$$

This inequality leads to

$$\|K_j\|_{L^2\to L^2} \le C\cdot 2^j,$$

because

$$\|K_j f\|_{L^2} = \|\mathscr{F}(\Psi_j * f)\|_{L^2} = C\|\widehat{\Psi_j} \cdot \widehat{f}\|_{L^2} \leq \|\widehat{\Psi_j}\|_{L^\infty}\|\widehat{f}\|_{L^2} \leq C \cdot 2^j \|f\|_{L^2}.$$

On the other hand, due to the estimate of the fundamental solution at infinity we can obtain that $|\Psi_j(x)| \leq C \cdot 2^{-j \cdot \frac{n-1}{2}}$ and

$$\|K_j\|_{L^1 \to L^\infty} \leq C \cdot 2^{-j \cdot \frac{n-1}{2}}.$$

We have used here two facts:

$$\left| H^{(1)}_{\frac{n-2}{2}}(|x|) \right| \leq \frac{C}{|x|^{\frac{1}{2}}}, \quad |x| > 1,$$

and $\operatorname{supp} \Psi_j(x) \subset \{x : 2^{j-1} < |x| < 2^{j+1}\}$. Interpolating these estimates, we obtain the self-dual estimates

$$\|K_j\|_{L^p \to L^{p'}} \leq C(2^j)^{2\left(1-\frac{1}{p}\right) - \frac{n-1}{2}\left(\frac{2}{p}-1\right)}.$$

For convergence of this series we need the condition $2(1-\frac{1}{p}) - \frac{n-1}{2}(\frac{2}{p}-1) < 0$, or $\frac{1}{p} - \frac{1}{p'} > \frac{2}{n+1}$. If we want to get the sharper inequality $\frac{1}{p} - \frac{1}{p'} \geq \frac{2}{n+1}$, we have to use Stein's theorem on interpolation [37]. Thus, Theorem 23.1 is proved. □

It follows from Theorem 23.1 that if we consider the values of p from the interval

$$\frac{2n}{n+2} \leq p \leq \frac{2n+2}{n+3}, \quad n \geq 3,$$

$$1 < p \leq 6/5, \quad n = 2,$$

then we have the self-dual estimate

$$\|\widehat{G}_k\|_{L^p \to L^{p'}} \leq \frac{C}{|k|^{2-n\left(\frac{1}{p}-\frac{1}{p'}\right)}}.$$

But we would like to extend the estimates for $\widehat{G}_k$ for $\frac{2n}{n+2} \leq p \leq 2$, $n \geq 3$, and $1 < p \leq 2$, $n = 2$. In order to do so, we use interpolation of Agmon's estimate and the latter estimate for $p = \frac{2n+2}{n+3}$. This process leads to the estimate

$$\|\widehat{G}_k\|_{L^p_\delta \to L^{p'}_{-\delta}} \leq \frac{C}{|k|^{1-(n-1)\left(\frac{1}{p}-\frac{1}{2}\right)}},$$

where $\frac{2n+2}{n+3} < p \leq 2$, $n \geq 2$, and $\delta > \frac{1}{2} - (n+1)\left(\frac{1}{2p} - \frac{1}{4}\right)$.

Theorem 23.5. *Assume that the potential $q(x)$ belongs to $L^p_\sigma(\mathbb{R}^n)$, $n \ge 2$, with $\frac{n}{2} < p \le \infty$ and $\sigma = 0$ for $\frac{n}{2} < p \le \frac{n+1}{2}$ and $\sigma > 1 - \frac{n+1}{2p}$ for $\frac{n+1}{2} < p \le +\infty$. Then for all $k \neq 0$, the limit*

$$\widehat{G}_q := \lim_{\varepsilon \to 0+} (H - k^2 - i\varepsilon)^{-1}$$

exists in the uniform operator topology from $L^{\frac{2p}{p+1}}_{\frac{\sigma}{2}}(\mathbb{R}^n)$ to $L^{\frac{2p}{p-1}}_{-\frac{\sigma}{2}}(\mathbb{R}^n)$ with the norm estimate

$$\|\widehat{G}_q f\|_{L^{\frac{2p}{p-1}}_{-\sigma/2}} \le C|k|^{-\gamma}\|f\|_{L^{\frac{2p}{p+1}}_{\sigma/2}}$$

for large k with p and σ as above and with $\gamma = 2 - \frac{n}{2}$ for $\frac{n}{2} < p \le \frac{n+1}{2}$ and $\gamma = 1 - \frac{n-1}{2p}$ for $\frac{n+1}{2} < p \le \infty$.

Proof. Let us prove first that the integral operator $\widehat{K}$ with kernel

$$K(x,y) := |q|^{\frac{1}{2}}(x) G^+_k(|x-y|) q_{\frac{1}{2}}(y),$$

where $q_{\frac{1}{2}}(y) = |q(y)|^{\frac{1}{2}} \operatorname{sgn} q(y)$ maps from $L^2(\mathbb{R}^n)$ to $L^2(\mathbb{R}^n)$ with the same norm estimate as in Theorem 23.5. Indeed, if $f \in L^2(\mathbb{R}^n)$ and $q \in L^p_\sigma(\mathbb{R}^n)$, then $|q|^{\frac{1}{2}} \in L^{2p}_{\frac{\sigma}{2}}(\mathbb{R}^n)$, and therefore, $f|q|^{\frac{1}{2}} \in L^{\frac{2p}{p+1}}_{\frac{\sigma}{2}}(\mathbb{R}^n)$. Applying Theorem 23.1, we obtain

$$\|\widehat{G}_k(|q|^{\frac{1}{2}} f)\|_{L^{\frac{2p}{p-1}}_{-\sigma/2}} \le C|k|^{-\gamma}\|f\|_{L^{\frac{2p}{p+1}}_{\sigma/2}},$$

where γ is as in Theorem 23.5. Then by Hölder's inequality we have $|q|^{\frac{1}{2}}\widehat{G}_k(q_{\frac{1}{2}} f) \in L^2(\mathbb{R}^n)$ as asserted.

Let us consider now the operator $\widehat{G}_q$. This operator satisfies the resolvent equation

$$\widehat{G}_q = \widehat{G}_k - \widehat{G}_k q \widehat{G}_q,$$

which follows easily from $(H - k^2)\widehat{G}_q = I$. We denote by $\widehat{G}_l$ and $\widehat{G}_r$ the integral operators having kernels $G^+_k(|x-y|) q_{\frac{1}{2}}(y)$ and $|q(x)|^{\frac{1}{2}} G^+_k(|x-y|)$, respectively. Then one can show that

$$\widehat{G}_q = \widehat{G}_k - \widehat{G}_l (1 + \widehat{K})^{-1} \widehat{G}_r$$

for large k. Since $\widehat{K} : L^2 \to L^2$, $\widehat{G}_r : L^{\frac{2p}{p+1}}_{\frac{\sigma}{2}} \to L^2$, and $\widehat{G}_l : L^2 \to L^{\frac{2p}{p-1}}_{-\frac{\sigma}{2}}$, Theorem 23.5 is proved. □

The fundamental solution of the Helmholtz operator that was considered in the previous chapter can be effectively used for the following *scattering problem*: find

$u \in H^2_{\mathrm{loc}}(\mathbb{R}^n), n \geq 2$ that satisfies

$$\begin{aligned} &-\Delta u + qu = k^2 u, \quad x \in \mathbb{R}^n, \quad k > 0 \\ &u = u_0 + u_{\mathrm{sc}}, \quad u_0 = \mathrm{e}^{\mathrm{i}k(x,\theta)}, \quad \theta \in \mathbb{S}^{n-1} \\ &\lim_{r\to\infty} r^{(n-1)/2}\left(\frac{\partial u_{\mathrm{sc}}}{\partial r} - \mathrm{i}ku_{\mathrm{sc}}\right) = 0, \quad r = |x|. \end{aligned} \tag{23.2}$$

The latter condition is called the *Sommerfeld radiation condition* at infinity. The problem (23.2) is called the scattering problem.

Theorem 23.6. *Assume that* $q \in L^1(\mathbb{R}^n) \cap L^p_\sigma(\mathbb{R}^n)$, $n/2 < p \leq \infty$, $\sigma > \max\{0, 1-(n+1)/(2p)\}$, *is real-valued. Then there exists a unique solution u of* (23.2) *such that* $u_{\mathrm{sc}} \in L^\infty(\mathbb{R}^n)$, *and this solution u necessarily satisfies the* Lippmann–Schwinger equation

$$u = u_0 - \int_{\mathbb{R}^n} G_k^+(|x-y|)q(y)u(y)\mathrm{d}y. \tag{23.3}$$

Proof. Let us show first that there is a constant $C > 0$ such that

$$\lim_{R\to\infty} \int_{|y|=R} |u_{\mathrm{sc}}(y)|^2 \mathrm{d}\sigma(y) \leq C. \tag{23.4}$$

Indeed, the Sommerfeld radiation condition at infinity and Green's identity imply that

$$\begin{aligned} 2\mathrm{i}k\int_{|y|=R} |u_{\mathrm{sc}}(y)|^2 \mathrm{d}\sigma(y) &= \int_{|y|=R} [\overline{u_{\mathrm{sc}}}(y)\mathrm{i}ku_{\mathrm{sc}}(y) - u_{\mathrm{sc}}(y)(-\mathrm{i}k\overline{u_{\mathrm{sc}}}(y))]\mathrm{d}\sigma(y) \\ &= \int_{|y|=R} [\overline{u_{\mathrm{sc}}}(y)\frac{\partial}{\partial r}u_{\mathrm{sc}}(y) - u_{\mathrm{sc}}(y)\frac{\partial}{\partial r}\overline{u_{\mathrm{sc}}}(y)]\mathrm{d}\sigma(y) \\ &\quad + o(1/R^{(n-1)/2})\int_{|y|=R} (\overline{u_{\mathrm{sc}}}(y) + u_{\mathrm{sc}}(y))\mathrm{d}\sigma(y) \\ &= \int_{|y|\leq R} [\overline{u_{\mathrm{sc}}}(y)\Delta u_{\mathrm{sc}}(y) - u_{\mathrm{sc}}(y)\Delta\overline{u_{\mathrm{sc}}}(y)]\mathrm{d}y \\ &\quad + o(1)\left(\int_{|y|=R} |u_{\mathrm{sc}}(y)|^2 \mathrm{d}\sigma(y)\right)^{1/2} \\ &= \int_{|y|\leq R} q(y)[\overline{u_{\mathrm{sc}}}(y)u_0(y) - u_{\mathrm{sc}}(y)\overline{u_0}(y)]\mathrm{d}y \\ &\quad + o(1)\left(\int_{|y|=R} |u_{\mathrm{sc}}(y)|^2 \mathrm{d}\sigma(y)\right)^{1/2}, \quad R \to \infty. \end{aligned}$$

This equality leads to the inequality

$$2k\int_{|y|=R}|u_{sc}(y)|^2\mathrm{d}\sigma(y)$$
$$\leq 2\|q\|_{L^1(\mathbb{R}^n)}\|u_{sc}\|_{L^\infty(\mathbb{R}^n)}+o(1)\left(\int_{|y|=R}|u_{sc}(y)|^2\mathrm{d}\sigma(y)\right)^{1/2}, \quad R\to\infty.$$

This inequality clearly implies (23.4). The next observation is that G_k^+ clearly satisfies the Sommerfeld radiation condition at infinity. Fixing now $x\in\mathbb{R}^n$ and $R>0$ sufficiently large that $x\in B_R=\{y:|y|<R\}$ and applying Green's identity to $u_{sc}(y)$ and $G_k^+(|x-y|)$, we obtain (using the fact that on the sphere $|y|=r$ we have $\partial_{\nu_y}=\frac{\partial}{\partial r}$)

$$\int_{|y|=R}[u_{sc}(y)\frac{\partial}{\partial r}G_k^+(|x-y|)-G_k^+(|x-y|)\frac{\partial}{\partial r}u_{sc}(y)]\mathrm{d}\sigma(y)$$
$$=\int_{|y|\leq R}[u_{sc}(y)(\Delta_y+k^2)G_k^+(|x-y|)-G_k^+(|x-y|)(\Delta+k^2)u_{sc}(y)]\mathrm{d}y.$$

The usual procedure of allocation of the singularity of G_k^+ allows us to obtain

$$u_{sc}(x)=-\int_{|y|\leq R}G_k^+(|x-y|)q(y)u(y)\mathrm{d}y$$
$$-\int_{|y|=R}\left(u_{sc}(y)\left(\frac{\partial}{\partial r}-\mathrm{i}k\right)G_k^+(|x-y|)-G_k^+(|x-y|)\left(\frac{\partial}{\partial r}-\mathrm{i}k\right)u_{sc}(y)\right)\mathrm{d}\sigma(y).$$

The integral over the sphere $|y|=R$ can be estimated from above by

$$\left(\int_{|y|=R}|u_{sc}(y)|^2\mathrm{d}\sigma(y)\right)^{1/2}\left(\int_{|y|=R}\left|\left(\frac{\partial}{\partial r}-\mathrm{i}k\right)G_k^+(|x-y|)\right|^2\mathrm{d}\sigma(y)\right)^{1/2}$$
$$+\left(\int_{|y|=R}|G_k^+(|x-y|)|^2\mathrm{d}\sigma(y)\right)^{1/2}\left(\int_{|y|=R}\left|\left(\frac{\partial}{\partial r}-\mathrm{i}k\right)u_{sc}(y)\right|^2\mathrm{d}\sigma(y)\right)^{1/2}.$$

Since for fixed x and $R\to\infty$ we have $|G_k^+(|x-y|)|\leq C/R^{(n-1)/2}$ and since G_k^+ and u_{sc} both satisfy the radiation condition, we may estimate the latter sum from above using (23.4) as

$$Co(1/R^{(n-1)/2})\left(\int_{|y|=R}\mathrm{d}\sigma(y)\right)^{1/2}$$
$$+\frac{C}{R^{(n-1)/2}}\left(\int_{|y|=R}\mathrm{d}\sigma(y)\right)^{1/2}o(1/R^{(n-1)/2})\left(\int_{|y|=R}\mathrm{d}\sigma(y)\right)^{1/2}.$$

But this sum tends to zero as $R\to\infty$. Thus we have

$$u_{sc}(x) = -\int_{\mathbb{R}^n} G_k^+(|x-y|)q(y)u(y)dy.$$

So u from (23.2) necessarily satisfies the Lippmann–Schwinger equation (23.3). For $k > 0$ sufficiently large we can prove the unique solvability of (23.3) (and (23.2) as well) as follows. Theorem 23.5 allows us to rewrite equation (23.3) in the form

$$v = v_0 - \widehat{K}v, \tag{23.5}$$

where $v = |q|^{1/2}u$, $v_0 = |q|^{1/2}u_0$, and $\widehat{K}$ is as in Theorem 23.5. Since the conditions on q and u imply that $v, v_0 \in L^2(\mathbb{R}^n)$ and

$$\left\|\widehat{K}\right\|_{L^2(\mathbb{R}^n)\to L^2(\mathbb{R}^n)} \le \frac{C_0}{k^\gamma},$$

where C_0 as in Theorem 23.5 and $\gamma = 2 - n/2$ for $n/2 < p \le (n+1)/2$ and $\gamma = 1-(n-1)/(2p)$ for $(n+1)/2 < p \le \infty$, we obtain for $k > C_0^{1/\gamma}$ that there is a unique solution v of (23.5), namely

$$v = \sum_{j=0}^{\infty} \widehat{K}^j v_0.$$

Moreover, the estimate

$$\|v - v_0\|_{L^2(\mathbb{R}^n)} \le \frac{2C_0}{k^\gamma}\|v_0\|_{L^2(\mathbb{R}^n)}$$

holds uniformly in $k \ge (2C_0)^{1/\gamma}$. This is equivalent to the estimate

$$\left\||q|^{1/2}u_{sc}\right\|_{L^2(\mathbb{R}^n)} \le \frac{2C_0}{k^\gamma}\|q\|_{L^1(\mathbb{R}^n)}^{1/2}. \tag{23.6}$$

For the values of k from the interval $0 < k \le (C_0)^{1/\gamma}$ we proceed as follows.

Exercise 23.4. Show that the integral operator $\widehat{G_k} \circ q$ for all $k > 0$ is a compact operator in $L^\infty(\mathbb{R}^n)$, where q satisfies the conditions of Theorem 23.6.

This exercise implies that the integral operators $\widehat{K}$ and $\widehat{G_q} \circ q$ are also compact in $L^2(\mathbb{R}^n)$ and $L^\infty(\mathbb{R}^n)$, respectively. Next, using Agmon's estimate and Theorems 23.1 and 23.5, we conclude that for all $k > 0$ the operator $\widehat{G_q} = (-\Delta - k^2 + q - \mathrm{i}0)^{-1}$ exists in the appropriate operator topology (see Theorem 23.5), and therefore for the solution u of (23.2) (or equivalently, (23.3)) the representation

$$u = (I - \widehat{G_q} \circ q)u_0 \tag{23.7}$$

holds. The Lippmann–Schwinger equation can be rewritten in the operator form

$$u_{sc} - Tu_{sc} = \widetilde{u_0}, \tag{23.8}$$

where $\widetilde{u_0} = -(\widehat{G_k} \circ q)u_0 \in L^\infty(\mathbb{R}^n)$ and T is a compact operator in $L^\infty(\mathbb{R}^n)$. By Riesz theory (see Chapter 34) we shall obtain the unique solvability of (23.8) if we are able to show that $I - T$ is injective. But injectivity follows immediately from (23.7). The theorem is therefore completely proved. □

Remark 23.7. For $k > 0$ large enough, the unique solvability in Theorem 23.6 holds for a complex-valued potential q.

Corollary 23.8. *Let v be the outgoing solution of the inhomogeneous Schrödinger equation*

$$(H - k^2)v = f,$$

i.e.,

$$v = (H - k^2 - \mathrm{i}0)^{-1} f,$$

where $f \in S(\mathbb{R}^n)$. Then the following representation holds:

$$v(x) = \widehat{G}_k(f - q\widehat{G}_q(f))(x).$$

Moreover, for $|x| \to \infty$ and fixed positive k,

$$v(x,k) = C_n \frac{\mathrm{e}^{\mathrm{i}k|x|} k^{\frac{n-3}{2}}}{|x|^{\frac{n-1}{2}}} A_f(k,\theta') + o\left(\frac{1}{|x|^{\frac{n-1}{2}}}\right),$$

where $\theta' = \frac{x}{|x|}$ and the function A_f, called the scattering amplitude, is defined by

$$A_f(k,\theta') := \int_{\mathbb{R}^n} \mathrm{e}^{-\mathrm{i}k(\theta',y)}(f(y) - q(y)\widehat{G}_q(f))\mathrm{d}y.$$

Proof. The first representation follows immediately from the definition of $\widehat{G}_q$. Indeed, since $v = \widehat{G}_q f$, we must have $\widehat{G}_k f = v + \widehat{G}_k q v$, or $v = \widehat{G}_k f - \widehat{G}_k q v = \widehat{G}_k(f - q\widehat{G}_q f)$.

In order to prove the asymptotic behavior for v let us assume that q and f have compact support, say in the ball $\{x : |x| \le R\}$. We will use the following asymptotic behavior of $G_k^+(|x|)$:

(1) $k|x| < 1$:

  (a) $G_k^+(|x|) \sim C|x|^{2-n}, \quad n \ge 3,$
  (b) $G_k^+(|x|) \sim C\log(k|x|), \quad n = 2.$

(2) $k|x| > 1$:

$$G_k^+(|x|) \sim C\frac{k^{\frac{n-3}{2}}}{|x|^{\frac{n-1}{2}}} \mathrm{e}^{\mathrm{i}k|x|}, \quad n \ge 2.$$

Since k is fixed, $|y| \le R$, and $|x| \to +\infty$, we may assume that $k|x-y| > 1$ for x large enough. Therefore, as $|x| \to \infty$, we have

$$v(x) = C\frac{e^{ik|x|}k^{\frac{n-3}{2}}}{|x|^{\frac{n-1}{2}}}\int_{|y|\le R} e^{ik(|x-y|-|x|)}(f - q\widehat{G}_q f)dy$$
$$+\int_{|y|\le R} o\left(\frac{1}{|x-y|^{\frac{n-1}{2}}}\right)(f - q\widehat{G}_q f)dy =: I_1 + I_2.$$

It is clear that for I_2 the following is true:

$$I_2 = o\left(\frac{1}{|x|^{\frac{n-1}{2}}}\int_{|y|\le R}(f(y) - q(y)\widehat{G}_q f(y))dy\right) = o\left(\frac{1}{|x|^{\frac{n-1}{2}}}\right), \quad |x| \to \infty,$$

because $f - q\widehat{G}_q f$ is an integrable function. Next, let us note that

$$|x-y| - |x| = \frac{|x-y|^2 - |x|^2}{|x-y| + |x|} = \frac{y^2 - 2(x,y)}{|x-y|+|x|} = -\left(\frac{x}{|x|}, y\right) + O\left(\frac{1}{|x|}\right)$$

as $|x| \to +\infty$. We can therefore rewrite the integral appearing in I_1 as follows:

$$\int_{|y|\le R} e^{-ik\left(\frac{x}{|x|},y\right)+O\left(\frac{1}{|x|}\right)}(f - q\widehat{G}_q f)dy$$
$$= \int_{|y|\le R} e^{-ik(\theta',y)}(f - q\widehat{G}_q f)dy + O\left(\frac{1}{|x|}\right)\int_{|y|\le R} e^{-ik(\theta',y)}(f - q\widehat{G}_q f)dy$$
$$= \int_{|y|\le R} e^{-ik(\theta',y)}(f - q\widehat{G}_q f)dy + O\left(\frac{1}{|x|}\right), \quad |x| \to \infty,$$

where $\theta' = \frac{x}{|x|} \in \mathbb{S}^{n-1}$. Thus, Corollary 23.8 is proved when q and f have compact support. The proof in the general case is much more difficult and is therefore omitted. □

Remark 23.9. Hint for the general case: The integral over $\mathbb{R}^n$ might be divided into two parts: $|y| < |x|^{\varepsilon}$ and $|y| > |x|^{\varepsilon}$, where $\varepsilon > 0$ is chosen appropriately.

Lemma 23.10 (Optical lemma). *For the function $A_f(k,\theta')$ the following equality holds:*

$$\int_{\mathbb{S}^{n-1}} |A_f(k,\theta')|^2 d\theta' = -\frac{1}{C^2 k^{n-2}}\int_{\mathbb{R}^n} \operatorname{Im}(f\overline{v})dx,$$

where C is the constant from the asymptotic representation of $v = (H - k^2 - i0)^{-1} f$.

Proof Let ρ be a smooth real-valued function on $[0,+\infty)$ such that $0 \le \rho \le 1$ and $\rho(r) = 1$ for $0 \le r < 1$ and $\rho(r) = 0$ for $r \ge 2$. We set $\rho_m(r) = \rho\left(\frac{r}{m}\right)$. Multiplying f by $\overline{v}\rho_m(|x|)$, integrating over $\mathbb{R}^n$, and taking the imaginary parts leads to

$$\mathrm{Im}\int_{\mathbb{R}^n} f(x)\rho_m(|x|)\overline{v}(x)\mathrm{d}x = \mathrm{Im}\int_{\mathbb{R}^n} (-\Delta v)\rho_m(|x|)\overline{v}(x)\mathrm{d}x.$$

As m tends to infinity, the left-hand side converges to $\mathrm{Im}\int_{\mathbb{R}^n} f(x)\overline{v}(x)\mathrm{d}x$. To get the desired limit for the right-hand side, we integrate by parts and obtain

$$\begin{aligned}
\mathrm{Im}\int_{\mathbb{R}^n} (-\Delta v)\rho_m(|x|)\overline{v}(x)\mathrm{d}x &= \mathrm{Im}\int_{\mathbb{R}^n} \frac{x}{|x|}\cdot\nabla v\rho_m'(|x|)\overline{v}(x)\mathrm{d}x \\
&= \mathrm{Im}\int_{\mathbb{R}^n} \left[(\theta'\cdot\nabla v - \mathrm{i}kv)\overline{v}(x)\rho_m'(|x|) + \mathrm{i}k\rho_m'(|x|)|v|^2\right]\mathrm{d}x \\
&= \mathrm{Im}\int_{\mathbb{R}^n} (\theta'\cdot\nabla v - \mathrm{i}kv)\overline{v}(x)\rho_m'(x)\mathrm{d}x + k\int_{\mathbb{R}^n} \rho_m'(|x|)|v(x)|^2\mathrm{d}x =: I_1 + I_2.
\end{aligned}$$

Since $v = (H-k^2-\mathrm{i}0)^{-1}f$, using the asymptotic representation we may conclude that v satisfies the *Sommerfeld radiation condition*

$$\frac{\partial v}{\partial r} - \mathrm{i}kv = o\left(\frac{1}{r^{\frac{n-1}{2}}}\right), \quad r = |x|,$$

at infinity. Hence $I_1 \to 0$ as $m \to \infty$. By Corollary 23.8, the second term I_2 is equal to

$$\begin{aligned}
k\int_{\mathbb{R}^n} \rho_m'(|x|)|v(x)|^2\mathrm{d}x &= k\int_{\mathbb{R}^n} \rho_m'(|x|)\cdot C^2\frac{k^{n-3}}{|x|^{n-1}}|A_f(k,\theta')|^2\mathrm{d}x \\
&\quad + k\int_{\mathbb{R}^n} \rho_m'(|x|)o\left(\frac{1}{|x|^{n-1}}\right)\mathrm{d}x \\
&= C^2k^{n-2}\int_{\mathbb{S}^{n-1}} |A_f(k,\theta')|^2\mathrm{d}\theta'\int_m^{2m} \frac{r^{n-1}}{r^{n-1}}\rho_m'(r)\mathrm{d}r \\
&\quad + k\int_{\mathbb{S}^{n-1}}\mathrm{d}\theta\int_m^{2m} r^{n-1}o\left(\frac{1}{r^{n-1}}\right)\rho_m'(r)\mathrm{d}r \\
&= C^2k^{n-2}\int_{\mathbb{S}^{n-1}} |A_f(k,\theta')|^2\mathrm{d}\theta'\int_m^{2m} \rho_m'(r)\mathrm{d}r + o(1) \\
&= C^2k^{n-2}\int_{\mathbb{S}^{n-1}} |A_f(k,\theta')|^2\mathrm{d}\theta'[\rho(2)-\rho(1)] + o(1) \\
&= -C^2k^{n-2}\int_{\mathbb{S}^{n-1}} |A_f(k,\theta')|^2\mathrm{d}\theta' + o(1), \quad m \to \infty.
\end{aligned}$$

Letting $m \to \infty$, we obtain

$$\mathrm{Im}\int_{\mathbb{R}^n} f(x)\overline{v}(x)\mathrm{d}x = -C^2k^{n-2}\int_{\mathbb{S}^{n-1}} |A_f(k,\theta')|^2\mathrm{d}\theta'.$$

Thus, Lemma 23.10 is proved. □

Exercise 23.5. Let $n=2$ or $n=3$. Assume that $q \in L^p(\mathbb{R}^n) \cap L^1(\mathbb{R}^n)$ with $1 < p \leq \infty$ if $n=2$ and $3 < p \leq \infty$ if $n=3$. Prove that the generalized eigenfunctions $u(x, \vec{k})$, that is, the solutions of the problem (23.2) with $(\vec{k}, \vec{k}) = k^2$, are uniformly bounded with respect to $x \in \mathbb{R}^n$ and $|\vec{k}|$ sufficiently large.

We will obtain very important corollaries from the optical lemma. Let $A_q(k)$ denote the linear mapping that takes the inhomogeneity f to the corresponding scattering amplitude

$$A_q(k) : f(x) \to A_f(k, \theta').$$

Lemma 23.11. *Let the potential $q(x)$ satisfy the conditions from Theorem 23.5. Then A_q is a well defined bounded operator from $L^{\frac{2p}{p+1}}_{\sigma/2}(\mathbb{R}^n)$ to $L^2(\mathbb{S}^{n-1})$ with the operator norm estimate*

$$\|A_q\|_{L^{\frac{2p}{p+1}}_{\sigma/2} \to L^2} \leq \frac{C}{|k|^{\frac{\gamma}{2} + \frac{n-2}{2}}},$$

where p, σ, and γ are as in Theorem 23.5.

Proof. By Lemma 23.10 and the definition of $A_q f$ we have that

$$\begin{aligned}\|A_q f\|^2_{L^2(\mathbb{S}^{n-1})} &= \int_{\mathbb{S}^{n-1}} |A_f(k, \theta')|^2 \mathrm{d}\theta' = -\frac{1}{C^2 |k|^{n-2}} \int_{\mathbb{R}^n} \operatorname{Im}(f \cdot \overline{v}) \mathrm{d}x \\ &\leq \frac{1}{C^2 |k|^{n-2}} \|v\|_{L^{\frac{2p}{p-1}}_{-\sigma/2}(\mathbb{R}^n)} \|f\|_{L^{\frac{2p}{p+1}}_{\sigma/2}(\mathbb{R}^n)}.\end{aligned}$$

Further, since $v = \widehat{G}_q f$, we obtain from Theorem 23.5 that

$$\|A_q f\|^2_{L^2(\mathbb{S}^{n-1})} \leq \frac{C}{|k|^{n-2}} \cdot |k|^{-\gamma} \|f\|^2_{L^{\frac{2p}{p+1}}_{\sigma/2}(\mathbb{R}^n)}.$$

Thus, Lemma 23.11 is proved. □

Let us denote by $A_0(k)$ the operator $A_q(k)$ that corresponds to the potential $q \equiv 0$, i.e.,

$$A_0 f(\theta') = \int_{\mathbb{R}^n} \mathrm{e}^{-\mathrm{i}k(\theta', y)} f(y) \mathrm{d}y.$$

It is not difficult to see that

$$A_q f(\theta') = A_f(k, \theta') = \int_{\mathbb{R}^n} f(y) \overline{u(y, k, \theta')} \mathrm{d}y,$$

where $u(\cdot,k,\theta')$ is the solution of the Lippmann–Schwinger equation. Indeed, by Corollary 23.8 we have

$$\begin{aligned} A_f(k,\theta') &= \int_{\mathbb{R}^n} \mathrm{e}^{-\mathrm{i}k(\theta',y)}(f(y)-q(y)\widehat{G}_q(f))\mathrm{d}y = ((I-q\widehat{G}_q)f,\mathrm{e}^{\mathrm{i}k(\theta',y)})_{L^2(\mathbb{R}^n)} \\ &= (f,(I-\widehat{G}_q(q))\mathrm{e}^{\mathrm{i}k(\theta',\cdot)})_{L^2(\mathbb{R}^n)} = \int_{\mathbb{R}^n} f(y)\overline{(I-\widehat{G}_q(q))(\mathrm{e}^{\mathrm{i}k(\theta',\cdot)})(y)}\mathrm{d}y \\ &= \int_{\mathbb{R}^n} f(y)\overline{u(y,k,\theta')}\mathrm{d}y, \end{aligned}$$

since $\widehat{G}_q$ is a self-adjoint operator.

Let us prove now that

$$u(y,k,\theta') := (I-\widehat{G}_q(q))(\mathrm{e}^{\mathrm{i}k(\theta',\cdot)})(y)$$

is the solution of the Lippmann–Schwinger equation. Indeed,

$$\begin{aligned} (H-k^2)u &= (H-k^2)(\mathrm{e}^{\mathrm{i}k(\theta',y)}) - (H-k^2)\widehat{G}_q(q)\cdot(\mathrm{e}^{\mathrm{i}k(\theta',\cdot)})(y) \\ &= (-\Delta-k^2)\mathrm{e}^{\mathrm{i}k(\theta',y)} + q\mathrm{e}^{\mathrm{i}k(\theta',y)} - q\mathrm{e}^{\mathrm{i}k(\theta',y)} = 0, \end{aligned}$$

since $(-\Delta-k^2)\mathrm{e}^{\mathrm{i}k(\theta',y)} = 0$ and $(H-k^2)\widehat{G}_q = I$. This means that this $u(y,k,\theta')$ is the solution of the equation $(H-k^2)u = 0$.

Remark 23.12. Let us consider the Lippmann–Schwinger equation

$$u(x,k,\theta) = \mathrm{e}^{\mathrm{i}k(x,\theta)} - \int_{\mathbb{R}^n} G_k^+(|x-y|)q(y)u(y,k,\theta)\mathrm{d}y.$$

Then for fixed $k>0$ and $|x|\to\infty$, the solution $u(x,k,\theta)$ admits the asymptotic representation

$$u(x,k,\theta) = \mathrm{e}^{\mathrm{i}k(x,\theta)} + C_n \frac{\mathrm{e}^{\mathrm{i}k|x|}k^{\frac{n-3}{2}}}{|x|^{\frac{n-1}{2}}} A(k,\theta',\theta) + o\left(\frac{1}{|x|^{\frac{n-1}{2}}}\right),$$

where $\theta' = \frac{x}{|x|}$ and the function $A(k,\theta',\theta)$ is called the *scattering amplitude* and has the form

$$A(k,\theta',\theta) = \int_{\mathbb{R}^n} \mathrm{e}^{-\mathrm{i}k(\theta',y)}q(y)u(y,k,\theta)\mathrm{d}y.$$

For $k<0$ we set

$$A(k,\theta',\theta) = \overline{A(-k,\theta',\theta)}, \quad u(x,k,\theta) = \overline{u(x,-k,\theta)}.$$

Proof. If $(H-k^2)u = 0$ and $u = \mathrm{e}^{\mathrm{i}k(\theta,x)} + u_{\mathrm{sc}}(x,k,\theta)$, then $u_{\mathrm{sc}}(x,k,\theta)$ satisfies the equation

$$(H-k^2)u_{sc} = -q\mathrm{e}^{\mathrm{i}k(\theta,x)}.$$

We may therefore apply Corollary 23.8 with $v := u_{\mathrm{sc}}$ and $f := -q\mathrm{e}^{\mathrm{i}k(\theta,x)}$ to obtain

$$u_{\mathrm{sc}}(x,k,\theta) = C_n \frac{\mathrm{e}^{\mathrm{i}k|x|}k^{\frac{n-3}{2}}}{|x|^{\frac{n-1}{2}}} A_f(k,\theta') + o\left(\frac{1}{|x|^{\frac{n-1}{2}}}\right),$$

where

$$\begin{aligned} A_f(k,\theta') &= \int_{\mathbb{R}^n} \mathrm{e}^{-\mathrm{i}k(\theta',y)}(-q\mathrm{e}^{\mathrm{i}k(\theta,y)} + q\widehat{G}_q(q\mathrm{e}^{\mathrm{i}k(\theta,\cdot)})(y))\mathrm{d}y \\ &= -\int_{\mathbb{R}^n} \mathrm{e}^{-\mathrm{i}k(\theta',y)} q(y)(\mathrm{e}^{\mathrm{i}k(\theta,y)} - \widehat{G}_q(q\mathrm{e}^{\mathrm{i}k(\theta,\cdot)})\mathrm{d}y. \end{aligned}$$

But we have proved that $\mathrm{e}^{\mathrm{i}k(\theta,y)} - \widehat{G}_q(q\mathrm{e}^{\mathrm{i}k(\theta,\cdot)})(y)$ is a solution of the equation $(H-k^2)u = 0$. We conclude that

$$A_f(k,\theta') = -\int_{\mathbb{R}^n} \mathrm{e}^{-\mathrm{i}k(\theta',y)} q(y)u(y,k,\theta)\mathrm{d}y =: -A(k,\theta',\theta).$$

This proves the remark. □

Now let $\Phi_0(k)$ and $\Phi(k)$ be the operators defined for $f \in L^2(\mathbb{S}^{n-1})$ as

$$(\Phi_0(k)f)(x) := |q(x)|^{\frac{1}{2}} \int_{\mathbb{S}^{n-1}} \mathrm{e}^{\mathrm{i}k(x,\theta)} f(\theta)\mathrm{d}\theta \tag{23.9}$$

and

$$(\Phi(k)f)(x) := |q(x)|^{\frac{1}{2}} \int_{\mathbb{S}^{n-1}} u(x,k,\theta)f(\theta)\mathrm{d}\theta. \tag{23.10}$$

Lemma 23.13. *The operators $\Phi_0(k)$ and $\Phi(k)$ are bounded from $L^2(\mathbb{S}^{n-1})$ to $L^2(\mathbb{R}^n)$ with the norm estimates*

$$\|\Phi_0(k)\|, \|\Phi(k)\| \le \frac{C}{k^{\frac{\gamma}{2}+\frac{n-2}{2}}}, \quad k>0,$$

where γ is as in Theorem 23.5.

Proof. Let us prove that

$$(\Phi_0(k)f)(x) = |q(x)|^{\frac{1}{2}}(A_0^* f)(x) \tag{23.11}$$

and

$$(\Phi(k)f)(x) = |q(x)|^{\frac{1}{2}}(A_q^* f)(x), \tag{23.12}$$

where A_0^* and A_q^* are the adjoint operators for A_0 and A_q, respectively. Indeed, if $f \in L^2(\mathbb{S}^{n-1})$ and $g \in L^2(\mathbb{R}^n)$, then

$$\begin{aligned}\int_{\mathbb{S}^{n-1}} f(\theta)\overline{(A_0 g)(\theta)}\mathrm{d}\theta &= \int_{\mathbb{S}^{n-1}} f(\theta)\mathrm{d}\theta \int_{\mathbb{R}^n} \mathrm{e}^{ik(\theta,y)}\overline{g(y)}\mathrm{d}y \\ &= \int_{\mathbb{R}^n} \overline{g(y)}\mathrm{d}y \int_{\mathbb{S}^{n-1}} \mathrm{e}^{ik(\theta,y)} f(\theta)\mathrm{d}\theta \\ &= \int_{\mathbb{R}^n} \left(\int_{\mathbb{S}^{n-1}} \mathrm{e}^{ik(\theta,y)} f(\theta)\mathrm{d}\theta \right) \overline{g(y)}\mathrm{d}y.\end{aligned}$$

This means that

$$A_0^* f(y) = \int_{\mathbb{S}^{n-1}} \mathrm{e}^{ik(\theta,y)} f(\theta)\mathrm{d}\theta,$$

and (23.11) is immediate. Similarly one proves (23.12). Since (see Lemma 23.11)

$$\|A_0\|, \|A_q\|_{L^{\frac{2p}{p+1}}_{\sigma/2} \to L^2(\mathbb{S}^{n-1})} \le \frac{C}{k^{\frac{\gamma}{2}+\frac{n-2}{2}}},$$

we have that

$$\|A_0^*\|, \|A_q^*\|_{L^2(\mathbb{S}^{n-1}) \to L^{\frac{2p}{p-1}}_{-\sigma/2}(\mathbb{R}^n)} \le \frac{C}{k^{\frac{\gamma}{2}+\frac{n-2}{2}}}.$$

The proof is finished by

$$\begin{aligned}\|\Phi_0(k)f\|_{L^2(\mathbb{R}^n)} &= \||q|^{\frac{1}{2}}(A_0^* f)\|_{L^2(\mathbb{R}^n)} \le \|q\|^{\frac{1}{2}}_{L^p_\sigma(\mathbb{R}^n)} \|A_0^* f\|_{L^{\frac{2p}{p-1}}_{-\sigma/2}(\mathbb{R}^n)} \\ &\le \frac{C}{k^{\frac{\gamma}{2}+\frac{n-2}{2}}} \|q\|^{\frac{1}{2}}_{L^p_\sigma(\mathbb{R}^n)} \|f\|_{L^2(\mathbb{S}^{n-1})},\end{aligned}$$

where we have made use of Hölder's inequality in the first estimate. It is clear that the same is true for $\Phi(k)$. □

Agmon's estimate (23.1) can be applied to the magnetic Schrödinger operator. In fact, in the work [2], Agmon proved a more general estimate than (23.1). Namely, it was proved that for all $g \in H^2_{-\delta}(\mathbb{R}^n)$ and $|k| \ge 1$,

$$\frac{1}{|k|}\|g\|_{H^2_{-\delta}(\mathbb{R}^n)} + \|g\|_{H^1_{-\delta}(\mathbb{R}^n)} + |k|\,\|g\|_{L^2_{-\delta}(\mathbb{R}^n)} \le C\left\|(\Delta + k^2)g\right\|_{L^2_\delta(\mathbb{R}^n)},$$

where $\delta > 1/2$ and $H^s_{-\delta}(\mathbb{R}^n)$, $s = 0,1,2$, denotes the weighted Sobolev space (see below for a precise definition). As a consequence of this estimate, for all $f \in L^2_\delta(\mathbb{R}^n)$, $\delta > 1/2$, one has the estimates

$$\begin{aligned} \|(-\Delta-k^2-\mathrm{i}0)^{-1}f\|_{L^2_{-\delta}(\mathbb{R}^n)} &\le \frac{\beta}{|k|}\|f\|_{L^2_\delta(\mathbb{R}^n)} \\ \|(-\Delta-k^2-\mathrm{i}0)^{-1}f\|_{H^1_{-\delta}(\mathbb{R}^n)} &\le \beta\|f\|_{L^2_\delta(\mathbb{R}^n)}. \end{aligned} \tag{23.13}$$

Here $(-\Delta-k^2-\mathrm{i}0)^{-1}$ is the integral operator with kernel $G_k^+(|x-y|)$, see (22.1), and the weighted Sobolev spaces $W^1_{p,\sigma}(\mathbb{R}^n)$ (or $H^1_\sigma(\mathbb{R}^n)$ if $p=2$) are understood so that f belongs to $W^1_{p,\sigma}(\mathbb{R}^n)$ if and only if f and ∇f belong to the weighted Lebesgue space $L^p_\sigma(\mathbb{R}^n)$ (see Example 18.17).

Since the integral operator $(-\Delta-k^2-\mathrm{i}0)^{-1}$ is of convolution type, using duality we can conclude that it maps $H^{-1}_\delta(\mathbb{R}^n)$ to $L^2_{-\delta}(\mathbb{R}^n)$ with the norm estimate

$$\|(-\Delta-k^2-\mathrm{i}0)^{-1}f\|_{L^2_{-\delta}(\mathbb{R}^n)} \le \beta\|f\|_{H^{-1}_\delta(\mathbb{R}^n)}, \quad |k|\ge 1, \tag{23.14}$$

where $H^{-1}_\delta(\mathbb{R}^n)$ denotes the dual space of the Sobolev space $H^1_{-\delta}(\mathbb{R}^n)$ and the constant β is the same as in (23.13).

We will consider now the scattering problem for the *magnetic Schrödinger operator* in $\mathbb{R}^n$, $n\ge 2$, of the form

$$H_{\mathrm{m}} := -(\nabla+\mathrm{i}\vec{W}(x))^2\cdot+V(x)\cdot, \tag{23.15}$$

where the coefficients $\vec{W}(x)$ and $V(x)$ are assumed to be real and are from the spaces

$$\vec{W}\in W^1_{p,\sigma}(\mathbb{R}^n), \quad V\in L^p_\sigma(\mathbb{R}^n), \quad n<p\le\infty, \quad \sigma>n/p', \quad 1/p+1/p'=1. \tag{23.16}$$

We are looking for the solutions to the equation $H_{\mathrm{m}}u=k^2u$, $k\ne 0$, with H_{m} from (23.15) in the form

$$\begin{cases} u(x)=u_0(x)+u_{\mathrm{sc}}(x), \quad u_0(x)=\mathrm{e}^{\mathrm{i}k(x,\theta)}, \theta\in\mathbb{S}^{n-1}, \\ \lim_{r\to\infty} r^{(n-1)/2}\left(\frac{\partial u_{\mathrm{sc}}(x)}{\partial r}-\mathrm{i}ku_{\mathrm{sc}}(x)\right)=0, \quad r=|x|. \end{cases} \tag{23.17}$$

Using the same procedure as for the Schrödinger operator (see Theorem 23.6), we conclude that the solution (23.17) necessarily satisfies the Lippmann–Schwinger integral equation

$$u(x)=u_0(x)+\int_{\mathbb{R}^n} G_k^+(|x-y|)(2\mathrm{i}\nabla(\vec{W}(y)u(y))-q(y)u(y))\mathrm{d}y,$$

where $q=\mathrm{i}\nabla\vec{W}+|\vec{W}|^2+V$. This equation can be rewritten as the following integral equation:

$$u_{\mathrm{sc}}(x)=\widetilde{u_0}(x)+L_k(u_{\mathrm{sc}})(x), \quad \widetilde{u_0}(x)=L_k(u_0)(x), \tag{23.18}$$

with the integral operator L_k defined as

$$L_k f(x) = \int_{\mathbb{R}^n} G_k^+(|x-y|)(2i\nabla(\vec{W}(y)f(y)) - q(y)f(y))dy. \tag{23.19}$$

Lemma 23.14. *Suppose that the conditions* (23.16) *are fulfilled. Then* $\widetilde{u_0}$ *belongs to* $L^2_{-\sigma/2}(\mathbb{R}^n)$*, and* L_k *from* (23.19) *maps* $L^2_{-\sigma/2}(\mathbb{R}^n)$ *into itself with* σ *as* (23.16)*. Moreover, for* $|k| \geq 1$ *uniformly,*

$$\begin{aligned} \|\widetilde{u_0}\|_{L^2_{-\sigma/2}(\mathbb{R}^n)} &\leq \beta\left(2\left\|\vec{W}\right\|_{L^2_{\sigma/2}(\mathbb{R}^n)} + \|q\|_{L^2_{\sigma/2}(\mathbb{R}^n)}\right), \\ \|L_k f\|_{L^2_{-\sigma/2}(\mathbb{R}^n)} &\leq \beta\left(2\left\|\vec{W}\right\|_{L^\infty_\sigma(\mathbb{R}^n)} + C_p \|q\|_{L^p_\sigma(\mathbb{R}^n)}\right)\|f\|_{L^2_{-\sigma/2}(\mathbb{R}^n)}, \end{aligned} \tag{23.20}$$

where β *is the same as in* (23.13) *and the constant* C_p *is equal to*

$$C_p = \left(\frac{1}{(2\sqrt{\pi})^n}\frac{\Gamma((p-n)/2)}{\Gamma(p/2)}\right)^{1/p}.$$

Proof. Conditions (23.16) imply that $\sigma/2 > 1/2$ and

$$L^p_\sigma(\mathbb{R}^n) \hookrightarrow L^2_{\sigma/2}(\mathbb{R}^n).$$

It is therefore true that under these conditions the functions $V, \nabla\vec{W}$ and $|\vec{W}|^2$ belong to $L^2_{\sigma/2}(\mathbb{R}^n)$ and $\vec{W} \in L^\infty_\sigma(\mathbb{R}^n)$. Using the first Agmon's estimate (23.13), one can easily obtain

$$\|\widetilde{u_0}\|_{L^2_{-\sigma/2}(\mathbb{R}^n)} \leq \frac{\beta}{|k|}\left(2|k|\left\|\vec{W}\right\|_{L^2_{\sigma/2}(\mathbb{R}^n)} + \|q\|_{L^2_{\sigma/2}(\mathbb{R}^n)}\right).$$

Hence the first inequality in (23.20) is proved. Next, applying now (23.14), we obtain that

$$\begin{aligned} \|L_k f\|_{L^2_{-\sigma/2}(\mathbb{R}^n)} &\leq \beta\left(2\left\|\nabla(\vec{W}f)\right\|_{H^{-1}_{\sigma/2}(\mathbb{R}^n)} + \|qf\|_{H^{-1}_{\sigma/2}(\mathbb{R}^n)}\right) \\ &\leq \beta\left(2\left\|\vec{W}f\right\|_{L^2_{\sigma/2}(\mathbb{R}^n)} + \|qf\|_{H^{-1}_{\sigma/2}(\mathbb{R}^n)}\right) \\ &\leq \beta\left(2\left\|\vec{W}\right\|_{L^\infty_\sigma(\mathbb{R}^n)}\|f\|_{L^2_{-\sigma/2}(\mathbb{R}^n)} + \|qf\|_{H^{-1}_{\sigma/2}(\mathbb{R}^n)}\right). \end{aligned}$$

To estimate the second term $\|qf\|_{H^{-1}_{\sigma/2}(\mathbb{R}^n)}$ we proceed using Hölder's inequality and the Hausdorff–Young inequalities as follows:

$$\|qf\|_{H^{-1}_{\sigma/2}(\mathbb{R}^n)} = \left\|\widetilde{q}\widetilde{f}\right\|_{H^{-1}(\mathbb{R}^n)} = \left\|\mathscr{F}(\widetilde{q}\widetilde{f})\right\|_{L^2_{-1}(\mathbb{R}^n)} \leq C_0 \left\|\mathscr{F}(\widetilde{q}\widetilde{f})\right\|_{L^{2p/(p-2)}(\mathbb{R}^n)}$$
$$\leq C_0(2\pi)^{-n/p} \left\|\widetilde{q}\widetilde{f}\right\|_{L^{2p/(p+2)}(\mathbb{R}^n)} \leq C_0(2\pi)^{-n/p} \|\widetilde{q}\|_{L^p(\mathbb{R}^n)} \left\|\widetilde{f}\right\|_{L^2(\mathbb{R}^n)},$$

where $p > n$, $\widetilde{q}(x) = (1+|x|^2)^{\sigma/2}q(x)$, $\widetilde{f}(x) = (1+|x|^2)^{-\sigma/4}f(x)$, and C_0 is equal to

$$C_0 = \left(\int_{\mathbb{R}^n} \frac{\mathrm{d}x}{(1+|x|^2)^{p/2}}\right)^{1/p} = \left(\frac{(\sqrt{\pi})^n}{\Gamma(n/2)} \int_0^\infty r^{(n-2)/2}(1+r)^{-p/2}\mathrm{d}r\right)^{1/p}.$$

Combining this constant C_0 with the latter inequality, we obtain C_p from this Lemma and (23.20). □

We denote by α and γ the following constants:

$$\alpha = 2\left\|\vec{W}\right\|_{L^\infty_\sigma(\mathbb{R}^n)} + C_p\|q\|_{L^p_\sigma(\mathbb{R}^n)}, \quad \gamma = 2\left\|\vec{W}\right\|_{L^2_{\sigma/2}(\mathbb{R}^n)} + \|q\|_{L^2_{\sigma/2}(\mathbb{R}^n)}. \tag{23.21}$$

Theorem 23.15. *Assume that the conditions* (23.16) *are satisfied and assume that* $\beta\alpha < 1$ *with* β *and* α *from* (23.13) *and* (23.21), *respectively. Then the integral equation* (23.18) *has a unique solution* u_{sc} *from the space* $L^2_{-\sigma/2}(\mathbb{R}^n)$, *and uniformly in* k, $|k| \geq 1$, *the following estimate holds:*

$$\|u_{\mathrm{sc}}\|_{L^2_{-\sigma/2}(\mathbb{R}^n)} \leq \frac{\beta\gamma}{1-\beta\alpha}. \tag{23.22}$$

Proof. Lemma 23.14 says that L_k maps in $L^2_{-\sigma/2}(\mathbb{R}^n)$ and

$$\|L_k\|_{L^2_{-\sigma/2}(\mathbb{R}^n)\to L^2_{-\sigma/2}(\mathbb{R}^n)} \leq \beta\alpha < 1.$$

Since $\widetilde{u_0}$ belongs to $L^2_{-\sigma/2}(\mathbb{R}^n)$ with the norm estimate $\beta\gamma$, the integral equation (23.18) has a unique solution u_{sc} from $L^2_{-\sigma/2}(\mathbb{R}^n)$ that can be obtained by the iterations

$$u_{\mathrm{sc}} = (I-L_k)^{-1}(\widetilde{u_0}) = \sum_{j=0}^\infty L_k^{j+1}(u_0).$$

The estimate (23.22) follows now from Lemma 23.14 and from the latter representation for u_{sc}. □

Corollary 23.16. *If the constant* α *from* (23.21) *is small enough, then for fixed* k, $|k| \geq 1$, $u_{\mathrm{sc}}(x,k,\theta)$ *belongs to* $L^\infty(\mathbb{R}^n)$ *in* $x \in \mathbb{R}^n$ *and uniformly in* $\theta \in \mathbb{S}^{n-1}$.

Proof. For α small enough, $\widetilde{u_0} \in L^\infty(\mathbb{R}^n)$ and L_k maps in $L^\infty(\mathbb{R}^n)$ with the norm estimate

$$\|L_k\|_{L^\infty(\mathbb{R}^n)\to L^\infty(\mathbb{R}^n)} \leq c(k)\alpha. \tag{23.23}$$

These two facts yield the proof. □

Lemma 23.17. *Under the assumptions of Theorem 23.15, for fixed $k \geq 1$ and for $f \in L^\infty(\mathbb{R}^n)$ the following asymptotic representation holds:*

$$L_k f(x) = C \frac{e^{ik|x|} k^{(n-3)/2}}{|x|^{(n-1)/2}} \int_{\mathbb{R}^n} e^{-ik(\theta',y)} (2k\theta'\vec{W} + q) f(y) dy + o\left(\frac{1}{|x|^{(n-1)/2}}\right), \tag{23.24}$$

as $|x| \to \infty$, where $\theta' = x/|x|$ and

$$C = \frac{1}{2} \frac{e^{-i\frac{\pi}{4}(n+1)}}{(2\pi)^{(n-1)/2}}.$$

Proof. In this proof we assume (for simplicity) that $n \geq 3$. Since $f \in L^\infty(\mathbb{R}^n)$ and $\vec{W}$ vanishes at infinity, integration by parts leads to

$$L_k f(x) = -2i \int_{\mathbb{R}^n} \nabla_y G_k^+(|x-y|) \vec{W}(y) f(y) dy - \int_{\mathbb{R}^n} G_k^+(|x-y|) q(y) f(y) dy.$$

In view of this, one must study the behavior as $|x| \to \infty$ of the functions

$$G_k^+(|x-y|) = \frac{i}{4} \left(\frac{k}{2\pi|x-y|}\right)^{(n-2)/2} H^{(1)}_{(n-2)/2}(k|x-y|)$$

and

$$\nabla_y G_k^+(|x-y|) = k \frac{x-y}{|x-y|} \frac{i}{4} \left(\frac{k}{2\pi|x-y|}\right)^{(n-2)/2} H^{(1)}_{n/2}(k|x-y|),$$

where $H^{(1)}_\nu$ denotes the Hankel function of the first kind of order ν. The behavior of the latter integrals can be studied by dividing them into two cases: $|y| \leq |x|^a$ and $|y| > |x|^a$, where $a > 0$ is a parameter that we can adjust to our liking. In the first case we have for $a < 1/2$ that

$$|x-y| = |x| - (\theta', y) + O(|x|^{2a-1})$$

and (as a consequence of it) $k|x-y| \to \infty$ for $|x| \to \infty$. Thus, we use the behavior of $H^{(1)}_\nu$ for large argument (see [23])

$$\begin{aligned} H^{(1)}_{(n-2)/2}(z) &= C_n \frac{e^{iz}}{\sqrt{z}} + O\left(\frac{1}{z^{3/2}}\right), \\ H^{(1)}_{n/2}(z) &= -iC_n \frac{e^{iz}}{\sqrt{z}} + O\left(\frac{1}{z^{3/2}}\right), \end{aligned} \tag{23.25}$$

as $|z| \to \infty$, where $C_n = \sqrt{\frac{2}{\pi}} e^{-i\frac{\pi}{4}(n-1)}$, $n \geq 2$. Hence we obtain in this case that

$$G_k^+(|x-y|) = \frac{iC_n}{4(2\pi)^{(n-2)/2}} \frac{e^{ik|x-y|}}{|x-y|^{(n-1)/2}} k^{(n-3)/2} + O\left(\frac{1}{|x|^{(n+1)/2}}\right)$$

$$\nabla_y G_k^+(|x-y|) = \frac{C_n}{4(2\pi)^{(n-2)/2}} \theta' k \frac{e^{ik|x-y|}}{|x-y|^{(n-1)/2}} k^{(n-3)/2} + O\left(\frac{1}{|x|^{(n+1)/2}}\right).$$

Since for $|y| \leq |x|^a$ we have in addition that

$$|x-y|^{-(n-1)/2} = |x|^{-(n-1)/2} + O(|x|^{-(n-1)/2+a-1}), \quad \frac{x-y}{|x-y|} = \frac{x}{|x|} + O(|x|^{a-1}),$$

and

$$e^{ik|x-y|} = e^{ik|x|} e^{-ik(\theta',y)} + O\left(\frac{1}{|x|^{1-2a}}\right),$$

it follows that the first part ($|y| \leq |x|^a$) of $L_k f(x)$ is equal to

$$\widetilde{C_n} \frac{e^{ik|x|}}{|x|^{(n-1)/2}} \int_{|y| \leq |x|^a} e^{-ik(\theta',y)} (2k\theta' \vec{W}(y) + q(y)) f(y) dy + O\left(\frac{1}{|x|^{(n-1)/2+1-2a}}\right),$$

where $\widetilde{C_n} = -\frac{iC_n}{4(2\pi)^{(n-2)/2}}$. We have used here the fact that the conditions (23.16) guarantee that $\vec{W}$ and V belong to $L^1(\mathbb{R}^n)$. This means that $L_k f$ is of the desired form as $|x| \to \infty$ in this case.

Turning now to $L_k f$, where $|y| > |x|^a$, we have two more possibilities: $|x|^a < |y| \leq |x|/2$ and $|y| > |x|/2$. In the first case we have that $|x-y| \geq |x|/2$. Using again the asymptotic (23.25) we may estimate $L_k f$ from above by

$$\frac{C}{|x|^{(n-1)/2}} \int_{|x|<|y| \leq |x|/2} (|\vec{W}(y)| + |V(y)|) dy \, \|f\|_{L^\infty(\mathbb{R}^n)} = o\left(\frac{1}{|x|^{(n-1)/2}}\right),$$

since $\vec{W}$ and V belong to $L^1(\mathbb{R}^n)$. For the case $|y| > |x|/2$ we have two subcases: $k|x-y| < 1$ and $k|x-y| > 1$. For the first subcase we use the behavior of $H_\nu^{(1)}$ for small argument, see [23],

$$H_\nu^{(1)}(z) = c_\nu z^{-\nu} + o(z^{-\nu}), \quad z \to 0+,$$

and obtain that

$$G_k^+(|x-y|) = c(k)|x-y|^{2-n} + o(|x-y|^{2-n}),$$

$$\nabla_y G_k^+(|x-y|) = c(k)|x-y|^{1-n} + o(|x-y|^{1-n}).$$

Hence, $L_k f(x)$ in this subcase can be estimated from above by

$$
\begin{aligned}
C\int_{k|x-y|<1,|y|>|x|/2}&\frac{|\vec{W}(y)||f(y)|\mathrm{d}y}{|x-y|^{n-1}}+C\int_{k|x-y|<1,|y|>|x|/2}\frac{|V(y)||f(y)|\mathrm{d}y}{|x-y|^{n-2}}\\
&\leq C\|f\|_{L^\infty(\mathbb{R}^n)}\left(\left\|\vec{W}\right\|_{L^\infty_\sigma(\mathbb{R}^n)}\int_{|y|>|x|/2}\frac{\mathrm{d}y}{|x-y|^{n-1}|y|^\sigma}\right.\\
&\quad\left.+\|V\|_{L^p_\sigma(\mathbb{R}^n)}\left(\int_{|y|>|x|/2}\frac{\mathrm{d}y}{|x-y|^{(n-2)p'}|y|^{\sigma p'}}\right)^{1/p'}\right)\\
\leq C\|f\|_{L^\infty(\mathbb{R}^n)}&\left(\frac{\left\|\vec{W}\right\|_{L^\infty_\sigma(\mathbb{R}^n)}}{|x|^{n-1+\sigma-n}}+\frac{\|V\|_{L^p_\sigma(\mathbb{R}^n)}}{|x|^{n-2+\sigma-n/p'}}\right)=o\left(\frac{1}{|x|^{(n-1)/2}}\right),
\end{aligned}
$$

since $\sigma > n/p'$, $n < p \leq \infty$, and $n \geq 3$. We have used here the estimates for the convolution of the weak singularities (see, for example, Lemma 34.3).

For the second subcase we can use (23.25) and estimate this part of $L_k f(x)$ from above by

$$
\begin{aligned}
&C\|f\|_{L^\infty(\mathbb{R}^n)}\int_{k|x-y|>1,|y|>|x|/2}\frac{(|\vec{W}|+|V|)(1+|y|)^\sigma}{|x-y|^{(n-1)/2}(1+|y|)^\sigma}\mathrm{d}y\\
&\leq C\|f\|_{L^\infty(\mathbb{R}^n)}\left(\left\|\vec{W}\right\|_{L^p_\sigma(\mathbb{R}^n)}+\|V\|_{L^p_\sigma(\mathbb{R}^n)}\right)\left(\int_{|y|>|x|/2}\frac{\mathrm{d}y}{|x-y|^{(n-1)p'/2}|y|^{\sigma p'}}\right)^{1/p'}\\
&\qquad=o\left(\frac{1}{|x|^{(n-1)/2}}\right)
\end{aligned}
$$

as $|x| \to \infty$, using the estimates for convolution of weak singularities and conditions (23.16). □

Since $u_{\text{sc}}(x,k,\theta)$ for fixed $k \geq 1$ is an L^∞-function in x, Lemma 23.17 yields the asymptotic representation for u as

$$
u(x,k,\theta)=\mathrm{e}^{\mathrm{i}k(x,\theta)}+\frac{1}{2}\frac{\mathrm{e}^{-\mathrm{i}\frac{\pi}{4}(n+1)}}{(2\pi)^{(n-1)/2}}\frac{\mathrm{e}^{\mathrm{i}k|x|}k^{(n-3)/2}}{|x|^{(n-1)/2}}A(k,\theta',\theta)+o\left(\frac{1}{|x|^{(n-1)/2}}\right)
$$

as $|x| \to \infty$, where the function $A(k,\theta',\theta)$ is called the *scattering amplitude* for the magnetic Schrödinger operator and it is defined as

$$
A(k,\theta',\theta)=\int_{\mathbb{R}^n}\mathrm{e}^{-\mathrm{i}k(\theta',y)}(2k\theta'\vec{W}(y)+q(y))u(y,k,\theta)\mathrm{d}y. \tag{23.26}
$$

Substituting $u = u_0 + u_{\text{sc}}$ into (23.26) implies

$$\begin{aligned}A(k,\theta',\theta) &= \int_{\mathbb{R}^n} \mathrm{e}^{-\mathrm{i}k(\theta',y)+\mathrm{i}k(\theta,y)}(2k\theta'\vec{W}(y)+q(y))\mathrm{d}y \\ &\quad + \int_{\mathbb{R}^n} \mathrm{e}^{-\mathrm{i}k(\theta',y)}(2k\theta'\vec{W}(y)+q(y))u_{\mathrm{sc}}(y,k,\theta)\mathrm{d}y \\ &=: A_{\mathrm{B}}(k,\theta',\theta)+R(k,\theta',\theta). \end{aligned} \tag{23.27}$$

The function $A_{\mathrm{B}}(k,\theta',\theta)$ is called the *direct Born approximation*. It can be easily checked that

$$\begin{aligned}A_{\mathrm{B}}(k,\theta',\theta) &= 2k\theta'\mathscr{F}(\vec{W})(k(\theta-\theta'))+\mathscr{F}(q)(k(\theta-\theta')) \\ &= k(\theta+\theta')\mathscr{F}(\vec{W})(k(\theta-\theta'))+\mathscr{F}(|\vec{W}|^2+V)(k(\theta-\theta')), \end{aligned} \tag{23.28}$$

where $\mathscr{F}$ denotes the usual n-dimensional Fourier transform.

The direct Born approximation allows us to obtain the approximation $u_{\mathrm{B}}(x,k,\theta)$ for the solution $u(x,k,\theta)$ of the equation $H_{\mathrm{m}}u = k^2 u$ as

$$u_{\mathrm{B}}(x,k,\theta) = \mathrm{e}^{\mathrm{i}k(x,\theta)} + \frac{1}{2}\frac{\mathrm{e}^{-\mathrm{i}\frac{\pi}{4}(n+1)}}{(2\pi)^{(n-1)/2}}\frac{\mathrm{e}^{\mathrm{i}k|x|}k^{(n-3)/2}}{|x|^{(n-1)/2}}A_{\mathrm{B}}(k,\theta',\theta) \tag{23.29}$$

and secondly, to prove the following very practical statement.

Proposition 23.18. *The Fourier transforms of* $|\vec{W}|^2+V$ *and* $\vec{W}$ *can be evaluated as*

$$\begin{aligned}\mathscr{F}(|\vec{W}|^2+V)(\xi) &= \frac{1}{2}(A_{\mathrm{B}}(k,\theta',\theta)+A_{\mathrm{B}}(k,-\theta,-\theta')) \\ \sqrt{4k^2-\xi^2}(\hat{\xi}_\perp,\mathscr{F}(\vec{W}))(\xi) &= \frac{1}{2}(A_{\mathrm{B}}(k,\theta',\theta)-A_{\mathrm{B}}(k,-\theta,-\theta')),\end{aligned}$$

where $\xi \neq 0$, $\hat{\xi}_\perp$ *is any unit vector that is orthogonal to* ξ, *and* k,θ',θ *are defined by*

$$\theta = \frac{\xi}{2k}+\frac{\hat{\xi}_\perp}{2k}\sqrt{4k^2-\xi^2}, \theta' = -\frac{\xi}{2k}+\frac{\hat{\xi}_\perp}{2k}\sqrt{4k^2-\xi^2}$$

so that $\xi = k(\theta-\theta')$ *and* $k^2 \leq \xi^2/4$.

Proof. The result follows straightforwardly from (23.28). □

All these results, in particular the direct Born approximation, are valid also for the Schrödinger operator ($\vec{W}=0$) as well as the approximation for the backscattering amplitude (see results below).

One may have interest in the particular case $\theta' = -\theta$. This case leads to the so-called direct backscattering Born approximation, i.e.,

$$A(k,-\theta,\theta) \approx A_{\mathrm{B}}^{\mathrm{b}}(k,-\theta,\theta) := \mathscr{F}(|\vec{W}|^2+V)(2k\theta). \tag{23.30}$$

But the approximation for the backscattering amplitude admits more terms than just the Born backscattering approximation. Namely, the following theorem holds.

Theorem 23.19. *Under the conditions of Theorem 23.15 the backscattering amplitude $A(k,-\theta,\theta)$ admits the following representation:*

$$A(k,-\theta,\theta) = \mathscr{F}(|\vec{W}|^2+V)(2k\theta) - \frac{1}{(2\pi)^n}\int_{\mathbb{R}^n}\frac{\mathscr{F}(\overline{q})(k\theta+\eta)\mathscr{F}(q)(k\theta-\eta)}{\eta^2-k^2-\mathrm{i}0}\mathrm{d}\eta$$
$$+\frac{4k}{(2\pi)^n}\int_{\mathbb{R}^n}\frac{\theta\mathscr{F}(\vec{W})(k\theta+\eta)\eta\mathscr{F}(\vec{W})(k\theta-\eta)}{\eta^2-k^2-\mathrm{i}0}\mathrm{d}\eta + h_{\mathrm{rest}}(k\theta), \tag{23.31}$$

where $\overline{q}$ denotes the complex conjugate of $q = \mathrm{i}\nabla\vec{W} + |\vec{W}|^2 + V$ and where h_{rest} belongs to $L^\infty(\mathbb{R}^n)$ and

$$\|h_{\mathrm{rest}}\|_{L^\infty(\mathbb{R}^n)} \le 3\frac{\beta^2\alpha\gamma^2}{1-\beta\alpha}. \tag{23.32}$$

Proof. The formulas (23.27) and (23.28) for the case $\theta' = -\theta$ show that we need to investigate only

$$R(k,-\theta,\theta) = -2k\theta\int_{\mathbb{R}^n}\mathrm{e}^{\mathrm{i}k(\theta,y)}\vec{W}(y)u_{\mathrm{sc}}(y,k,\theta)\mathrm{d}y + \int_{\mathbb{R}^n}\mathrm{e}^{\mathrm{i}k(\theta,y)}q(y)u_{\mathrm{sc}}(y,k,\theta)\mathrm{d}y. \tag{23.33}$$

But since $u_{\mathrm{sc}} = \sum_{j=1}^\infty L_k^j(u_0)$, we see that (23.33) can be rewritten as

$$\begin{aligned}R(k,-\theta,\theta) &= -2k\theta\int_{\mathbb{R}^n}\mathrm{e}^{\mathrm{i}k(\theta,y)}\vec{W}(y)L_k u_0(y,k,\theta)\mathrm{d}y\\
&\quad+\int_{\mathbb{R}^n}\mathrm{e}^{\mathrm{i}k(\theta,y)}q(y)L_k u_0(y,k,\theta)\mathrm{d}y\\
&\quad-2k\theta\int_{\mathbb{R}^n}\mathrm{e}^{\mathrm{i}k(\theta,y)}\vec{W}(y)\sum_{j=2}^\infty L_k^j u_0(y,k,\theta)\mathrm{d}y\\
&\quad+\int_{\mathbb{R}^n}\mathrm{e}^{\mathrm{i}k(\theta,y)}q(y)\sum_{j=2}^\infty L_k^j u_0(y,k,\theta)\mathrm{d}y =: R_1+R_2.\end{aligned}$$

The definition of $L_k u_0$ allows us to obtain

$$
\begin{aligned}
R_1 = 4k^2 \int_{\mathbb{R}^n}\int_{\mathbb{R}^n} & \mathrm{e}^{\mathrm{i}k(\theta,y+z)} G_k^+(|y-z|)\theta\vec{W}(y)\theta\vec{W}(z)\mathrm{d}y\mathrm{d}z \\
&+ 2k \int_{\mathbb{R}^n}\int_{\mathbb{R}^n} \mathrm{e}^{\mathrm{i}k(\theta,y+z)} G_k^+(|y-z|)\theta\vec{W}(y)(\overline{q}(z)-q(z))\mathrm{d}y\mathrm{d}z \\
&- \int_{\mathbb{R}^n}\int_{\mathbb{R}^n} \mathrm{e}^{\mathrm{i}k(\theta,y+z)} G_k^+(|y-z|)q(y)\overline{q}(z)\mathrm{d}y\mathrm{d}z =: I_1 + I_2 + I_3.
\end{aligned}
$$

Using now the facts

$$
\mathscr{F}(G_k^+(\cdot))(\eta) = \frac{1}{\eta^2 - k^2 - \mathrm{i}0}
$$

and $\mathscr{F}(\varphi\psi) = (2\pi)^{-n}\mathscr{F}(\varphi) * \mathscr{F}(\psi)$, we obtain that I_j, $j = 1,2,3$, can be written as

$$
\begin{aligned}
I_1 &= \frac{4k^2}{(2\pi)^n}\int_{\mathbb{R}^n} \frac{\theta\mathscr{F}(\vec{W})(k\theta+\eta)\theta\mathscr{F}(\vec{W})(k\theta-\eta)}{\eta^2-k^2-\mathrm{i}0}\mathrm{d}\eta, \\
I_2 &= -\frac{4\mathrm{i}k}{(2\pi)^n}\int_{\mathbb{R}^n} \frac{\theta\mathscr{F}(\vec{W})(k\theta+\eta)\mathscr{F}(\nabla\vec{W})(k\theta-\eta)}{\eta^2-k^2-\mathrm{i}0}\mathrm{d}\eta, \\
&= -I_1 + \frac{4k}{(2\pi)^n}\int_{\mathbb{R}^n} \frac{\theta\mathscr{F}(\vec{W})(k\theta+\eta)\eta\mathscr{F}(\vec{W})(k\theta-\eta)}{\eta^2-k^2-\mathrm{i}0}\mathrm{d}\eta, \\
I_3 &= -\frac{1}{(2\pi)^n}\int_{\mathbb{R}^n} \frac{\mathscr{F}(\overline{q})(k\theta+\eta)\mathscr{F}(q)(k\theta-\eta)}{\eta^2-k^2-\mathrm{i}0}\mathrm{d}\eta.
\end{aligned}
$$

Thus, the second terms in (23.31) are proved. It remains to estimate h_{rest} (or R_2). Indeed, the definition of R_2 allows us to obtain (using integration by parts) that

$$
\begin{aligned}
R_2(k,-\theta,\theta) = &- \int_{\mathbb{R}^n} \mathrm{e}^{\mathrm{i}k(\theta,y)}\overline{q}(y)\sum_{j=2}^{\infty} L_k^j u_0(y,k,\theta)\mathrm{d}y \\
&+ 2\mathrm{i}\int_{\mathbb{R}^n} \mathrm{e}^{\mathrm{i}k(\theta,y)}\vec{W}(y)\cdot\nabla\left(\sum_{j=2}^{\infty} L_k^j u_0(y,k,\theta)\right)\mathrm{d}y.
\end{aligned}
$$

Since

$$
\|L_k\|_{H^1_{-\sigma/2}(\mathbb{R}^n)\to L^2_{-\sigma/2}(\mathbb{R}^n)} \le \beta\alpha,
$$

it follows using duality that $R_2(k,-\theta,\theta)$ can be estimated as

$$
\begin{aligned}
&|R_2(k,-\theta,\theta)| \\
&\le \|q\|_{L^2_{\sigma/2}(\mathbb{R}^n)} \left\|\sum_{j=2}^{\infty} L_k^j u_0\right\|_{L^2_{-\sigma/2}(\mathbb{R}^n)} + 2\left\|\vec{W}\right\|_{H^1_{\sigma/2}(\mathbb{R}^n)} \left\|\nabla\left(\sum_{j=2}^{\infty} L_k^j u_0\right)\right\|_{H^{-1}_{-\sigma/2}(\mathbb{R}^n)} \\
&\qquad \le \|q\|_{L^2_{\sigma/2}(\mathbb{R}^n)}\frac{\beta\alpha\beta\gamma}{1-\beta\alpha} + 2\left\|\vec{W}\right\|_{H^1_{\sigma/2}(\mathbb{R}^n)}\frac{\beta\alpha\beta\gamma}{1-\beta\alpha} \le 3\frac{\beta^2\alpha\gamma^2}{1-\beta\alpha}.
\end{aligned}
$$

Thus, Theorem 23.19 is completely proved. □

Remark 23.20. This theorem (as well as Theorem 23.15) is a generalization of the corresponding results for the Schrödinger operator. But the difference is that compared with the Schrödinger operator, the magnetic Schrödinger operator is not a "small" perturbation of the Laplacian. This is a reason for the smallness of norms in Theorem 23.15. For the Schrödinger operator we do not need this requirement.

Remark 23.21 (One-dimensional case). There is one interesting remark that should be made here. The asymptotic representations (see Theorem 23.15 of formulas (23.24)) and (23.26) coincide with well known formulas in the one-dimensional case. Moreover, the definition (23.26) of the scattering amplitude defines the reflection and transmission coefficients for the one-dimensional "magnetic" Schrödinger operator.

Part III
Operator Theory and Integral Equations

Chapter 24
Introduction

Despite the fact that this part is devoted to Hilbert spaces, it is assumed that the following concepts are known (they are necessary mainly for examples and exercises):

(1) the Lebesgue integral in a bounded domain $\Omega \subset \mathbb{R}^n$ and in $\mathbb{R}^n$;
(2) functions of bounded variation $BV[a,b]$ on an interval $[a,b]$ (see Part I for details);
(3) the Stieltjes integral of continuous functions on $[a,b]$;
(4) a complete normed space $C^k(\overline{\Omega})$, $k = 0,1,2,\ldots$, on a closed bounded domain $\overline{\Omega} \subset \mathbb{R}^n$ defined by

$$C^k(\overline{\Omega}) := \{f : \overline{\Omega} \to \mathbb{C} : \|f\|_{C^k(\overline{\Omega})} := \max_{\overline{\Omega}} \sum_{|\alpha| \le k} |\partial^\alpha f(x)| < \infty\},$$

where α is an n-dimensional multi-index, i.e., $\alpha = (\alpha_1,\ldots,\alpha_n)$, $\alpha_j \in \mathbb{N} \cup \{0\}$, $j = 1,2,\ldots,n$ with $|\alpha| = \alpha_1 + \alpha_2 + \cdots + \alpha_n$ and $\partial^\alpha f = \frac{\partial^{|\alpha|} f}{\partial x_1^{\alpha_1} \cdots \partial x_n^{\alpha_n}}$;

(5) a complete normed space $L^\infty(\Omega)$ defined by

$$L^\infty(\Omega) := \{f : \Omega \to \mathbb{C} : \|f\|_{L^\infty(\Omega)} := \operatorname{ess\,sup}_{\Omega} |f(x)| < \infty\};$$

(6) a complete normed space $L^1(\Omega)$ for an open set $\Omega \subset \mathbb{R}^n$ defined by

$$L^1(\Omega) := \{f : \Omega \to \mathbb{C} : \|f\|_{L^1(\Omega)} := \int_\Omega |f(x)| \mathrm{d}x < \infty\};$$

(7) the generalized (in the L^2 sense) derivatives $\partial^\alpha f(x)$, $\alpha = (\alpha_1,\ldots,\alpha_n)$ (see Part II for details);

V. Serov, *Fourier Series, Fourier Transform and Their Applications to Mathematical Physics*, Applied Mathematical Sciences 197,
DOI 10.1007/978-3-319-65262-7_24

(8) Lebesgue's dominated convergence theorem: let $\Omega \subset \mathbb{R}^n$ be measurable and let $\{f_k(x)\}_{k=1}^{\infty}$ be a sequence of measurable functions converging to $f(x)$ pointwise in Ω; if there exists a function $g(x) \in L^1(\Omega)$ such that $|f_k(x)| \leq g(x)$, $k = 1, 2, \ldots$, then $f(x) \in L^1(\Omega)$ and

$$\lim_{k\to\infty} \int_{\Omega} f_k(x)\mathrm{d}x = \int_{\Omega} f(x)\mathrm{d}x;$$

(9) Fubini's theorem on interchanging the order of integration: if $f(x,y)$ is integrable on $X \times Y$, then

$$\int_X \mathrm{d}x \left(\int_Y f(x,y)\mathrm{d}y \right) = \int_Y \mathrm{d}y \left(\int_X f(x,y)\mathrm{d}x \right) = \int_{X\times Y} f(x,y)\mathrm{d}x\mathrm{d}y;$$

(10) the *uniform boundedness principle* in Hilbert space (*Banach–Steinhaus theorem*): let H be a Hilbert space; suppose that F is a collection of bounded (continuous) linear operators in H; if for all $x \in H$, then one has

$$\sup_{A\in F} \|Ax\|_H < \infty,$$

whence

$$\sup_{A\in F, \|x\|=1} \|Ax\|_H = \sup_{A\in F} \|A\|_{H\to H} < \infty,$$

see Theorem 26.3.

Chapter 25
Inner Product Spaces and Hilbert Spaces

A collection of elements H is called a complex (real) *vector space* (*linear space*) if the following axioms are satisfied:

(1) To every pair $x, y \in H$ there corresponds a vector $x+y$, called the sum, with the following properties:

  (a) $x+y=y+x$;
  (b) $x+(y+z)=(x+y)+z \equiv x+y+z$;
  (c) there exists a unique element $0 \in H$ such that $x+0=x$;
  (d) for every $x \in H$ there exists a unique element $y_1 \in H$ such that $x+y_1=0$. We set $y_1 := -x$.

(2) For every $x \in H$ and every $\lambda, \mu \in \mathbb{C}$ there corresponds a vector $\lambda \cdot x$ such that

  (a) $\lambda(\mu x) = (\lambda\mu)x \equiv \lambda\mu x$;
  (b) $(\lambda+\mu)x = \lambda x + \mu x$;
  (c) $\lambda(x+y) = \lambda x + \lambda y$;
  (d) $1 \cdot x = x$.

Definition 25.1. For a linear space H, a mapping $(\cdot,\cdot) : H \times H \to \mathbb{C}$ is called an *inner product* or a *scalar product* if for every $x, y, z \in H$ and $\lambda \in \mathbb{C}$ the following conditions are satisfied:

(1) $(x,x) \geq 0$ and $(x,x) = 0$ if and only if $x = 0$;
(2) $(x, y+z) = (x,y) + (x,z)$;
(3) $(\lambda x, y) = \lambda(x,y)$;
(4) $(x,y) = \overline{(y,x)}$.

A linear space equipped with an inner product is called an *inner product space*.

V. Serov, *Fourier Series, Fourier Transform and Their Applications to Mathematical Physics*, Applied Mathematical Sciences 197,
DOI 10.1007/978-3-319-65262-7_25

An immediate consequence of this definition is that

$$(\lambda x+\mu y,z)=\lambda(x,z)+\mu(y,z),$$
$$(x,\lambda y)=\overline{\lambda}(x,y),$$

for every $x,y,z\in H$ and $\lambda,\mu\in\mathbb{C}$.

Example 25.2. In the complex Euclidean space $H=\mathbb{C}^n$ the standard inner product is

$$(x,y)=\sum_{j=1}^{n}x_j\overline{y_j},$$

where $x=(x_1,\dots,x_n)\in\mathbb{C}^n$ and $y=(y_1,\dots,y_n)\in\mathbb{C}^n$.

Example 25.3. In the linear space $C[a,b]$ of continuous complex-valued functions, the formula

$$(f,g)=\int_a^b f(x)\overline{g(x)}\mathrm{d}x$$

defines an inner product.

Definition 25.4. Suppose H is an inner product space. Then

(1) $x\in H$ is *orthogonal* to $y\in H$ if $(x,y)=0$.

(2) A system $\{x_\alpha\}_{\alpha\in I}\subset H$ is *orthonormal* if $(x_\alpha,x_\beta)=\delta_{\alpha,\beta}=\begin{cases}1, & \alpha=\beta,\\ 0, & \alpha\neq\beta,\end{cases}$ where I is some index set.

(3) $\|x\|:=\sqrt{(x,x)}$ is called the *length* of $x\in H$.

Exercise 25.1. Prove the *Pythagorean theorem*: If $\{x_j\}_{j=1}^k$, $k\in\mathbb{N}$, is an orthonormal system in an inner product space H, then

$$\|x\|^2=\sum_{j=1}^{k}|(x,x_j)|^2+\left\|x-\sum_{j=1}^{k}(x,x_j)x_j\right\|^2$$

for every $x\in H$.

Exercise 25.2. Prove *Bessel's inequality*: if $\{x_j\}_{j=1}^k$, $k\le\infty$, is an orthonormal system, then

$$\sum_{j=1}^{k}|(x,x_j)|^2\le\|x\|^2,$$

for every $x\in H$.

Exercise 25.3. Prove the *Cauchy–Bunyakovsky–Schwarz inequality*:

$$|(x,y)| \leq \|x\| \, \|y\|, \quad x,y \in H.$$

Prove also that $(\cdot,\cdot)$ is continuous as a map from $H \times H$ to $\mathbb{C}$.

If H is an inner product space, then

$$\|x\| := \sqrt{(x,x)}$$

has the following properties:

(1) $\|x\| \geq 0$ for every $x \in H$ and $\|x\| = 0$ if and only if $x = 0$.
(2) $\|\lambda x\| = |\lambda| \, \|x\|$ for every $x \in H$ and $\lambda \in \mathbb{C}$.
(3) $\|x+y\| \leq \|x\| + \|y\|$ for every $x,y \in H$. This is the *triangle inequality*.

The function $\|\cdot\| = \sqrt{(\cdot,\cdot)}$ is thus a *norm* on H. It is called the *norm induced by the inner product*.

Every inner product space H is a normed space under the induced norm. The *neighborhood* of $x \in H$ is the *open ball* $B_r(x) = \{y \in H : \|x-y\| < r\}$. This system of neighborhoods defines the *norm topology* on H such that the following conditions are satisfied:

(1) Addition $x+y$ is a continuous map $H \times H \to H$.
(2) Scalar multiplication $\lambda \cdot x$ is a continuous map $\mathbb{C} \times H \to H$.
(3) The inner product $(x,y) : H \times H \to \mathbb{C}$ is continuous.

Definition 25.5. (1) A sequence $\{x_j\}_{j=1}^{\infty} \subset H$ is called a *Cauchy sequence* if for every $\varepsilon > 0$ there exists $n_0 \in \mathbb{N}$ such that $\|x_k - x_j\| < \varepsilon$ for $k, j \geq n_0$.
(2) A sequence $\{x_j\}_{j=1}^{\infty} \subset H$ is said to be *convergent* if there exists $x \in H$ such that for every $\varepsilon > 0$ there exists $n_0 \in \mathbb{N}$ such that $\|x - x_j\| < \varepsilon$ whenever $j \geq n_0$.
(3) An inner product space H is a *complete space* if every Cauchy sequence in H converges.

Corollary 25.6. *(1) Every convergent sequence is a Cauchy sequence.*
(2) If $\{x_j\}_{j=1}^{\infty}$ converges to $x \in H$, then

$$\lim_{j\to\infty} \|x_j\| = \|x\|.$$

Definition 25.7 (J. von Neumann, 1925). A *Hilbert space* is an inner product space that is complete (with respect to its norm topology).

Exercise 25.4. Prove that in an inner product space the norm induced by this inner product satisfies the *parallelogram law*

$$\|x+y\|^2 + \|x-y\|^2 = 2\|x\|^2 + 2\|y\|^2.$$

Exercise 25.5. Prove that if in a normed space H the parallelogram law holds, then there is an inner product on H such that $\|x\|^2 = (x,x)$ and that this inner product is defined by the *polarization identity*

$$(x,y) := \frac{1}{4}\left(\|x+y\|^2 - \|x-y\|^2 + \mathrm{i}\,\|x+\mathrm{i}y\|^2 - \mathrm{i}\,\|x-\mathrm{i}y\|^2\right).$$

Exercise 25.6. Prove that on $C[a,b]$ the norm

$$\|f\| = \max_{x\in[a,b]} |f(x)|$$

is not induced by an inner product.

Exercise 25.7. Give an example of an inner product space that is not complete.

Next we list some examples of Hilbert spaces.

(1) The Euclidean spaces $\mathbb{R}^n$ and $\mathbb{C}^n$.
(2) The matrix space $M_n(\mathbb{C})$ consisting of $n \times n$ matrices whose elements are complex numbers. For $A, B \in M_n(\mathbb{C})$ the inner product is given by

$$(A,B) = \sum_{k,j=1}^{n} a_{kj}\overline{b_{kj}} = \mathrm{Tr}(AB^*),$$

where $B^* = \overline{B}^T$.
(3) The *sequence space* $l^2(\mathbb{C})$ defined by

$$l^2(\mathbb{C}) := \left\{ \{x_j\}_{j=1}^{\infty}, x_j \in \mathbb{C} : \sum_{j=1}^{\infty} |x_j|^2 < \infty \right\}.$$

The estimates

$$|x_j + y_j|^2 \le 2\left(|x_j|^2 + |y_j|^2\right), \quad |\lambda x_j|^2 = |\lambda|^2 |x_j|^2$$

and

$$|x_j y_j| \le \frac{1}{2}\left(|x_j|^2 + |y_j|^2\right)$$

imply that $l^2(\mathbb{C})$ is a linear space. Let us define the inner product by

$$(x,y) := \sum_{j=1}^{\infty} x_j \overline{y_j}$$

and prove that $l^2(\mathbb{C})$ is complete. Suppose that $\{x^{(k)}\}_{k=1}^{\infty} \in l^2(\mathbb{C})$ is a Cauchy sequence. Then for every $\varepsilon > 0$ there exists $n_0 \in \mathbb{N}$ such that

$$\left\|x^{(k)} - x^{(m)}\right\|^2 = \sum_{j=1}^{\infty} |x_j^{(k)} - x_j^{(m)}|^2 < \varepsilon^2$$

for $k, m \geq n_0$. This implies that

$$|x_j^{(k)} - x_j^{(m)}| < \varepsilon, \quad j = 1, 2, \ldots,$$

or that $\{x_j^{(k)}\}_{k=1}^{\infty}$ is a Cauchy sequence in $\mathbb{C}$ for every $j = 1, 2, \ldots$. Since $\mathbb{C}$ is a complete space, it follows that $\{x_j^{(k)}\}_{k=1}^{\infty}$ converges for every fixed $j = 1, 2, \ldots$, i.e., there exists $x_j \in \mathbb{C}$ such that

$$x_j = \lim_{k\to\infty} x_j^{(k)}.$$

This fact and

$$\sum_{j=1}^{l} |x_j^{(k)} - x_j^{(m)}|^2 < \varepsilon^2, \quad l \in \mathbb{N},$$

imply that

$$\lim_{m\to\infty} \sum_{j=1}^{l} |x_j^{(k)} - x_j^{(m)}|^2 = \sum_{j=1}^{l} |x_j^{(k)} - x_j|^2 \leq \varepsilon^2$$

for all $k \geq n_0$ and $l \in \mathbb{N}$. Therefore, the sequence

$$s_l := \sum_{j=1}^{l} |x_j^{(k)} - x_j|^2, \quad k \geq n_0,$$

is a monotonically increasing sequence that is bounded from above by ε^2. Hence this sequence has a limit with the same upper bound, i.e.,

$$\sum_{j=1}^{\infty} |x_j^{(k)} - x_j|^2 = \lim_{l\to\infty} \sum_{j=1}^{l} |x_j^{(k)} - x_j|^2 \leq \varepsilon^2,$$

from which we conclude that

$$\|x\| \leq \left\|x^{(k)}\right\| + \left\|x^{(k)} - x\right\| \leq \left\|x^{(k)}\right\| + \varepsilon$$

and $x \in l^2(\mathbb{C})$.

(4) The *Lebesgue space* $L^2(\Omega)$, where $\Omega \subset \mathbb{R}^n$ is an open set. The space $L^2(\Omega)$ consists of all Lebesgue measurable functions f that are square integrable, i.e.,

$$\int_{\Omega} |f(x)|^2 \mathrm{d}x < \infty.$$

This space is a linear space with the inner product

$$(f,g) = \int_\Omega f(x)\overline{g(x)}\mathrm{d}x$$

and the Riesz–Fischer theorem reads as follows: $L^2(\Omega)$ is a Hilbert space.

(5) The *Sobolev spaces* $W_2^k(\Omega)$ consisting of functions $f \in L^2(\Omega)$ whose weak or distributional derivatives $\partial^\alpha f$ also belong to $L^2(\Omega)$ up to order $|\alpha| \le k$, $k = 1, 2, \ldots$. On the space $W_2^k(\Omega)$ the natural inner product is

$$(f,g) = \sum_{|\alpha|\le k} \int_\Omega \partial^\alpha f(x)\overline{\partial^\alpha g(x)}\mathrm{d}x.$$

Definition 25.8. Let H be an inner product space. For a linear subspace $M \subset H$ the *orthogonal complement* of M is defined as

$$M^\perp := \{y \in H : (y,x) = 0, \text{for all} x \in M\}.$$

Remark 25.9. It is clear that $M^\perp$ is a linear subspace of H. Moreover, $M \cap M^\perp = \{0\}$, since we always have $0 \in M$.

Definition 25.10. A *closed subspace* of a Hilbert space H is a linear subspace of H that is closed (i.e., $\overline{M} = M$) with respect to the induced norm.

Remark 25.11. The subspace $M^\perp$ is closed if M is a subset of a Hilbert space.

Theorem 25.12 (Projection theorem). *Suppose M is a closed subspace of a Hilbert space H. Then every $x \in H$ has a unique representation as*

$$x = u + v,$$

where $u \in M$ and $v \in M^\perp$, or equivalently,

$$H = M \oplus M^\perp.$$

Moreover, one has that

$$\|v\| = \inf_{y\in M} \|x - y\| =: d(x,M).$$

Proof. Let $x \in H$. Then

$$d := d(x,M) \equiv \inf_{y\in M} \|x - y\| \le \|x - u\|$$

for all $u \in M$. The definition of infimum implies that there exists a sequence $\{u_j\}_{j=1}^\infty \subset M$ such that

$$d = \lim_{j\to\infty} \|x - u_j\|.$$

The parallelogram law implies that

$$\|u_j - u_k\|^2 = \|(u_j - x) + (x - u_k)\|^2 = 2\|u_j - x\|^2 + 2\|x - u_k\|^2 - 4\left\|x - \frac{u_j + u_k}{2}\right\|^2.$$

Since $(u_j + u_k)/2 \in M$, it follows that

$$\|u_j - u_k\|^2 \le 2\|u_j - x\|^2 + 2\|x - u_k\|^2 - 4d^2 \to 2d^2 + 2d^2 - 4d^2 = 0$$

as $j,k \to \infty$. Hence $\{u_j\}_{j=1}^{\infty} \subset M$ is a Cauchy sequence in the Hilbert space H. This means that there exists $u \in H$ such that

$$u = \lim_{j\to\infty} u_j.$$

But $M = \overline{M}$ implies that $u \in M$. By construction one has that

$$d = \lim_{j\to\infty} \|x - u_j\| = \|x - u\|.$$

Let us set $v := x - u$ and show that $v \in M^\perp$. For all $y \in M$, $y \ne 0$, we introduce the number

$$\alpha = -\frac{(v,y)}{\|y\|^2}.$$

Since $u - \alpha y \in M$, we have

$$d^2 \le \|x - (u - \alpha y)\|^2 = \|v + \alpha y\|^2 = \|v\|^2 + (v, \alpha y) + (\alpha y, v) + |\alpha|^2 \|y\|^2$$
$$= d^2 - \frac{\overline{(v,y)}(v,y)}{\|y\|^2} - \frac{(v,y)(y,v)}{\|y\|^2} + \frac{|(v,y)|^2}{\|y\|^2} = d^2 - \frac{|(v,y)|^2}{\|y\|^2}.$$

This inequality implies that $(y,v) = 0$, which means that $v \in M^\perp$. In order to prove uniqueness, assume that $x = u_1 + v_1 = u_2 + v_2$, where $u_1, u_2 \in M$ and $v_1, v_2 \in M^\perp$. It follows that

$$u_1 - u_2 = v_2 - v_1 \in M \cap M^\perp.$$

But $M \cap M^\perp = \{0\}$, so that $u_1 = u_2$ and $v_1 = v_2$. □

Remark 25.13. In the framework of this theorem we have that

$$\|x\|^2 = \|u\|^2 + \|v\|^2, \quad \|v\|^2 = (x,v), \quad \|u\|^2 = (x,u).$$

Corollary 25.14 (Riesz-Fréchet theorem). *If T is a linear continuous functional on the Hilbert space H, then there exists a unique $h \in H$ such that $T(x) = (x,h)$ for all $x \in H$. Moreover, $\|T\|_{H\to\mathbb{C}} = \|h\|$.*

Proof. If $T \equiv 0$, then $h = 0$ will do. If $T \neq 0$, then there exists $v_0 \in H$ such that $T(v_0) \neq 0$. Let

$$M := \{u \in H : T(u) = 0\}.$$

Then $v_0 \in M^\perp$, $v_0 \neq 0$, and $T(v_0) \neq 0$. Since T is linear and continuous, M is a closed subspace. It follows from Theorem 25.12 that

$$H = M \oplus M^\perp,$$

i.e., every $x \in H$ has a unique representation as $x = \widetilde{u} + \widetilde{v}$. Therefore, for every $x \in H$, we can define

$$u := x - \frac{T(x)}{T(v_0)} v_0.$$

Then $T(u) = 0$, i.e., $u \in M$. It follows that

$$(x, v_0) = (u, v_0) + \frac{T(x)}{T(v_0)} \|v_0\|^2 = \frac{T(x)}{T(v_0)} \|v_0\|^2,$$

or

$$T(x) = \frac{T(v_0)}{\|v_0\|^2}(x, v_0) = \left(x, \frac{\overline{T(v_0)}}{\|v_0\|^2} v_0\right),$$

which is of the desired form. The uniqueness of h can be seen as follows. If $T(x) = (x,h) = (x,\widetilde{h})$, then $(x, h - \widetilde{h}) = 0$ for all $x \in H$. In particular, $\left\|h - \widetilde{h}\right\|^2 = (h - \widetilde{h}, h - \widetilde{h}) = 0$, i.e., $h = \widetilde{h}$. It remains to prove the statement about the norm $\|T\|_{H\to\mathbb{C}} = \|T\|$. Firstly,

$$\|T\| = \sup_{\|x\|\le 1} |T(x)| = \sup_{\|x\|\le 1} |(x,h)| \le \|h\|.$$

On the other hand, $T(h/\|h\|) = \|h\|$ implies that $\|T\| \ge \|h\|$. Thus $\|T\| = \|h\|$. This completes the proof. □

Corollary 25.15. *If M is a linear subspace of a Hilbert space H, then*

$$M^{\perp\perp} := \left(M^\perp\right)^\perp = \overline{M}.$$

Proof. It is not difficult to check that $M^\perp = (\overline{M})^\perp$. Therefore,

$$M^{\perp\perp} = \left((\overline{M})^\perp\right)^\perp,$$

and Theorem 25.12 implies that

$$H = \overline{M} \oplus (\overline{M})^{\perp}, \quad H = (\overline{M})^{\perp} \oplus M^{\perp\perp}.$$

The uniqueness of this representation guarantees that $M^{\perp\perp} = \overline{M}$. □

Definition 25.16. Let $A \subset H$ be a subset of an inner product space. The subset

$$\operatorname{span} A := \left\{ x \in H : x = \sum_{j=1}^{k} \lambda_j x_j, x_j \in A, \lambda_j \in \mathbb{C} \right\}$$

is called the *linear span* of A.

Definition 25.17. Let H be a Hilbert space.

(1) A subset $B \subset H$ is called a *basis* of H if B is linearly independent in H and

$$\overline{\operatorname{span} B} = H,$$

i.e., for every $x \in H$ and every $\varepsilon > 0$ there exist $k \in \mathbb{N}$ and $\{c_j\}_{j=1}^{k} \subset \mathbb{C}$ such that

$$\left\| x - \sum_{j=1}^{k} c_j x_j \right\| < \varepsilon, \quad x_j \in B.$$

(2) H is called *separable* if it has a countable or finite basis.
(3) An orthonormal system $B = \{x_\alpha\}_{\alpha \in A}$ in H that is a basis is called an *orthonormal basis*.

By Gram–Schmidt orthonormalization we may conclude that every separable Hilbert space has an orthonormal basis.

Theorem 25.18 (Characterization of an orthonormal basis). *Let $B = \{x_j\}_{j=1}^{\infty}$ be an orthonormal system in a separable Hilbert space H. Then the following statements are equivalent:*

(1) B is maximal, i.e., it is not a proper subset of any other orthonormal system.
(2) For every $x \in H$ the condition $(x, x_j) = 0$, $j = 1, 2, \ldots$, implies that $x = 0$.
(3) Every $x \in H$ has the Fourier expansion

$$x = \sum_{j=1}^{\infty} (x, x_j) x_j,$$

i.e.,

$$\left\| x - \sum_{j=1}^{k} (x, x_j) x_j \right\| \to 0, \quad k \to \infty.$$

This means that B is an orthonormal basis.

(4) Every pair $x,y \in H$ satisfies the completeness relation

$$(x,y) = \sum_{j=1}^{\infty}(x,x_j)\overline{(y,x_j)}.$$

(5) Every $x \in H$ satisfies Parseval's equality

$$\|x\|^2 = \sum_{j=1}^{\infty}|(x,x_j)|^2.$$

Proof. (1)⇒(2) Suppose that there is $z \in H, z \neq 0$ such that $(z,x_j) = 0$ for all $j = 1,2,\ldots$. Then

$$B' := \left\{\frac{z}{\|z\|}, x_1, x_2, \ldots\right\}$$

is an orthonormal system in H. This fact implies that B is not maximal, which contradicts (1) and proves (2).

(2)⇒(3) Given $x \in H$, we introduce the sequence

$$x^{(k)} = \sum_{j=1}^{k}(x,x_j)x_j.$$

The Pythagorean theorem and Bessel's inequality (Exercises 25.1 and 25.2) imply that

$$\left\|x^{(k)}\right\|^2 = \sum_{j=1}^{k}|(x,x_j)|^2 \leq \|x\|^2.$$

It follows that

$$\sum_{j=1}^{\infty}|(x,x_j)|^2$$

converges. Therefore, for $m < k$,

$$\left\|x^{(k)} - x^{(m)}\right\|^2 = \sum_{j=m+1}^{k}|(x,x_j)|^2 \to 0$$

as $k,m \to \infty$. Hence $x^{(k)}$ is a Cauchy sequence in H. Thus there exists $y \in H$ such that

$$y = \lim_{k\to\infty} x^{(k)} = \sum_{j=1}^{\infty}(x,x_j)x_j.$$

Next, since the inner product is continuous, we deduce that

$$(y,x_j) = \lim_{k\to\infty}(x^{(k)},x_j) = (x,x_j)$$

for all $j = 1,2,\ldots$. Therefore, $(y-x,x_j) = 0$ for all $j = 1,2,\ldots$. Part (2) implies that $y = x$ and part (3) follows.

(3)⇒(4) Let $x,y \in H$. We know from part (3) that

$$x = \sum_{j=1}^{\infty}(x,x_j)x_j, \quad y = \sum_{k=1}^{\infty}(y,x_k)x_k.$$

The continuity of the inner product and the orthonormality of $\{x_j\}_{j=1}^{\infty}$ allow us to conclude that

$$(x,y) = \sum_{j=1}^{\infty}\sum_{k=1}^{\infty}(x,x_j)\overline{(y,x_k)}(x_j,x_k) = \sum_{j=1}^{\infty}(x,x_j)\overline{(y,x_j)}.$$

(4)⇒(5) Take $y = x$ in part (4).

(5)⇒(1) Suppose that B is not maximal. Then we can add a unit vector $z \in H$ to it that is orthogonal to B. Parseval's equality gives then

$$1 = \|z\|^2 = \sum_{j=1}^{\infty}|(z,x_j)|^2 = 0.$$

This contradiction proves the result. □

Exercise 25.8. Let $\{x_j\}_{j=1}^{\infty}$ be an orthonormal system in an inner product space H. Let $x \in H$, $\{c_j\}_{j=1}^{k} \subset \mathbb{C}$, and $k \in \mathbb{N}$. Prove that

$$\left\|x - \sum_{j=1}^{k}(x,x_j)x_j\right\| \le \left\|x - \sum_{j=1}^{k}c_jx_j\right\|.$$

Chapter 26
Symmetric Operators in Hilbert Spaces

Assume that H is a Hilbert space. A *linear operator* from H to H is a mapping

$$A : D(A) \subset H \to H,$$

where $D(A)$ is a linear subspace of H and A satisfies the condition

$$A(\lambda x + \mu y) = \lambda Ax + \mu Ay$$

for all $\lambda, \mu \in \mathbb{C}$ and $x, y \in D(A)$. The space $D(A)$ is called the *domain* of A. The space

$$N(A) := \{x \in D(A) : Ax = 0\}$$

is called the *null space* (or *kernel*) of A. The space

$$R(A) := \{y \in H : y = Ax \text{ for some } x \in D(A)\}$$

is called the *range* of A. Both $N(A)$ and $R(A)$ are linear subspaces of H. We say that A is *bounded* if there exists $M > 0$ such that

$$\|Ax\| \le M\|x\|, \quad x \in D(A).$$

We say that A is *densely defined* if $\overline{D(A)} = H$. In such a case, A can be extended to A_{ex}, which will be defined on the whole H with the same norm estimate, and we may define

$$\|A\|_{H \to H} := \inf\{M : \|Ax\| \le M\|x\|, x \in D(A)\},$$

or equivalently,

$$\|A\|_{H \to H} = \sup_{\|x\|=1} \|Ax\|.$$

V. Serov, *Fourier Series, Fourier Transform and Their Applications to Mathematical Physics*, Applied Mathematical Sciences 197,
DOI 10.1007/978-3-319-65262-7_26

Example 26.1 (Integral operator in L^2). Suppose that $K(s,t) \in L^2(\Omega \times \Omega), \Omega \subset \mathbb{R}^n$. Let us show that the integral operator $\widehat{K}$ defined as

$$\widehat{K}f(s) = \int_\Omega K(s,t)f(t)\mathrm{d}t, \quad f \in L^2(\Omega)$$

is bounded. Indeed,

$$\begin{aligned}
\left\|\widehat{K}f\right\|^2_{L^2(\Omega)} &= \int_\Omega |\widehat{K}f(s)|^2 \mathrm{d}s = \int_\Omega \left|\int_\Omega K(s,t)f(t)\mathrm{d}t\right|^2 \mathrm{d}s \\
&= \int_\Omega \left|(K(s,\cdot),\overline{f})_{L^2}\right|^2 \mathrm{d}s \le \int_\Omega \|K(s,\cdot)\|^2_{L^2} \|\overline{f}\|^2_{L^2} \mathrm{d}s \\
&= \int_\Omega \left(\int_\Omega |K(s,t)|^2 \mathrm{d}t \int_\Omega |f(t)|^2 \mathrm{d}t\right) \mathrm{d}s = \|K\|^2_{L^2(\Omega\times\Omega)} \|f\|^2_{L^2(\Omega)},
\end{aligned}$$

where we have made use of the Cauchy–Bunyakovsky–Schwarz inequality. We therefore have

$$\left\|\widehat{K}\right\|_{L^2 \to L^2} \le \|K\|_{L^2(\Omega\times\Omega)}.$$

The norm

$$\left\|\widehat{K}\right\|_{HS} := \|K\|_{L^2(\Omega\times\Omega)}$$

is called the *Hilbert–Schmidt norm* of $\widehat{K}$.

Example 26.2 (Schur test). Assume that p and q are positive measurable functions on $\Omega \subset \mathbb{R}^n$ and α and β are positive numbers such that

$$\int_\Omega |K(x,y)|p(y)\mathrm{d}y \le \alpha q(x), \quad \text{a.e. in } \Omega$$

and

$$\int_\Omega |K(x,y)|q(x)\mathrm{d}x \le \beta p(y), \quad \text{a.e. in } \Omega.$$

Then $\widehat{K}$ is bounded and

$$\left\|\widehat{K}\right\|_{L^2 \to L^2} \le \sqrt{\alpha\beta}.$$

Proof. For all $f \in L^2(\Omega)$ we have

$$\int_\Omega\left(\int_\Omega |K(x,y)|\cdot|f(y)|\mathrm{d}y\right)^2\mathrm{d}x$$
$$= \int_\Omega\left(\int_\Omega \sqrt{|K(x,y)|}\sqrt{p(y)}\sqrt{\frac{|K(x,y)|}{p(y)}}|f(y)|\mathrm{d}y\right)^2\mathrm{d}x$$
$$\leq \int_\Omega\left(\int_\Omega |K(x,y)|p(y)\mathrm{d}y\right)\left(\int_\Omega \frac{|K(x,y)|}{p(y)}|f(y)|^2\mathrm{d}y\right)\mathrm{d}x$$
$$\leq \alpha\int_\Omega\left(\int_\Omega |K(x,y)|q(x)\mathrm{d}x\right)\frac{|f(y)|^2}{p(y)}\mathrm{d}y \leq \alpha\beta\int_\Omega |f(y)|^2\mathrm{d}y$$

by the Cauchy–Bunyakovsky–Schwarz inequality and Fubini's theorem. □

Exercise 26.1. Assume that α and β are positive constants such that

$$\int_\Omega |K(x,y)|\mathrm{d}y \leq \alpha, \quad \text{a.e. in } \Omega$$

and

$$\int_\Omega |K(x,y)|\mathrm{d}x \leq \beta, \quad \text{a.e. in } \Omega$$

for some measurable function $K(x,y)$ on $\Omega\times\Omega, \Omega\subset\mathbb{R}^n$. Show that the integral operator $\widehat{K}$ with kernel K is bounded in $L^p(\Omega)$ for all $1\leq p\leq\infty$ and

$$\left\|\widehat{K}\right\|_{L^p\to L^p} \leq \alpha^{\frac{1}{p'}}\beta^{\frac{1}{p}}, \quad \frac{1}{p}+\frac{1}{p'}=1.$$

The following fundamental result can be used in the theory of bounded linear operators (see [29]).

Theorem 26.3 (Uniform boundedness principle). *Suppose that a sequence $A_n : H\to H$ of bounded linear operators satisfies the property (pointwise boundedness)*

$$\sup_n \|A_n u\|_H \leq C_u. \tag{26.1}$$

Then there is a constant $C>0$ such that

$$\sup_n \|A_n\|_{H\to H} \leq C.$$

Proof. Let us assume to the contrary that

$$\sup_n \|A_n\|_{H\to H} = +\infty. \tag{26.2}$$

Then for each $k \geq 1$ there exist A_{n_k} and $u_k \in H$ such that

$$\|u_k\| = 4^{-k}, \quad \|A_{n_k}u_k\| \geq \frac{2}{3}\|A_{n_k}\|\,\|u_k\|, \quad \|A_{n_k}u_k\| \geq 2(M_{k-1}+k), \tag{26.3}$$

where $M_0 = 1$ and $M_k = \sup_n \|A_n(u_1 + \cdots + u_k)\|$. Indeed, by (26.2) there exists A_{n_1} with $\|A_{n_1}\| \geq 24$. The definition of the norm of a linear operator allows us to choose $\widetilde{u_1}$ such that $\|\widetilde{u_1}\| = 1$ and $\|A_{n_1}\widetilde{u_1}\| \geq \frac{2}{3}\|A_{n_1}\|$. Setting $u_1 = \widetilde{u_1}/4$ shows that all conditions (26.3) are satisfied for $k = 1$ and with $M_0 = 1$.

Assuming that $u_1, u_2, \ldots, u_{k-1}, A_{n_1}, A_{n_2}, \ldots, A_{n_{k-1}}$ have been defined, choose A_{n_k} such that

$$\|A_{n_k}\| \geq 3 \cdot 4^k(M_{k-1}+k),$$

which is possible by hypothesis (26.2). With this choice of A_{n_k} there exists $\widetilde{u_k}$ such that $\|\widetilde{u_k}\| = 1$ and $\|A_{n_k}\widetilde{u_k}\| \geq \frac{2}{3}\|A_{n_k}\|$. Setting $u_k = \widetilde{u_k}/4^k$, we have again that $\|u_k\| = 4^{-k}$ and

$$\|A_{n_k}u_k\| \geq \frac{2}{3} \cdot 4^{-k}\|A_{n_k}\| \geq \frac{2}{3} \cdot 4^{-k} 3 \cdot 4^k(M_{k-1}+k) = 2(M_{k-1}+k).$$

To complete the proof we put $u := \sum_{k=1}^{\infty} u_k$, which is well defined in H. But then we have

$$\left\|A_{n_k}\sum_{j=k+1}^{\infty} u_j\right\| \leq \|A_{n_k}\| \sum_{j=k+1}^{\infty} \|u_j\| \leq \|A_{n_k}\| \sum_{j=k+1}^{\infty} 4^{-j} = \frac{1}{3}\|A_{n_k}\|\,\|u_k\|.$$

By the triangle inequality and the definition of M_k we have

$$\begin{aligned}\|A_{n_k}u\| &\geq \|A_{n_k}u_k\| - \left\|A_{n_k}\sum_{j=1}^{k-1} u_j\right\| - \left\|A_{n_k}\sum_{j=k+1}^{\infty} u_j\right\| \\ &\geq \|A_{n_k}u_k\| - M_{k-1} - \frac{1}{2}\|A_{n_k}u_k\| \geq k.\end{aligned}$$

This contradiction with (26.1) proves the theorem. □

Remark 26.4. The uniform boundedness principle holds not only in Hilbert spaces but also in Banach spaces (complete normed spaces). This fact follows straightforwardly from the proof.

Corollary 26.5 (Banach–Steinhaus). *Under the conditions of Theorem 26.3 it is true that for pointwise convergence $A_n u \to Au$, $n \to \infty$, for all $u \in H$ it is necessary and sufficient that $\sup_n \|A_n\|_{H \to H} \leq C$ and that $A_n u \to Au, n \to \infty$ for all $u \in U$, where U is some dense subset of H.*

Corollary 26.6 (Trigonometric Fourier series). *There exists a continuous function whose Fourier partial sums $S_n f(x)$ do not remain uniformly bounded. For every*

$x \in [-\pi, \pi]$ there exists a continuous function whose Fourier partial sums $S_n f(x)$ are unbounded at x.

Proof. Let us consider on the Banach space $C[-\pi, \pi]$ of continuous and periodic functions on the interval $[-\pi, \pi]$ the linear operators

$$f \mapsto S_n f(x) = \sum_{n \le N} c_n(f) \mathrm{e}^{inx},$$

where $c_n(f)$ are the trigonometric Fourier coefficients of f. Since we have the sharp estimate

$$\|S_n f\|_{L^\infty(-\pi,\pi)} \le \frac{1}{2\pi} \int_{-\pi}^{\pi} |D_N(x)| \mathrm{d}x \, \|f\|_{L^\infty(-\pi,\pi)},$$

where $D_N(x)$ is the Dirichlet kernel (see Chapter 10), choosing the sequence $f_n(x) := \sigma_n(f_0)$ defined by Fejér means with $f_0(x) = \operatorname{sgn} D_N(x)$, we obtain

$$\begin{aligned} 2\pi S_N f_n(0) &= \int_{-\pi}^{\pi} f_n(x) D_N(x) \mathrm{d}x \\ &\to \int_{-\pi}^{\pi} f_0(x) D_N(x) \mathrm{d}x = \int_{-\pi}^{\pi} |D_N(x)| \mathrm{d}x = \frac{8 \log N}{\pi} + O(1), \quad N \to \infty, \end{aligned}$$

as is stated in Exercise 10.3. Thus, the linear operators $f \mapsto S_N f(x)$ are bounded (for each fixed N) with operator norms

$$\|S_N\| = \frac{4 \log N}{\pi^2} + O(1).$$

Therefore, by the uniform boundedness principle, there exists a continuous function satisfying the present corollary. □

Exercise 26.2 (Hellinger–Toeplitz). Suppose that $D(A) = H$ and

$$(Ax, y) = (x, Ay), \quad x, y \in H.$$

Prove that A is bounded.

Exercise 26.3. Suppose that $f \in L^1(-\pi, \pi)$ is periodic. Prove that if for some x there exists

$$\lim_{y \to 0} \frac{f(x+y) + f(x-y)}{2},$$

then $S_N f(x) = o(\log N)$ as $N \to \infty$.

Example 26.7 (Differential operator in L^2). Consider the differential operator

$$A := \mathrm{i}\frac{\mathrm{d}}{\mathrm{d}t}$$

of order 1 in $L^2(0,1)$ with domain

$$D(A) = \{f \in C^1[0,1] : f(0) = f(1) = 0\}.$$

First of all, we have that $\overline{D(A)} = L^2(0,1)$; see, e.g., Lemma 17.2. Moreover, integration by parts gives

$$(Af,g) = \int_0^1 \mathrm{i}f'(t)\overline{g(t)}\mathrm{d}t = \mathrm{i}f\overline{g}|_0^1 - \int_0^1 \mathrm{i}f(t)\overline{g'(t)}\mathrm{d}t = \int_0^1 f(t)\overline{\mathrm{i}g'(t)}\mathrm{d}t = (f,Ag)$$

for all $f,g \in D(A)$. Let us now consider the sequence

$$u_n(t) := \sin(n\pi t), \quad n = 1,2,\ldots.$$

Clearly, $u_n \in D(A)$ and

$$\|u_n\|_{L^2}^2 = \int_0^1 |\sin(n\pi t)|^2 \mathrm{d}t = \frac{1}{2}.$$

But

$$\|Au_n\|_{L^2}^2 = \int_0^1 \left|\mathrm{i}\frac{d}{\mathrm{d}t}\sin(n\pi t)\right|^2 \mathrm{d}t = (n\pi)^2 \int_0^1 |\cos(n\pi t)|^2 \mathrm{d}t = \frac{(n\pi)^2}{2} = (n\pi)^2 \|u_n\|_{L^2}^2.$$

Therefore, A is unbounded. This shows that $D(A) = H$ is an essential assumption in Exercise 26.2.

Example 26.8 (Differential operator in L^2). Consider the differential operator

$$A := p_0\frac{\mathrm{d}^2}{\mathrm{d}t^2} + \mathrm{i}p_1\frac{\mathrm{d}}{\mathrm{d}t} + p_2$$

of order 2 in $L^2(0,1)$ with domain

$$D(A) = \{f \in C^2[0,1] : f(0) = f(1) = 0\}$$

and with real nonzero constant coefficients p_0, p_1, and p_2. The fact $\overline{D(A)} = L^2$ and integration by parts gives

$$\begin{aligned}
(Af,g) &= p_0\int_0^1 f''\cdot\overline{g}\mathrm{d}t + \mathrm{i}p_1\int_0^1 f'\cdot\overline{g}\mathrm{d}t + p_2\int_0^1 f\cdot\overline{g}\mathrm{d}t \\
&= p_0\left[f'\overline{g}|_0^1 - \int_0^1 f'\cdot\overline{g}'\mathrm{d}t\right] + \mathrm{i}p_1\left[f\overline{g}|_0^1 - \int_0^1 f\cdot\overline{g}'\mathrm{d}t\right] + p_2(f,g)_{L^2} \\
&= -p_0\int_0^1 f'\cdot\overline{g}'\mathrm{d}t - \mathrm{i}p_1\int_0^1 f\cdot\overline{g}'\mathrm{d}t + (f,p_2 g)_{L^2} \\
&= -p_0\left[f\overline{g'}|_0^1 - \int_0^1 f\cdot\overline{g''}\mathrm{d}t\right] + (f,\mathrm{i}p_1 g')_{L^2} + (f,p_2 g)_{L^2} \\
&= p_0\int_0^1 f\cdot\overline{g''}\mathrm{d}t + (f,\mathrm{i}p_1 g')_{L^2} + (f,p_2 g)_{L^2} = (f,Ag)_{L^2}
\end{aligned}$$

for all $f,g\in D(A)$. Moreover, for the sequence $u_n(t)=\sin(n\pi t)$ we have (for sufficiently large n) that

$$\begin{aligned}
\|Au_n\|_{L^2}^2 &= \int_0^1 |p_0(\sin(n\pi t))'' + \mathrm{i}p_1(\sin(n\pi t))' + p_2\sin(n\pi t)|^2\mathrm{d}t \\
&= \int_0^1 \left[(p_0(n\pi)^2 - p_2)^2\sin^2(n\pi t) + (n\pi)^2 p_1^2\cos^2(n\pi t)\right]\mathrm{d}t \\
&\geq \int_0^1 \left[\frac{(n\pi)^4}{2}p_0^2\sin^2(n\pi t) + (n\pi)^2 p_1^2\cos^2(n\pi t)\right]\mathrm{d}t \\
&\geq (n\pi)^2 p_1^2\int_0^1 (\sin^2(n\pi t)+\cos^2(n\pi t))\,\mathrm{d}t \\
&= 2(n\pi)^2 p_1^2\frac{1}{2} = 2(n\pi)^2 p_1^2\,\|u_n\|_{L^2}^2 .
\end{aligned}$$

So A is unbounded, since

$$\|A\|_{L^2\to L^2}^2 \geq 2(n\pi)^2 p_1^2$$

for $n\to\infty$.

From now on we assume that $\overline{D(A)}=H$, i.e., that A is densely defined in any case.

Definition 26.9. The *graph* $\Gamma(A)$ of a linear operator A in a Hilbert space H is defined as

$$\Gamma(A) := \{(x;y)\in H\times H : x\in D(A)\,\text{and}\,y=Ax\}.$$

Remark 26.10. The graph $\Gamma(A)$ is a linear subspace of a Hilbert space $H\times H$. The inner product in $H\times H$ can be defined as

$$((x_1;y_1),(x_2;y_2))_{H\times H} := (x_1,x_2)_H + (y_1,y_2)_H$$

for all $(x_1;y_1),(x_2;y_2)\in H\times H$.

Definition 26.11. The operator A is said to be *closed* if $\Gamma(A) = \overline{\Gamma(A)}$. We denote this fact by $A = \overline{A}$.

By definition, the *criterion for closedness* is that

$$\begin{cases} x_n \in D(A), \\ x_n \to x, \\ Ax_n \to y \end{cases} \quad \Rightarrow \quad \begin{cases} x \in D(A), \\ y = Ax. \end{cases}$$

The reader is asked to verify that it is also possible to use a seemingly weaker, but equivalent, criterion:

$$\begin{cases} x_n \in D(A), \\ x_n \overset{w}{\to} x, \\ Ax_n \overset{w}{\to} y \end{cases} \quad \Rightarrow \quad \begin{cases} x \in D(A), \\ y = Ax, \end{cases}$$

where $x_n \overset{w}{\to} x$ indicates *weak convergence* in the sense that

$$(x_n, y) \to (x, y)$$

for all $y \in H$.

Remark 26.12. It is important from the point of view of applications (in particular, for numerical procedures) that the closedness of an operator guarantees the convergence of some process to the "correct" result.

Definition 26.13. Let A and A_1 be two linear operators in a Hilbert space H. The operator A_1 is called an *extension* of A (or A is a *restriction* of A_1) if $D(A) \subset D(A_1)$ and $Ax = A_1x$ for all $x \in D(A)$. We denote this fact by $A \subset A_1$ and $A = A_1|_{D(A)}$.

Definition 26.14. An operator A is called *closable* if A has an extension A_1 and $A_1 = \overline{A_1}$. The *closure* of A, denoted by $\overline{A}$, is the smallest closed extension of A if it exists, i.e.,

$$\overline{A} = \bigcap_{\substack{A \subset A_1 \\ A_1 = \overline{A_1}}} A_1.$$

Here, by $A_1 \cap \widetilde{A_1}$ we mean the operator whose domain is $D(A_1 \cap \widetilde{A_1}) := D(A_1) \cap D(\widetilde{A_1})$ and

$$(A_1 \cap \widetilde{A_1})x := A_1x = \widetilde{A_1}x, \quad x \in D(A_1 \cap \widetilde{A_1}),$$

whenever $A \subset A_1 = \overline{A_1}$ and $A \subset \widetilde{A_1} = \overline{\widetilde{A_1}}$.

If A is closable, then $\Gamma\left(\overline{A}\right) = \overline{\Gamma(A)}$.

Definition 26.15. Consider the subspace

$$D^* := \{v \in H : \text{there exists}\, h \in H\, \text{such that}\, (Ax, v) = (x, h)\, \text{for all}\, x \in D(A)\}.$$

The operator A^* with domain $D(A^*) := D^*$ and mapping $A^*v = h$ is called the *adjoint operator* of A.

Exercise 26.4. Prove that A^* exists as a unique linear operator.

Remark 26.16. The adjoint operator is maximal among all linear operators B (in the sense that $B \subset A^*$) that satisfy

$$(Ax, y) = (x, By)$$

for all $x \in D(A)$ and $y \in D(B)$.

Example 26.17. Consider the operator

$$Af(x) := x^{-\alpha} f(x), \quad \alpha > 0$$

in the Hilbert space $H = L^2(0,1)$. Let us define

$$D(A) := \{f \in L^2(0,1) : f(x) = \chi_n(x)g(x), g \in L^2 \text{ for some } n \in \mathbb{N}\},$$

where

$$\chi_n(x) = \begin{cases} 0, & 0 \le x \le 1/n, \\ 1, & 1/n < x \le 1. \end{cases}$$

It is clear that $\overline{D(A)} = L^2(0,1)$. For $v \in D(A^*)$ we have

$$(Af, v) = \int_0^1 x^{-\alpha} \chi_n(x) g(x) \overline{v(x)} \mathrm{d}x = \int_0^1 f(x) \overline{x^{-\alpha} v(x)} \mathrm{d}x = (f, A^* v).$$

We conclude that

$$D(A^*) = \{v \in L^2 : x^{-\alpha} v \in L^2\}.$$

Let us show that A is not closed. To see this, we take the sequence

$$f_n(x) = \begin{cases} x^\alpha, & 1/n < x \le 1, \\ 0, & 0 \le x \le 1/n. \end{cases}$$

Then $f_n \in D(A)$ and

$$Af_n(x) = \begin{cases} 1, & 1/n < x \le 1, \\ 0, & 0 \le x \le 1/n. \end{cases}$$

If we assume that $A = \overline{A}$, then

$$\begin{cases} f_n \in D(A), \\ f_n \to x^\alpha, \\ Af_n \to 1 \end{cases} \Rightarrow \begin{cases} x^\alpha \in D(A), \\ 1 = Ax^\alpha. \end{cases}$$

But $x^\alpha \notin D(A)$. This contradiction shows that A is not closed. It is not bounded either, since $\alpha > 0$.

Theorem 26.18. *Let A be a linear and densely defined operator. Then*

(1) $A^* = \overline{A^*}$.
(2) A is closable if and only if $\overline{D(A^*)} = H$. In this case $A^{**} := (A^*)^* = \overline{A}$.
(3) If A is closable, then $\left(\overline{A}\right)^* = A^*$.

Proof. (1) Let us define in $H \times H$ the linear and bounded operator V as the mapping

$$V : (u; v) \to (v; -u).$$

It has the property $V^2 = -I$. The equality $(Au, v) = (u, A^*v)$ for $u \in D(A)$ and $v \in D(A^*)$ can be rewritten as

$$(V(u; Au), (v; A^*v))_{H\times H} = 0.$$

This implies that $\Gamma(A^*) \perp V\Gamma(A)$ and $\Gamma(A^*) \perp \overline{V\Gamma(A)}$, which in turn means that $\Gamma(A^*) \subset \left(\overline{V\Gamma(A)}\right)^\perp$. Let us check that the criterion for closedness holds, i.e.,

$$\begin{cases} v_n \in D(A^*), \\ v_n \to v, \\ A^*v_n \to y \end{cases} \Rightarrow \begin{cases} v \in D(A^*), \\ y = A^*v. \end{cases}$$

Indeed, for all $u \in D(A)$ we have

$$(Au, v_n) \to (Au, v).$$

On the other hand,

$$(Au, v_n) = (u, A^*v_n) \to (u, y).$$

Hence $(Au, v) = (u, y)$. Thus $v \in D(A^*)$ and $y = A^*v$. This proves (1).
(2) Assume $\overline{D(A^*)} = H$. Then A^{**} exists and due to part (1) we may conclude that

$$\Gamma(A^*) \subset \overline{V\Gamma(A)}^\perp.$$

Then

$$V\Gamma(A) \subset \Gamma(A^*)^{\perp}.$$

It follows that

$$\Gamma(A) \subset (-V\Gamma(A^*))^{\perp},$$

since $V^2 = -I$. Here

$$-V\Gamma(A^*) = \{(-A^*u; u), u \in D(A^*)\}.$$

Thus

$$(-V\Gamma(A^*))^{\perp} = \{(e_1; e_2)\},$$

so that

$$(-A^*u, e_1)_H + (u, e_2)_H = 0,$$

or

$$(A^*u, e_1)_H = (u, e_2)_H.$$

Therefore, $e_1 \in D(A^{**})$ and $A^{**}e_1 = e_2$. This shows that $(e_1; e_2) \in \Gamma(A^{**})$ and hence

$$(-V\Gamma(A^*))^{\perp} \subset \Gamma(A^{**}).$$

Therefore,

$$\Gamma(A) \subset \Gamma(A^{**}),$$

which means that $A \subset A^{**}$, and since A^{**} is closed, A is closable and $\overline{A} \subset A^{**}$. Let us show that in this case, in fact, $\overline{A} = A^{**}$. Indeed, if $u \in D(A^{**})$, then

$$(v, A^{**}u) = (A^*v, u), \quad v \in D(A^*),$$

or

$$(u, A^*v) = (A^{**}u, v), \quad v \in D(A^*),$$

or

$$(Au, v) = (A^{**}u, v), \quad v \in D(A^*).$$

Since $\overline{D(A^*)} = H$, we obtain $Au = A^{**}u$ on $D(A^{**})$. It follows that $A^{**} \subset A$ and furthermore $A^{**} \subset \overline{A}$. Hence $\overline{A} = A^{**}$.
This proves (2) in one direction. Let us assume now that A is closable (i.e., $\overline{A}$ exists and is minimal among all closed extensions) but $\overline{D(A^*)} \neq H$. Then there exists $u_0 \neq 0$ such that $u_0 \perp \overline{D(A^*)}$. So $u_0 \perp D(A^*)$ also. Then

$$(u_0; 0) \perp (v; A^*v), \quad v \in D(A^*).$$

It follows that

$$(u_0, v) = (0, A^*v),$$

or

$$(A^*v, 0) = (v, u_0).$$

In part (1) it is shown that $\Gamma(A) \perp (-V\Gamma(A^*))$. Then

$$\overline{\Gamma(A)} \perp (-V\Gamma(A^*))$$

or

$$\Gamma(\overline{A}) \perp (-V\Gamma(A^*))$$

since $\overline{A}$ exists. Since also $(0; u_0) \perp (-V\Gamma(A^*))$ then $(0; u_0) \in \Gamma(\overline{A})$ i.e. $0 = \overline{A}(0) = u_0 \neq 0$. This contradiction proves (2).

(3) Since A is closable, (1) and (2) imply

$$A^* = \overline{A^*} = (A^*)^{**} = (A)^{***} = (A^{**})^* = \left(\overline{A}\right)^*.$$

This completes the proof. □

Example 26.19. Consider the Hilbert space $H = L^2(\mathbb{R})$ and the operator

$$Au(x) = (u, f_0)u_0(x),$$

where $u_0 \not\equiv 0$, $u_0 \in L^2(\mathbb{R})$, is fixed and $f_0 \neq 0$ is an arbitrary but fixed constant. We consider A on the domain

$$D(A) = \left\{ u \in L^2(\mathbb{R}) : \int_{\mathbb{R}} |f_0 u(x)| \mathrm{d}x < \infty \right\} = L^2(\mathbb{R}) \cap L^1(\mathbb{R}).$$

It is known that $\overline{L^2(\mathbb{R}) \cap L^1(\mathbb{R})} = L^2(\mathbb{R})$. Thus A is densely defined. Let v be an element of $D(A^*)$. Then

$$(Au, v) = ((u, f_0)u_0, v) = (u, f_0)(u_0, v) = \left(u, \overline{(u_0, v)} f_0\right) = (u, (v, u_0) f_0).$$

It means that

$$A^*v = (v, u_0) f_0.$$

But $(v, u_0) f_0$ must belong to $L^2(\mathbb{R})$. Since $(v, u_0) f_0$ is a constant and $f_0 \neq 0$, it follows that (v, u_0) must be equal to 0. Thus

$$u_0 \perp D(A^*),$$

which implies that

$$u_0 \perp \overline{D(A^*)}.$$

Since $u_0 \neq 0$, we have $\overline{D(A^*)} \neq H$. Thus A^* exists but is not densely defined. So A is not closable.

Exercise 26.5. Assume that A is closable. Prove that $D(\overline{A})$ can be obtained as the closure of $D(A)$ by the norm

$$\left(\|Au\|^2 + \|u\|^2\right)^{1/2}.$$

Theorem 26.20 (Closed graph theorem). *If $A : H \to H$ is a linear operator whose graph $\Gamma(A)$ is closed in $H \times H$, then A is bounded.*

Proof. As a closed subspace of the Hilbert space $H \times H$, the graph $\Gamma(A)$ is a Hilbert space (see Exercise 26.5). Let us define the projection mappings P_1 and P_2 as follows:

$$P_1 : \Gamma(A) \to H, \quad P_1(u,v) = u,$$
$$P_2 : \Gamma(A) \to H, \quad P_2(u,v) = v.$$

Since A is linear, both P_1 and P_2 are linear. Moreover, P_1 is injective and surjective and P_1 and P_2 are continuous, since

$$\|P_1(u,v)\|_H = \|u\|_H \leq \|u\|_H + \|v\|_H,$$
$$\|P_2(u,v)\|_H = \|v\|_H \leq \|u\|_H + \|v\|_H.$$

Hence P_1 is a bijective continuous (bounded) linear map of $\Gamma(A)$ onto H and has a continuous (bounded) inverse, since it is open; see [5]. But at the same time,

$$Au = P_2(P_1)^{-1}(u), \quad u \in H,$$

and therefore as a superposition of two bounded linear operators, A is also bounded. □

Definition 26.21. An operator $A : H \to H$ with $\overline{D(A)} = H$ is called

(1) *symmetric* if $A \subset A^*$;
(2) *self-adjoint* if $A = A^*$;
(3) *essentially self-adjoint* if $\left(\overline{A}\right)^* = \overline{A}$.

Remark 26.22. A symmetric operator is always closable, and its closure is also symmetric. Indeed, if $A \subset A^*$, then $D(A) \subset D(A^*)$. Hence

$$H = \overline{D(A)} \subset \overline{D(A^*)} \subset H$$

implies that $\overline{D(A^*)} = H$. Therefore, A is closable. Since $\overline{A}$ is the smallest closed extension of A, we have

$$A \subset \overline{A} \subset A^* = \left(\overline{A}\right)^*,$$

i.e., $\overline{A}$ is also symmetric.

Some properties of a symmetric operator A are as follows:

(1) $A \subset \overline{A} = A^{**} \subset A^*$,
(2) $A = \overline{A} = A^{**} \subset A^*$ if A is closed,
(3) $A = \overline{A} = A^{**} = A^*$ if A is self-adjoint,
(4) $A \subset \overline{A} = A^{**} = A^*$ if A is essentially self-adjoint.

Example 26.23. Consider the operator

$$A := \frac{\mathrm{d}^2}{\mathrm{d}x^2}$$

in the Hilbert space $H = L^2(0,1)$ with domain

$$D(A) = \{f \in C^2[0,1] : f(0) = f(1) = f'(0) = f'(1) = 0\}.$$

It is clear that $\overline{D(A)} = L^2(0,1)$ and A is not closed. Moreover, integration by parts gives

$$(Af,g)_{L^2} = (f,Ag)_{L^2}$$

for every $f \in D(A)$ and $g \in W_2^2(0,1)$. That is, A is symmetric such that $A \subset A^*$ and $D(A^*) = W_2^2(0,1)$. As we know, $A^* = \overline{A^*}$ always. Now we will show that $\overline{A}$ is the same differential operator of order 2 with $D(\overline{A}) = \overset{\circ}{W}{}_2^2(0,1)$, where $\overset{\circ}{W}{}_2^2(0,1)$ denotes the closure of $D(A)$ with respect to the norm of the Sobolev space $W_2^2(0,1)$. Indeed, for every $f \in D(A)$ we have

$$\|Af\|_{L^2}^2 + \|f\|_{L^2}^2 \le \|f\|_{W_2^2}^2$$

and

$$\begin{aligned}\|f\|_{W_2^2}^2 &= \|Af\|_{L^2}^2 + \|f\|_{L^2}^2 + \int_0^1 |f'|^2 \mathrm{d}x \\ &= \|Af\|_{L^2}^2 + \|f\|_{L^2}^2 - \int_0^1 f\overline{f''}\mathrm{d}x \le \frac{3}{2}\|Af\|_{L^2}^2 + \frac{3}{2}\|f\|_{L^2}^2 .\end{aligned}$$

This means that

$$\|Af\|_{L^2}^2 + \|f\|_{L^2}^2 \asymp \|f\|_{W_2^2}^2 .$$

Exercise 26.5 gives now that

$$D(\overline{A}) = \overset{\circ}{W}{}_2^2(0,1).$$

So we have finally

$$D(A) \subsetneq D(\overline{A}) = \overset{\circ}{W}{}_2^2(0,1) = D(A^{**}) \subsetneq W_2^2(0,1).$$

The closure $\overline{A}$ is symmetric but not self-adjoint, since

$$\overset{\circ}{W}{}_2^2(0,1) = D(\overline{A}) \neq D(\overline{A}^*) = D(A^*) = W_2^2(0,1).$$

Theorem 26.24 (J. von Neumann). *Assume that $A \subset A^*$.*

(1) If $D(A) = H$, then $A = A^*$ and A is bounded.
(2) If $R(A) = H$, then $A = A^*$ and A^{-1} exists and is bounded.
(3) If A^{-1} exists, then $A = A^*$ if and only if $A^{-1} = \left(A^{-1}\right)^*$.

Proof. (1) Since $A \subset A^*$, we have $H = D(A) \subset D(A^*) \subset H$ and hence $D(A) = D(A^*) = H$. Thus $A = A^*$, and the Hellinger–Toeplitz theorem (Exercise 26.2) says that A is bounded.

(2), (3) Let us assume that $u_0 \in D(A)$ and $Au_0 = 0$. Then for all $v \in D(A)$ we obtain that

$$0 = (Au_0, v) = (u_0, Av).$$

This means that $u_0 \perp H$ and therefore $u_0 = 0$. It follows that A^{-1} exists and $D(A^{-1}) = R(A) = H$. Hence $\left(A^{-1}\right)^*$ exists. Let us prove that $(A^*)^{-1}$ exists too and $(A^*)^{-1} = \left(A^{-1}\right)^*$. Indeed, if $u \in D(A)$ and $v \in D\left(\left(A^{-1}\right)^*\right)$, then

$$(u,v) = (A^{-1}Au, v) = \left(Au, \left(A^{-1}\right)^* v\right).$$

This equality implies that

$$\left(A^{-1}\right)^* v \in D(A^*)$$

and

$$A^* \left(A^{-1}\right)^* v = v. \tag{26.4}$$

Similarly, if $u \in D(A^{-1})$ and $v \in D(A^*)$, then

$$(u,v) = (AA^{-1}u, v) = (A^{-1}u, A^*v)$$

and therefore

$$A^*v \in D\left(\left(A^{-1}\right)^*\right)$$

and

$$\left(A^{-1}\right)^* A^* v = v. \tag{26.5}$$

It follows from (26.4) and (26.5) that $(A^*)^{-1}$ exists and $(A^*)^{-1} = \left(A^{-1}\right)^*$. The boundedness of A^{-1} follows from part (1).

Exercise 26.6. Let A and B be injective operators. Prove that if $A \subset B$, then $A^{-1} \subset B^{-1}$.

Since $A \subset A^*$, we have by Exercise 26.6 that

$$A^{-1} \subset (A^*)^{-1} = \left(A^{-1}\right)^*,$$

i.e., A^{-1} is also symmetric. But $D(A^{-1}) = H$. We conclude that $H = D(A^{-1}) \subset D\left(\left(A^{-1}\right)^*\right) \subset H$ and hence $D(A^{-1}) = D\left(\left(A^{-1}\right)^*\right) = H$. Thus A^{-1} is self-adjoint and bounded (Hellinger–Toeplitz theorem; see Exercise 26.2). Finally,

$$A^{-1} = \left(A^{-1}\right)^* = (A^*)^{-1}$$

if and only if $A = A^*$.
This completes the proof. □

Theorem 26.25 (Basic criterion of self-adjointness). *If $A \subset A^*$, then the following statements are equivalent:*

(1) $A = A^*$.
(2) $A = \overline{A}$ and $N(A^* \pm \mathrm{i}I) = \{0\}$.
(3) $R(A \pm \mathrm{i}I) = H$.

Proof. (1) ⇒ (2) Since $A = A^*$, it follows that A is closed. Suppose that $u_0 \in N(A^* - \mathrm{i}I)$, i.e., $u_0 \in D(A^*) = D(A)$ and $Au_0 = \mathrm{i}u_0$. Then

$$\mathrm{i}(u_0, u_0) = (\mathrm{i}u_0, u_0) = (Au_0, u_0) = (u_0, Au_0) = (u_0, \mathrm{i}u_0) = -\mathrm{i}(u_0, u_0).$$

This implies that $u_0 = 0$, i.e., $N(A^* - \mathrm{i}I) = \{0\}$. The proof of $N(A^* + \mathrm{i}I) = \{0\}$ is left to the reader.

(2) ⇒ (3) Since $A = \overline{A}$ and $N(A^* \pm \mathrm{i}I) = \{0\}$, it follows, for example, that the equation $A^*u = -\mathrm{i}u$ has only the trivial solution $u = 0$. This implies that $\overline{R(A - \mathrm{i}I)} = H$. For otherwise, there exists $u_0 \neq 0$ such that $u_0 \perp \overline{R(A - \mathrm{i}I)}$. This means that for all $u \in D(A)$ we have

$$((A - \mathrm{i}I)u, u_0) = 0$$

and therefore $u_0 \in D(A^* + \mathrm{i}I)$ and $(A^* + \mathrm{i}I)u_0 = 0$, or $A^*u_0 = -\mathrm{i}u_0, u_0 \neq 0$. This contradiction proves that $\overline{R(A - \mathrm{i}I)} = H$. Next, since A is closed, $\Gamma(A)$ is also closed, and due to the fact that A is symmetric, we have

$$\begin{aligned}\|(A-\mathrm{i}I)u\|^2 &= ((A-\mathrm{i}I)u,(A-\mathrm{i}I)u)\\ &= \|Au\|^2-\mathrm{i}(u,Au)+\mathrm{i}(Au,u)+\|u\|^2=\|Au\|^2+\|u\|^2\end{aligned}$$

for $u \in D(A)$. It follows that if $(A-\mathrm{i}I)u_n \to v_0$, then Au_n and u_n are convergent, i.e., $Au_n \to v_0'$, $u_n \to u_0'$, and $u_n \in D(A)$. The closedness of A implies that $u_0' \in D(A)$ and $v_0' = Au_0'$, i.e., $(A-\mathrm{i}I)u_n \to Au_0' - \mathrm{i}u_0' = v_0$. This means that $R(A-\mathrm{i}I)$ is a closed set, i.e., $\overline{R(A-\mathrm{i}I)} = R(A-\mathrm{i}I) = H$. The proof of $R(A+\mathrm{i}I) = H$ is left to the reader.

(3) $\Rightarrow$ (1) Assume that $R(A\pm\mathrm{i}I) = H$. Since $A \subset A^*$, it suffices to show that $D(A^*) \subset D(A)$. For every $u \in D(A^*)$ we have $(A^*-\mathrm{i}I)u \in H$. Part (3) implies that there exists $v_0 \in D(A)$ such that

$$(A-\mathrm{i}I)v_0 = (A^*-\mathrm{i}I)u.$$

It is clear that $u - v_0 \in D(A^*)$ (since $A \subset A^*$) and

$$\begin{aligned}(A^*-\mathrm{i}I)(u-v_0) &= (A^*-\mathrm{i}I)u-(A^*-\mathrm{i}I)v_0 = (A^*-\mathrm{i}I)u-(A-\mathrm{i}I)v_0\\ &= (A-\mathrm{i}I)v_0-(A-\mathrm{i}I)v_0 = 0.\end{aligned}$$

Hence $u - v_0 \in N(A^*-\mathrm{i}I)$.

Exercise 26.7. Let A be a linear and densely defined operator in the Hilbert space H. Prove that

$$H = N(A^*) \oplus \overline{R(A)}.$$

By this exercise we know that

$$H = N(A^*-\mathrm{i}I) \oplus \overline{R(A+\mathrm{i}I)}.$$

But in our case $R(A+\mathrm{i}I) = H$. Hence $N(A^*-\mathrm{i}I) = \{0\}$ and therefore $u = v_0$. Thus $D(A) = D(A^*)$.
This concludes the proof. □

Example 26.26. Assume that an operator A is symmetric and closed in a Hilbert space H. Consider the operator A^*A on the domain

$$D(A^*A) = \{f \in D(A) : Af \in D(A^*)\}.$$

This operator is self-adjoint. Indeed, since $(A^*A)^* = A^*A^{**} = A^*\overline{A} = A^*A$, we have that A^*A is symmetric. At the same time, for all $f \in D(A)$, we have

$$(A^*Af,f) = (Af,A^{**}f) = (Af,\overline{A}f) = \|Af\|_H^2.$$

This fact leads to $R(A^*A\pm\mathrm{i}I) = H$, since $A^*A \pm \mathrm{i}I$ is invertible in this case. Thus, Theorem 26.25 gives us that A^*A is self-adjoint. The same is true for the operator

AA^* on the domain

$$D(AA^*) = \{f \in D(A^*) : A^*f \in D(A)\}.$$

It is clear that in general,

$$AA^* \neq A^*A.$$

If equality holds here, the operator A is said to be *normal.*

Exercise 26.8. Let $H = L^2(0,1)$ and $A := \mathrm{i}\dfrac{\mathrm{d}}{\mathrm{d}x}$.

(1) Prove that A is closed and symmetric on the domain

$$D(A) = \{f \in L^2(0,1) : f' \in L^2(0,1), f(0) = f(1) = 0\} \equiv \overset{\circ}{W}{}_2^1(0,1).$$

(2) Prove that A is self-adjoint on the domain

$$D_\gamma(A) = \left\{f \in L^2(0,1) : f' \in L^2(0,1), f(0) = f(1)\mathrm{e}^{\mathrm{i}\gamma}, \gamma \in \mathbb{R}\right\}.$$

Chapter 27
John von Neumann's Spectral Theorem

Definition 27.1. A bounded linear operator P on a Hilbert space H that is self-adjoint and *idempotent*, i.e., $P^2 = P$, is called an orthogonal projection operator or a *projector*.

Proposition 27.2. *Let P be a projector. Then*

(1) $\|P\| = 1$ *if* $P \neq 0$.
(2) P *is a projector if and only if* $P^{\perp} := I - P$ *is a projector.*
(3) $H = R(P) \oplus R(P^{\perp})$, $P|_{R(P)} = I$, *and* $P|_{R(P^{\perp})} = 0$.
(4) There is a one-to-one correspondence between projectors on H and closed linear subspaces of H. More precisely, if $M \subset H$ is a closed linear subspace, then there exists a projector $P_M : H \to M$, and conversely, if $P : H \to H$ is a projector, then $R(P)$ is a closed linear subspace.
(5) If $\{e_j\}_{j=1}^N, N \le \infty$ is an orthonormal system, then

$$P_N x := \sum_{j=1}^{N} (x, e_j) e_j, \quad x \in H,$$

is a projector.

Proof. (1) Since $P = P^*$ and $P = P^2$, we have $P = P^*P$. Hence $\|P\| = \|P^*P\|$. But $\|P^*P\| = \|P\|^2$. Indeed,

$$\|P^*P\| \le \|P^*\| \, \|P\| \le \|P\|^2$$

and

$$\|P\|^2 = \sup_{\|x\|=1} \|Px\|^2 = \sup_{\|x\|=1} (Px, Px)$$
$$= \sup_{\|x\|=1} (P^*Px, x) \le \sup_{\|x\|=1} \|P^*Px\| = \|P^*P\|.$$

V. Serov, *Fourier Series, Fourier Transform and Their Applications to Mathematical Physics*, Applied Mathematical Sciences 197,
DOI 10.1007/978-3-319-65262-7_27

Therefore, $\|P\| = \|P\|^2$, or $\|P\| = 1$, if $P \neq 0$.

(2) Since P is linear and bounded, the same is true about $I - P$. Moreover,

$$(I-P)^* = I - P^* = I - P$$

and

$$(I-P)^2 = (I-P)(I-P) = I - 2P + P^2 = I - P.$$

(3) It follows immediately from $I = P + P^{\perp}$ that every $x \in H$ is of the form $u + v$, where $u \in R(P)$ and $v \in R(P^{\perp})$. Let us prove that $R(P) = \left(R(P^{\perp})\right)^{\perp}$. First assume that $w \in \left(R(P^{\perp})\right)^{\perp}$, i.e., $(w, (I-P)x) = 0$ for all $x \in H$. This is equivalent to

$$(w,x) = (w, Px) = (Pw, x), \quad x \in H,$$

or $Pw = w$. Hence $w \in R(P)$, and so we have proved that $\left(R(P^{\perp})\right)^{\perp} \subset R(P)$. For the opposite embedding we let $w \in R(P)$. Then there exists $x_w \in H$ such that $w = Px_w$. If $z \in R(P^{\perp})$, then $z = P^{\perp}x_z = (I-P)x_z$ for some $x_z \in H$. Thus

$$(w,z) = (Px_w, (I-P)x_z) = (Px_w, x_z) - (Px_w, Px_z) = 0,$$

since P is a projector. Therefore, $w \in \left(R(P^{\perp})\right)^{\perp}$, and we may conclude that $R(P) = \left(R(P^{\perp})\right)^{\perp}$. This fact allows us to conclude that $R(P) = \overline{R(P)}$ and $H = R(P) \oplus R(P^{\perp})$. Moreover, it is easy to check from the definition that $P|_{R(P)} = I$ and $P|_{R(P^{\perp})} = 0$.

(4) If $M \subset H$ is a closed subspace, then Theorem 25.12 implies that $x = u + v \in H$, where $u \in M$ and $v \in M^{\perp}$. In that case, let us define $P_M : H \to M$ as

$$P_M x = u.$$

It is clear that $P_M^2 x = P_M u = u = P_M x$, i.e., $P_M^2 = P_M$. Moreover, if $y \in H$, then $y = u_1 + v_1, u_1 \in M, v_1 \in M^{\perp}$ and

$$(P_M x, y) = (u, u_1 + v_1) = (u, u_1) = (u + v, u_1) = (u + v, P_M y) = (x, P_M y),$$

i.e., $P_M^* = P_M$. Hence P_M is a projector. If P is a projector, then we know from part (3) that $M := R(P)$ is a closed subspace of H.

(5) Let us assume that $N = \infty$. Define M as

$$M := \left\{ x \in H : x = \sum_{j=1}^{\infty} c_j e_j, \ \sum_{j=1}^{\infty} |c_j|^2 < \infty \right\}.$$

Then M is a closed subspace of H. If we define a linear operator P_M as

$$P_M x := \sum_{j=1}^{\infty} (x, e_j) e_j, \quad x \in H,$$

then by Bessel's inequality we obtain that $P_M x \in M$ and

$$\|P_M x\| \leq \|x\| .$$

This means that P_M is a bounded linear operator into M. But $P_M e_j = e_j$ and thus $P_M^2 x = P_M x$ for all $x \in H$. Next, for all $x, y \in H$ we have

$$\begin{aligned}(P_M x, y) &= \left(\sum_{j=1}^{\infty} (x, e_j) e_j, y \right) = \sum_{j=1}^{\infty} (x, e_j) \, (e_j, y) = \sum_{j=1}^{\infty} (x, (y, e_j) e_j) \\ &= \left(x, \sum_{j=1}^{\infty} (y, e_j) e_j \right) = (x, P_M y),\end{aligned}$$

i.e., $P_M^* = P_M$. The case of finite N requires no convergence questions and is left to the reader.

This completes the proof. □

Definition 27.3. A bounded linear operator A on a Hilbert space H is said to be smaller than or equal to a bounded operator B on H if

$$(Ax, x) \leq (Bx, x), \quad x \in H.$$

We denote this fact by $A \leq B$. The operator A is *nonnegative* if $A \geq 0$; A is *positive*, denoted by $A > 0$, if $A \geq c_0 I$ for some $c_0 > 0$.

Remark 27.4. In the framework of this definition, (Ax, x) and (Bx, x) must be real for all $x \in H$.

Proposition 27.5. *For two projectors P and Q the following statements are equivalent:*

(1) $P \leq Q$.
(2) $\|Px\| \leq \|Qx\|$ *for all* $x \in H$.
(3) $R(P) \subset R(Q)$.
(4) $P = PQ = QP$.

Proof. (1) $\Rightarrow$ (2) Follows directly from $(Px, x) = (P^2 x, x) = (Px, Px) = \|Px\|^2$.
(3) $\Rightarrow$ (4) Assume $R(P) \subset R(Q)$. Then $QPx = Px$ or $QP = P$. Conversely, if $QP = P$, then clearly $R(P) \subset R(Q)$. Finally, $P = QP = P^* = (QP)^* = P^* Q^* = PQ$.

(2) ⇒ (4) If (4) holds, then $Px = PQx$ and $\|Px\| = \|PQx\| \le \|Qx\|$ for all $x \in H$. Conversely, if $\|Px\| \le \|Qx\|$, then $Px = QPx + Q^{\perp}Px$ implies that

$$\|Px\|^2 = \|QPx\|^2 + \left\|Q^{\perp}Px\right\|^2 \le \|QPx\|^2.$$

Hence

$$\left\|Q^{\perp}Px\right\|^2 = 0,$$

i.e., $Q^{\perp}Px = 0$ for all $x \in H$. Hence $P = QP = PQ$.

This completes the proof. □

Exercise 27.1. Let $\{P_j\}_{j=1}^{\infty}$ be a sequence of projectors with $P_j \le P_{j+1}$ for each $j = 1, 2, \ldots$. Prove that $\lim_{j\to\infty} P_j := P$ exists and that P is a projector.

Definition 27.6. A linear map $A : H \to H$ with the property

$$\|Ax\| = \|x\|, \quad x \in H,$$

is called an *isometry*.

Exercise 27.2. Prove that

(1) A is an isometry if and only if $A^*A = I$.
(2) Every isometry A has an inverse $A^{-1} : R(A) \to H$ and $A^{-1} = A^*|_{R(A)}$.
(3) If A is an isometry, then AA^* is a projector on $R(A)$.

Definition 27.7. A surjective isometry $U : H \to H$ is called a *unitary operator*.

Remark 27.8. It follows that U is unitary if and only if it is surjective and $U^*U = UU^* = I$, i.e., $(Ux, Uy) = (x, y)$ for all $x, y \in H$.

Definition 27.9. Let H be a Hilbert space. The family of operators $\{E_\lambda\}_{\lambda=-\infty}^{\infty}$ is called a *spectral family* if the following conditions are satisfied:

(1) E_λ is a projector for all $\lambda \in \mathbb{R}$.
(2) $E_\lambda \le E_\mu$ for all $\lambda < \mu$.
(3) $\{E_\lambda\}$ is right continuous with respect to the strong operator topology, i.e.,

$$\lim_{s\to t+} \|E_s x - E_t x\| = 0$$

for all $x \in H$.

(4) $\{E_\lambda\}$ is normalized as follows:

$$\lim_{\lambda\to-\infty}\|E_\lambda x\| = 0, \quad \lim_{\lambda\to+\infty}\|E_\lambda x\| = \|x\|$$

for all $x \in H$. The latter condition can also be formulated as

$$\lim_{\lambda\to+\infty}\|E_\lambda x - x\| = 0.$$

Remark 27.10. It follows from the previous definition and Proposition 27.5 that

$$E_\lambda E_\mu = E_{\min\{\lambda,\mu\}}.$$

Proposition 27.11. *For every fixed $x, y \in H$, $(E_\lambda x, y)$ is a function of bounded variation with respect to $\lambda \in \mathbb{R}$.*

Proof. Let us define

$$E(\alpha,\beta] := E_\beta - E_\alpha, \quad \alpha < \beta.$$

Then $E(\alpha,\beta]$ is a projector. Indeed,

$$E(\alpha,\beta]^* = E_\beta^* - E_\alpha^* = E_\beta - E_\alpha = E(\alpha,\beta],$$

i.e., $E(\alpha,\beta]$ is self-adjoint. It is also idempotent due to

$$\begin{aligned}(E(\alpha,\beta])^2 &= (E_\beta - E_\alpha)(E_\beta - E_\alpha) = E_\beta^2 - E_\alpha E_\beta - E_\beta E_\alpha + E_\alpha^2 \\ &= E_\beta - E_\alpha - E_\alpha + E_\alpha = E(\alpha,\beta].\end{aligned}$$

Another property is that

$$E(\alpha_1,\beta_1]x \perp E(\alpha,\beta]y, \quad x,y \in H$$

if $\beta_1 \le \alpha$ or $\beta \le \alpha_1$. To see this for $\beta_1 \le \alpha$ we calculate

$$\begin{aligned}(E(\alpha_1,\beta_1]x, E(\alpha,\beta]y) &= (E_{\beta_1}x - E_{\alpha_1}x, E_\beta y - E_\alpha y) \\ &= (E_{\beta_1}x, E_\beta y) - (E_{\alpha_1}x, E_\beta y) - (E_{\beta_1}x, E_\alpha y) + (E_{\alpha_1}x, E_\alpha y) \\ &= (x, E_{\beta_1}y) - (x, E_{\alpha_1}y) - (x, E_{\beta_1}y) + (x, E_{\alpha_1}y) = 0.\end{aligned}$$

Let now

$$\lambda_0 < \lambda_1 < \cdots < \lambda_n.$$

Then

$$
\begin{aligned}
\sum_{j=1}^{n} |(E_{\lambda_j}x,y) - (E_{\lambda_{j-1}}x,y)| &= \sum_{j=1}^{n} \left|(E(\lambda_{j-1},\lambda_j]x,y)\right| \\
&= \sum_{j=1}^{n} \left|(E(\lambda_{j-1},\lambda_j]x, E(\lambda_{j-1},\lambda_j]y)\right| \\
&\leq \sum_{j=1}^{n} \|E(\lambda_{j-1},\lambda_j]x\| \, \|E(\lambda_{j-1},\lambda_j]y\| \\
&\leq \left(\sum_{j=1}^{n} \|E(\lambda_{j-1},\lambda_j]x\|^2\right)^{1/2} \left(\sum_{j=1}^{n} \|E(\lambda_{j-1},\lambda_j]y\|^2\right)^{1/2} \\
&= \left\|\sum_{j=1}^{n} E(\lambda_{j-1},\lambda_j]x\right\| \left\|\sum_{j=1}^{n} E(\lambda_{j-1},\lambda_j]y\right\| \\
&= \|E(\lambda_0,\lambda_n]x\| \, \|E(\lambda_0,\lambda_n]y\| \leq \|x\| \, \|y\|.
\end{aligned}
$$

Here we have made use of orthogonality, normalization, and the Cauchy–Bunyakovsky–Schwarz inequality. □

By Proposition 27.11 we can define a Stieltjes integral. Indeed, for every continuous function $f(\lambda)$ we may conclude the equality of limits

$$
\lim_{\Delta\to 0} \sum_{j=1}^{n} f(\lambda_j^*) \left(E(\lambda_{j-1},\lambda_j]x,y\right) = \lim_{\Delta\to 0} \left(\sum_{j=1}^{n} f(\lambda_j^*) E(\lambda_{j-1},\lambda_j]x, y\right),
$$

where $\lambda_j^* \in [\lambda_{j-1},\lambda_j]$, $\alpha = \lambda_0 < \lambda_1 < \cdots < \lambda_n = \beta$, and $\Delta = \max_{1\leq j\leq n} |\lambda_{j-1} - \lambda_j|$ exists, and by definition this limit is

$$
\int_{\alpha}^{\beta} f(\lambda) \mathrm{d}(E_\lambda x, y), \quad x,y \in H.
$$

It can be shown that this is equivalent to the existence of the limit in H

$$
\lim_{\Delta\to 0} \sum_{j=1}^{n} f(\lambda_j^*) E(\lambda_{j-1},\lambda_j]x,
$$

which we denote by

$$
\int_{\alpha}^{\beta} f(\lambda) \mathrm{d}E_\lambda x.
$$

Thus

$$
\int_{\alpha}^{\beta} f(\lambda) \mathrm{d}(E_\lambda x, y) = \left(\int_{\alpha}^{\beta} f(\lambda) \mathrm{d}E_\lambda x, y\right), \quad x,y \in H.
$$

For the spectral representation of self-adjoint operators one needs integrals not only over finite intervals but also over the whole line, which is naturally defined as the limit

$$\int_{-\infty}^{\infty} f(\lambda)\mathrm{d}(E_\lambda x,y) = \lim_{\substack{\alpha\to-\infty\\ \beta\to\infty}} \int_{\alpha}^{\beta} f(\lambda)\mathrm{d}(E_\lambda x,y) = \left(\int_{-\infty}^{\infty} f(\lambda)\mathrm{d}E_\lambda x,y\right)$$

if it exists. Deriving first some basic properties of the integral just defined, one can check that

$$\begin{aligned}\int_{-\infty}^{\infty} f(\lambda)\mathrm{d}(E_\lambda E_\beta x,y) &= \int_{-\infty}^{\beta} f(\lambda)\mathrm{d}(E_\lambda x,y)\\ &:= \lim_{\alpha\to-\infty}\int_{\alpha}^{\beta} f(\lambda)\mathrm{d}(E_\lambda x,y), \quad x,y\in H.\end{aligned}$$

Theorem 27.12. *Let $\{E_\lambda\}_{\lambda=-\infty}^{\infty}$ be a spectral family on a Hilbert space H and let f be a real-valued continuous function on the line. Define*

$$D := \left\{x\in H : \int_{-\infty}^{\infty} |f(\lambda)|^2\mathrm{d}(E_\lambda x,x) < \infty\right\}$$

(or $D := \left\{x\in H : \int_{-\infty}^{\infty} f(\lambda)\mathrm{d}E_\lambda x \, exists\right\}$). Let us define on this domain an operator A as

$$(Ax,y) = \int_{-\infty}^{\infty} f(\lambda)\mathrm{d}(E_\lambda x,y), \quad x\in D(A) := D, y\in H$$

(or $Ax = \int_{-\infty}^{\infty} f(\lambda)\mathrm{d}E_\lambda x, x\in D(A)$). Then A is self-adjoint and satisfies

$$E(\alpha,\beta]A \subset AE(\alpha,\beta], \quad \alpha<\beta.$$

Proof. It can be shown that the integral

$$\int_{-\infty}^{\infty} f(\lambda)\mathrm{d}(E_\lambda x,y)$$

exists for $x\in D$ and $y\in H$. Thus (Ax,y) is well defined. Let v be an element of H and let $\varepsilon>0$. Then by normalization, there exist $\alpha<-R$ and $\beta>R$ with R sufficiently large such that

$$\|v - E(\alpha,\beta]v\| = \|v - E_\beta v + E_\alpha v\| \le \|(I-E_\beta)v\| + \|E_\alpha v\| < \varepsilon.$$

On the other hand,

$$\begin{aligned}\int_{-\infty}^{\infty}|f(\lambda)|^2\mathrm{d}(E_\lambda E(\alpha,\beta]v,E(\alpha,\beta]v) &= \int_{-\infty}^{\infty}|f(\lambda)|^2\mathrm{d}(E_\lambda E(\alpha,\beta]v,v)\\ &= \int_{-\infty}^{\infty}|f(\lambda)|^2\mathrm{d}(E_\lambda E_\beta v,v) - \int_{-\infty}^{\infty}|f(\lambda)|^2\mathrm{d}(E_\lambda E_\alpha v,v)\\ &= \int_{-\infty}^{\beta}|f(\lambda)|^2\mathrm{d}(E_\lambda v,v) - \int_{-\infty}^{\alpha}|f(\lambda)|^2\mathrm{d}(E_\lambda v,v)\\ &= \int_{\alpha}^{\beta}|f(\lambda)|^2\mathrm{d}(E_\lambda v,v) < \infty.\end{aligned}$$

These two facts mean that $E(\alpha,\beta]v \in D$ and $\overline{D} = H$. Since $f(\lambda) = \overline{f(\lambda)}$, it follows that A is symmetric. Indeed,

$$\begin{aligned}(Ax,y) &= \int_{-\infty}^{\infty} f(\lambda)\mathrm{d}(E_\lambda x,y) = \lim_{\substack{\alpha\to-\infty\\ \beta\to\infty}}\int_{\alpha}^{\beta} f(\lambda)\mathrm{d}(E_\lambda x,y)\\ &= \lim_{\substack{\alpha\to-\infty\\ \beta\to\infty}}\int_{\alpha}^{\beta} f(\lambda)\mathrm{d}(x,E_\lambda y) = \lim_{\substack{\alpha\to-\infty\\ \beta\to\infty}}\left(x, \int_{\alpha}^{\beta} f(\lambda)\mathrm{d}E_\lambda y\right)\\ &= \left(x, \lim_{\substack{\alpha\to-\infty\\ \beta\to\infty}}\int_{\alpha}^{\beta} f(\lambda)\mathrm{d}E_\lambda y\right) = (x,Ay).\end{aligned}$$

In order to prove that $A = A^*$, it remains to show that $D(A^*) \subset D(A)$. Let $u \in D(A^*)$. Then

$$(E(\alpha,\beta]z,A^*u) = (AE(\alpha,\beta]z,u) = \int_{\alpha}^{\beta} f(\lambda)\mathrm{d}(E_\lambda z,u)$$

for all $z \in H$. This equality implies that

$$\begin{aligned}(z,A^*u) &= \lim_{\substack{\alpha\to-\infty\\ \beta\to\infty}}\int_{\alpha}^{\beta} f(\lambda)\mathrm{d}(E_\lambda z,u) = \int_{-\infty}^{\infty} f(\lambda)\mathrm{d}(E_\lambda z,u)\\ &= \int_{-\infty}^{\infty} f(\lambda)\mathrm{d}(z,E_\lambda u) = \overline{\int_{-\infty}^{\infty} f(\lambda)\mathrm{d}(E_\lambda u,z)} = \overline{(Au,z)} = (z,Au),\end{aligned}$$

where the integral exists because (z,A^*u) exists. Hence $u \in D(A)$ and $A^*u = Au$. For the second claim we first calculate

$$
\begin{aligned}
E(\alpha,\beta]Ax &= (E_\beta - E_\alpha)Ax = (E_\beta - E_\alpha)\int_{-\infty}^{\infty} f(\lambda)\mathrm{d}E_\lambda x \\
&= \int_{-\infty}^{\infty} f(\lambda)\mathrm{d}E_\lambda E_\beta x - \int_{-\infty}^{\infty} f(\lambda)\mathrm{d}E_\lambda E_\alpha x \\
&= \int_{-\infty}^{\beta} f(\lambda)\mathrm{d}E_\lambda x - \int_{-\infty}^{\alpha} f(\lambda)\mathrm{d}E_\lambda x \\
&= \int_{\alpha}^{\beta} f(\lambda)\mathrm{d}E_\lambda x = \int_{-\infty}^{\infty} f(\lambda)\mathrm{d}E_\lambda (E_\beta - E_\alpha)x \\
&= A(E_\beta - E_\alpha)x = AE(\alpha,\beta]x
\end{aligned}
$$

for all $x \in D(A)$. Since the left-hand side is defined on $D(A)$ and the right-hand side on all of H, the latter is an extension of the former. □

Exercise 27.3. Let A be as in Theorem 27.12. Prove that

$$\|Au\|^2 = \int_{-\infty}^{\infty} |f(\lambda)|^2 \mathrm{d}(E_\lambda u, u)$$

if $u \in D(A)$.

Exercise 27.4. Let $H = L^2(\mathbb{R})$ and $Au(t) = tu(t)$, $t \in \mathbb{R}$. Define $D(A)$ on which $A = A^*$ and evaluate the spectral family $\{E_\lambda\}_{\lambda=-\infty}^{\infty}$.

Theorem 27.13 (John von Neumann's spectral theorem). *Every self-adjoint operator A on a Hilbert space H has a unique spectral representation, i.e., there is a unique spectral family $\{E_\lambda\}_{\lambda=-\infty}^{\infty}$ such that*

$$Ax = \int_{-\infty}^{\infty} \lambda \mathrm{d}E_\lambda x, \quad x \in D(A)$$

(i.e., $(Ax,y) = \int_{-\infty}^{\infty} \lambda \mathrm{d}(E_\lambda x, y), x \in D(A), y \in H$), where $D(A)$ is defined as

$$D(A) = \left\{ x \in H : \int_{-\infty}^{\infty} \lambda^2 \mathrm{d}(E_\lambda x, x) < \infty \right\}.$$

Proof. First we assume that this theorem holds when A is bounded, that is, that there is a unique spectral family $\{F_\mu\}_{\mu=-\infty}^{\infty}$ such that

$$Au = \int_{-\infty}^{\infty} \mu \mathrm{d}F_\mu u, \quad u \in H,$$

since $D(A) = H$ in this case. But $F_\mu = 0$ for $\mu < m$ and $F_\mu = I$ for $\mu > M$, where

$$m = \inf_{\|x\|=1} (Ax,x), \quad M = \sup_{\|x\|=1} (Ax,x).$$

The spectral representation therefore can be written in the form

$$Au = \int_m^M \mu \mathrm{d}F_\mu u, \quad u \in H.$$

Let us consider now an unbounded operator that is semibounded from below, i.e.,

$$(Au, u) \geq m_0(u, u), \quad u \in D(A)$$

with some constant m_0. We assume without loss of generality that $(Au, u) \geq (u, u)$. This condition implies that A^{-1} exists, it is defined over all of H, and $\|A^{-1}\| \leq 1$. Indeed, A^{-1} exists and is bounded because $Au = 0$ if and only if $u = 0$. The norm estimate follows from

$$(v, A^{-1}v) \geq \|A^{-1}v\|^2, \quad v \in D(A^{-1}).$$

Since A^{-1} is bounded, $D(A^{-1})$ is a closed subspace in H. The self-adjointness of A means that $A^{-1} = (A^{-1})^*$. Therefore, A^{-1} is closed and $\overline{D(A^{-1})} = H$, i.e., A^{-1} is densely defined. Therefore, $D(A^{-1}) = H$ and $R(A) = H$. Since

$$0 \leq (A^{-1}v, v) \leq \|v\|^2, \quad v \in H,$$

we may conclude in this case that $m \geq 0$, $M \leq 1$, and

$$A^{-1}v = \int_0^1 \mu \mathrm{d}F_\mu v, \quad v \in H,$$

where $\{F_\mu\}$ is the spectral family of A^{-1}. Let us note that $F_1 = I$ and $F_0 = 0$, which follows from the spectral theorem for bounded operators and from the fact that $A^{-1}v = 0$ if and only if $v = 0$. Next, let us define the operator B_ε, $\varepsilon > 0$, as

$$B_\varepsilon u := \int_\varepsilon^1 \frac{1}{\mu} \mathrm{d}F_\mu u, \quad u \in D(A).$$

For every $v \in H$ we have

$$\begin{aligned} B_\varepsilon A^{-1}v &= \int_\varepsilon^1 \frac{1}{\mu} \mathrm{d}F_\mu (A^{-1}v) = \int_\varepsilon^1 \frac{1}{\mu} \mathrm{d}F_\mu \left(\int_0^1 \lambda \mathrm{d}F_\lambda v \right) \\ &= \int_\varepsilon^1 \frac{1}{\mu} d \left(\int_0^1 \lambda \mathrm{d}(F_\mu F_\lambda v) \right) = \int_\varepsilon^1 \frac{1}{\mu} \mathrm{d} \left(\int_\varepsilon^\mu \lambda \mathrm{d}F_\lambda v \right) = \int_\varepsilon^1 \frac{1}{\mu} \mu \mathrm{d}F_\mu v \\ &= \int_\varepsilon^1 \mathrm{d}F_\mu v = F_1 v - F_\varepsilon v = v - F_\varepsilon v. \end{aligned}$$

Since every spectral family is right continuous, it follows that

$$\lim_{\varepsilon\to 0+} B_\varepsilon A^{-1}v = v$$

exists. For every $u \in D(A)$ we have similarly

$$A^{-1}B_\varepsilon u = \int_0^1 \mu \mathrm{d}F_\mu(B_\varepsilon u) = \int_\varepsilon^1 \mu \mathrm{d}\left(\int_\varepsilon^\mu \frac{1}{\lambda}\mathrm{d}F_\lambda u\right) = u - F_\varepsilon u,$$

and hence

$$\lim_{\varepsilon\to 0+} A^{-1}B_\varepsilon u = u$$

exists. These two equalities mean that

$$\lim_{\varepsilon\to 0+} B_\varepsilon = \left(A^{-1}\right)^{-1} = A$$

exists and the spectral representation

$$A = \int_0^1 \frac{1}{\mu}\mathrm{d}F_\mu = \lim_{\varepsilon\to 0+}\int_\varepsilon^1 \frac{1}{\mu}\mathrm{d}F_\mu$$

holds. If we define $E_\lambda = I - F_{\frac{1}{\lambda}}, 1 \le \lambda < \infty$, then

$$A = -\int_0^1 \frac{1}{\mu}\mathrm{d}E_{\frac{1}{\mu}} = \int_1^\infty \lambda \mathrm{d}E_\lambda.$$

Exercise 27.5. Prove that this $\{E_\lambda\}$ is a spectral family which is left-continuous.

The domain $D(A)$ can be characterized as

$$D(A) = \left\{u \in H : \int_1^\infty \lambda^2 \mathrm{d}(E_\lambda u, u) < \infty\right\} = \left\{u \in H : \int_0^1 \frac{1}{\mu^2}\mathrm{d}(F_\mu u, u) < \infty\right\}.$$

This proves the theorem for self-adjoint operators that are semibounded from below. For bounded operators we will only sketch the proof.

Step 1. If $A = A^*$ and A is bounded, then we can define

$$p_N(A) := a_0 I + a_1 A + \cdots + a_N A^N, \quad N \in \mathbb{N},$$

where $a_j \in \mathbb{R}$ for $j = 0, 1, \ldots, N$. Then $p_N(A)$ is also self-adjoint and bounded with

$$\|p_N(A)\| \le \sup_{|t|\le\|A\|} |p_N(t)|.$$

Step 2. For every continuous real-valued function f on $[m,M]$, where m and M are as above, we can define $f(A)$ as an approximation by $p_N(A)$, i.e., we can prove that for every $\varepsilon > 0$ there exists $p_N(A)$ such that

$$\|f(A) - p_N(A)\| < \varepsilon.$$

Step 3. For every $u, v \in H$ let us define the functional L as

$$L(f) := (f(A)u, v).$$

Then

$$|L(f)| = |(f(A)u, v)| \leq \|f(A)\| \, \|u\| \, \|v\|,$$

that is, $L(f)$ is a bounded linear functional on $C[m,M]$.

Step 4. (Riesz's theorem) A continuous positive linear functional L on $C[a,b]$ can be represented in the form

$$L(f) = \int_a^b f(x) \mathrm{d}\nu(x),$$

where ν is a measure that satisfies the conditions

(1) $L(f) \geq 0$ for $f \geq 0$;
(2) $|L(f)| \leq \nu(K) \|f\|_K$, where $K \subset [a,b]$ is compact and

$$\|f\|_K = \max_{x \in K} |f(x)|.$$

Step 5. It follows from Step 4 that

$$(Au, v) = \int_m^M \lambda \mathrm{d}\nu(\lambda; u, v).$$

Step 6. It is possible to prove that $\nu(\lambda; u, v)$ is a self-adjoint sesquilinear form, from which we conclude that there exists a self-adjoint and bounded operator E_λ such that

$$\nu(\lambda; u, v) = (E_\lambda u, v).$$

This operator is idempotent, and we may define $E_\lambda \equiv 0$ for $\lambda < m$ and $E_\lambda \equiv I$ for $\lambda \geq M$. Thus $\{E_\lambda\}_{\lambda=-\infty}^{\infty}$ is the required spectral family, and the theorem is proved. See [4] for an alternative proof of this theorem. □

Let $A : H \to H$ be a self-adjoint operator in a Hilbert space H. Then by von Neumann's spectral theorem we can write

$$Au = \int_{-\infty}^{\infty} \lambda \mathrm{d}E_\lambda u, \quad u \in D(A).$$

For every continuous function f we can define

$$D_f := \left\{ u \in H : \int_{-\infty}^{\infty} |f(\lambda)|^2 \mathrm{d}(E_\lambda u, u) < \infty \right\}.$$

This set is a linear subspace of H. For every $u \in D_f$ and $v \in H$ let us define the linear functional

$$L(v) := \int_{-\infty}^{\infty} f(\lambda) \mathrm{d}(E_\lambda u, v) = \left(\int_{-\infty}^{\infty} f(\lambda) \mathrm{d}E_\lambda u, v \right).$$

This functional is continuous because it is bounded. Indeed,

$$|L(v)|^2 \leq \left\| \int_{-\infty}^{\infty} f(\lambda) \mathrm{d}E_\lambda u \right\|^2 \|v\|^2 = \int_{-\infty}^{\infty} |f(\lambda)|^2 \mathrm{d}(E_\lambda u, u) \, \|v\|^2 = c(u) \, \|v\|^2 .$$

By the Riesz–Fréchet theorem this functional can be expressed in the form of an inner product, i.e., there exists $z \in H$ such that

$$\int_{-\infty}^{\infty} f(\lambda) \mathrm{d}(E_\lambda u, v) = (z, v), \quad v \in H.$$

We set

$$z := f(A)u, \quad u \in D_f,$$

i.e.,

$$(f(A)u, v) = \int_{-\infty}^{\infty} f(\lambda) \mathrm{d}(E_\lambda u, v).$$

Remark 27.14. Since in general f is not real-valued, $f(A)$ is not a self-adjoint operator in general.

Example 27.15. Consider

$$f(\lambda) = \frac{\lambda - \mathrm{i}}{\lambda + \mathrm{i}}, \quad \lambda \in \mathbb{R},$$

and a self-adjoint operator A with spectral family E_λ. Define

$$U_A := f(A) = \int_{-\infty}^{\infty} \frac{\lambda - \mathrm{i}}{\lambda + \mathrm{i}} \mathrm{d}E_\lambda .$$

The operator U_A is called the *Cayley transform* of A. Since $|f(\lambda)| = 1$, we have $D_f = D(U_A) = H$ and

$$\|U_A u\|^2 = \int_{-\infty}^{\infty} |f(\lambda)|^2 \mathrm{d}(E_\lambda u,u) = \lim_{\substack{\alpha\to-\infty\\ \beta\to\infty}} \int_\alpha^\beta \mathrm{d}(E_\lambda u,u)$$

$$= \lim_{\substack{\alpha\to-\infty\\ \beta\to\infty}} \left((E_\beta u,u)-(E_\alpha u,u)\right) = \lim_{\substack{\alpha\to-\infty\\ \beta\to\infty}} \left(\|E_\beta u\|^2 - \|E_\alpha u\|^2\right) = \|u\|^2$$

by normalization of $\{E_\lambda\}$. Hence U_A is an isometry. There is a one-to-one correspondence between self-adjoint operators and their Cayley transforms. Indeed,

$$U_A = (A-\mathrm{i}I)(A+\mathrm{i}I)^{-1}$$

is equivalent to

$$\begin{cases} I-U_A = 2\mathrm{i}(A+\mathrm{i}I)^{-1}, \\ I+U_A = 2A(A+\mathrm{i}I)^{-1}, \end{cases}$$

or

$$A = \mathrm{i}(I+U_A)(I-U_A)^{-1}.$$

Example 27.16. Consider

$$f(\lambda) = \frac{1}{\lambda - z}, \quad \lambda\in\mathbb{R}, z\in\mathbb{C}, \operatorname{Im} z\neq 0.$$

Define

$$R_z := (A-zI)^{-1} = \int_{-\infty}^{\infty} \frac{1}{\lambda - z} \mathrm{d}E_\lambda.$$

The operator R_z is called the *resolvent* of A. Since

$$\left|\frac{1}{\lambda - z}\right| \le \frac{1}{|\operatorname{Im} z|}$$

for all $\lambda\in\mathbb{R}$, we have that R_z is bounded and defined on the whole of H.

Example 27.17. Suppose that $K(x,y)\in L^2(\Omega\times\Omega)$. Define an integral operator on $L^2(\Omega)$ as

$$Af(x) = \int_\Omega K(x,y)f(y)\mathrm{d}y.$$

Then

$$A^*f(x) = \int_\Omega \overline{K(y,x)} f(y)\mathrm{d}y$$

and therefore

$$A^*Af(x) = \int_\Omega \left(\int_\Omega K(y,z)\overline{K(y,x)}\mathrm{d}y\right) f(z)\mathrm{d}z.$$

As we know from Example 26.26, A^*A is self-adjoint on $L^2(\Omega)$. This fact can also be checked directly, since

$$\int_\Omega K(y,z)\overline{K(y,x)}\mathrm{d}y = \overline{\int_\Omega K(y,x)\overline{K(y,z)}\mathrm{d}y}.$$

Von Neumann's spectral theorem gives us for this operator and for all $s \geq 0$ that

$$(A^*A)^s = \int_0^{\|A\|^2_{L^2\to L^2}} \lambda^s \mathrm{d}E_\lambda,$$

since A^*A is positive and bounded by $\|A\|^2_{L^2\to L^2}$.

Exercise 27.6. Let $A = A^*$ with spectral family E_λ. Let $u \in D(f(A))$ and $v \in D(g(A))$. Prove that

$$(f(A)u, g(A)v) = \int_{-\infty}^{\infty} f(\lambda)\overline{g(\lambda)}\mathrm{d}(E_\lambda u, v).$$

Exercise 27.7. Let $A = A^*$ with spectral family E_λ. Let $u \in D(f(A))$. Prove that $f(A)u \in D(g(A))$ if and only if $u \in D((gf)(A))$ and that

$$(gf)(A)u = \int_{-\infty}^{\infty} g(\lambda)f(\lambda)\mathrm{d}E_\lambda u.$$

Remark 27.18. It follows from Exercise 27.7 that

$$(gf)(A) = (fg)(A)$$

on the domain $D((fg)(A)) \cap D((gf)(A))$.

Chapter 28
Spectra of Self-Adjoint Operators

Definition 28.1. Given a linear operator A on a Hilbert space H with domain $D(A)$, $\overline{D(A)} = H$, the set

$$\rho(A) = \left\{z \in \mathbb{C} : (A - zI)^{-1} \text{exists as a bounded operator from} \, H \, \text{to} \, D(A)\right\}$$

is called the *resolvent set* of A. Its complement

$$\sigma(A) = \mathbb{C} \setminus \rho(A)$$

is called the *spectrum* of A.

Theorem 28.2.

(1) If $A = \overline{A}$ then the resolvent set is open and the resolvent operator $R_z := (A - zI)^{-1}$ is an analytic function from $\rho(A)$ to $B(H;H)$, the set of all linear bounded operators in H. Furthermore, the resolvent identity

$$R_z - R_\xi = (z - \xi) R_z R_\xi, \quad z, \xi \in \rho(A)$$

holds and $R_z' = (R_z)^2$.

(2) If $A = A^*$ then $z \in \rho(A)$ if and only if there exists $C_z > 0$ such that

$$\|(A - zI)u\| \geq C_z \|u\|$$

for all $u \in D(A)$.

Proof. (1) Assume that $z_0 \in \rho(A)$. Then R_{z_0} is a bounded linear operator from H to $D(A)$ and thus $r := \left\|R_{z_0}\right\|^{-1} > 0$. Let us define for $|z - z_0| < r$ the operator

$$G_{z_0} := (z - z_0) R_{z_0}.$$

V. Serov, *Fourier Series, Fourier Transform and Their Applications to Mathematical Physics*, Applied Mathematical Sciences 197,
DOI 10.1007/978-3-319-65262-7_28

Then G_{z_0} is bounded with $\|G_{z_0}\| < 1$. Hence it defines the operator

$$(I - G_{z_0})^{-1} = \sum_{j=0}^{\infty} (G_{z_0})^j,$$

because this Neumann series converges. But for $|z - z_0| < r$ we have

$$A - zI = (A - z_0 I)(I - G_{z_0}),$$

or

$$(A - zI)^{-1} = (I - G_{z_0})^{-1} R_{z_0}.$$

Hence R_z exists with $D(R_z) = H$ and is bounded. It remains to show that $R(R_z) \subset D(A)$. For $x \in H$ we know that

$$y := (A - zI)^{-1} x \in H.$$

We claim that $y \in D(A)$. Indeed,

$$\begin{aligned} y = (A - zI)^{-1} x = (I - G_{z_0})^{-1} R_{z_0} x &= \sum_{j=0}^{\infty} (z - z_0)^j (R_{z_0})^{j+1} x \\ = \lim_{n \to \infty} \sum_{j=0}^{n} (z - z_0)^j (R_{z_0})^{j+1} x. \end{aligned}$$

It follows from this representation that $R_z = (A - zI)^{-1}$ is an analytic function from $\rho(A)$ to $B(H;H)$. Next we define

$$s_n x := \sum_{j=0}^{n} (z - z_0)^j (R_{z_0})^{j+1} x.$$

It is clear that $s_n x \in D(A)$ and that $\lim_{n\to\infty} s_n x = y$. Moreover,

$$\lim_{n \to \infty} (A - zI) s_n x = x.$$

Writing $y_n := s_n x$ we conclude from the criterion for closedness that

$$\begin{cases} y_n \in D(A), \\ y_n \to y, \\ (A - zI) y_n \to x \end{cases} \Rightarrow \begin{cases} y \in D(A), \\ x = (A - zI) y. \end{cases}$$

Hence $y = (A - zI)^{-1} x \in D(A)$, and therefore $\rho(A)$ is open. The resolvent identity is proved by a straightforward calculation:

$$R_z - R_\xi = R_z(A-\xi I)R_\xi - R_z(A-zI)R_\xi = R_z\left[(A-\xi I)-(A-zI)\right]R_\xi$$
$$= (z-\xi)R_zR_\xi.$$

Finally, the limit

$$\lim_{z\to\xi}\frac{R_z - R_\xi}{z-\xi} = \lim_{z\to\xi} R_zR_\xi = (R_z)^2$$

exists, and hence $R_z' = (R_z)^2$ exists, which proves this part.

(2) Assume that $A = A^*$. If $z \in \rho(A)$, then by definition R_z maps from H to $D(A)$. Hence there exists $M_z > 0$ such that

$$\|R_zv\| \le M_z\|v\|, \quad v\in H.$$

Since $u = R_z(A-zI)u$ for all $u \in D(A)$, we get

$$\|u\| \le M_z\|(A-zI)u\|, \quad u\in D(A).$$

This is equivalent to

$$\|(A-zI)u\| \ge \frac{1}{M_z}\|u\|, \quad u\in D(A).$$

Conversely, if there exists $C_z > 0$ such that

$$\|(A-zI)u\| \ge C_z\|u\|, \quad u\in D(A),$$

then $(A-zI)^{-1}$ is bounded. Since A is self-adjoint, $(A-zI)^{-1}$ is defined over all of H. Indeed, if $R(A-zI) \ne H$, then there exists $v_0 \ne 0$ such that $v_0 \perp R(A-zI)$. This means that

$$(v_0,(A-zI)u) = 0, \quad u\in D(A),$$

or

$$(Au,v_0) = (zu,v_0),$$

or

$$(u,A^*v_0) = (u,\bar{z}v_0).$$

Thus $v_0 \in D(A^*)$ and $A^*v_0 = \bar{z}v_0$. Since $A = A^*$, it follows that $v_0 \in D(A)$ and $Av_0 = \bar{z}v_0$, or

$$(A-\bar{z}I)v_0 = 0.$$

It is easy to check that $\|(A-\bar{z}I)u\|^2 = \|(A-zI)u\|^2$ for all $u\in D(A)$. Therefore,

$$\|(A-\bar{z}I)v_0\| = \|(A-zI)v_0\| \ge C_z\|v_0\|.$$

Hence $v_0 = 0$ and $D\left((A-zI)^{-1}\right) = R(A-zI) = H$. This means that $z \in \rho(A)$.
□

Corollary 28.3. *If $A = A^*$, then $\sigma(A) \neq \emptyset, \sigma(A) = \overline{\sigma(A)}$ and $\sigma(A) \subset \mathbb{R}$.*

Proof. If $z = \alpha + i\beta \in \mathbb{C}$ with $\operatorname{Im} z = \beta \neq 0$, then

$$\|(A-zI)x\|^2 = \|(A-\alpha I)x - i\beta x\|^2 = \|(A-\alpha I)x\|^2 + |\beta|^2 \|x\|^2 \geq |\beta|^2 \|x\|^2.$$

This implies (see part (2) of Theorem 28.2) that $z \in \rho(A)$, which means that $\sigma(A) \subset \mathbb{R}$. Since $A = A^*$ and is therefore closed, the spectrum $\sigma(A)$ is closed as the complement of an open set (see part (1) of Theorem 28.2).

It remains to prove that $\sigma(A) \neq \emptyset$. Assume to the contrary that $\sigma(A) = \emptyset$. Then the resolvent R_z is an entire analytic function. Let us prove that $\|R_z\|$ is uniformly bounded with respect to $z \in \mathbb{C}$. We introduce the functional

$$T_z(y) := (R_z x, y), \quad \|x\| = 1, y \in H.$$

Then $T_z(y)$ is a linear functional on the Hilbert space H. Moreover, since R_z is bounded for every (fixed) $z \in \mathbb{C}$, it follows that

$$|T_z(y)| \leq \|R_z x\| \|y\| \leq \|R_z\| \|y\| = C_z \|y\|.$$

Therefore, $T_z(y)$ is continuous, i.e., $\{T_z, z \in \mathbb{C}\}$ is a pointwise bounded family of continuous linear functionals. By the Banach–Steinhaus theorem (or the uniform boundedness principle) we conclude that

$$\sup_{z \in \mathbb{C}} \|T_z\| = c_0 < \infty.$$

We therefore have

$$|T_z(y)| = |(R_z x, y)| \leq c_0 \|y\|, \quad \|x\| = 1, z \in \mathbb{C},$$

which implies that $\|R_z x\| \leq c_0$, i.e., $\|R_z\| \leq c_0$. By Liouville's theorem we may conclude now that R_z is constant with respect to z. But by von Neumann's spectral theorem,

$$R_z = \int_{-\infty}^{\infty} \frac{1}{\lambda - z} dE_\lambda,$$

where $\{E_\lambda\}$ is the spectral family of $A = A^*$. Due to the estimate

$$\|R_z\| \leq \frac{1}{|\operatorname{Im} z|},$$

we may conclude that $\|R_z\| \to 0$ as $|\operatorname{Im} z| \to \infty$. Hence $R_z \equiv 0$. This contradiction completes the proof. □

Exercise 28.1. Consider $A = \frac{\mathrm{d}}{\mathrm{d}x}$ defined in $L^2(0,1)$ with domain

$$D(A) = \{u \in W_2^1(0,1) : u(1) = 0\}.$$

Show that $A \neq A^*$ and $\sigma(A) = \emptyset$.

Exercise 28.2. [Weyl's criterion] Let $A = A^*$. Prove that $\lambda \in \sigma(A)$ if and only if there exists $x_n \in D(A)$, $\|x_n\| = 1$, such that

$$\lim_{n\to\infty} \|(A - \lambda I)x_n\| = 0.$$

Definition 28.4. Let us assume that $A = \overline{A}$. The *point spectrum* $\sigma_{\mathrm{p}}(A)$ of A is the set of eigenvalues of A, i.e.,

$$\sigma_{\mathrm{p}}(A) = \{\lambda \in \sigma(A) : N(A - \lambda I) \neq \{0\}\}.$$

This means that $(A - \lambda I)^{-1}$ does not exist, i.e., there exists a nontrivial $u \in D(A)$ such that $Au = \lambda u$. The complement $\sigma(A) \setminus \sigma_{\mathrm{p}}(A)$ is called the *continuous spectrum* $\sigma_{\mathrm{c}}(A)$. The *discrete spectrum* is the set

$$\sigma_{\mathrm{d}}(A) = \left\{\lambda \in \sigma_{\mathrm{p}}(A) : \dim N(A - \lambda I) < \infty \,\text{and}\, \lambda \,\text{is isolated in}\, \sigma(A)\right\}.$$

The set $\sigma_{\mathrm{ess}}(A) := \sigma(A) \setminus \sigma_{\mathrm{d}}(A)$ is called the *essential spectrum* of A.

In the framework of this definition, the complex plane can be divided into regions according to

$$\mathbb{C} = \rho(A) \cup \sigma(A),$$

$$\sigma(A) = \sigma_{\mathrm{p}}(A) \cup \sigma_{\mathrm{c}}(A),$$

and

$$\sigma(A) = \sigma_{\mathrm{d}}(A) \cup \sigma_{\mathrm{ess}}(A),$$

with all the unions disjoint.

Remark 28.5. If $A = A^*$, then

(1) $\lambda \in \sigma_{\mathrm{c}}(A)$ means that $(A - \lambda I)^{-1}$ exists but is not bounded.

(2)

$$\begin{aligned}\sigma_{\mathrm{ess}}(A) = {} & \overline{\sigma_{\mathrm{c}}(A)} \\ & \cup \{\text{eigenvalues of infinite multiplicity and their accumulation points}\} \\ & \cup \{\text{accumulation points of}\, \sigma_{\mathrm{d}}(A)\}.\end{aligned}$$

Exercise 28.3. Let $A = A^*$ and $\lambda_1, \lambda_2 \in \sigma_{\mathrm{p}}(A)$. Prove that if $\lambda_1 \neq \lambda_2$, then

$$N(A-\lambda_1 I) \perp N(A-\lambda_2 I).$$

Exercise 28.4. Let $\{e_j\}_{j=1}^{\infty}$ be an orthonormal basis in H and let $\{s_j\}_{j=1}^{\infty} \subset \mathbb{C}$ be some sequence. Introduce the set

$$D = \left\{ x \in H : \sum_{j=1}^{\infty} |s_j|^2 |(x, e_j)|^2 < \infty \right\}.$$

Define

$$Ax = \sum_{j=1}^{\infty} s_j (x, e_j) e_j, \quad x \in D.$$

Prove that $A = \overline{A}$ and that $\sigma(A) = \overline{\{s_j : j = 1, 2, \ldots\}}$. Prove also that

$$(A - zI)^{-1} x = \sum_{j=1}^{\infty} \frac{1}{s_j - z} (x, e_j) e_j$$

for all $z \in \rho(A)$ and $x \in D$.

Exercise 28.5. Prove that the spectrum $\sigma(U)$ of a unitary operator U lies on the unit circle in $\mathbb{C}$.

Theorem 28.6. *Let $A = A^*$ and let $\{E_\lambda\}_{\lambda \in \mathbb{R}}$ be its spectral family. Then*

(1) $\mu \in \sigma(A)$ if and only if $E_{\mu+\varepsilon} - E_{\mu-\varepsilon} \neq 0$ for every $\varepsilon > 0$.
(2) $\mu \in \sigma_{\mathrm{p}}(A)$ if and only if $E_\mu - E_{\mu-0} \neq 0$. Here $E_{\mu-0} := \lim_{\varepsilon \to 0+} E_{\mu-\varepsilon}$ in the sense of the strong operator topology.

Proof. (1) Suppose that $\mu \in \sigma(A)$ but there exists $\varepsilon > 0$ such that $E_{\mu+\varepsilon} - E_{\mu-\varepsilon} = 0$. Then by the spectral theorem we obtain for every $x \in D(A)$ that

$$\begin{aligned}
\|(A - \mu I)x\|^2 &= \int_{-\infty}^{\infty} (\lambda - \mu)^2 \mathrm{d}(E_\lambda x, x) \geq \int_{|\lambda - \mu| \geq \varepsilon} (\lambda - \mu)^2 \mathrm{d}(E_\lambda x, x) \\
&\geq \varepsilon^2 \int_{|\lambda - \mu| \geq \varepsilon} \mathrm{d}(E_\lambda x, x) = \varepsilon^2 \left[\int_{-\infty}^{\mu-\varepsilon} + \int_{\mu+\varepsilon}^{\infty} \right] \mathrm{d}(E_\lambda x, x) \\
&= \varepsilon^2 \left[(E_{\mu-\varepsilon} x, x) + \|x\|^2 - (E_{\mu+\varepsilon} x, x) \right] = \varepsilon^2 \|x\|^2.
\end{aligned}$$

This inequality means (see part (2) of Theorem 28.2) that $\mu \notin \sigma(A)$ but $\mu \in \rho(A)$. This contradiction proves (1) in one direction. Conversely, if

$$P_n := E_{\mu+\frac{1}{n}} - E_{\mu-\frac{1}{n}} \neq 0$$

for all $n \in \mathbb{N}$, then there is a sequence $\{x_n\}_{n=1}^{\infty}$ such that $x_n \in R(P_n)$, i.e., $x_n = P_n x_n$, i.e., $x_n \in D(A)$ and $\|x_n\| = 1$. For this sequence it is true that

$$\begin{aligned}\|(A-\mu I)x_n\|^2 &= \int_{-\infty}^{\infty}(\lambda-\mu)^2 \mathrm{d}(E_\lambda P_n x_n, P_n x_n)\\ &= \int_{|\lambda-\mu|\le 1/n}(\lambda-\mu)^2 \mathrm{d}(E_\lambda x_n, x_n)\\ &\le \frac{1}{n^2}\int_{-\infty}^{\infty}\mathrm{d}(E_\lambda x_n, x_n) = \frac{1}{n^2}\|x_n\|^2 = \frac{1}{n^2} \to 0\end{aligned}$$

as $n \to \infty$. Hence, this sequence satisfies Weyl's criterion (see Exercise 28.2) and therefore $\mu \in \sigma(A)$.

(2) Suppose $\mu \in \mathbb{R}$ is an eigenvalue of A. Then there exists $x_0 \in D(A)$, $x_0 \neq 0$, such that

$$0 = \|(A-\mu I)x_0\|^2 = \int_{-\infty}^{\infty}(\lambda-\mu)^2 \mathrm{d}(E_\lambda x_0, x_0).$$

In particular, for all $n \in \mathbb{N}$ large enough and $\varepsilon > 0$ we have that

$$\begin{aligned}0 &= \int_{\mu+\varepsilon}^{n}(\lambda-\mu)^2 \mathrm{d}(E_\lambda x_0, x_0) \ge \varepsilon^2 \int_{\mu+\varepsilon}^{n}\mathrm{d}(E_\lambda x_0, x_0) = \varepsilon^2((E_n - E_{\mu+\varepsilon})x_0, x_0)\\ &= \varepsilon^2\left\|(E_n - E_{\mu+\varepsilon})x_0\right\|^2.\end{aligned}$$

Thus we may conclude that

$$0 = E_n x_0 - E_{\mu+\varepsilon}x_0.$$

Similarly we can get that

$$0 = E_{-n}x_0 - E_{\mu-\varepsilon}x_0.$$

Letting $n \to \infty$ and $\varepsilon \to 0$, we obtain

$$x_0 = E_\mu x_0, \quad 0 = E_{\mu-0}x_0.$$

Hence

$$x_0 = (E_\mu - E_{\mu-0})x_0$$

and therefore

$$E_\mu - E_{\mu-0} \neq 0.$$

Conversely, define the projector

$$P := E_\mu - E_{\mu-0}.$$

If $P \neq 0$, then there exists $y \in H$, $y \neq 0$, such that $y = Py$ (e.g., any $y \in R(P) \neq \{0\}$ will do). For $\lambda > \mu$ it follows that

$$E_\lambda y = E_\lambda P y = E_\lambda E_\mu y - E_\lambda E_{\mu-0} y = P y = y.$$

For $\lambda < \mu$ we have that

$$E_\lambda y = E_\lambda E_\mu y - E_\lambda E_{\mu-0} y = E_\lambda y - E_\lambda y = 0.$$

Hence

$$\|(A - \mu I)y\|^2 = \int_{-\infty}^{\infty} (\lambda - \mu)^2 \mathrm{d}(E_\lambda y, y) = \int_{\mu}^{\infty} (\lambda - \mu)^2 \mathrm{d}_\lambda (y, y) = 0.$$

Therefore, $Ay = \mu y$ and $y \in D(A)$, $y \neq 0$, i.e., μ is an eigenvalue of A, or $\mu \in \sigma_{\mathrm{p}}(A)$.

□

Remark 28.7. The statements of Theorem 28.6 can be reformulated as follows:

(1) $\mu \in \sigma_{\mathrm{p}}(A)$ if and only if $E_\mu - E_{\mu-0} \neq 0$.
(2) $\mu \in \sigma_{\mathrm{c}}(A)$ if and only if $E_\mu - E_{\mu-0} = 0$.

Definition 28.8. Let H and H_1 be two Hilbert spaces. A bounded linear operator $K : H \to H_1$ is called *compact* or *completely continuous* if it maps bounded sets in H into *precompact* sets in H_1, i.e., for every bounded sequence $\{x_n\}_{n=1}^{\infty} \subset H$ the sequence $\{Kx_n\}_{n=1}^{\infty} \subset H_1$ contains a convergent subsequence.

If $K : H \to H_1$ is compact, then the following statements hold.

(1) K maps every weakly convergent sequence in H into a norm convergent sequence in H_1. This condition is also sufficient.
(2) If $H = H_1$ is separable, then every compact operator is a norm limit of a sequence of operators of *finite rank* (i.e., operators with finite-dimensional ranges).
(3) The norm limit of a sequence of compact operators is compact.

Let us prove (2). Let K be a compact operator. Since H is separable, it has an orthonormal basis $\{e_j\}_{j=1}^{\infty}$. Consider for $n = 1, 2, \ldots$ the projector

$$P_n x := \sum_{j=1}^{n} (x, e_j) e_j, \quad x \in H.$$

Then $P_n \leq P_{n+1}$ and $\|(I - P_n)x\| \to 0$ as $n \to \infty$. Define

$$d_n := \sup_{\|x\|=1} \|K(I - P_n)x\| \equiv \|K(I - P_n)\|.$$

Since $R(I - P_n) \supset R(I - P_{n+1})$ (see Proposition 27.5), it follows that $\{d_n\}_{n=1}^{\infty}$ is a monotonically decreasing sequence of positive numbers. Hence the limit

$$\lim_{n\to\infty} d_n := d \geq 0$$

exists. Let us choose $y_n \in R(I-P_n)$, $\|y_n\| = 1$, such that

$$\|K(I-P_n)y_n\| = \|Ky_n\| \geq \frac{d}{2}.$$

Then

$$|(y_n,x)| = |((I-P_n)y_n,x)| = |(y_n,(I-P_n)x)| \leq \|y_n\| \, \|(I-P_n)x\| \to 0$$

as $n \to \infty$ for all $x \in H$. This means that $y_n \xrightarrow{w} 0$. The compactness of K implies that $Ky_n \to 0$. Thus $d = 0$. Therefore,

$$d_n = \|K - KP_n\| \to 0.$$

Since P_n is of finite rank, so is KP_n, i.e., K is a norm limit of finite-rank operators.

Lemma 28.9. *Suppose $A = A^*$ is compact. Then at least one of the two numbers $\pm\|A\|$ is an eigenvalue of A.*

Proof. Since

$$\|A\| = \sup_{\|x\|=1} |(Ax,x)|,$$

there exists a sequence x_n with $\|x_n\| = 1$ such that

$$\|A\| = \lim_{n\to\infty} |(Ax_n,x_n)|.$$

In fact, we can assume that $\lim_{n\to\infty}(Ax_n,x_n)$ exists and equals, say, a. Otherwise, we would take a subsequence of $\{x_n\}$. Since $A = A^*$, it follows that a is real and $\|A\| = |a|$. Due to the fact that every bounded set of a Hilbert space is weakly relatively compact (the unit ball in our case), we can choose a subsequence of $\{x_n\}$, say $\{x_{k_n}\}$, that converges weakly, i.e., $x_{k_n} \xrightarrow{w} x$. The compactness of A implies that $Ax_{k_n} \to y$. Next we observe that

$$\begin{aligned}\|Ax_{k_n} - ax_{k_n}\|^2 &= \|Ax_{k_n}\|^2 - 2a(Ax_{k_n},x_{k_n}) + a^2 \leq \|A\|^2 - 2a(Ax_{k_n},x_{k_n}) + a^2 \\ &= 2a^2 - 2a(Ax_{k_n},x_{k_n}) \to 2a^2 - 2a^2 = 0\end{aligned}$$

as $n \to \infty$. Hence

$$\begin{cases} Ax_{k_n} - ax_{k_n} \to 0, \\ Ax_{k_n} \to y, \\ x_{k_n} \xrightarrow{w} x \end{cases} \Rightarrow \quad \begin{cases} x_{k_n} \to x, \\ Ax = ax. \end{cases}$$

Since $\|x_{k_n}\| = 1$, we have $\|x\| = 1$ also. Hence $x \neq 0$, and a is an eigenvalue of A. □

Remark 28.10. It is not difficult to show that the statement of Lemma 28.9 remains true if A is just bounded and self-adjoint.

Theorem 28.11 (Riesz–Schauder). *Suppose $A = A^*$ is compact. Then*

(1) A has a sequence of real eigenvalues $\lambda_j \neq 0$ that can be enumerated in such a way that

$$|\lambda_1| \geq |\lambda_2| \geq \cdots \geq |\lambda_j| \geq \cdots.$$

(2) If there are infinitely many eigenvalues, then $\lim_{j\to\infty} \lambda_j = 0$ and 0 is the only accumulation point of $\{\lambda_j\}$.
(3) The multiplicity of λ_j is finite.
(4) If e_j is the normalized eigenvector for λ_j, then $\{e_j\}_{j=1}^{\infty}$ is an orthonormal system and

$$Ax = \sum_{j=1}^{\infty} \lambda_j (x, e_j) e_j = \sum_{j=1}^{\infty} (Ax, e_j) e_j, \quad x \in H.$$

This means that $\{e_j\}_{j=1}^{\infty}$ is an orthonormal basis of $R(A)$.
(5) $\sigma(A) = \{0, \lambda_1, \lambda_2, \ldots, \lambda_j, \ldots\}$, while 0 is not necessarily an eigenvalue of A.

Proof. Lemma 28.9 gives the existence of an eigenvalue $\lambda_1 \in \mathbb{R}$ with $|\lambda_1| = \|A\|$ and a normalized eigenvector e_1. Introduce $H_1 = e_1^{\perp}$. Then H_1 is a closed subspace of H, and A maps H_1 into itself. Indeed,

$$(Ax, e_1) = (x, Ae_1) = (x, \lambda_1 e_1) = \lambda_1 (x, e_1) = 0$$

for every $x \in H_1$. The restriction of the inner product of H to H_1 makes H_1 a Hilbert space (since H_1 is closed), and the restriction of A to H_1, denoted by $A_1 = A|_{H_1}$, is again a self-adjoint compact operator that maps in H_1. Clearly, its norm is bounded by the norm of A, i.e., $\|A_1\| \leq \|A\|$. Applying Lemma 28.9 to A_1 on H_1, we get an eigenvalue λ_2 with $|\lambda_2| = \|A_1\|$ and a normalized eigenvector e_2 with $e_2 \perp e_1$. It is clear that $|\lambda_2| \leq |\lambda_1|$. Next introduce the closed subspace $H_2 = (\text{span}\{e_1, e_2\})^{\perp}$. Again, A leaves H_2 invariant, and thus $A_2 := A_1|_{H_2} = A|_{H_2}$ is a self-adjoint compact operator in H_2. Applying Lemma 28.9 to A_2 on H_2, we obtain λ_3 with $|\lambda_3| = \|A_2\|$ and a normalized eigenvector e_3 with $e_3 \perp e_2$ and $e_3 \perp e_1$. This process in an infinite-dimensional Hilbert space leads us to the sequence $\{\lambda_j\}_{j=1}^{\infty}$ such that $|\lambda_{j+1}| \leq |\lambda_j|$ and corresponding normalized eigenvectors. Since $|\lambda_j| > 0$ and the sequence is monotonically decreasing, there is a limit

$$\lim_{j\to\infty} |\lambda_j| = r.$$

Clearly $r \geq 0$. Let us prove that $r = 0$. If $r > 0$, then $|\lambda_j| \geq r > 0$ for each $j = 1, 2, \ldots$, or

$$\frac{1}{|\lambda_j|} \le \frac{1}{r} < \infty.$$

Hence the sequence of vectors

$$y_j := \frac{e_j}{\lambda_j}$$

is bounded, and therefore there is a weakly convergent subsequence $y_{j_k} \xrightarrow{w} y$. The compactness of A implies the strong convergence of $Ay_{j_k} \equiv e_{j_k}$. But for $k \neq m$ we have $\left\|e_{j_k} - e_{j_m}\right\| = \sqrt{2}$. This contradiction proves (1) and (2).

Exercise 28.6. Prove that if H is an infinite-dimensional Hilbert space, then the identity operator I is not compact, and the inverse of a compact operator (if it exists) is unbounded.

Exercise 28.7. Prove part (3) of Theorem 28.11.

Consider now the projector

$$P_n x := \sum_{j=1}^{n} (x, e_j) e_j, \quad x \in H.$$

Then $I - P_n$ is a projector onto $(\text{span}\{e_1, \ldots, e_n\})^{\perp} \equiv H_n$ and hence

$$\|A(I-P_n)x\| \le \|A\|_{H_n} \|(I-P_n)x\| \le |\lambda_{n+1}| \|x\| \to 0$$

as $n \to \infty$. Since

$$AP_n x = \sum_{j=1}^{n} (x, e_j) A e_j = \sum_{j=1}^{n} \lambda_j (x, e_j) e_j$$

and

$$\|A(I-P_n)x\| = \|Ax - AP_n x\| \to 0, \quad n \to \infty,$$

we have

$$Ax = \sum_{j=1}^{\infty} \lambda_j (x, e_j) e_j,$$

and part (4) follows. Finally, Exercise 28.4 gives immediately that

$$\sigma(A) = \{0, \lambda_1, \lambda_2, \ldots, \lambda_j, \ldots\}.$$

This completes the proof. □

Corollary 28.12 (Hilbert–Schmidt theorem). *An orthonormal system of eigenvectors $\{e_j\}_{j=1}^{\infty}$ of a compact self-adjoint operator A in a Hilbert space H is an orthonormal basis if and only if $N(A) = \{0\}$.*

Proof. Recall from Exercise 26.7 that

$$H = N(A^*) \oplus \overline{R(A)} = N(A) \oplus \overline{R(A)}.$$

If $N(A) = \{0\}$, then $H = \overline{R(A)}$. This means that for every $x \in H$ and $\varepsilon > 0$ there exists $y_\varepsilon \in R(A)$ such that

$$\|x - y_\varepsilon\| < \varepsilon/2.$$

But by the Riesz–Schauder theorem,

$$y_\varepsilon = Ax_\varepsilon = \sum_{j=1}^{\infty} \lambda_j (x_\varepsilon, e_j) e_j.$$

Hence

$$\|x - y_\varepsilon\| = \left\| x - \sum_{j=1}^{\infty} \lambda_j (x_\varepsilon, e_j) e_j \right\| < \varepsilon/2.$$

Making use of the Pythagorean theorem, Bessel's inequality, and Exercise 25.8 yields

$$\begin{aligned}
\left\| x - \sum_{j=1}^{n} (x, e_j) e_j \right\| &\le \left\| x - \sum_{j=1}^{n} \lambda_j (x_\varepsilon, e_j) e_j \right\| \\
&= \left\| x - \sum_{j=1}^{\infty} \lambda_j (x_\varepsilon, e_j) e_j + \sum_{j=n+1}^{\infty} \lambda_j (x_\varepsilon, e_j) e_j \right\| \\
&< \varepsilon/2 + \left\| \sum_{j=n+1}^{\infty} \lambda_j (x_\varepsilon, e_j) e_j \right\| \\
&\le \varepsilon/2 + \left(\sum_{j=n+1}^{\infty} |\lambda_j|^2 |(x_\varepsilon, e_j)|^2 \right)^{1/2} \\
&\le \varepsilon/2 + |\lambda_{n+1}| \left(\sum_{j=n+1}^{\infty} |(x_\varepsilon, e_j)|^2 \right)^{1/2} \\
&\le \varepsilon/2 + |\lambda_{n+1}| \, \|x_\varepsilon\| < \varepsilon
\end{aligned}$$

for n sufficiently large. This means that $\{e_j\}_{j=1}^{\infty}$ is a basis of H, and moreover, it is an orthonormal basis.

Conversely, if $\{e_j\}_{j=1}^{\infty}$ is complete in H, then $\overline{R(A)} = H$ (Riesz–Schauder) and therefore $N(A) = \{0\}$. □

Remark 28.13. The condition $N(A) = \{0\}$ means that A^{-1} exists and H must be separable in this case.

Proposition 28.14 (Riesz). *If A is a compact operator on H and $\mu \in \mathbb{C}$, then the null space of $I - \mu A$ is a finite-dimensional subspace.*

Proof. The null space $N(I-\mu A)$ is a closed subspace of H, since $I-\mu A$ is bounded. Indeed, for each sequence $f_n \to f$ and $f_n - \mu A f_n = 0$ we have that $f - \mu A f = 0$, since A is continuous.

The operator A is compact on H and therefore also compact from $N(I-\mu A)$ onto $N(I-\mu A)$, since $N(I-\mu A)$ is closed. Hence, for every $f \in N(I-\mu A)$ we have

$$If = (I-\mu A)f + \mu A f = \mu A f,$$

and I is compact on $N(I-\mu A)$. Thus $N(I-\mu A)$ is finite-dimensional. □

Theorem 28.15 (Riesz's lemma). *If A is a compact operator on H and $\mu \in \mathbb{C}$, then $R(I-\mu A)$ is closed in H.*

Proof. If $\mu = 0$, then $R(I-\mu A) = H$. If $\mu \neq 0$, then we assume without loss of generality that $\mu = 1$. Let $f \in \overline{R(I-A)}$, $f \neq 0$. Then there exists a sequence $\{g_n\} \subset H$ such that

$$f = \lim_{n\to\infty} (I-A)g_n.$$

We will prove that $f \in R(I-A)$, i.e., there exists $g \in H$ such that $f = (I-A)g$. Since $f \neq 0$, we can assume by the decomposition $H = N(I-A) \oplus N(I-A)^{\perp}$ that $g_n \in N(I-A)^{\perp}$ and $g_n \neq 0$ for all $n \in \mathbb{N}$.

Suppose that g_n is bounded. Then there is a subsequence $\{g_{k_n}\}$ such that

$$g_{k_n} \overset{w}{\to} g.$$

The compactness of A implies that

$$Ag_{k_n} \to h = Ag.$$

Next,

$$g_{k_n} = (I-A)g_{k_n} + Ag_{k_n} \to f + h.$$

Hence $g = f + Ag$, i.e., $f = (I-A)g$.

Suppose that g_n is not bounded. Then we can assume without loss of generality that $\|g_n\| \to \infty$. Let us introduce a new sequence

$$u_n := \frac{g_n}{\|g_n\|}.$$

Since $\|u_n\| = 1$, there exists a subsequence $u_{k_n} \overset{w}{\to} u$. The compactness of A gives $Au_{k_n} \to Au$. Since $(I-A)g_n \to f$, we have

$$(I-A)u_{k_n} = \frac{1}{\|g_{k_n}\|}(I-A)g_{k_n} \to 0.$$

This means again that

$$u_{k_n} = (I-A)u_{k_n} + Au_{k_n} \to Au$$

and $u = Au$, i.e., $u \in N(I-A)$. But $g_n \in N(I-A)^{\perp}$. Hence $u_{k_n} \in N(I-A)^{\perp}$ and further $u \in N(I-A)^{\perp}$, because $N(I-A)^{\perp}$ is closed. Since $\|u_{k_n}\| = 1$, we have $\|u\| = 1$. Therefore, $u \neq 0$, while

$$u \in N(I-A) \cap N(I-A)^{\perp}.$$

This contradiction shows that unbounded g_n cannot occur. □

We are now ready to derive the following fundamental result of Riesz theory.

Theorem 28.16 (Riesz). *Let $A : H \to H$ be a compact linear operator on a Hilbert space H. Then for every $\mu \in \mathbb{C}$ the operator $I - \mu A$ is* injective *(i.e., $(I-\mu A)^{-1}$ exists) if and only if it is* surjective *(i.e., $R(I-\mu A) = H$). Moreover, in this case the inverse operator $(I-\mu A)^{-1} : H \to H$ is bounded.*

Proof. If $(I-\mu A)^{-1}$ exists, then $(I-\overline{\mu}A^*)^{-1}$ exists too and therefore $N(I-\overline{\mu}A^*) = 0$. Then Riesz's lemma (Theorem 28.15) and Exercise 26.7 imply $H = R(I-\mu A)$, i.e., $I-\mu A$ is surjective.

Conversely, if $I-\mu A$ is surjective, then $N(I-\overline{\mu}A^*) = 0$, i.e., $I-\overline{\mu}A^*$ is injective and so is $I-\mu A$.

It remains to show that $(I-\mu A)^{-1}$ is bounded on H if $I-\mu A$ is injective. Assume that $(I-\mu A)^{-1}$ is not bounded. Then there exists a sequence $f_n \in H$ with $\|f_n\| = 1$ such that

$$\left\|(I-\mu A)^{-1} f_n\right\| \geq n.$$

Define

$$g_n := \frac{f_n}{\|(I-\mu A)^{-1} f_n\|}, \quad \varphi_n := \frac{(I-\mu A)^{-1} f_n}{\|(I-\mu A)^{-1} f_n\|}.$$

Then $g_n \to 0$ as $n \to \infty$ and $\|\varphi_n\| = 1$. Since A is compact, we can select a subsequence φ_{k_n} such that $A\varphi_{k_n} \to \varphi$ as $k_n \to \infty$. But

$$\varphi_n - \mu A \varphi_n = g_n,$$

and we observe that $\varphi_{k_n} \to \mu\varphi$ and $\varphi \in N(I-\mu A)$. Hence $\varphi = 0$, and this contradicts $\|\varphi_n\| = 1$. □

Theorem 28.17 (Fredholm alternative). *Suppose $A = A^*$ is compact. For given $g \in H$ either the equation*

$$(I-\mu A)f = g$$

has the unique solution ($\mu^{-1} \notin \sigma(A)$), in which case $f = (I-\mu A)^{-1} g$, or else $\mu^{-1} \in \sigma(A)$, and this equation has a solution if and only if $g \in R(I-\mu A)$, i.e., $g \perp N(I-\mu A)$. In this case, the general solution of the equation is of the form $f = f_0 + u$, *where f_0 is a particular solution and $u \in N(I-\mu A)$ (u is the general solution of the corresponding homogeneous equation), and the set of all solutions is a finite-dimensional affine subspace of H.*

Proof. Riesz's lemma (Theorem 28.15) gives

$$R(I-\mu A) = N(I-\overline{\mu}A)^{\perp}.$$

If $\mu^{-1} \notin \sigma(A)$, then $(\overline{\mu})^{-1} \notin \sigma(A)$ also. Thus

$$R(I-\mu A) = N(I-\overline{\mu}A)^{\perp} = \{0\}^{\perp} = H.$$

Since $A = A^*$, this means that $(I-\mu A)^{-1}$ exists, and the unique solution is $f = (I-\mu A)^{-1}g$.

If $\mu^{-1} \in \sigma(A)$, then $R(I-\mu A)$ is a proper subspace of H, and the equation $(I-\mu A)f = g$ has a solution if and only if $g \in R(I-\mu A)$. Since the equation is linear, every solution is of the form

$$f = f_0 + u, \quad u \in N(I-\mu A),$$

and the dimension of $N(I-\mu A)$ is finite. □

Exercise 28.8. Let $A = A^*$ be compact and injective. Prove that $\sigma_{\text{p}}(A) = \sigma_{\text{d}}(A) = \sigma(A) \setminus \{0\}$ and $0 \in \sigma_{\text{ess}}(A)$.

Exercise 28.9. Consider the Hilbert space $H = l^2(\mathbb{C})$ and

$$A(x_1, x_2, \ldots, x_n, \ldots) = (0, x_1, \frac{x_2}{2}, \ldots, \frac{x_n}{n}, \ldots)$$

for $(x_1, x_2, \ldots, x_n, \ldots) \in l^2(\mathbb{C})$. Show that A is compact and has no eigenvalues (moreover, $\sigma(A) = \{0\}$) and is not self-adjoint.

Exercise 28.10. Consider the Hilbert space $H = L^2(\mathbb{R})$ and

$$(Af)(t) = tf(t).$$

Show that the equation $Af = f$ has no nontrivial solutions and that $(I-A)^{-1}$ does not exist. This means that the Fredholm alternative does not hold for a noncompact but self-adjoint operator.

Exercise 28.11. Let $H = L^2(\mathbb{R}^n)$ and let

$$Af(x) = \int_{\mathbb{R}^n} K(x,y) f(y) \mathrm{d}y,$$

where $K(x,y) \in L^2(\mathbb{R}^n \times \mathbb{R}^n)$ is such that $K(x,y) = \overline{K(y,x)}$. Prove that $A = A^*$ and that A is compact.

Theorem 28.18 **(Weyl).** *If $A = A^*$, then $\lambda \in \sigma_{\text{ess}}(A)$ if and only if there exists an orthonormal system $\{x_n\}_{n=1}^{\infty}$ such that*

$$\|(A - \lambda I)x_n\| \to 0$$

as $n \to \infty$.

Proof. We will provide only a partial proof. See [5] for a full proof. Suppose that $\lambda \in \sigma_{\text{ess}}(A)$. If λ is an eigenvalue of infinite multiplicity, then there is an infinite orthonormal system of eigenvectors $\{x_n\}_{n=1}^{\infty}$, because $\dim(E_\lambda - E_{\lambda-0})H = \infty$ in this case. Since $(A - \lambda I)x_n \equiv 0$, it is clear that

$$(A - \lambda I)x_n \to 0.$$

Next, suppose that λ is an accumulation point of $\sigma(A)$. This means that $\lambda \in \sigma(A)$ and

$$\lambda = \lim_{n \to \infty} \lambda_n,$$

where $\lambda_n \neq \lambda_m$, $n \neq m$, and $\lambda_n \in \sigma(A)$. Hence for each $n = 1, 2, \ldots$ we have that

$$E_{\lambda_n + \varepsilon} - E_{\lambda_n - \varepsilon} \neq 0$$

for all $\varepsilon > 0$. Therefore, there exists a sequence $r_n \to 0$ such that

$$E_{\lambda_n + r_n} - E_{\lambda_n - r_n} \neq 0.$$

We can therefore find a normalized vector $x_n \in R(E_{\lambda_n + r_n} - E_{\lambda_n - r_n})$. Since $\lambda_n \neq \lambda_m$ for $n \neq m$, we can find $\{x_n\}_{n=1}^{\infty}$ as an orthonormal system. By the spectral theorem we have

$$\begin{aligned}
\|(A - \lambda I)x_n\|^2 &= \int_{-\infty}^{\infty} (\lambda - \mu)^2 \mathrm{d}(E_\mu x_n, x_n) \\
&= \int_{-\infty}^{\infty} (\lambda - \mu)^2 \mathrm{d}(E_\mu (E_{\lambda_n + r_n} - E_{\lambda_n - r_n})x_n, x_n) \\
&= \int_{\lambda_n - r_n}^{\lambda_n + r_n} (\lambda - \mu)^2 \mathrm{d}(E_\mu x_n, x_n) \\
&\leq \max_{\lambda_n - r_n \leq \mu \leq \lambda_n + r_n} (\lambda - \mu)^2 \int_{-\infty}^{\infty} \mathrm{d}(E_\mu x_n, x_n) \\
&= \max_{\lambda_n - r_n \leq \mu \leq \lambda_n + r_n} (\lambda - \mu)^2 \to 0, \quad n \to \infty.
\end{aligned}$$

This completes the proof. □

Theorem 28.19 **(Weyl).** *Let A and B be two self-adjoint operators in a Hilbert space. If there is $z \in \rho(A) \cap \rho(B)$ such that*

$$T := (A - zI)^{-1} - (B - zI)^{-1}$$

is a compact operator, then $\sigma_{\text{ess}}(A) = \sigma_{\text{ess}}(B)$.

Proof. We show first that $\sigma_{\text{ess}}(A) \subset \sigma_{\text{ess}}(B)$. Take any $\lambda \in \sigma_{\text{ess}}(A)$. Then there is an orthonormal system $\{x_n\}_{n=1}^{\infty}$ such that

$$\|(A - \lambda I)x_n\| \to 0, \quad n \to \infty.$$

Define the sequence y_n as

$$y_n := (A - zI)x_n \equiv (A - \lambda I)x_n + (\lambda - z)x_n.$$

Due to Bessel's inequality, every orthonormal system in the Hilbert space converges weakly to 0. Hence $y_n \xrightarrow{w} 0$. We also have

$$\|y_n\| \geq |\lambda - z| \, \|x_n\| - \|(A - \lambda I)x_n\| = |\lambda - z| - \|(A - \lambda I)x_n\| > \frac{|\lambda - z|}{2} > 0$$

for all $n \geq n_0 \gg 1$. Next we take the identity

$$\left[(B - zI)^{-1} - (\lambda - z)^{-1}\right] y_n = -Ty_n - (\lambda - z)^{-1}(A - \lambda I)x_n.$$

Since T is compact and $y_n \xrightarrow{w} 0$, we deduce that

$$\left[(B - zI)^{-1} - (\lambda - z)^{-1}\right] y_n \to 0.$$

Introduce

$$z_n := (B - zI)^{-1} y_n.$$

Then

$$z_n - (\lambda - z)^{-1} y_n \to 0,$$

or

$$y_n + (z - \lambda) z_n \to 0.$$

This fact and $\|y_n\| > \frac{|\lambda - z|}{2}$ imply that $\|z_n\| \geq \frac{|\lambda - z|}{3}$ for all $n \geq n_0 \gg 1$. But

$$(B - \lambda I)z_n \equiv (B - zI)z_n + (z - \lambda)z_n = y_n + (z - \lambda)z_n \to 0.$$

Due to $\|z_n\| > \frac{|\lambda - z|}{3} > 0$, the sequence $\{z_n\}_{n=1}^{\infty}$ can be chosen as an orthonormal system. Thus $\lambda \in \sigma_{\text{ess}}(B)$. This proves that $\sigma_{\text{ess}}(A) \subset \sigma_{\text{ess}}(B)$. Finally, since $-T$ is compact too, we can interchange the roles of A and B and obtain the opposite embedding. $\square$

Chapter 29
Quadratic Forms. Friedrichs Extension.

Definition 29.1. Let D be a linear subspace of a Hilbert space H. A function $Q : D \times D \to \mathbb{C}$ is called a *quadratic form* if

(1) $Q(\alpha_1 x_1 + \alpha_2 x_2, y) = \alpha_1 Q(x_1, y) + \alpha_2 Q(x_2, y)$,
(2) $Q(x, \beta_1 y_1 + \beta_2 y_2) = \overline{\beta_1} Q(x, y_1) + \overline{\beta_2} Q(x, y_2)$,

for all $\alpha_1, \alpha_2, \beta_1, \beta_2 \in \mathbb{C}$ and $x_1, x_2, x, y_1, y_2, y \in D$. The space $D(Q) := D$ is called the *domain* of Q. Then Q is

(1) *densely defined* if $\overline{D(Q)} = H$.
(2) *symmetric* if $Q(x,y) = \overline{Q(y,x)}$.
(3) *semibounded from below* if there exists $\lambda \in \mathbb{R}$ such that $Q(x,x) \geq -\lambda \|x\|^2$ for all $x \in D(Q)$.
(4) *closed* (and semibounded) if $D(Q)$ is complete with respect to the norm

$$\|x\|_Q := \sqrt{Q(x,x) + (\lambda + 1)\|x\|^2}.$$

(5) *bounded* (continuous) if there exists $M > 0$ such that

$$|Q(x,y)| \leq M \|x\| \|y\|$$

for all $x, y \in D(Q)$.

Exercise 29.1. Prove that $\|\cdot\|_Q$ is a norm and that

$$(x,y)_Q := Q(x,y) + (\lambda + 1)(x,y)$$

is an inner product.

V. Serov, *Fourier Series, Fourier Transform and Their Applications to Mathematical Physics*, Applied Mathematical Sciences 197,
DOI 10.1007/978-3-319-65262-7_29

Theorem 29.2. *Let Q be a densely defined, closed, semibounded, and symmetric quadratic form in a Hilbert space H such that*

$$Q(x,x) \geq -\lambda \|x\|^2, \quad x \in D(Q).$$

Then there exists a unique self-adjoint operator A defined by the quadratic form Q as

$$Q(x,y) = (Ax,y), \quad x \in D(A), y \in D(Q)$$

that is semibounded from below, i.e.,

$$(Ax,x) \geq -\lambda \|x\|^2, \quad x \in D(A),$$

and $D(A) \subset D(Q)$.

Proof. Let us introduce an inner product on $D(Q)$ by

$$(x,y)_Q := Q(x,y) + (\lambda+1)(x,y), \quad x,y \in D(Q)$$

(see Exercise 29.1). Since Q is closed, $D(Q) = \overline{D(Q)}$ is a closed subspace of H with respect to the norm $\|\cdot\|_Q$. This means that $D(Q)$ with this inner product defines a new Hilbert space H_Q. It is clear also that

$$\|x\|_Q \geq \|x\|$$

for all $x \in H_Q$. Thus, for fixed $x \in H$,

$$L(y) := (y,x), \quad y \in H_Q$$

defines a continuous (bounded) linear functional on the Hilbert space H_Q. Applying the Riesz–Fréchet theorem to H_Q, we obtain an element $x^* \in H_Q$ ($x^* \in D(Q)$) such that

$$(y,x) \equiv L(y) = (y,x^*)_Q.$$

It is clear that the map

$$H \ni x \mapsto x^* \in H_Q$$

defines a linear operator J such that

$$J : H \to H_Q, \quad Jx = x^*.$$

Hence

$$(y,x) = (y,Jx)_Q, \quad x \in H,\ y \in H_Q.$$

Next we prove that J is self-adjoint and that it has an inverse operator J^{-1}. For all $x, y \in H$ we have

$$(Jy,x) = (Jy,Jx)_Q = \overline{(Jx,Jy)_Q} = \overline{(Jx,y)} = (y,Jx).$$

Hence $J = J^*$. It is bounded by the Hellinger–Toeplitz theorem (Exercise 26.2). Suppose that $Jx = 0$. Then

$$(y,x) = (y,Jx)_Q = 0$$

for every $y \in D(Q)$. Since $\overline{D(Q)} = H$, the last equality implies that $x = 0$, and therefore $N(J) = \{0\}$ and J^{-1} exists. Moreover,

$$H = N(J) \oplus \overline{R(J^*)} = \overline{R(J)}$$

and $R(J) \subset H_Q$. Now we can define a linear operator A on the domain $D(A) \equiv R(J)$ as

$$Ax := J^{-1}x - (\lambda+1)x, \quad \lambda \in \mathbb{R}.$$

It is clear that A is densely defined and $A = A^*$ (J^{-1} is self-adjoint, since J is). If now $x \in D(A)$ and $y \in D(Q) \equiv H_Q$, then

$$Q(x,y) = (x,y)_Q - (\lambda+1)(x,y) = (J^{-1}x,y) - (\lambda+1)(x,y) = (Ax,y).$$

The semiboundedness of A from below follows from that of Q. It remains to prove that this representation for A is unique. Assume that we have two such representations, A_1 and A_2. Then for every $x \in D(A_1) \cap D(A_2)$ and $y \in D(Q)$ we have that

$$Q(x,y) = (A_1x,y) = (A_2x,y).$$

It follows that

$$((A_1 - A_2)x,y) = 0.$$

Since $\overline{D(Q)} = H$, we must have $A_1x = A_2x$. This completes the proof. □

Corollary 29.3. *Under the same assumptions as in Theorem 29.2, there exists $\sqrt{A+\lambda I}$ that is self-adjoint on $D(\sqrt{A+\lambda I}) \equiv D(Q) = H_Q$. Moreover,*

$$Q(x,y) + \lambda(x,y) = (\sqrt{A+\lambda I}x, \sqrt{A+\lambda I}y)$$

for all $x, y \in D(Q)$.

Proof. Since $A + \lambda I$ is self-adjoint and nonnegative, there exists a spectral family $\{E_\mu\}_{\mu=0}^{\infty}$ such that

$$A + \lambda I = \int_0^\infty \mu \, \mathrm{d}E_\mu.$$

We can therefore define the operator

$$\sqrt{A+\lambda I} := \int_0^\infty \sqrt{\mu}\mathrm{dE}_\mu,$$

which is also self-adjoint and nonnegative. Then

$$Q(x,y)+\lambda(x,y) = ((A+\lambda I)x,y) = (\sqrt{A+\lambda I}x, \left(\sqrt{A+\lambda I}\right)^* y)$$

for all $x \in D(A)$ and $y \in D(Q)$. This fact means that $x \in D(\sqrt{A+\lambda I})$ and $y \in D(\left(\sqrt{A+\lambda I}\right)^*)$. But $\sqrt{A+\lambda I}$ is self-adjoint, and therefore,

$$D(\sqrt{A+\lambda I}) = D(\left(\sqrt{A+\lambda I}\right)^*) = D(Q) \equiv H_Q.$$

This completes the proof. □

Theorem 29.4 ***(Friedrichs extension).*** *Let A be a nonnegative symmetric linear operator in a Hilbert space H. Then there exists a self-adjoint extension A_F of A that is the smallest among all nonnegative self-adjoint extensions of A in the sense that its corresponding quadratic form has the smallest domain. This extension A_F is called the* Friedrichs extension *of A.*

Proof Let A be a nonnegative symmetric operator with $\overline{D(A)} = H$. Its associated quadratic form

$$Q(x,y) := (Ax,y), \quad x,y \in D(Q) \equiv D(A),$$

is densely defined, nonnegative, and symmetric. Let us define a new inner product

$$(x,y)_Q = Q(x,y)+(x,y), \quad x,y \in D(Q).$$

Then $D(Q)$ becomes an inner product space. This inner product space has a completion H_Q with respect to the norm

$$\|x\|_Q := \sqrt{Q(x,x)+\|x\|^2}.$$

Moreover, the quadratic form $Q(x,y)$ has an extension $Q_1(x,y)$ to this Hilbert space H_Q defined by

$$Q_1(x,y) = \lim_{n\to\infty} Q(x_n,y_n)$$

whenever $x \overset{H_Q}{=} \lim_{n\to\infty} x_n, y \overset{H_Q}{=} \lim_{n\to\infty} y_n, x_n, y_n \in D(Q)$ and these limits exist. The quadratic form Q_1 is densely defined, closed, nonnegative, and symmetric. Therefore, Theorem 29.2, applied to Q_1, gives a unique nonnegative self-adjoint operator A_F such that

$$Q_1(x,y) = (A_{\mathrm{F}}x,y), \quad x \in D(A_{\mathrm{F}}) \subset H_Q, y \in D(Q_1) \equiv H_Q.$$

Since for $x, y \in D(A)$ one has

$$(Ax,y) = Q(x,y) = Q_1(x,y) = (A_{\mathrm{F}}x,y),$$

it follows that A_{F} is a self-adjoint extension of A.

It remains to prove that A_{F} is the smallest nonnegative self-adjoint extension of A. Suppose that $B \geq 0$, $B = B^*$, is such that $A \subset B$. The associated quadratic form $Q_B(x,y) := (Bx,y)$ is an extension of $Q \equiv Q_A$. Hence

$$\overline{Q_B} \supset \overline{Q} = Q_1.$$

This completes the proof. □

Chapter 30
Elliptic Differential Operators

Let Ω be a domain in $\mathbb{R}^n$, i.e., an open and connected set. We introduce the following notation:

(1) $x = (x_1, \dots, x_n) \in \Omega$;
(2) $|x| = \sqrt{x_1^2 + \cdots + x_n^2}$;
(3) $\alpha = (\alpha_1, \dots, \alpha_n)$ is a *multi-index*, i.e., $\alpha_j \in \mathbb{N}_0 \equiv \mathbb{N} \cup \{0\}$:

(a) $|\alpha| = \alpha_1 + \cdots + \alpha_n$,
(b) $\alpha \geq \beta$ if $\alpha_j \geq \beta_j$ for all $j = 1, 2, \dots, n$,
(c) $\alpha + \beta = (\alpha_1 + \beta_1, \dots, \alpha_n + \beta_n)$,
(d) $\alpha - \beta = (\alpha_1 - \beta_1, \dots, \alpha_n - \beta_n)$ if $\alpha \geq \beta$,
(e) $x^\alpha = x_1^{\alpha_1} \cdots x_n^{\alpha_n}$ with $0^0 = 1$,
(f) $\alpha! = \alpha_1! \cdots \alpha_n!$ with $0! = 1$;

(4) $\partial_j = \frac{\partial}{\partial x_j}$ and $\partial^\alpha = \partial_1^{\alpha_1} \cdots \partial_n^{\alpha_n}$.

Definition 30.1. An *elliptic partial differential operator* $A(x, \partial)$ of order m on Ω is an operator of the form

$$A(x, \partial) = \sum_{|\alpha| \leq m} a_\alpha(x) \partial^\alpha,$$

where $a_\alpha(x) \in C^\infty(\Omega)$, whose *principal symbol*

$$a(x, \xi) = \sum_{|\alpha| = m} a_\alpha(x) \xi^\alpha, \quad \xi \in \mathbb{R}^n$$

is invertible for all $x \in \Omega$ and $\xi \in \mathbb{R}^n \setminus \{0\}$, that is, $a(x, \xi) \neq 0$ for all $x \in \Omega$ and $\xi \in \mathbb{R}^n \setminus \{0\}$.

Assumption 30.2. We assume that $a_\alpha(x)$ are real for $|\alpha| = m$.

V. Serov, *Fourier Series, Fourier Transform and Their Applications to Mathematical Physics*, Applied Mathematical Sciences 197,
DOI 10.1007/978-3-319-65262-7_30

Under Assumption 30.2 either $a(x,\xi) > 0$ or $a(x,\xi) < 0$ for all $x \in \Omega$ and $\xi \in \mathbb{R}^n \setminus \{0\}$. Without loss of generality we assume that $a(x,\xi) > 0$. Assumption 30.2 implies also that m is even and that for every compact set $K \subset \Omega$ there exists $C_K > 0$ such that

$$a(x,\xi) \geq C_K |\xi|^m, \quad x \in \Omega, \xi \in \mathbb{R}^n.$$

Assumption 30.3. We assume that $A(x,\partial)$ is *formally self-adjoint*, i.e.,

$$A(x,\partial) = A'(x,\partial) := \sum_{|\alpha|\leq m} (-1)^{|\alpha|} \partial^\alpha (\overline{a_\alpha(x)} \cdot).$$

Exercise 30.1. Prove that $A(x,\partial) = A'(x,\partial)$ if and only if

$$a_\alpha(x) = \sum_{\substack{\alpha\leq\beta \\ |\beta|\leq m}} (-1)^{|\beta|} C_\beta^\alpha \partial^{\beta-\alpha} \overline{a_\beta(x)},$$

where

$$C_\beta^\alpha = \frac{\beta!}{\alpha!(\beta-\alpha)!}.$$

Hint: Make use of the *generalized Leibniz formula*

$$\partial^\alpha (fg) = \sum_{\beta\leq\alpha} C_\alpha^\beta \partial^{\alpha-\beta} f \partial^\beta g.$$

Assumption 30.4. We assume that $A(x,\partial)$ has a divergence form

$$A(x,\partial) \equiv \sum_{|\alpha|=|\beta|\leq m/2} (-1)^{|\alpha|} \partial^\alpha (a_{\alpha\beta}(x) \partial^\beta),$$

where $a_{\alpha\beta} = a_{\beta\alpha}$ and this value is real for all α and β. We assume also the *generalized ellipticity condition*

$$\int_\Omega \sum_{|\alpha|=|\beta|=m/2} a_{\alpha\beta}(x) \partial^\alpha f \partial^\beta \overline{f} \mathrm{d}x \geq \nu \int_\Omega \sum_{|\alpha|=m/2} |\partial^\alpha f|^2 \mathrm{d}x, \quad f \in C_0^\infty(\Omega),$$

where $\nu > 0$ is called the *constant of ellipticity*.

Remark 30.5. If the coefficients $a_{\alpha\beta}$ of $A(x,\partial)$ are constants, then this generalized ellipticity condition reads

$$\sum_{|\alpha|=|\beta|=m/2} a_{\alpha\beta} \xi^{\alpha+\beta} \geq \nu \sum_{|\alpha|=m/2} \xi^{2\alpha}.$$

Exercise 30.2. Prove that

$$\sum_{|\alpha|=m/2} \xi^{2\alpha} \asymp |\xi|^m,$$

i.e.,

$$c|\xi|^m \leq \sum_{|\alpha|=m/2} \xi^{2\alpha} \leq C|\xi|^m,$$

where c and C are some constants.

Example 30.6. Let us consider

$$A(x,\partial) = -\sum_{j=1}^{n} \partial_j^2 = -\Delta, \quad x \in \Omega \subset \mathbb{R}^n$$

in $H = L^2(\Omega)$ and prove that $A \subset A^*$ with

$$D(A) = C_0^\infty(\Omega) = \left\{ f \in C^\infty(\Omega) : \operatorname{supp} f = \overline{\{x : f(x) \neq 0\}} \text{ is compact in } \Omega \right\}.$$

Let $u, v \in C_0^\infty(\Omega)$. Then

$$\begin{aligned}
(Au,v)_{L^2} &= -\sum_{j=1}^{n} \int_\Omega (\partial_j^2 u)\, \overline{v} \mathrm{d}x \\
&= -\sum_{j=1}^{n} \int_\Omega \partial_j ((\partial_j u)\overline{v}) \,\mathrm{d}x + \sum_{j=1}^{n} \int_\Omega (\partial_j u) \left(\overline{\partial_j v}\right) \mathrm{d}x \\
&= -\int_{\partial\Omega} (\overline{v}\nabla u, n_x) \mathrm{d}x + (\nabla u, \nabla v)_{L^2} = (\nabla u, \nabla v)_{L^2},
\end{aligned}$$

where $\partial\Omega$ is the boundary of Ω and n_x is the unit outward normal vector at $x \in \partial\Omega$. Here we have made use of the divergence theorem. In a similar fashion we obtain

$$(\nabla u, \nabla v)_{L^2} = -\sum_{j=1}^{n} \int_\Omega u \partial_j^2 \overline{v} \mathrm{d}x = (u, -\Delta v)_{L^2} = (u, Av)_{L^2}.$$

Hence $A \subset A^*$ and A is closable.

Example 30.7. Recall from Example 30.6 that

$$(-\Delta u, v)_{L^2} = (\nabla u, \nabla v)_{L^2}, \quad u, v \in C_0^\infty(\Omega).$$

Hence

$$(-\Delta u, u)_{L^2} = \|\nabla u\|_{L^2}^2 \leq \|u\|_{L^2} \|\Delta u\|_{L^2}, \quad u \in C_0^\infty(\Omega).$$

Therefore,

$$\begin{aligned}\|u\|_{W_2^2}^2 &= \|u\|_{L^2}^2 + \|\nabla u\|_{L^2}^2 + \|\Delta u\|_{L^2}^2 \\ &\le \|u\|_{L^2}^2 + \|u\|_{L^2}\|\Delta u\|_{L^2} + \|\Delta u\|_{L^2}^2 \\ &\le \frac{3}{2}\|u\|_{L^2}^2 + \frac{3}{2}\|\Delta u\|_{L^2}^2 \equiv \frac{3}{2}\|u\|_A^2,\end{aligned}$$

where $\|\cdot\|_A$ is a norm that corresponds to the operator $A = -\Delta$ as follows:

$$\|u\|_A^2 := \|u\|_{L^2}^2 + \|-\Delta u\|_{L^2}^2.$$

It is also clear that $\|u\|_A \le \|u\|_{W_2^2}$. Combining these inequalities gives

$$\sqrt{\frac{2}{3}}\|u\|_{W_2^2} \le \|u\|_A \le \|u\|_{W_2^2}$$

for all $u \in C_0^\infty(\Omega)$. A completion of $C_0^\infty(\Omega)$ with respect to these norms leads us to the statement

$$D(\overline{A}) = \overset{\circ}{W}{}_2^2(\Omega).$$

Thus $\overline{A} = -\Delta$ on $D(\overline{A}) = \overset{\circ}{W}{}_2^2(\Omega)$. Let us determine $D(A^*)$ in this case. By the definition of $D(A^*)$ we have

$$D((-\Delta)^*) = \{v \in L^2(\Omega) : \text{there exists}\, v^* \in L^2(\Omega)\, \text{such that} \\ (-\Delta u, v) = (u, v^*)\, \text{for all}\, u \in C_0^\infty(\Omega)\}.$$

If we assume that $v \in W_2^2(\Omega)$, then this is equivalent to

$$(u, (-\Delta)^* v) = (u, v^*),$$

i.e., $(-\Delta)^* v = v^*$ and $D((-\Delta)^*) = W_2^2(\Omega)$. Finally, for $\Omega \subset \mathbb{R}^n$ with $\Omega \neq \mathbb{R}^n$ we obtain that

$$A \subset \overline{A} \subset A^* \equiv (\overline{A})^*$$

and $A \neq \overline{A}$ and $\overline{A} \neq (\overline{A})^*$, that is, the closure of A does not lead us to a self-adjoint operator.

Remark 30.8. If $\Omega = \mathbb{R}^n$, then $\overset{\circ}{W}{}_2^2(\mathbb{R}^n) \equiv W_2^2(\mathbb{R}^n)$ and therefore

$$\overline{A} = A^* = (\overline{A})^*.$$

Hence the closure of A is self-adjoint in that case, i.e., A is essentially self-adjoint.

Example 30.9. Consider again $A = -\Delta$ on $D(A) = C_0^\infty(\Omega)$ with $\Omega \neq \mathbb{R}^n$. Since

$$(-\Delta u, u)_{L^2} = \|\nabla u\|_{L^2}^2 \geq 0,$$

it follows that $-\Delta$ is nonnegative with lower bound $\lambda = 0$. Therefore,

$$Q(u,v) := (\nabla u, \nabla v)_{L^2}$$

is a densely defined nonnegative quadratic form with $D(Q) \equiv D(A) = C_0^\infty(\Omega)$. A new inner product is defined as

$$(u,v)_Q := (\nabla u, \nabla v)_{L^2} + (u,v)_{L^2}$$

and

$$\|u\|_Q^2 \equiv \|u\|_{W_2^1(\Omega)}^2.$$

If we apply now the procedure from Theorem 29.4, then we obtain the existence of $Q_1 = \overline{Q}$ with respect to the norm $\|\cdot\|_Q$, which will also be nonnegative and closed with $D(Q_1) \equiv \overset{\circ}{W}{}_2^1(\Omega)$. The next step is to obtain the Friedrichs extension A_{F} as

$$A_{\mathrm{F}} = J^{-1} - I$$

with $D(A_{\mathrm{F}}) \equiv R(J) \subset \overset{\circ}{W}{}_2^1(\Omega)$. A more careful examination of Theorem 29.2 leads us to the fact

$$D(A_{\mathrm{F}}) = \overset{\circ}{W}{}_2^1(\Omega) \cap D(A^*) = \overset{\circ}{W}{}_2^1(\Omega) \cap W_2^2(\Omega).$$

Remark 30.10. In general, for a symmetric operator we have

$$D(A_{\mathrm{F}}) = \{u \in H_Q : Au \in H\},$$

which is equivalent to

$$D(A_{\mathrm{F}}) = \{u \in H_Q : u \in D(A^*)\}.$$

Exercise 30.3. Let $H = L^2(\Omega)$ and $A(x,D) = -\Delta + q(x)$, where $q(x) = \overline{q(x)}$ and $q(x) \in L^\infty(\Omega)$. Define $\overline{A}$, A^*, and A_{F}.

Exercise 30.4. Let $H = L^2(\Omega)$ and

$$A(x,\partial) = -(\nabla + \mathrm{i}\vec{W}(x))^2 + q(x),$$

where $\vec{W}$ is an n-dimensional real-valued vector from $W_\infty^1(\Omega)$ and q is a real-valued function from $L^\infty(\Omega)$. Define $\overline{A}$, A^*, and A_{F}.

Consider now a bounded domain $\Omega \subset \mathbb{R}^n$ and an elliptic operator $A(x,\partial)$ in Ω of the form

$$A(x,\partial) = \sum_{|\alpha|=|\beta|\le m/2} (-1)^{|\alpha|}\partial^\alpha(a_{\alpha\beta}(x)\partial^\beta),$$

where $a_{\alpha\beta}(x) = a_{\beta\alpha}(x)$ are real. Assume that there exists $C_0 > 0$ such that

$$|a_{\alpha\beta}(x)| \le C_0, \quad |\alpha|,|\beta| < \frac{m}{2},$$

for all $x \in \Omega$. Assume also that $A(x,\partial)$ is elliptic, that is,

$$\int_\Omega \sum_{|\alpha|=|\beta|=m/2} a_{\alpha\beta}(x)\partial^\alpha f \partial^\beta \overline{f} \mathrm{d}x \ge \nu \int_\Omega \sum_{|\alpha|=m/2} |\partial^\alpha|^2 \mathrm{d}x, \quad \nu > 0.$$

Theorem 30.11 **(Gårding's inequality)**. *Suppose that $A(x,\partial)$ is as above. Then for every $\varepsilon > 0$ there is $C_\varepsilon > 0$ such that*

$$(Af,f)_{L^2(\Omega)} \ge (\nu-\varepsilon)\|f\|^2_{W_2^{m/2}(\Omega)} - C_\varepsilon \|f\|^2_{L^2(\Omega)}$$

for all $f \in C_0^\infty(\Omega)$.

Proof. Let $f \in C_0^\infty(\Omega)$. Then integration by parts yields

$$\begin{aligned}
(Af,f)_{L^2(\Omega)} &= \sum_{|\alpha|=|\beta|\le m/2} (-1)^{|\alpha|} \int_\Omega \partial^\alpha(a_{\alpha\beta}(x)\partial^\beta f)\overline{f}\mathrm{d}x \\
&= \sum_{|\alpha|=|\beta|=m/2} \int_\Omega a_{\alpha\beta}(x)\overline{\partial^\alpha f}\partial^\beta f \mathrm{d}x \\
&+ \sum_{|\alpha|=|\beta|<m/2} \int_\Omega a_{\alpha\beta}(x)\overline{\partial^\alpha f}\partial^\beta f \mathrm{d}x \\
&\ge \nu \sum_{|\alpha|=m/2} \int_\Omega |\partial^\alpha f|^2 \mathrm{d}x - C_0 \sum_{|\alpha|=|\beta|<m/2} \int_\Omega |\partial^\alpha f||\partial^\beta f| \mathrm{d}x \\
&\ge \nu \sum_{|\alpha|\le m/2} \int_\Omega |\partial^\alpha f|^2 \mathrm{d}x - (C_0+\nu) \sum_{|\alpha|<m/2} \int_\Omega |\partial^\alpha f|^2 \mathrm{d}x \\
&= \nu \|f\|^2_{W_2^{m/2}(\Omega)} - (C_0+\nu)\|f\|^2_{W_2^{m/2-1}(\Omega)}.
\end{aligned}$$

Next we make use of the following lemma.

Lemma 30.12. *For all $\varepsilon > 0$ and $0 < \delta \le m/2$ there is $C_\varepsilon(\delta) > 0$ such that*

$$(1+|\xi|^2)^{m/2-\delta} \le \varepsilon(1+|\xi|^2)^{m/2} + C_\varepsilon(\delta)$$

for all $\xi \in \mathbb{R}^n$.

Proof. Let $\varepsilon > 0$ and $0 < \delta \le m/2$. If $(1+|\xi|^2)^\delta \ge \frac{1}{\varepsilon}$, then

$$(1+|\xi|^2)^{-\delta} \le \varepsilon.$$

Hence

$$(1+|\xi|^2)^{m/2-\delta} \le \varepsilon(1+|\xi|^2)^{m/2},$$

i.e., the claim holds for every positive constant $C_\varepsilon(\delta)$. For $(1+|\xi|^2)^\delta < \frac{1}{\varepsilon}$ we can obtain

$$(1+|\xi|^2)^{m/2-\delta} < \left(\frac{1}{\varepsilon}\right)^{\frac{m/2-\delta}{\delta}} \equiv C_\varepsilon(\delta).$$

This proves the claim. □

Applying this lemma with $\delta = 1$ to the norm of the Sobolev spaces W_2^k, we conclude that

$$\|f\|^2_{W_2^{m/2-1}(\Omega)} \le \varepsilon_1 \|f\|^2_{W_2^{m/2}(\Omega)} + C_{\varepsilon_1} \|f\|^2_{L^2(\Omega)}$$

for all $\varepsilon_1 > 0$. Hence

$$\begin{aligned}(Af,f)_{L^2(\Omega)} &\ge \nu \|f\|^2_{W_2^{m/2}(\Omega)} - (C_0+\nu)\|f\|^2_{W_2^{m/2-1}(\Omega)}\\ &\ge \nu \|f\|^2_{W_2^{m/2}(\Omega)} - (C_0+\nu)\varepsilon_1 \|f\|^2_{W_2^{m/2}(\Omega)} - (C_0+\nu)C_{\varepsilon_1}\|f\|^2_{L^2(\Omega)}\\ &= (\nu-\varepsilon)\|f\|^2_{W_2^{m/2}(\Omega)} - C_\varepsilon \|f\|^2_{L^2(\Omega)}.\end{aligned}$$

This proves the theorem. □

Corollary 30.13. *There exists a self-adjoint Friedrichs extension A_F of A with domain $D(A_F) = \overset{\circ}{W}{}_2^{m/2}(\Omega) \cap W_2^m(\Omega)$.*

Proof. It follows from Gårding's inequality that

$$(Af,f)_{L^2(\Omega)} \ge -C_\varepsilon \|f\|^2_{L^2(\Omega)}, \quad f \in D(A).$$

This means that $A_\mu := A + \mu I$ is positive for $\mu > C_\varepsilon$, and therefore Theorem 29.4 gives us the existence of

$$(A_\mu)_F \equiv (A_F)_\mu = A_F + \mu I$$

with domain

$$D(A_F) = D((A_\mu)_F) = \overset{\circ}{W}{}_2^{m/2}(\Omega) \cap D(A^*),$$

where $\overset{\circ}{W}{}_2^{m/2}(\Omega)$ is the domain of the corresponding closed quadratic form (see Theorem 29.4). If Ω is bounded with smooth boundary $\partial\Omega$, then it can be proved that

$$D(A^*) = W_2^m(\Omega).$$

This concludes the proof. □

Gårding's inequality has two more consequences. Firstly,

$$\left\|(A_{\mathrm{F}})_\mu f\right\|_{L^2} \geq C_0 \|f\|_{L^2}, \quad C_0 > 0,$$

so that

$$(A_{\mathrm{F}})_\mu^{-1} : L^2(\Omega) \to L^2(\Omega).$$

Secondly,

$$\left\|(A_{\mathrm{F}})_\mu f\right\|_{W_2^{-m/2}(\Omega)} \geq C_0' \|f\|_{W_2^{m/2}(\Omega)}, \quad C_0' > 0,$$

so that

$$(A_{\mathrm{F}})_\mu^{-1} : L^2(\Omega) \to \overset{\circ}{W}{}_2^{m/2}(\Omega).$$

Corollary 30.14. *The spectrum $\sigma(A_{\mathrm{F}}) = \{\lambda_j\}_{j=1}^\infty$ is the sequence of eigenvalues of finite multiplicity with only one accumulation point at $+\infty$. In short, $\sigma(A_{\mathrm{F}}) = \sigma_{\mathrm{d}}(A_{\mathrm{F}})$. The corresponding orthonormal system $\{\psi_j\}_{j=1}^\infty$ of eigenfunctions forms an orthonormal basis and*

$$A_{\mathrm{F}} f \overset{L^2}{=} \sum_{j=1}^{\infty} \lambda_j (f, \psi_j) \psi_j$$

for all $f \in D(A_{\mathrm{F}})$.

Proof. We begin with a lemma.

Lemma 30.15. *The embedding*

$$\overset{\circ}{W}{}_2^{m/2}(\Omega) \hookrightarrow L^2(\Omega)$$

is compact.

Proof. It is enough to show that for every $\{\varphi_k\}_{k=1}^\infty \subset \overset{\circ}{W}{}_2^{m/2}(\Omega)$ with $\|\varphi_k\|_{W_2^{m/2}} \leq 1$ there exists $\{\varphi_{j_k}\}_{k=1}^\infty$ that is a Cauchy sequence in $L^2(\Omega)$. Since Ω is bounded, we have

$$|\widehat{\varphi_k}(\xi)| \leq \|\varphi_k\|_{L^2} |\Omega|^{1/2},$$

i.e., the Fourier transform $\widehat{\varphi_k}(\xi)$ (see Chapter 16) is uniformly bounded. Thus there exists $\widehat{\varphi_{j_k}}(\xi)$ that converges pointwise in $\mathbb{R}^n$. Next, using Parseval's equality and the definition of the Sobolev spaces $H^s(\mathbb{R}^n)$ (see Chapter 20), we have

$$\begin{aligned}
\|\varphi_{j_k}-\varphi_{j_m}\|_{L^2}^2 &= \int_{\mathbb{R}^n} |\widehat{\varphi_{j_k}}(\xi)-\widehat{\varphi_{j_m}}(\xi)|^2 d\xi \\
&= \int_{|\xi|<r} |\widehat{\varphi_{j_k}}(\xi)-\widehat{\varphi_{j_m}}(\xi)|^2 d\xi + \int_{|\xi|>r} |\widehat{\varphi_{j_k}}(\xi)-\widehat{\varphi_{j_m}}(\xi)|^2 d\xi \\
&\le \int_{|\xi|<r} |\widehat{\varphi_{j_k}}(\xi)-\widehat{\varphi_{j_m}}(\xi)|^2 d\xi \\
&\quad + \frac{1}{(1+r^2)^{m/2}} \int_{\mathbb{R}^n} (1+|\xi|^2)^{m/2} |\widehat{\varphi_{j_k}}(\xi)-\widehat{\varphi_{j_m}}(\xi)|^2 d\xi \\
&= \int_{|\xi|<r} |\widehat{\varphi_{j_k}}(\xi)-\widehat{\varphi_{j_m}}(\xi)|^2 d\xi + (1+r^2)^{-m/2} \|\varphi_{j_k}-\varphi_{j_m}\|_{W_2^{m/2}}^2 \\
&=: I_1 + I_2.
\end{aligned}$$

The first term I_1 tends to 0 as $k,m \to \infty$ by the Lebesgue dominated convergence theorem for every fixed $r > 0$. The second term converges to 0 as $r \to \infty$ because $\|\varphi_{j_k}-\varphi_{j_m}\|_{W_2^{m/2}} \le 2$. □

Lemma 30.15 gives us that

$$(A_\mu)_{\mathrm{F}}^{-1} : L^2(\Omega) \to L^2(\Omega)$$

is a compact operator. Applying the Riesz–Schauder and Hilbert–Schmidt theorems, we get the following statements:

(1) $\sigma((A_\mu)_{\mathrm{F}}^{-1}) = \{0, \mu_1, \mu_2, \ldots\}$ with $\mu_j \ge \mu_{j+1} > 0$ and $\mu_j \to 0$ as $j \to \infty$.
(2) μ_j is of finite multiplicity.
(3) $(A_\mu)_{\mathrm{F}}^{-1} \psi_j = \mu_j \psi_j$, where $\{\psi_j\}_{j=1}^\infty$ is an orthonormal system.
(4) $\{\psi_j\}_{j=1}^\infty$ forms an orthonormal basis in $L^2(\Omega)$.

Since $A_{\mathrm{F}}\psi_j = \lambda_j \psi_j$ with $\lambda_j = \frac{1}{\mu_j} - \mu$, we conclude that

$$\sigma(A_{\mathrm{F}}) = \{\lambda_j\}_{j=1}^\infty, \quad \lambda_j \le \lambda_{j+1}, \lambda_j \to \infty.$$

Moreover, λ_j has finite multiplicity and the ψ_j are the corresponding eigenfunctions. We have also the following representation:

$$(A_\mu)_{\mathrm{F}}^{-1} f = \sum_{j=1}^\infty \mu_j (f, \psi_j)\psi_j, \quad f \in L^2(\Omega).$$

Exercise 30.5. Prove that

$$A_{\mathrm{F}} f = \sum_{j=1}^\infty \lambda_j (f, \psi_j)\psi_j$$

for all $f \in D(A_F)$.

The corollary is proved. □

In some applications it is quite useful to deal with semigroups of operators. We consider these semigroups in Hilbert spaces. This approach allows us to characterize the domains of operators (when they are not bounded); see, e.g., [3].

Let A be a nonnegative self-adjoint operator in a Hilbert space H. By the spectral theorem we can characterize $D(A)$ as follows: $f \in D(A)$ if and only if

$$\int_0^\infty (1+\lambda^2)\mathrm{d}(E_\lambda f, f) < \infty$$

and we define a new norm

$$\|f\|_{D(A)} := \|f\|_H + \|Af\|_H.$$

Definition 30.16. Let $\{G(t)\}_{t>0}$ be a family of bounded linear operators from H to H. This family is called an *equi-bounded strongly continuous semigroup* if

(1) $G(t+s)f = G(t)(G(s)f)$ for $s,t > 0$ and $f \in H$,
(2) $\|G(t)f\|_H \le M\|f\|_H$ for $t > 0$ and $f \in H$ with $M > 0$ that does not depend on t of f,
(3) $\lim_{t\to 0+} \|G(t)f - f\|_H = 0$ for $f \in H$.

Remark 30.17. We can complete this definition by $G(0) := I$.

Definition 30.18. The *infinitesimal generator* A of the semigroup $\{G(t)\}_{t>0}$ is defined by the formula

$$\lim_{t\to 0}\left\|\frac{G(t)-I}{t} - Af\right\| = 0$$

with domain $D(A)$ consisting of all $f \in H$ such that

$$\lim_{t\to 0}\frac{G(t)-I}{t}f$$

exists in H.

Remark 30.19. In the sense of the previous definition we write $G'(0) = A$.

Example 30.20. Let $H = L^2(\mathbb{R}^n)$. Let $\omega(\xi)$ be an infinitely differentiable positive function on $\mathbb{R}^n \setminus \{0\}$ that is positively homogeneous of order $m > 0$, i.e., $\omega(t\xi) = |t|^m\omega(\xi)$. Let us define the family $\{G(t)\}_{t>0}$ by the formula

$$G(t)f := \mathscr{F}^{-1}(\mathrm{e}^{-t\omega(\xi)}\mathscr{F} f), \quad f \in L^2(\mathbb{R}^n).$$

It is clear that $G(t) : L^2(\mathbb{R}^n) \to L^2(\mathbb{R}^n)$. Moreover,

(1)

$$G(t+s)f = \mathscr{F}^{-1}(e^{-(t+s)\omega(\xi)}\mathscr{F}f)$$
$$= \mathscr{F}^{-1}(e^{-t\omega(\xi)}\mathscr{F}\mathscr{F}^{-1}(e^{-s\omega(\xi)}\mathscr{F}f)) = G(t)(G(s)f);$$

(2)

$$\|G(t)f\|_{L^2} = \left\|\mathscr{F}^{-1}(e^{-t\omega(\xi)}\mathscr{F}f)\right\|_{L^2} = \left\|e^{-t\omega(\xi)}\mathscr{F}f\right\|_{L^2} \le \|\mathscr{F}f\|_{L^2} = \|f\|_{L^2};$$

(3)

$$\|G(t)f - f\|_{L^2} = \left\|\mathscr{F}^{-1}(e^{-t\omega(\xi)}\mathscr{F}f - \mathscr{F}f)\right\|_{L^2} = \left\|(e^{-t\omega(\xi)} - 1)\mathscr{F}f\right\|_{L^2} \to 0$$

as $t \to 0$ by the Lebesgue dominated convergence theorem. Also by this theorem we have that

$$\lim_{t\to 0}\frac{G(t)f - f}{t} = \lim_{t\to 0}\mathscr{F}^{-1}\left(\frac{e^{-t\omega(\xi)} - 1}{t}\mathscr{F}f\right) = -\mathscr{F}^{-1}(\omega(\xi)\mathscr{F}f) \equiv Af.$$

The domain of A is

$$D(A) = \{f \in L^2 : \|\omega(\xi)\mathscr{F}f\|_{L^2} < \infty\}.$$

For example, if $\omega(\xi) = |\xi|^m$, then $A = -(-\Delta)^{m/2}$ and $D(A) = W_2^m(\mathbb{R}^n)$.

Example 30.21. Let $A = A^* \ge 0$. Define

$$G(t) := e^{ItA} \equiv \int_0^\infty e^{It\lambda} dE_\lambda.$$

Then

(1)

$$G(t+s) = \int_0^\infty e^{I(t+s)\lambda} dE_\lambda = \int_0^\infty e^{It\lambda}e^{Is\lambda} dE_\lambda$$
$$= \int_0^\infty e^{It\lambda} dE_\lambda \int_0^\infty e^{Is\mu} dE_\mu = G(t)G(s);$$

(2)

$$\|G(t)f\|^2 = \int_0^\infty |e^{It\lambda}|^2 d(E_\lambda f, f) = \|f\|^2;$$

(3)

$$\|G(t)f - f\|^2 = \int_0^\infty |e^{It\lambda} - 1|^2 d(E_\lambda f, f) \to 0, \quad t \to 0,$$

and

$$\frac{G(t)f - f}{t} = \int_0^\infty \frac{e^{It\lambda} - 1}{t} dE_\lambda f \to I \int_0^\infty \lambda dE_\lambda f \equiv IAf, \quad t \to 0,$$

and

$$\lim_{t \to 0} \left\| \frac{G(t) - I}{t} f - IAf \right\|_H = 0.$$

These examples reveal a one-to-one correspondence between the infinitesimal generators of semigroups and self-adjoint operators in Hilbert space.

Chapter 31
Spectral Functions

Let us consider a bounded domain $\Omega \subset \mathbb{R}^n$ and an elliptic differential operator $A(x,\partial)$ in Ω of the form

$$A(x,\partial) = \sum_{|\alpha|=|\beta|\le m/2} (-1)^{|\alpha|}\partial^\alpha (a_{\alpha\beta}(x)\partial^\beta),$$

where $a_{\alpha\beta} = a_{\beta\alpha}$ are real, in $C^\infty(\Omega)$, and bounded for all α and β. We assume that

$$\int_\Omega \sum_{|\alpha|=|\beta|=m/2} a_{\alpha\beta}(x)\partial^\alpha f \partial^\beta \overline{f} \mathrm{d}x \ge \nu \int_\Omega \sum_{|\alpha|=m/2} |\partial^\alpha f|^2 \mathrm{d}x, \quad \nu > 0.$$

As was proved above, there exists at least one self-adjoint extension of A with $D(A) = C_0^\infty(\Omega)$, namely, the Friedrichs extension A_F with

$$D(A_\mathrm{F}) = \overset{\circ}{W}{}_2^{m/2}(\Omega) \cap W_2^m(\Omega).$$

Let us consider an arbitrary self-adjoint extension $\widehat{A}$ of A. Without loss of generality we assume that $\widehat{A} \ge 0$. Therefore, $\widehat{A}$ has the spectral representation

$$\widehat{A} = \int_0^\infty \lambda \mathrm{d}E_\lambda$$

with domain

$$D\left(\widehat{A}\right) = \left\{ f \in L^2(\Omega) : \int_0^\infty \lambda^2 \mathrm{d}(E_\lambda f, f) < \infty \right\}.$$

In general, we have no formula for $D\left(\widehat{A}\right)$ like that for the Friedrichs extension A_F. But we can say that

V. Serov, *Fourier Series, Fourier Transform and Their Applications to Mathematical Physics*, Applied Mathematical Sciences 197,
DOI 10.1007/978-3-319-65262-7_31

$$\overset{\circ}{W}{}_2^m(\Omega) \subset D\left(\widehat{A}\right).$$

Indeed, since $a_{\alpha\beta} \in C^\infty(\Omega)$ and $a_{\alpha\beta}$ is bounded, $A(x,\partial)$ can be rewritten in the usual form

$$A(x,\partial) = \sum_{|\gamma|\le m} \tilde{a_\gamma}(x)\partial^\gamma$$

with bounded coefficients. Hence

$$\|Af\|_{L^2(\Omega)} \le c \sum_{|\gamma|\le m} \|\partial^\gamma f\|_{L^2(\Omega)} \equiv c\|f\|_{W_2^m(\Omega)}.$$

This proves the embedding.

But even in this general case, one can obtain more significant results than just the previous embedding into the domain of the operator. The basis for these results is the following classical theorem of L. Gårding, which is given here without proof (see, e.g., [14, 15]). In this theorem it is assumed that $\widehat{A}$ is an arbitrary (semibounded from below) self-adjoint extension of an elliptic differential operator $A(x,D)$ with smooth and bounded coefficients.

Theorem 31.1 (Gårding). *If $\widehat{A} = \widehat{A}^*$, then E_λ is an integral operator in $L^2(\Omega)$ such that*

$$E_\lambda f(x) = \int_\Omega \theta(x,y,\lambda) f(y)\mathrm{d}y,$$

where $\theta(x,y,\lambda)$ is called the spectral function *and has the properties*

(1) $\theta(x,y,\lambda) = \overline{\theta(y,x,\lambda)}$,

(2)

$$\theta(x,y,\lambda) = \int_\Omega \theta(x,z,\lambda)\theta(z,y,\lambda)\mathrm{d}z$$

and

$$\theta(x,x,\lambda) = \int_\Omega |\theta(x,z,\lambda)|^2 \mathrm{d}z \ge 0,$$

(3)

$$\sup_{x\in\Omega_1} \|\theta(x,\cdot,\lambda)\|_{L^2(\Omega)} \le c_1\lambda^k,$$

where $\overline{\Omega_1} = \Omega_1 \subset \Omega, k \in \mathbb{N}$ with $k > \frac{n}{2m}$ and $c_1 = c(\Omega_1)$.

Remark 31.2. It was proved by L. Hörmander that in fact,

$$\theta(x,x,\lambda) \le c_1\lambda^{n/m}.$$

Corollary 31.3. *Let* $z \in \rho\left(\widehat{A}\right)$*. Then* $(\widehat{A}-zI)^{-1}$ *is an integral operator whose kernel* $G(x,y,z)$ *is called the* Green's function *corresponding to* $\widehat{A}$ *and that has the properties*

(1)

$$G(x,y,z) = \int_0^\infty \frac{\mathrm{d}_\lambda \theta(x,y,\lambda)}{\lambda - z},$$

(2) $\overline{G(x,y,z)} = G(y,x,\overline{z})$.

Proof. Since $z \in \rho\left(\widehat{A}\right)$, von Neumann's spectral theorem gives us

$$(\widehat{A}-zI)^{-1}f = \int_0^\infty (\lambda - z)^{-1}\mathrm{d}E_\lambda f.$$

Next, by Theorem 31.1 we get

$$\begin{aligned}(\widehat{A}-zI)^{-1}f &= \int_0^\infty (\lambda - z)^{-1}\mathrm{d}_\lambda\left(\int_\Omega \theta(x,y,\lambda)f(y)\mathrm{d}y\right)\\ &= \int_\Omega\left(\int_0^\infty (\lambda-z)^{-1}\mathrm{d}_\lambda\theta(x,y,\lambda)\right)f(y)\mathrm{d}y = \int_\Omega G(x,y,z)f(y)\mathrm{d}y,\end{aligned}$$

where $G(x,y,z)$ is as in (1). Since

$$\overline{G(x,y,z)} = \int_0^\infty \frac{\overline{\mathrm{d}\theta(x,y,\lambda)}}{\lambda - \overline{z}} = \int_0^\infty \frac{\mathrm{d}\theta(y,x,\lambda)}{\lambda - \overline{z}} = G(y,x,\overline{z}),$$

(2) is also proved. □

Exercise 31.1. Prove that $\theta(x,x,\lambda)$ is a monotonically increasing function with respect to λ and

(1) $|\theta(x,y,\lambda)|^2 \le \theta(x,x,\lambda)\theta(y,y,\lambda)$,

(2) $|E_\lambda f(x)| \le \theta(x,x,\lambda)^{1/2}\|f\|_{L^2(\Omega)}$.

Exercise 31.2. Prove that

$$|E_\lambda f(x) - E_\mu f(x)| \le \left\|E_\lambda f - E_\mu f\right\|_{L^2(\Omega)} |\theta(x,x,\lambda) - \theta(x,x,\mu)|^{1/2}$$

for all $\lambda > 0$ and $\mu > 0$.

Exercise 31.3. Let us assume that $n < m$. Prove that

$$G(x,y,z) = \int_0^\infty \frac{\theta(x,y,\lambda)\mathrm{d}\lambda}{(\lambda - z)^2}$$

and that $G(\cdot,y,z) \in L^2(\Omega)$.

In the case of the Friedrichs extension for a bounded domain, the spectral function $\theta(x,y,\lambda)$ and the Green's function have a special form. We know from Corollary 7.14 that the spectrum $\sigma(A_{\mathrm{F}})$ is the sequence $\{\lambda_j\}_{j=1}^{\infty}$ of eigenvalues with only one accumulation point at $+\infty$, and the corresponding orthonormal system $\{\psi_j\}_{j=1}^{\infty}$ forms an orthonormal basis in $L^2(\Omega)$ such that

$$A_{\mathrm{F}}f = \sum_{j=1}^{\infty} \lambda_j (f,\psi_j)\psi_j \quad \text{in } L^2.$$

This fact implies that

$$\begin{aligned} E_\lambda f &= \sum_{\lambda_j<\lambda} (f,\psi_j)\psi_j = \sum_{\lambda_j<\lambda} \int_\Omega f(y)\overline{\psi_j(y)}\mathrm{d}y\,\psi_j(x) \\ &= \int_\Omega \left(\sum_{\lambda_j<\lambda} \psi_j(x)\overline{\psi_j(y)} \right) f(y)\mathrm{d}y = \int_\Omega \theta(x,y,\lambda) f(y)\mathrm{d}y, \end{aligned}$$

i.e., the spectral function $\theta(x,y,\lambda)$ has the following form:

$$\theta(x,y,\lambda) = \sum_{\lambda_j<\lambda} \psi_j(x)\overline{\psi_j(y)}.$$

Hence (see Corollary 31.3) the Green's function has the form

$$G(x,y,z) = \sum_{j=1}^{\infty} \frac{\psi_j(x)\overline{\psi_j(y)}}{\lambda_j - z} \quad \text{in } L^2.$$

If we assume now that $n<m$, then we obtain that the Green's function $G(x,y,z)$ is uniformly bounded in $(x,y)\in\Omega\times\Omega$. Let us assume for simplicity that $z = \mathrm{i}z_2$ and $A_{\mathrm{F}} \geq I$. Then applying Hörmander's estimate (see Remark 31.2) for the spectral function, we obtain

$$\begin{aligned} |G(x,y,z)| &\leq \sum_{j=1}^{\infty} \frac{|\psi_j(x)||\psi_j(y)|}{\sqrt{\lambda_j^2+z_2^2}} = \sum_{k=0}^{\infty} \sum_{2^k\leq\lambda_j<2^{k+1}} \frac{|\psi_j(x)||\psi_j(y)|}{\sqrt{\lambda_j^2+z_2^2}} \\ &\leq \sum_{k=0}^{\infty} \frac{1}{(2^{2k}+z_2^2)^{1/2}} \left(\sum_{2^k\leq\lambda_j<2^{k+1}} |\psi_j(x)|^2 \right)^{\frac{1}{2}} \left(\sum_{2^k\leq\lambda_j<2^{k+1}} |\psi_j(y)|^2 \right)^{\frac{1}{2}} \\ &\leq \sum_{k=0}^{\infty} \frac{(2^{k+1})^{n/m}}{(2^{2k}+z_2^2)^{1/2}}. \end{aligned}$$

Since $n<m$, this series converges for all z_2.

Chapter 32
The Schrödinger Operator

There are certain physical problems that are connected with the reconstruction of the quantum-mechanical potential in the Schrödinger operator $H = -\Delta + q(x)$. This operator is defined in $\mathbb{R}^n$. Here and throughout we assume that q is real-valued.

First of all we have to define H as a self-adjoint operator in $L^2(\mathbb{R}^n)$. Our basic assumption is that the potential $q(x)$ belongs to $L^p(\mathbb{R}^n)$ for $\frac{n}{2} < p \le \infty$ and has the following special behavior at infinity:

$$|q(x)| \le c|x|^{-\mu}, \quad |x| > R,$$

with some $\mu \ge 0$ and $R > 0$ sufficiently large. The parameter μ will be specified later, depending on the situation. We would like to construct the self-adjoint extension of this operator by Friedrichs's method, because formally our operator is defined now only for smooth functions, say for functions from $C_0^\infty(\mathbb{R}^n)$. In order to construct such an extension let us consider the Hilbert space H_1 defined as follows:

$$H_1 = \{f \in L^2(\mathbb{R}^n) : \nabla f(x) \in L^2(\mathbb{R}^n) \text{ and } \int_{\mathbb{R}^n} |q(x)||f(x)|^2 \mathrm{d}x < \infty\}.$$

The inner product in H_1 is defined by

$$(f,g)_{H_1} = (\nabla f, \nabla g)_{L^2} + \int_{\mathbb{R}^n} q(x) f(x) \overline{g(x)} \mathrm{d}x + \mu_0 (f,g)_{L^2},$$

with $\mu_0 > 0$ sufficiently large and fixed.

Lemma 32.1. *Assume that $f \in W_2^1(\mathbb{R}^n)$ and $q \in L^p(\mathbb{R}^n)$ for $\frac{n}{2} < p \le \infty$, $n \ge 2$. Then for every $0 < \varepsilon < 1$ there exists $c_\varepsilon > 0$ such that*

$$|(qf,f)_{L^2}| \le \varepsilon \|\nabla f\|^2_{L^2(\mathbb{R}^n)} + c_\varepsilon \|f\|^2_{L^2(\mathbb{R}^n)}.$$

V. Serov, *Fourier Series, Fourier Transform and Their Applications to Mathematical Physics*, Applied Mathematical Sciences 197,
DOI 10.1007/978-3-319-65262-7_32

Proof. If $p = \infty$, then

$$
\begin{aligned}
|(qf, f)_{L^2}| &\le \int_{\mathbb{R}^n} |q(x)||f(x)|^2 \mathrm{d}x \le \|q\|_{L^\infty(\mathbb{R}^n)} \|f\|_{L^2(\mathbb{R}^n)}^2 \\
&\le \varepsilon \|\nabla f\|_{L^2(\mathbb{R}^n)}^2 + \|q\|_{L^\infty(\mathbb{R}^n)} \|f\|_{L^2(\mathbb{R}^n)}^2 .
\end{aligned}
$$

If $\frac{n}{2} < p < \infty$, then we estimate

$$
\begin{aligned}
|(qf, f)_{L^2}| &\le \int_{|q(x)|<A} |q(x)||f(x)|^2 \mathrm{d}x + \int_{|q(x)|>A} |q(x)||f(x)|^2 \mathrm{d}x \\
&\le \int_{|q(x)|>A} |q(x)||f(x)|^2 \mathrm{d}x + A \|f\|_{L^2(\mathbb{R}^n)}^2 .
\end{aligned}
$$

Let us consider the integral appearing in the last estimate. For $n \ge 3$ it follows from Hölder's inequality that

$$
\begin{aligned}
\int_{|q(x)|>A} |q(x)||f(x)|^2 \mathrm{d}x &\le \left(\int_{|q(x)|>A} |q(x)|^{\frac{n}{2}} \mathrm{d}x \right)^{\frac{2}{n}} \left(\int_{|q(x)|>A} |f(x)|^{\frac{2n}{n-2}} \mathrm{d}x \right)^{\frac{n-2}{n}} \\
&\le A^{\left(\frac{n}{2}-p\right)\frac{2}{n}} \left(\int_{|q(x)|>A} |q(x)|^p \mathrm{d}x \right)^{\frac{2}{n}} c_1 \|f\|_{W_2^1(\mathbb{R}^n)}^2 \\
&\le c_1 A^{1-\frac{2p}{n}} \|q\|_{L^p(\mathbb{R}^n)}^{\frac{2p}{n}} \|f\|_{W_2^1(\mathbb{R}^n)}^2 .
\end{aligned}
$$

To obtain the last inequality we used the fact that $\frac{n}{2} < p < \infty$ and a well known [1, 3] embedding: $W_2^1(\mathbb{R}^n) \subset L^{\frac{2n}{n-2}}(\mathbb{R}^n)$, $n \ge 3$, with the norm estimate

$$
\|f\|_{L^{\frac{2n}{n-2}}(\mathbb{R}^n)} \le \sqrt{c_1} \|f\|_{W_2^1(\mathbb{R}^n)} .
$$

Collecting these estimates, we obtain

$$
\begin{aligned}
|(qf, f)_{L^2}| &\le c_1 A^{1-\frac{2p}{n}} \|q\|_{L^p(\mathbb{R}^n)}^{\frac{2p}{n}} \|f\|_{W_2^1(\mathbb{R}^n)}^2 + A \|f\|_{L^2(\mathbb{R}^n)}^2 \\
&= c_1 A^{1-\frac{2p}{n}} \|q\|_{L^p(\mathbb{R}^n)}^{\frac{2p}{n}} \|\nabla f\|_{L^2(\mathbb{R}^n)}^2 + \left(A + c_1 A^{1-\frac{2p}{n}} \|q\|_{L^p(\mathbb{R}^n)}^{\frac{2p}{n}} \right) \|f\|_{L^2(\mathbb{R}^n)}^2 .
\end{aligned}
$$

The claim follows now from the last inequality, since $A^{1-\frac{2p}{n}}$ can be chosen sufficiently small for $\frac{n}{2} < p < \infty$. □

Exercise 32.1. Prove Lemma 32.1 for $n = 2$.

Exercise 32.2. Let us assume that $q(x)$ satisfies the conditions

(1) $|q| \le c_1 |x|^{-\gamma_1}$, $|x| < 1$, and
(2) $|q| \le c_2 |x|^{-\gamma_2}$, $|x| > 1$.

Find the conditions on γ_1 and γ_2 that ensure the statement of Lemma 32.1.

Remark 32.2. Lemma 32.1 holds for every potential $q \in L^p(\mathbb{R}^n) + L^\infty(\mathbb{R}^n)$ for $p > \frac{n}{2}, n \geq 2$.

Using Lemma 32.1, we obtain

$$\begin{aligned}
\|f\|_{H_1}^2 &= \|\nabla f\|_{L^2(\mathbb{R}^n)}^2 + \mu_0 \|f\|_{L^2(\mathbb{R}^n)}^2 + (qf, f)_{L^2} \\
&\geq \|\nabla f\|_{L^2(\mathbb{R}^n)}^2 + \mu_0 \|f\|_{L^2(\mathbb{R}^n)}^2 - \varepsilon \|\nabla f\|_{L^2(\mathbb{R}^n)}^2 - c_\varepsilon \|f\|_{L^2(\mathbb{R}^n)}^2 \\
&= (1-\varepsilon) \|\nabla f\|_{L^2(\mathbb{R}^n)}^2 + (\mu_0 - c_\varepsilon) \|f\|_{L^2(\mathbb{R}^n)}^2 .
\end{aligned}$$

We choose here $0 < \varepsilon < 1$ and $\mu_0 > c_\varepsilon$. On the other hand,

$$\|f\|_{H_1}^2 \leq (1+\varepsilon) \|\nabla f\|_{L^2(\mathbb{R}^n)}^2 + (\mu_0 + c_\varepsilon) \|f\|_{L^2(\mathbb{R}^n)}^2 .$$

These two inequalities mean that the new Hilbert space H_1 is equivalent to the space $W_2^1(\mathbb{R}^n)$ up to equivalent norms. Thus we may conclude that for every $f \in H_1$ our operator is well defined by

$$(f, (H+\mu_0)f)_{L^2(\mathbb{R}^n)} = \|f\|_{H_1}^2 .$$

Moreover, since $H + \mu_0$ is positive, we must have

$$\|f\|_{H_1}^2 = \left\| (H+\mu_0)^{\frac{1}{2}} f \right\|_{L^2(\mathbb{R}^n)}^2 ,$$

and the following statements hold:

(1) the domain of $(H+\mu_0)^{\frac{1}{2}}$ is $W_2^1(\mathbb{R}^n)$;
(2) $D(H+\mu_0) \equiv D(H) \subset W_2^1(\mathbb{R}^n)$;
(3) $D(H) = \{f \in W_2^1(\mathbb{R}^n) : Hf \in L^2(\mathbb{R}^n)\}$.

Remark 32.3. $(H+\mu_0)f = (H+\mu_0)^{\frac{1}{2}}(H+\mu_0)^{\frac{1}{2}} f$ is equivalent to

$$D(H) = \{f \in W_2^1(\mathbb{R}^n) : g := (H+\mu_0)^{\frac{1}{2}} f \in W_2^1(\mathbb{R}^n)\}.$$

Remark 32.4. Let us consider this extension procedure from another point of view. The inequality

$$(f, (H+\mu_0)f)_{L^2} \geq (1-\varepsilon) \|\nabla f\|_{L^2(\mathbb{R}^n)}^2 + (\mu_0 - c_\varepsilon) \|f\|_{L^2(\mathbb{R}^n)}^2$$

allows us to conclude that

(1) $(f, (H+\mu_0)f)_{L^2} \geq c' \|f\|_{L^2(\mathbb{R}^n)}^2$ and
(2) $(f, (H+\mu_0)f)_{L^2} \geq c'' \|f\|_{W_2^1(\mathbb{R}^n)}^2$

for every $f \in C_0^\infty(\mathbb{R}^n)$. This means that there exists $(H+\mu_0)^{-1}$ that is also defined for $g \in C_0^\infty(\mathbb{R}^n)$ and satisfies the inequality

(1) $\left\|(H+\mu_0)^{-1}g\right\|_{L^2(\mathbb{R}^n)} \le \frac{1}{c'}\|g\|_{L^2(\mathbb{R}^n)}$ or even
(2) $\left\|(H+\mu_0)^{-1}g\right\|_{W_2^1(\mathbb{R}^n)} \le \frac{1}{c''}\|g\|_{W_2^{-1}(\mathbb{R}^n)}$, where $W_2^{-1}(\mathbb{R}^n)$ is the dual space of $W_2^1(\mathbb{R}^n)$.

Since $(H+\mu_0)^{-1}$ is a bounded operator and $\overline{C_0^\infty(\mathbb{R}^n)} \overset{L^2}{=} L^2(\mathbb{R}^n)$ and $\overline{C_0^\infty(\mathbb{R}^n)} \overset{W_2^{-1}}{=} W_2^{-1}(\mathbb{R}^n)$, we can extend $(H+\mu_0)^{-1}$ as a bounded operator onto $L^2(\mathbb{R}^n)$ in the first case and onto $W_2^{-1}(\mathbb{R}^n)$ in the second. The extension for the differential operator is $H+\mu_0 = ((H+\mu_0)^{-1})^{-1}$ and $D(H+\mu_0) = R((H+\mu_0)^{-1})$ in both cases. It is also clear that $H+\mu_0$ and $(H+\mu_0)^{-1}$ are self-adjoint operators.

Lemma 32.5. *Let us assume that $q \in L^p(\mathbb{R}^n)$ for $2 \le p \le \infty$ if $n = 2,3$ and $q \in L^p(\mathbb{R}^n)$ for $\frac{n}{2} < p \le \infty$ if $n \ge 4$. Then*

$$W_2^2(\mathbb{R}^n) \subset D(H).$$

Proof. Since $H = -\Delta + q$ and $D(H) = \{f \in W_2^1(\mathbb{R}^n) : Hf \in L^2(\mathbb{R}^n)\}$, it is enough to show for the required embedding that for $f \in W_2^2(\mathbb{R}^n)$ it follows that $qf \in L^2(\mathbb{R}^n)$.

If $p = \infty$, then

$$\int_{\mathbb{R}^n} |qf|^2 \mathrm{d}x \le \|q\|^2_{L^\infty(\mathbb{R}^n)} \|f\|^2_{L^2(\mathbb{R}^n)} < \infty$$

for every $f \in W_2^2(\mathbb{R}^n)$, $n \ge 2$.

For finite p let us consider first the case $n = 2,3$. Since $W_2^2(\mathbb{R}^n) \subset C(\mathbb{R}^n) \cap L^\infty(\mathbb{R}^n)$ (Sobolev embedding), we must have

$$\begin{aligned}
\int_{\mathbb{R}^n} |qf|^2 \mathrm{d}x &= \int_{|q|<A} |qf|^2 \mathrm{d}x + \int_{|q|>A} |qf|^2 \mathrm{d}x \\
&\le A^2 \int_{|q|<A} |f|^2 \mathrm{d}x + \|f\|^2_{L^\infty(\mathbb{R}^n)} \int_{|q|>A} |q|^p |q|^{2-p} \mathrm{d}x \\
&\le A^2 \|f\|^2_{L^2(\mathbb{R}^n)} + C \|f\|^2_{W_2^2(\mathbb{R}^n)} A^{2-p} \|q\|^p_{L^p(\mathbb{R}^n)} < \infty.
\end{aligned}$$

We will apply the following embeddings:

$$\begin{aligned}
&f \in W_2^2(\mathbb{R}^4) \subset L^p(\mathbb{R}^4), \quad p < \infty. \\
&f \in W_2^2(\mathbb{R}^n) \subset L^{\frac{2n}{n-4}}(\mathbb{R}^n), \quad \text{if } n \ge 5.
\end{aligned}$$

Therefore, on applying Hölder's inequality we obtain

$$\begin{aligned}
\int_{\mathbb{R}^n} |qf|^2 \mathrm{d}x &= \int_{|q|<A} |qf|^2 \mathrm{d}x + \int_{|q|>A} |qf|^2 \mathrm{d}x \\
&\le A^2 \|f\|^2_{L^2(\mathbb{R}^n)} + \left(\int_{|q|>A} |q|^{\frac{n}{2}} \mathrm{d}x\right)^{\frac{4}{n}} \left(\int_{|q|>A} |f|^{\frac{2n}{n-4}} \mathrm{d}x\right)^{\frac{n-4}{n}}
\end{aligned}$$

$$\leq A^2 \|f\|^2_{L^2(\mathbb{R}^n)} + CA^{(\frac{n}{2}-p)\frac{4}{n}} \left(\int_{|q|>A} |q|^p \mathrm{d}x\right)^{\frac{4}{n}} \|f\|^2_{W_2^2(\mathbb{R}^n)} < \infty.$$

if $n \geq 5$ and

$$\int_{\mathbb{R}^4} |qf|^2 \mathrm{d}x \leq \left(\int_{\mathbb{R}^4} |q|^p \mathrm{d}x\right)^{\frac{2}{p}} \left(\int_{\mathbb{R}^4} |f|^{p'} \mathrm{d}x\right)^{\frac{2}{p'}} < \infty$$

if $n = 4$ for $2 < p < \infty$ and $p' < \infty$. □

Exercise 32.3. Prove this lemma for $q \in L^p(\mathbb{R}^n) + L^\infty(\mathbb{R}^n)$, $\frac{n}{2} < p \leq \infty$, if $n \geq 4$ and for $q \in L^2(\mathbb{R}^n) + L^\infty(\mathbb{R}^n)$ if $n = 2, 3$.

Remark 32.6. For $n \geq 5$ we may consider $q \in L^{\frac{n}{2}}(\mathbb{R}^n)$.

Lemma 32.7. *Let us assume that $q \in L^n(\mathbb{R}^n)$, $n \geq 3$. Then*

$$D(H) = W_2^2(\mathbb{R}^n).$$

Proof. The embedding $W_2^2(\mathbb{R}^n) \subset D(H)$ was proved in Lemma 32.5. Let us now assume that $f \in D(H)$, i.e., $f \in W_2^1(\mathbb{R}^n)$ and $Hf \in L^2(\mathbb{R}^n)$. Note that for $g := Hf \in L^2$ we have the following representation:

$$\begin{aligned} -f &= (-\Delta+1)^{-1}(q-1)f - (-\Delta+1)^{-1}g \\ &= (-\Delta+1)^{-1}(qf) - (-\Delta+1)^{-1}g - (-\Delta+1)^{-1}f. \end{aligned}$$

It is therefore enough to show that $qf \in L^2(\mathbb{R}^n)$. We use the same arguments as in Lemmas 32.1 and 32.5. So it suffices to show that $qf \in L^2(\mathbb{R}^n)$ for every $f \in W_2^1(\mathbb{R}^n)$. From the embedding $W_2^1(\mathbb{R}^n) \subset L^{\frac{2n}{n-2}}(\mathbb{R}^n)$ for $n \geq 3$ we have by Hölder's inequality

$$\begin{aligned} \int_{\mathbb{R}^n} |q(x)|^2 |f(x)|^2 \mathrm{d}x &= \int_{|q|<A} |q(x)|^2 |f(x)|^2 \mathrm{d}x + \int_{|q|>A} |q(x)|^2 |f(x)|^2 \mathrm{d}x \\ &\leq A^2 \|f\|^2_{L^2(\mathbb{R}^n)} + \left(\int_{|q|>A} |q|^n \mathrm{d}x\right)^{\frac{2}{n}} \left(\int_{|q|>A} |f|^{\frac{2n}{n-2}} \mathrm{d}x\right)^{\frac{n-2}{n}} < \infty. \end{aligned}$$

Thus the lemma is proved. □

Exercise 32.4. Describe the domain of H for the case $\frac{n}{2} < p < n$, $n \geq 3$. Hint: Prove that $D(H) \subset W_2^2(\mathbb{R}^n) + W_s^2(\mathbb{R}^n)$ with some $s = s(p)$.

Let us consider now the Laplacian $H_0 = -\Delta$ in $\mathbb{R}^n$, $n \geq 1$. Since $(-\Delta f, f)_{L^2} = \|\nabla f\|^2_{L^2(\mathbb{R}^n)} \geq 0$ for every $f \in W_2^1(\mathbb{R}^n)$, it follows that H_0 is a nonnegative operator. Moreover, $H_0 = H_0^*$ with domain $D(H_0) = W_2^2(\mathbb{R}^n)$, and this operator has the spectral representation

$$H_0 f = \int_0^\infty \lambda \mathrm{d}E_\lambda f.$$

It follows that $\sigma(H_0) \subset [0,+\infty)$, but in fact, $\sigma(H_0) = [0,+\infty)$ and even $\sigma(H_0) = \sigma_c(H_0) = \sigma_{ess}(H_0) = [0,+\infty)$. In order to understand this fact it is enough to observe that for every $\lambda \in [0,+\infty)$ the homogeneous equation $(H_0 - \lambda)u = 0$ has a solution of the form $u(x,\vec{k}) = \mathrm{e}^{\mathrm{i}(\vec{k},x)}$, where $(\vec{k},\vec{k}) = \lambda$ and $\vec{k} \in \mathbb{R}^n$. These solutions $u(x,\vec{k})$ are called *generalized eigenfunctions*, but $u(x,\vec{k}) \notin L^2(\mathbb{R}^n)$. These solutions are bounded and correspond to the continuous spectrum of H_0. Consequently, $u(x,\vec{k})$ are not eigenfunctions, but generalized eigenfunctions. If we consider the solutions of the equation $(H_0 - \lambda)u = 0$ for $\lambda < 0$, then these solutions will be exponentially increasing at the infinity. This implies that $\lambda < 0$ does not belong to $\sigma(H_0)$.

For the spectral representation of H_0 we have two forms:

(1) the *Neumann spectral representation*

$$-\Delta f = \int_0^\infty \lambda \mathrm{d}E_\lambda f, \quad f \in W_2^2(\mathbb{R}^n);$$

(2) the *scattering theory* representation

$$-\Delta f = \mathscr{F}^{-1}(|\xi|^2 \widehat{f}) = (2\pi)^{-n} \int_{\mathbb{R}^n} |\xi|^2 \mathrm{e}^{\mathrm{i}(\xi,x)} \mathrm{d}\xi \int_{\mathbb{R}^n} \mathrm{e}^{-\mathrm{i}(\xi,y)} f(y) \mathrm{d}y.$$

Exercise 32.5. Determine the connection between these two representations.

There are some important remarks to be made about the resolvent $(-\Delta - z)^{-1}$ for $z \notin [0,+\infty)$. A consequence of the spectral theorem is that

$$(-\Delta - z)^{-1} = \int_0^\infty (\lambda - z)^{-1} \mathrm{d}E_\lambda, \quad z \in \mathbb{C} \setminus [0,+\infty),$$

and for such z the operator $(-\Delta - z)^{-1}$ is a bounded operator in $L^2(\mathbb{R}^n)$. Moreover, with respect to $z \notin [0,+\infty)$, the operator $(-\Delta - z)^{-1}$ as an operator-valued function is a *holomorphic* function. This fact follows immediately from

$$((-\Delta - z)^{-1})'_z = \int_0^\infty (\lambda - z)^{-2} \mathrm{d}E_\lambda = (-\Delta - z)^{-2}.$$

The last integral converges as well as the previous one (even better). Now we are in a position to formulate a theorem about the spectrum of $H = -\Delta + q$.

Theorem 32.8. *Assume that $q \in L^p(\mathbb{R}^n)$, $\frac{n}{2} < p \le \infty$, $n \ge 2$, and $q(x) \to 0$ as $|x| \to +\infty$. Then*

(1) $\sigma_c(H) \supset (0,+\infty)$;

(2) $\sigma_p(H) \subset [-c_0, 0]$ is of finite multiplicity with its only accumulation point at $\{0\}$ with c_0 such that $-\Delta + q \ge -c_0$.

In order to prove this theorem we will prove two lemmas.

Lemma 32.9. *Assume that the potential $q(x)$ satisfies the assumptions of Theorem 32.8. Assume in addition that $q(x) \in L^2(\mathbb{R}^n)$ for $n = 2,3$. Then*

$$(-\Delta - z)^{-1} \circ q : L^2(\mathbb{R}^n) \to L^2(\mathbb{R}^n)$$

is a compact operator for $z \notin [0, +\infty)$.

Proof. Due to our assumptions on the potential $q(x)$, it can be represented as the sum $q(x) = q_1(x) + q_2(x)$, where $q_1 \in L^p(|x| < R)$ with the same p and $q_2 \to 0$ as $|x| \to \infty$. We may assume (without loss of generality) that q_2 is supported in $\{x \in \mathbb{R}^n : |x| > R\}$ and that it is a continuous function. Let us consider first the cases $n = 2,3$. If $f \in L^2(\mathbb{R}^n)$, then $q_1 f \in L^1(|x| < R)$ and $(-\Delta - z)^{-1}(q_1 f) \in W_1^2(\mathbb{R}^n)$ (by the Fourier transform). By the embedding theorem for Sobolev spaces (see, e.g., [1, 3]) we have that

$$(-\Delta - z)^{-1}(q_1 f) \in W_1^2(\mathbb{R}^n) \subset W_2^{2-\frac{n}{2}}(\mathbb{R}^n), \quad n = 2,3,$$

with the norm estimate

$$\begin{aligned}
\left\|(-\Delta - z)^{-1}(q_1 f)\right\|_{L^2(\mathbb{R}^n)} &\le \left\|(-\Delta - z)^{-1}(q_1 f)\right\|_{W_2^{2-\frac{n}{2}}} \\
&\le c\left\|(-\Delta - z)^{-1}(q_1 f)\right\|_{W_1^2} \le c\left\|q_1 f\right\|_{L^1(\mathbb{R}^n)} \\
&\le c\left\|q_1\right\|_{L^2(|x|<R)}\left\|f\right\|_{L^2(|x|<R)},
\end{aligned}$$

or

$$\left\|(-\Delta - z)^{-1} \circ q_1\right\|_{L^2(|x|<R) \to L^2(\mathbb{R}^n)} \le c\left\|q_1\right\|_{L^2},$$

where c may depend only on z.

In the case $n \ge 4$ and $q \in L^p(|x| < R)$, $p > \frac{n}{2}$, we may obtain by Hölder's inequality that

$$q_1 f \in L^s(|x| < R), \quad s > \frac{2n}{n+4},$$

for $f \in L^2(\mathbb{R}^n)$, and therefore, $(-\Delta - z)^{-1}(q_1 f) \in W_s^2(\mathbb{R}^n)$. Again by the embedding theorem for Sobolev spaces we have

$$(-\Delta - z)^{-1}(q_1 f) \in W_2^{2-n\left(\frac{1}{s}-\frac{1}{2}\right)}(\mathbb{R}^n)$$

for some $s > \frac{2n}{n+4}$, with the norm estimate

$$\left\|(-\Delta - z)^{-1} \circ q_1\right\|_{L^2(|x|<R) \to L^2(\mathbb{R}^n)} \le c\left\|q_1\right\|_{L^p(|x|<R)}.$$

In order to prove that $(-\Delta - z)^{-1} \circ q_1$ is a compact operator we approximate it as follows:

$$A := (-\Delta - z)^{-1} \circ q_1, \quad A_j := \varphi_j(x) A,$$

where $\varphi_j(x) \in C_0^\infty(\mathbb{R}^n), |\varphi_j(x)| \le C$ and

$$\|A - A_j\|_{L^2 \to L^2} \to 0, \quad j \to \infty.$$

The reason is that $(-\Delta - z)^{-1} \circ q_1$ is actually an integral operator with kernel $K_z(x-y)$ that tends to 0 as $|x| \to \infty$ uniformly with respect to $|y| < R$ (note that q_1 is supported in $|y| < R$). We therefore can approximate this kernel K_z by the functions $\varphi_j \in C_0^\infty(\mathbb{R}^n)$. But A_j is a compact operator for each $j = 1, 2, \ldots$, because the embedding

$$W_2^\alpha(|x| < R) \subset L^2(|x| < R)$$

is compact for positive α. This implies that A is also a compact operator.

Next we consider q_2. Since for $f(x) \in L^2(\mathbb{R}^n)$ we know that $(-\Delta - z)^{-1} f \in W_2^2(\mathbb{R}^n)$, we conclude that $q_2(-\Delta - z)^{-1} f \in L^2(|x| > R)$. In fact,

$$q_2 \cdot : W_2^2(\mathbb{R}^n) \to L^2(|x| > R)$$

is a compact embedding. In order to establish this fact let us consider again $\varphi_j(x) \in C_0^\infty(\mathbb{R}^n), |\varphi_j(x)| \le c$ and $\varphi_j \to q_2$ as $j \to \infty$. We can state this because $\overline{C_0^\infty} \overset{L^\infty}{=} \dot{C}$. That is why we required such behavior of $q(x)$ at infinity ($q \to 0$ as $|x| \to +\infty$). If we set $A := q_2(-\Delta - z)^{-1}$ and $A_j := \varphi_j(-\Delta - z)^{-1}$, then we obtain

$$\begin{aligned} \|A - A_j\|_{L^2 \to L^2} &\le \sup_x |\varphi_j - q_2| \, \|(-\Delta - z)^{-1}\|_{L^2 \to L^2} \\ &\le c \sup_x |\varphi_j - q_2| \to 0, \; j \to +\infty. \end{aligned} \tag{32.1}$$

But we know that $W_{2,\mathrm{comp}}^2 \subset L_{\mathrm{comp}}^2$ is a compact embedding. This implies (together with (32.1)) that A is a compact operator. Since

$$(-\Delta - z)^{-1} \circ q_2 = (q_2(-\Delta - \bar{z})^{-1})^*,$$

the Lemma is proved. □

Lemma 32.10. *Let Q be an open and connected set in $\mathbb{C}$. Let $A(z)$ be a compact, operator-valued, and holomorphic function in Q and in $L^2(\mathbb{R}^n)$. If $(I + A(z_0))^{-1}$ exists for some $z_0 \in Q$, then $(I + A(z))^{-1}$ exists in all of Q except for finitely many points from Q with the only possible accumulation points on ∂Q.*

Proof. We will prove this lemma only for our concrete operator $A(z) := (-\Delta - z)^{-1} q(x)$ (see [22] for a full proof). Lemma 32.9 shows us that $A(z)$ is a compact operator for $z \notin [0, +\infty)$. The remarks about $R_z = (-\Delta - z)^{-1}$ show us that $A(z)$ is a holomorphic function in $\mathbb{C}\setminus[0,+\infty)$. Also we can prove that $(I+(-\Delta-z)^{-1}q)^{-1}$ exists for all $z\in\mathbb{C}\setminus\mathbb{R}$ and for real $z<-c_0$, where $-\Delta+q\geq -c_0$. Indeed, if $z\in\mathbb{C}$ with $\operatorname{Im} z\neq 0$, then $(I+(-\Delta-z)^{-1}q)u=0$, or $(-\Delta-z)u=-qu$, or $(\Delta u,u)+z(u,u)=(qu,u)$. This implies for z, $\operatorname{Im} z\neq 0$, that $(u,u)=0$ if and only if $u=0$. In the real case $z<-c_0$, the equality $(I+(-\Delta-z)^{-1}q)u=0$ implies

$$((-\Delta+q)u,u)-z(u,u)=0.$$

It follows that

$$(-c_0-z)\,\|u\|_{L^2}^2\leq 0$$

and thus $u=0$. These remarks show us that in $\mathbb{C}\setminus[0,+\infty)$ our operator $I+(-\Delta-z)^{-1}q$ may be noninvertible only on $[-c_0,0)$.

Let us consider an open and connected set Q in $\mathbb{C}\setminus[0,+\infty)$ such that $[-c_0,0)\subset Q$; see Figure 32.1.

It is easily seen that there exists $z_0\in Q$ such that $(I+(-\Delta-z_0)^{-1}q)^{-1}$ exists also. It is not difficult to show that there exists $\delta>0$ such that $(I+(-\Delta-z)^{-1}q)^{-1}$ exists in $U_\delta(z_0)$. Indeed, let us choose $\delta>0$ such that

$$\|A(z)-A(z_0)\|_{L^2\to L^2}<\frac{1}{\|(I+A(z_0))^{-1}\|_{L^2\to L^2}} \tag{32.2}$$

for all z such that $|z-z_0|<\delta$. Then

$$(I+A(z))^{-1}=(I+A(z_0))^{-1}(I+B)^{-1},$$

where $B:=(A(z)-A(z_0))(I+A(z_0))^{-1}$. But $\|B\|<1$ due to (32.2), and then

$$(I+B)^{-1}=I-B+B^2+\cdots+(-1)^nB^n+\cdots$$

exists in the strong topology from L^2 to L^2. We may therefore conclude that $I+A(z)$ may be noninvertible only for finitely many points in Q. This fact follows from the holomorphicity of $A(z)$ with respect to z by analogy with the theorem about the zeros of a holomorphic function in complex analysis. Moreover, since $A(z)$ is a compact operator, it follows by Fredholm's alternative that $\operatorname{Ker}(I+A(z))$ has finite

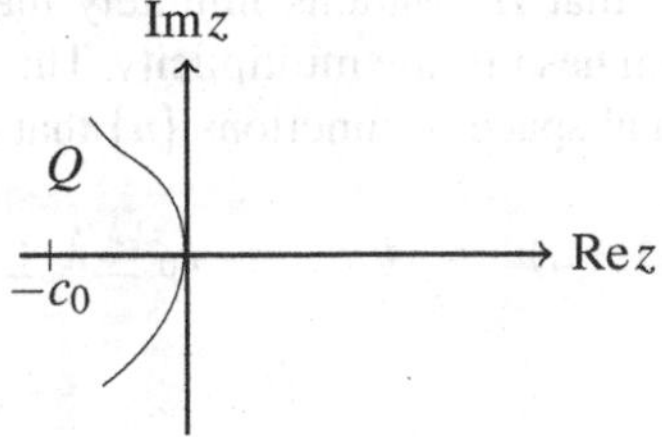

Fig. 32.1 The set Q.

dimension. We conclude that $(I+(-\Delta-z)^{-1}q)^{-1}$ does not exist at only a finite numbers of points (at most) on $[-c_0,-\varepsilon]$ for all $\varepsilon>0$, and these points are of finite multiplicity. This completes the proof. □

Let us return to the proof of Theorem 32.8.

Proof (Proof of Theorem. 32.8*).* Let μ be a positive number and $\mu+c_0>0$ ($H\geq -c_0I$). Let us consider for such μ the second resolvent equation

$$(H+\mu)^{-1}=(H_0+\mu)^{-1}-(H+\mu)^{-1}\circ q\circ(H_0+\mu)^{-1},$$

where $H_0=-\Delta$ and $H=-\Delta+q(x)$. It follows from Lemma 32.9 that $q\circ(H_0+\mu)^{-1}$ is a compact operator in $L^2(\mathbb{R}^n)$. This means that $(H+\mu)^{-1}$ is a compact perturbation of $(H_0+\mu)^{-1}$. Hence, by Theorem 28.18 above we have

$$\sigma_{\text{ess}}((H+\mu)^{-1})=\sigma_{\text{ess}}((H_0+\mu)^{-1}).$$

But $\sigma_{\text{ess}}((H_0+\mu)^{-1})=[0,\frac{1}{\mu}]=\sigma_{\text{c}}((H_0+\mu)^{-1})$, from which we conclude that

$$\sigma_{\text{ess}}(H+\mu)=[\mu,+\infty].$$

Outside of this set we have only points of the discrete spectrum with one possible accumulation point at μ. This statement is a simple corollary of Lemma 32.10. Moreover, these points of the discrete spectrum are located on $[\mu-c_0,\mu)$ and are of finite multiplicity. Hence the discrete spectrum $\sigma_{\text{d}}(H)$ of H belongs to $[-c_0,0)$ with only one possible accumulation point at $\{0\}$. And $(0,+\infty)$ is the continuous part of $\sigma(H)$. There is only one problem. Weyl's theorem states that the operators H and H_0 do not have the same spectrum but the same essential spectrum. Thus on $(0,+\infty)$ there can be eigenvalues of infinite multiplicity (see the definition of σ_{ess}). In order to eliminate such a possibility and to prove that $0\in\sigma_{\text{c}}(H)$ and $\sigma_{\text{d}}(H)$ is finite, let us assume additionally that our potential $q(x)$ has a special behavior at infinity:

$$|q(x)|\leq c|x|^{-\mu},\quad |x|\to+\infty,$$

where $\mu>2$. In that case we can prove that on the interval $[-c_0,0)$ the operator H has at most finitely many points of the discrete spectrum. And we prove also that $0\in\sigma_{\text{c}}(H)$.

Assume to the contrary that H contains infinitely many points of the discrete spectrum or that one of them has infinite multiplicity. This means that in $D(H)$ there exists an infinite-dimensional space of functions $\{u\}$ that satisfy the equation

$$(-\Delta+q)u=\lambda u,\qquad -c_0\leq\lambda\leq 0.$$

It follows that

$$\int_{\mathbb{R}^n}(|\nabla u(x)|^2+q^+(x)|u(x)|^2)\mathrm{d}x\leq\int_{\mathbb{R}^n}q^-(x)|u(x)|^2\mathrm{d}x,$$

where q^+ and q^- are the positive and negative parts of the potential $q(x)$, respectively. Let us consider an infinite sequence of functions $\{u(x)\}$ that are orthogonal with respect to the inner product $\int_{\mathbb{R}^n} q^-(x)u(x)\overline{v(x)}\mathrm{d}x$. This sequence is uniformly bounded in the metric $\int_{\mathbb{R}^n}(|\nabla u|^2+|q||u|^2)\mathrm{d}x$, and hence in the metric $\int_{\mathbb{R}^n}(|\nabla u|^2+|u|^2)\mathrm{d}x$. But for every eigenfunction $u(x)$ of the operator H with eigenvalue $\lambda \in [-c_0,0]$ the following inequality holds (see [9]):

$$|u(x)| \le c|\lambda| \int_{|x-y|\le 1} |u(y)|\mathrm{d}y,$$

where c does not depend on x. It follows from this inequality that

(1) $\lambda = 0$ is not an eigenvalue;
(2) this orthogonal sequence is uniformly bounded in every fixed ball.

Lemma 32.11. *Denote by U the set of functions $u(x) \in D(H)$ that are uniformly bounded in every fixed ball in $\mathbb{R}^n$. Then U is a precompact set in the metric*

$$\int_{\mathbb{R}^n} |q||u|^2\mathrm{d}x$$

if it is a bounded set in the metric

$$\int_{\mathbb{R}^n} (|\nabla u|^2+|u|^2)\mathrm{d}x.$$

Proof. Let $\{u_k(x)\}_{k=1}^\infty \subset U$ be an arbitrary sequence that is bounded in the second metric. Then for $u(x) := u_k(x) - u_m(x)$ we have for r sufficiently large that

$$\begin{aligned}\int_{\mathbb{R}^n} |q(x)||u(x)|^2\mathrm{d}x &\le c\int_{|x|>r} \frac{|u(x)|^2}{|x|^\mu}\mathrm{d}x + \int_{|x|\le r, |q(x)|\le A} |q(x)||u(x)|^2\mathrm{d}x \\ &+ \int_{|x|\le r, |q(x)|>A} |q(x)||u(x)|^2\mathrm{d}x =: I_0 + I_1 + I_2.\end{aligned}$$

For $n \ge 3$ (for $n = 2$ the proof needs some changes) and $\mu > 2$ we get

$$I_0 \le cr^{2-\mu} \int_{|x|>r} |x|^{-2}|u(x)|^2\mathrm{d}x \le cr^{2-\mu} \int_{\mathbb{R}^n} |\nabla u(x)|^2\mathrm{d}x, \quad u \in W_2^1(\mathbb{R}^n).$$

Due to the uniform boundedness of U in every ball, we conclude that

$$I_2 \le c\int_{|x|\le r, |q(x)|>A} |q(x)|\mathrm{d}x \to 0$$

as $A \to +\infty$ uniformly on U with fixed r. Since the embedding $W_2^1 \subset L^2$ for every ball is compact, the boundedness of the sequence in the second metric implies the precompactness in L^2 for every ball. We therefore have

$$I_1 \le A \int_{|x|\le r} |u(x)|^2 \mathrm{d}x \to 0, \quad m,k \to \infty$$

with r and A fixed. On passing to the limit, these inequalities for I_0, I_1, and I_2 show that

$$\int_{\mathbb{R}^n} |q(x)||u(x)|^2 \mathrm{d}x \to 0, \quad m,k \to \infty.$$

Thus the lemma is proved. □

Let us return to the proof of (1). By Lemma 32.11 we obtain that our sequence (which is orthogonal with respect to the inner product $\int_{\mathbb{R}^n} q^-(x)u(x)\overline{v(x)}\mathrm{d}x$) is a Cauchy sequence in the first metric. But this fact contradicts its orthogonality. Thus (1) is proved.

(2) Let us discuss (briefly) the situation with a positive eigenvalue on the continuous spectrum. If we consider the homogeneous equation

$$[I + (-\Delta - k^2 - \mathrm{i}0)^{-1} q] f = 0, \quad k^2 > 0,$$

in the space $\dot{C}(\mathbb{R}^n)$, then by Green's formula one can show (see [20] or [21]) that the solution $f(x)$ of this equation behaves at infinity as $o(|x|^{-\frac{n-1}{2}})$. We thus conclude [21] that $f(x) \equiv 0$ outside some ball in $\mathbb{R}^n$. By the unique continuation principle for the Schrödinger operator it follows that $f \equiv 0$ in the whole of $\mathbb{R}^n$. □

Let us consider now the spectral representation of the Schrödinger operator $H = -\Delta + q(x)$, with $q(x)$ as in Theorem 32.8 with the behavior $O(|x|^{-\mu})$, $\mu > 2$, at infinity (compare with the spectral representation that follows from von Neumann's spectral theorem, Theorem 27.13, for the self-adjoint operator $-\Delta + q$ in $L^2(\mathbb{R}^n)$). For all $f \in D(H)$, we have

$$Hf(x) = (2\pi)^{-n} \int_{\mathbb{R}^n} k^2 u(x,\vec{k}) \mathrm{d}\vec{k} \int_{\mathbb{R}^n} f(y)\overline{u(y,\vec{k})}\mathrm{d}y + \sum_{j=1}^{M} \lambda_j f_j u_j(x),$$

where $u(x,\vec{k})$ are the solutions of the equation $Hu = k^2 u$, $u_j(x)$ are the orthonormal eigenfunctions corresponding to the negative eigenvalues λ_j, taking into account the multiplicity of λ_j and $f_j = (f, u_j)_{L^2(\mathbb{R}^n)}$. The functions $u(x,\vec{k})$ are called *generalized eigenfunctions*. When $q \equiv 0$, the generalized eigenfunctions have the form $u(x,\vec{k}) = \mathrm{e}^{\mathrm{i}(x,\vec{k})}$. This follows by means of the Fourier transform. Indeed,

$$(-\Delta - k^2) u = 0$$

if and only if

$$(|\xi|^2 - k^2)\widehat{u} = 0,$$

or

$$\widehat{u} = \sum_{\alpha} c_\alpha \delta^{(\alpha)}(\xi - \vec{k}),$$

since

$$|\xi|^2 = k^2$$

if and only if

$$\xi - \vec{k} = 0.$$

Hence

$$\begin{aligned} u(x,\vec{k}) &= \sum_{\alpha} c_\alpha \mathscr{F}^{-1}(\delta^{(\alpha)}(\xi - \vec{k}))(x) \\ &= \sum_{\alpha} c_\alpha \mathrm{e}^{\mathrm{i}(x,\vec{k})} \mathscr{F}^{-1}(\delta^{(\alpha)}(\xi))(x) = \sum_{\alpha} c'_\alpha \mathrm{e}^{\mathrm{i}(x,\vec{k})} x^\alpha. \end{aligned}$$

But $u(x,\vec{k})$ must be bounded, and so $u(x,\vec{k}) = c'_0 \mathrm{e}^{\mathrm{i}(x,\vec{k})}$. We choose $c'_0 = 1$. If we have the Schrödinger operator $H = -\Delta + q$ with $q \neq 0$, then it is natural to look for the scattering solutions of $Hu = k^2 u$ of the form $u(x,\vec{k}) = \mathrm{e}^{\mathrm{i}(x,\vec{k})} + u_{\mathrm{sc}}(x,\vec{k})$. Due to this representation, we have

$$(-\Delta - k^2)(\mathrm{e}^{\mathrm{i}(x,\vec{k})} + u_{\mathrm{sc}}) = -qu,$$

or

$$(-\Delta - k^2) u_{\mathrm{sc}} = -qu.$$

In order to find u_{sc}, let us recall that from Chapter 22 we know the fundamental solution of the operator $-\Delta - k^2$. Therefore,

$$u(x,k) = \mathrm{e}^{\mathrm{i}(x,\vec{k})} - \int_{\mathbb{R}^n} G_k^+(|x-y|) q(y) u(y) \mathrm{d}y,$$

where

$$G_k^+(|x|) = \frac{\mathrm{i}}{4} \left(\frac{|k|}{2\pi |x|} \right)^{\frac{n-2}{2}} H^{(1)}_{\frac{n-2}{2}}(|k||x|)$$

is the fundamental solution for the operator $-\Delta - k^2$. This equation is called the *Lippmann–Schwinger integral equation.*

Chapter 33
The Magnetic Schrödinger Operator

As a continuation (and, in some sense, an extension) of the previous chapter, where the Schrödinger operator was considered, in this chapter we consider the *magnetic Schrödinger operator*

$$H_{\mathrm{m}}u := -(\nabla + \mathrm{i}\vec{W}(x))^2 u + V(x)u, \quad x \in \Omega \subset \mathbb{R}^n, n \geq 2, \tag{33.1}$$

where Ω is an open set (not necessarily bounded) in $\mathbb{R}^n$, $n \geq 2$, with smooth boundary. It is assumed that the electric potential $V(x)$ and the magnetic potential $\vec{W}(x)$ are real-valued and belong to the following spaces:

$$\begin{aligned} &V \in L^p(\Omega) \quad \text{with some } 1 < p \leq \infty \text{ for } n = 2 \text{ and } n/2 \leq p \leq \infty \text{ for } n \geq 3, \\ &\vec{W} \in L^s(\Omega) \quad \text{with some } 2 < s \leq \infty \text{ for } n = 2 \text{ and } n \leq s \leq \infty \text{ for } n \geq 3. \end{aligned} \tag{33.2}$$

The operator H_{m} of the form (33.1) is symmetric in the Hilbert space $L^2(\Omega)$ on the domain $C_0^\infty(\Omega)$. We want to construct the Friedrichs self-adjoint extension of this operator and to describe the domain of this extension.

Lemma 33.1. *Assume that the conditions* (33.2) *are satisfied for the coefficients of* H_{m}. *Then for all* $f \in C_0^\infty(\Omega)$ *the following double inequality holds:*

$$\gamma_1 \|\nabla f\|_{L^2(\Omega)}^2 - C_1 \|f\|_{L^2(\Omega)}^2 \leq (H_{\mathrm{m}} f, f)_{L^2(\Omega)} \leq \gamma_2 \|\nabla f\|_{L^2(\Omega)}^2 + C_2 \|f\|_{L^2(\Omega)}^2, \tag{33.3}$$

where $0 < \gamma_1 < 1 < \gamma_2$ *and* $C_1, C_2 > 0$.

Proof. For all $f \in C_0^\infty(\Omega)$ we have by integration by parts that

$$\begin{aligned} (H_{\mathrm{m}} f, f)_{L^2(\Omega)} &= \int_\Omega |(\nabla + \mathrm{i}\vec{W}) f|^2 \mathrm{d}x + \int_\Omega V(x)|f|^2 \mathrm{d}x \\ &= \int_\Omega |\nabla f|^2 \mathrm{d}x + \int_\Omega |\vec{W}|^2 |f|^2 \mathrm{d}x + \int_\Omega V(x)|f|^2 \mathrm{d}x - 2\int_\Omega \operatorname{Im}(\vec{W} f \cdot \nabla \overline{f}) \mathrm{d}x. \end{aligned}$$

V. Serov, *Fourier Series, Fourier Transform and Their Applications to Mathematical Physics*, Applied Mathematical Sciences 197,
DOI 10.1007/978-3-319-65262-7_33

Therefore, for $\varepsilon > 0$ sufficiently small we obtain the following double inequality:

$$(1-\varepsilon)\|\nabla f\|^2_{L^2(\Omega)} - (1/\varepsilon - 1)\left\|\vec{W} f\right\|^2_{L^2(\Omega)} - \left\||V|^{1/2} f\right\|^2_{L^2(\Omega)} \le (H_{\mathrm{m}} f, f)_{L^2(\Omega)}$$
$$\le (1+\varepsilon)\|\nabla f\|^2_{L^2(\Omega)} + (1/\varepsilon + 1)\left\|\vec{W} f\right\|^2_{L^2(\Omega)} + \left\||V|^{1/2} f\right\|^2_{L^2(\Omega)}.$$

Due to the conditions (33.2) the functions $\vec{W}$ and $|V|^{1/2}$ are from equivalent spaces (with respect to norm estimates). We shall therefore estimate only the norm $\||V|^{1/2} f\|^2_{L^2(\Omega)}$, and the norm $\left\|\vec{W} f\right\|^2_{L^2(\Omega)}$ can be estimated in the same manner. Let us consider first $n \ge 3$ and some p satisfying $n/2 \le p < \infty$. Then for $R > 0$, using the Hölder's inequality we obtain

$$\left\||V|^{1/2} f\right\|^2_{L^2(\Omega)} \le \int_{|V(x)|>R} |V||f|^2 \mathrm{d}x + \int_{|V(x)|\le R} |V||f|^2 \mathrm{d}x$$
$$\le \left(\int_{|V(x)|>R} |V|^{n/2} \mathrm{d}x\right)^{2/n} \left(\int_{|V(x)|>R} |f|^{2n/(n-2)} \mathrm{d}x\right)^{(n-2)/n} + R\|f\|^2_{L^2(\Omega)}$$
$$\le C_1 R^{1-2p/n} \|V\|^{2p/n}_{L^p(\Omega \cap \{x:|V(x)|>R\})} \|f\|^2_{W^1_2(\Omega)} + R\|f\|^2_{L^2(\Omega)}.$$

In obtaining the latter inequality we have used the fact that $n \ge 3$, $n/2 \le p < \infty$, and the well known embedding [1, 3] $W^1_2(\Omega) \hookrightarrow L^{2n/(n-2)}(\Omega)$ with the norm estimate

$$\|f\|_{L^{2n/(n-2)}(\Omega)} \le \sqrt{C_1}\, \|f\|_{W^1_2(\Omega)}.$$

Collecting all these estimates, we get

$$\left\||V|^{1/2} f\right\|^2_{L^2(\Omega)} \le C_1 R^{1-2p/n} \|V\|^{2p/n}_{L^p(\Omega \cap \{x:|V(x)|>R\})} \|f\|^2_{W^1_2(\Omega)} + R\|f\|^2_{L^2(\Omega)}$$
$$\le \delta(R)\|f\|^2_{W^1_2(\Omega)} + R\|f\|^2_{L^2(\Omega)},$$

where $\delta(R) > 0$ can be chosen as small as we want if R is sufficiently large. The same is true (with some evident changes) for $p = \infty$ and for $n = 2$. Hence, we have for arbitrarily small $\varepsilon > 0$ and for arbitrarily small $\delta(R) > 0$ that

$$[1 - \varepsilon - \delta(R) - (1/\varepsilon - 1)\delta(R)]\,\|\nabla f\|^2_{L^2(\Omega)}$$
$$- [R + \delta(R) + (1/\varepsilon - 1)(R + \delta(R))]\,\|f\|^2_{L^2(\Omega)} \le (H_{\mathrm{m}} f, f)_{L^2(\Omega)}$$
$$\le [1 + \varepsilon + \delta(R) + (1/\varepsilon - 1)\delta(R)]\,\|\nabla f\|^2_{L^2(\Omega)} + [(2 + 1/\varepsilon)(R + \delta(R))]\,\|f\|^2_{L^2(\Omega)}.$$

Choosing $\varepsilon > 0$ arbitrarily small and $R > 0$ such that $\delta(R) = \varepsilon^2$, we obtain the required estimate (33.3). □

Exercise 33.1. Prove the previous lemma in the cases $p = s = \infty$ for $n \geq 2$ and $p, s < \infty$ for $n = 2$.

This lemma implies that there exists $\mu_0 > 0$ such that $H_{\mathrm{m}} + \mu_0 I$ is positive and

$$((H_{\mathrm{m}} + \mu_0 I)f, f)_{L^2(\Omega)} \asymp \|f\|^2_{W_2^1(\Omega)}.$$

This fact implies that there is a Friedrichs self-adjoint extension of the positive operator $H_{\mathrm{m}} + \mu_0 I$, denoted by $(H_{\mathrm{m}} + \mu_0 I)_{\mathrm{F}}$ (see, for example [5]), with the domain

$$D((H_{\mathrm{m}} + \mu_0 I)_{\mathrm{F}}) = \{f \in \overset{\circ}{W}{}_2^1(\Omega) : (H_{\mathrm{m}} + \mu_0 I)f \in L^2(\Omega)\}, \tag{33.4}$$

where $\overset{\circ}{W}{}_2^1(\Omega)$ is the closure of $C_0^\infty(\Omega)$ with respect to the norm of the Sobolev space $W_2^1(\Omega)$. Hence, the Friedrichs extension $(H_{\mathrm{m}})_{\mathrm{F}}$ of H_{m} can be defined as $(H_{\mathrm{m}})_{\mathrm{F}} := (H_{\mathrm{m}} + \mu_0 I)_{\mathrm{F}} - \mu_0 I_{\mathrm{F}}$ with the same domain (33.4).

Exercise 33.2. Show that if $\vec{W} \in L^\infty(\Omega)$ and $\nabla \cdot \vec{W}, V \in L^p(\Omega)$ with some $n \leq p \leq \infty$ for $n \geq 3$ and with some $2 < p \leq \infty$ for $n = 2$, then

$$D((H_{\mathrm{m}})_{\mathrm{F}}) = \overset{\circ}{W}{}_2^1(\Omega) \cap W_2^2(\Omega).$$

In particular, for $\Omega = \mathbb{R}^n$ we obtain in this case that

$$D((H_{\mathrm{m}})_{\mathrm{F}}) = W_2^2(\mathbb{R}^n).$$

Hint. Represent first H_{m} in the form

$$H_{\mathrm{m}} u = -\Delta - 2\mathrm{i}\vec{W}(x)\nabla u + [|\vec{W}|^2 + V - \mathrm{i}\nabla \cdot \vec{W}]u$$

and then use the same technique and the same embedding theorems for Sobolev spaces as in the proof of Lemma 33.1.

Remark 33.2. The Friedrichs self-adjoint extension of the magnetic Schrödinger operator H_{m} exists under much "broader" assumptions for the coefficients V and $\vec{W}$ than in Lemma 33.1. Namely, if we just assume that $\vec{W} \in L^2(\Omega)$ and $V \in L^1(\Omega)$ but $V \geq 0$, then since for all $f \in C_0^\infty(\Omega)$ we have

$$(H_{\mathrm{m}} f, f)_{L^2(\Omega)} = \left\| (\nabla + \mathrm{i}\vec{W})f \right\|^2_{L^2(\Omega)} + (Vf, f)_{L^2(\Omega)} \geq 0,$$

we may conclude that for all $\mu_0 > 0$, $H_{\mathrm{m}} + \mu_0 I$ is positive, and thus the Friedrichs self-adjoint extension exists (see, for example, [5]). But in this so-called "general" case we cannot characterize the domain of $(H_{\mathrm{m}})_{\mathrm{F}}$ constructively. We can say only that

$$D((H_{\rm m})_{\rm F}) = \{f \in D(\nabla + {\rm i}\vec{W}) : V^{1/2} f \in L^2(\Omega) \text{ and } H_{\rm m} f \in L^2(\Omega)\}.$$

But even in this "general" case we may prove the *diamagnetic inequality* (see [35]).

For all $t > 0$ we may consider (using von Neumann's spectral theorem, see Theorem 27.13) the self-adjoint operators

$$\begin{aligned} {\rm e}^{-t(H_{\rm m})_{\rm F}} f(x) &:= \int_0^\infty {\rm e}^{-t\lambda} {\rm d}E_\lambda f(x), \\ {\rm e}^{-t(-\Delta)_{\rm F}} f(x) &:= \int_0^\infty {\rm e}^{-t\lambda} {\rm d}E_\lambda^{(0)} f(x), \end{aligned} \tag{33.5}$$

where E_λ and $E_\lambda^{(0)}$ are the spectral families corresponding to the self-adjoint operators $(H_{\rm m})_{\rm F}$ and $(-\Delta)_{\rm F}$, respectively.

Theorem 33.3. *Assume that $\vec{W} \in L^2(\Omega)$, $V \in L^1(\Omega)$, $V \ge 0$, and that these potentials are real-valued. Then for all $f \in L^2(\Omega)$, $t > 0$, and $\mu > 0$ we have that*

$$|{\rm e}^{-t(H_{\rm m}+\mu I)_{\rm F}} f(x)| \le {\rm e}^{-t(-\Delta+\mu)_{\rm F}} |f(x)| \tag{33.6}$$

holds for almost every $x \in \Omega$.

Proof. For brevity we denote $\nabla + {\rm i}\vec{W}$ by $D_{\vec{W}}$. We have the following two lemmas (which are also of independent interest).

Lemma 33.4. *For all $f \in C_0^\infty(\Omega)$ we have*

$$|D_{\vec{W}} f(x)| \ge |\nabla |f(x)|| \tag{33.7}$$

almost everywhere.

Proof. Indeed,

$$\nabla |f(x)|^2 = \nabla(\overline{f(x)} f(x)) = \overline{f} \nabla f + f \nabla \overline{f} = \overline{f} D_{\vec{W}} f + f \overline{D_{\vec{W}} f} = 2\,{\rm Re}(\overline{f} D_{\vec{W}} f).$$

This is equivalent to

$$2|f| \nabla |f| = 2\,{\rm Re}(\overline{f} D_{\vec{W}} f).$$

The latter equality implies that

$$|D_{\vec{W}} f||f| \ge |f||\nabla |f||.$$

Thus, the lemma is proved. □

Lemma 33.5. *For all $f \in C_0^\infty(\Omega)$ and $\varphi \ge 0$ sufficiently smooth we have that*

$$\operatorname{Re}\left(\overline{D_{\vec{W}}\left(\frac{f}{|f|}\varphi\right)} D_{\vec{W}} f\right) \ge \nabla \varphi \nabla |f| \tag{33.8}$$

almost everywhere.

Proof. Without loss of generality we may assume that $f \neq 0$; otherwise, we consider $f_\varepsilon = \sqrt{|f|^2+\varepsilon}$ and then take the limit $\varepsilon \to 0+$. We have

$$\begin{aligned} D_{\vec{W}}\left(\frac{f}{|f|}\varphi\right) &= \nabla\left(\frac{f}{|f|}\varphi\right) + \mathrm{i}\vec{W}\frac{f}{|f|}\varphi = \varphi\nabla\frac{f}{|f|} + \frac{f}{|f|}\nabla\varphi + \mathrm{i}\vec{W}\frac{f}{|f|}\varphi \\ &= \varphi\left[\frac{D_{\vec{W}}f}{|f|} - \frac{f\nabla|f|}{|f|^2}\right] + \frac{f}{|f|}\nabla\varphi. \end{aligned}$$

This equality implies that

$$\overline{D_{\vec{W}}\left(\frac{f}{|f|}\varphi\right)} D_{\vec{W}}f = \varphi\left[\frac{\overline{D_{\vec{W}}f}D_{\vec{W}}f}{|f|} - \frac{\overline{f}\nabla|f|}{|f|^2}D_{\vec{W}}f\right] + \frac{\overline{f}}{|f|}\nabla\varphi, D_{\vec{W}}f$$

so

$$\mathrm{Re}\left(\overline{D_{\vec{W}}\left(\frac{f}{|f|}\varphi\right)}D_{\vec{W}}f\right) = \varphi\frac{|D_{\vec{W}}f|^2}{|f|} - \varphi\frac{\nabla|f|}{|f|^2}\mathrm{Re}(\overline{f}D_{\vec{W}}f) + \frac{\mathrm{Re}(\overline{f}D_{\vec{W}}f)}{|f|}\nabla\varphi.$$

Calculating $\mathrm{Re}(\overline{f}D_{\vec{W}}f)$ for $f = f_1 + \mathrm{i}f_2$, we obtain

$$\mathrm{Re}(\overline{f}D_{\vec{W}}f) = \mathrm{Re}[(f_1 - \mathrm{i}f_2)(\nabla f_1 + \mathrm{i}\nabla f_2 + \mathrm{i}\nabla f_2 - \vec{W}f_2)] = f_1\nabla f_1 + f_2\nabla f_2 = |f|\nabla|f|.$$

Thus,

$$\begin{aligned} \mathrm{Re}\left(\overline{D_{\vec{W}}\left(\frac{f}{|f|}\varphi\right)}D_{\vec{W}}f\right) &= \varphi\frac{|D_{\vec{W}}f|^2}{|f|} - \varphi\frac{\nabla|f|}{|f|^2}|f|\nabla|f| + \frac{|f|\nabla|f|}{|f|}\nabla\varphi \\ &= \varphi\frac{|D_{\vec{W}}f|^2 - |\nabla|f||^2}{|f|} + \nabla\varphi\nabla|f| \geq \nabla\varphi\nabla|f| \end{aligned}$$

by Lemma 33.4 and the fact that $\varphi \geq 0$. □

To end the proof of Theorem 33.3 we consider $\mu > 0$. Using these two lemmas we obtain

$$\begin{aligned} \int_\Omega \nabla\varphi\nabla|f|\mathrm{d}x + \mu\int_\Omega \varphi|f|\mathrm{d}x &\leq \mathrm{Re}\int_\Omega \overline{D_{\vec{W}}\left(\frac{f}{|f|}\varphi\right)}D_{\vec{W}}f\mathrm{d}x + \mu\int_\Omega\varphi|f|\mathrm{d}x \\ &\leq \left|\int_\Omega \overline{D_{\vec{W}}\left(\frac{f}{|f|}\varphi\right)}D_{\vec{W}}f\mathrm{d}x + \mu\int_\Omega\varphi|f|\mathrm{d}x\right|, \end{aligned}$$

since $\mathrm{Re}\,z + a \leq |z+a|$ for real a. Using now integration by parts in both integrals we have that

$$\int_\Omega \varphi(-\Delta+\mu)|f|\mathrm{d}x \le \left|\int_\Omega \left(-D^2_{\vec{W}} f\frac{\overline{f}}{|f|}\varphi + \mu f\frac{\overline{f}}{|f|}\varphi\right)\mathrm{d}x\right|,$$

i.e.,

$$\begin{aligned}((-\Delta+\mu)|f|,f)_{L^2(\Omega)} &\le \left|\int_\Omega \frac{\overline{f}}{|f|}\varphi(-D^2_{\vec{W}}+\mu)f\mathrm{d}x\right| \\ &\le \int_\Omega \varphi|(-D^2_{\vec{W}}+\mu)f|\mathrm{d}x = (|(-D^2_{\vec{W}}+\mu)f|,\varphi)_{L^2(\Omega)}.\end{aligned}$$

Since $(-D^2_{\vec{W}}+\mu I)^{-1}$ exists, by introducing $f := (-D^2_{\vec{W}}+\mu I)^{-1}u$ we can rewrite the latter inequality as

$$\begin{aligned}((-\Delta+\mu)|f|,\varphi)_{L^2(\Omega)} &= (|f|,(-\Delta+\mu)\varphi)_{L^2(\Omega)} = (|(-D^2_{\vec{W}}+\mu I)^{-1}u|,\psi)_{L^2(\Omega)} \\ &\le (|u|,\varphi)_{L^2(\Omega)} = (|u|,(-\Delta+\mu I)^{-1}\psi)_{L^2(\Omega)} = ((-\Delta+\mu I)^{-1}|u|,\psi)_{L^2(\Omega)},\end{aligned}$$

where we have used the self-adjointness of all operators and the notation $\psi = (-\Delta+\mu I)\varphi$. Hence, for arbitrary $\psi \ge 0$ sufficiently smooth we obtain the inequality

$$(|(-D^2_{\vec{W}}+\mu I)^{-1}u|,\psi)_{L^2(\Omega)} \le ((-\Delta+\mu)^{-1}|u|,\psi)_{L^2(\Omega)}.$$

Since ψ is an arbitrary function of such type, we may conclude from here that for every $u \in L^2(\Omega)$ we have that

$$|(-D^2_{\vec{W}}+\mu I)^{-1}u(x)| \le (-\Delta+\mu I)^{-1}|u|(x)$$

almost everywhere. Iterating the latter inequality, we obtain for all $m \in \mathbb{N}$ that

$$|(-D^2_{\vec{W}}+\mu I)^{-m}u(x)| \le (-\Delta+\mu I)^{-m}|u|(x).$$

Hence, for every $u \in L^2(\Omega)$ we have

$$\begin{aligned}|\mathrm{e}^{-t(-D^2_{\vec{W}}+\mu)}u(x)| &\le \overline{\lim_{m\to\infty}}\left|\left(\frac{m}{t}(-D^2_{\vec{W}}+\mu+\frac{m}{t})\right)^{-m}u(x)\right| \\ &\le \overline{\lim_{m\to\infty}}\left(\frac{m}{t}(-\Delta+\mu+\frac{m}{t})\right)^{-m}|u(x)| \\ &= \lim_{m\to\infty}\left(\frac{m}{t}(-\Delta+\mu+\frac{m}{t})\right)^{-m}|u(x)| = \mathrm{e}^{-t(-\Delta+\mu)}|u(x)|\end{aligned}$$

almost everywhere. Thus (33.6) is proved in the case $V=0$. In order to add $V \ge 0$ to $-D^2_{\vec{W}}+\mu I$ we repeat the above procedure and easily obtain that for all $u \in L^2(\Omega)$ and $\psi \ge 0$ sufficiently smooth, we have

$$\begin{aligned}(|(-D_{\vec{W}}^2+\mu+V)^{-1}u|,\psi)_{L^2(\Omega)} &\le (|u|,(-\Delta+\mu+V)^{-1}\psi)_{L^2(\Omega)} \\ &\le (|u|,(-\Delta+\mu)^{-1}\psi)_{L^2(\Omega)} = ((-\Delta+\mu)^{-1}|u|,\psi)_{L^2(\Omega)},\end{aligned}$$

since

$$(-\Delta+\mu+V)^{-1}\psi \le (-\Delta+\mu)^{-1}\psi$$

almost everywhere. This completes the proof of Theorem 33.3. □

There are many applications of the diamagnetic inequality. We will consider some of them. If A is a nonnegative self-adjoint operator acting in $L^2(\Omega)$, its *heat kernel* $P(x,y,t)$ (if it exists) is defined to be a function such that for every $t>0$ the self-adjoint operator e^{-tA} is an integral operator with this kernel (see for comparison Definition 22.6 and Chapter 45), i.e., for all $f \in L^2(\Omega)$,

$$\mathrm{e}^{-tA}f(x) = \int_\Omega P(x,y,t)f(y)\mathrm{d}y. \tag{33.9}$$

Using this definition, we may conclude that if a heat kernel exists, then for every $\mu>0$ the inverse operator $(A+\mu I)^{-1}$ (which exists) is an integral operator with kernel

$$G(x,y,\mu) = \int_0^\infty \mathrm{e}^{-\mu t}P(x,y,t)f(y)\mathrm{d}t. \tag{33.10}$$

Indeed, by von Neumann's theorem for A (see Chapter 27), we have that for every $f \in L^2(\Omega)$,

$$\mathrm{e}^{-tA}f(x) = \int_0^\infty \mathrm{e}^{-t\lambda}\mathrm{d}E_\lambda f,$$

where $\{E_\lambda\}$ is the spectral family corresponding to A.

Since $(A+\mu I)^{-1}$ exists for every $\mu>0$ and it is self-adjoint, it follows that for every $f \in L^2(\Omega)$ we have

$$\begin{aligned}(A+\mu I)^{-1}f(x) &= \int_0^\infty \frac{1}{\lambda+\mu}\mathrm{d}E_\lambda f(x) = \int_0^\infty \mathrm{d}E_\lambda f(x)\int_0^\infty \mathrm{e}^{-t(\lambda+\mu)}\mathrm{d}t \\ &= \int_0^\infty \mathrm{e}^{-t\mu}\mathrm{d}t\int_0^\infty \mathrm{e}^{-t\lambda}\mathrm{d}E_\lambda f(x) = \int_0^\infty \mathrm{e}^{-t\mu}P(x,y,t)\mathrm{d}t.\end{aligned}$$

There is (at least) one quite general situation in which the heat kernel exists and has "good" estimates. Let us assume that $\Omega \subset \mathbb{R}^n$ is a bounded domain with smooth boundary (the smoothness is required for the Sobolev embedding theorem). We consider the magnetic Schrödinger operator H_{m} in Ω with electric potential $V \ge 0$ and with magnetic potential $\vec{W}$ satisfying all assumptions of Lemma 33.1. In this case H_{m} and $H_0 = -\Delta$ have Friedrichs self-adjoint extensions, which are denoted by the same symbols H_{m} and H_0, respectively. We have the following theorem.

Theorem 33.6. *Under the conditions of Lemma 33.1 for $\mu>0$, the resolvent $(H_{\mathrm{m}}+\mu I)^{-1}$ is an integral operator with kernel $G(x,y,\mu)$, called the* Green's function *corresponding to the Friedrichs extension of H_{m}. Moreover, the following estimates are valid:*

$$|G(x,y,\mu)| \le C \begin{cases} |x-y|^{2-n} \mathrm{e}^{-\sqrt{\mu}|x-y|}, & n \ge 3, \\ 1 + |\log(\sqrt{\mu}|x-y|)|, & n = 2. \end{cases} \tag{33.11}$$

Proof. Using Lemma 33.1, we conclude that for $\mu > 0$ sufficiently large and for all $f \in \overset{\circ}{W}{}_2^1(\Omega)$ we have

$$((H_{\mathrm{m}} + \mu I)f, f)_{L^2(\Omega)} \ge \gamma \|f\|^2_{W_2^1(\Omega)}, \quad \gamma > 0.$$

Since the embedding $\overset{\circ}{W}{}_2^1(\Omega) \hookrightarrow L^2(\Omega)$ is compact (see Lemma 30.15), we have that $(H_{\mathrm{m}} + \mu I)^{-1}$ is compact. Using now the Riesz–Schauder and Hilbert–Schmidt theorems (see Theorem 28.10), we conclude that the spectrum $\sigma(H_{\mathrm{m}}) = \{\lambda_j\}_{j=1}^{\infty}$ is discrete and of finite multiplicity with only one accumulation point at infinity. The corresponding normalized eigenfunctions $\{\varphi_j\}_{j=1}^{\infty}$ form an orthonormal basis in $L^2(\Omega)$ such that the spectral family for H_{m} is defined as

$$E_\lambda f(x) = \sum_{\lambda_j < \lambda} f_j \varphi_j(x), \quad f_j = (f, \varphi_j)_{L^2(\Omega)}.$$

Hence

$$\mathrm{e}^{-tH_{\mathrm{m}}} f(x) = \int_0^\infty \mathrm{e}^{-t\lambda} \mathrm{d}E_\lambda f(x) = \sum_{j=1}^{\infty} \mathrm{e}^{-t\lambda_j} f_j \varphi_j(x)$$

$$= \sum_{j=1}^{\infty} \mathrm{e}^{-t\lambda_j} \left(\int_\Omega f(y) \overline{\varphi_j(y)} \mathrm{d}y \right) \varphi_j(x) = \int_\Omega \left(\sum_{j=1}^{\infty} \mathrm{e}^{-t\lambda_j} \varphi_j(x) \overline{\varphi_j(y)} \right) f(y) \mathrm{d}y,$$

and the heat kernel of H_{m} will be equal in this case to

$$P(x,y,t) = \sum_{j=1}^{\infty} \mathrm{e}^{-t\lambda_j} \varphi_j(x) \overline{\varphi_j(y)}. \tag{33.12}$$

It must be mentioned here that all these equalities (and operations) are considered in the sense of $L^2(\Omega)$. The equality (33.12) implies that $(H_{\mathrm{m}} + \mu I)^{-1}$ is an integral operator with kernel $G(x,y,\mu)$ defined by

$$G(x,y,\mu) = \int_0^\infty \mathrm{e}^{-\mu t} P(x,y,t) \mathrm{d}t = \sum_{j=1}^{\infty} \varphi_j(x) \overline{\varphi_j(y)} \int_0^\infty \mathrm{e}^{-t(\lambda_j + \mu)} \mathrm{d}t = \sum_{j=1}^{\infty} \frac{\varphi_j(x) \overline{\varphi_j(y)}}{\lambda_j + \mu}. \tag{33.13}$$

This function $G(x,y,\mu)$ is called the Green's function of the Friedrichs extension of the magnetic Schrödinger operator H_{m} in the bounded domain. To obtain the estimates (33.11) we proceed as follows. It is known that the heat kernel $P_0(x,y,t)$

for H_0 in the whole of $\mathbb{R}^n$ (see, for example, Chapter 22) is equal to

$$P_0(x,y,t) = (4\pi t)^{-n/2} e^{-|x-y|^2/(4t)}. \tag{33.14}$$

At the same time, the heat kernel $\widetilde{P_0}(x,y,t)$ of H_0 in the bounded domain Ω satisfies the following boundary value problem:

$$\begin{cases} \partial_t \widetilde{P_0}(x,y,t) = \Delta_x \widetilde{P_0}(x,y,t), & x,y \in \Omega, t>0, \\ \widetilde{P_0}(x,y,t)\Big|_{\partial\Omega} = 0, & x \in \partial\Omega,\ y \in \Omega,\ t>0, \\ \widetilde{P_0}(x,y,0) = \delta(x-y). \end{cases}$$

If we define $\widetilde{P_0}(x,y,t) = P_0(x,y,t) + R(x,y,t)$, then $R(x,y,t)$ has to satisfy

$$\begin{cases} \partial_t R(x,y,t) = \Delta_x R(x,y,t), & x,y \in \Omega,\ t>0, \\ R(x,y,t)|_{\partial\Omega} = -P_0(x,y,t), & x \in \partial\Omega,\ y \in \Omega,\ t>0 \\ R(x,y,0) = 0, & x,y \in \Omega. \end{cases}$$

But $-P_0(x,y,t) < 0$ for $x,y \in \Omega, t>0$ (see (33.14)). Using then the maximum principle for the heat equation (see Theorem 45.7) we obtain that $R(x,y,t) \le 0$ for all $x,y \in \Omega$ and $t > 0$, i.e.,

$$0 \le \widetilde{P_0}(x,y,t) \le P_0(x,y,t).$$

The next step is as follows: the diamagnetic inequality (33.6) leads in this case to

$$\left| \int_\Omega P(x,y,t) f(y) dy \right| \le \int_\Omega P_0(x,y,t) |f(y)| dy,$$

which holds almost everywhere in $x \in \Omega$ and for all $f \in L^2(\Omega)$. Using the Hardy–Littlewood maximal function (see, e.g., [18]), we can obtain from the latter inequality that

$$|P(x,y,t)| \le (4\pi t)^{-n/2} e^{-|x-y|^2/(4t)}, \quad x,y \in \Omega,\ t>0.$$

Using this and (33.10), we get (see Example 22.8)

$$\begin{aligned} |G(x,y,\mu)| &\le \int_0^\infty e^{-\mu t} (4\pi t)^{-n/2} e^{-|x-y|^2/(4t)} dt \\ &= (2\pi)^{-n/2} \left(\frac{|x-y|}{\sqrt{\mu}} \right)^{(2-n)/2} K_{(n-2)/2}(\sqrt{\mu}|x-y|), \end{aligned}$$

where $K_\nu(z)$ is the Macdonald function of order ν. Using the asymptotic expansion for $K_\nu(z)$ for $z \to 0$ and $z \to \infty$ (see, for example, [23]), we can obtain the following inequalities (see also the straightforward calculations in Example 22.8):

$$|G(x,y,\mu)| \le C \begin{cases} |x-y|^{2-n} e^{-\sqrt{\mu}|x-y|}, & n \ge 3, \\ 1 + |\log(\sqrt{\mu}|x-y|)|, & n = 2, \end{cases}$$

where $x, y \in \Omega$ and the constant $C > 0$ depends only on the dimension n. Thus, Theorem 33.6 is completely proved. □

One more application of the diamagnetic inequality concerns the estimates of the normalized eigenfunctions of H_{m}.

Corollary 33.7. *Let φ be a normalized eigenfunction of H_{m} with corresponding eigenvalue $\lambda > 0$. Then*

$$\|\varphi\|_{L^\infty(\Omega)} \le \left(\frac{e}{2\pi n}\right)^{n/4} \lambda^{n/4}. \tag{33.15}$$

Proof. Applying (33.6) to $\varphi(x)$, we obtain ($\|\varphi\|_{L^2(\Omega)} = 1$)

$$\begin{aligned}
|e^{-t\lambda}\varphi(x)| = |e^{-tH_{\mathrm{m}}}\varphi| &\le \int_\Omega (4\pi t)^{-n/2} e^{-|x-y|^2/(4t)} |\varphi(y)| dy \\
&\le \|\varphi\|_{L^2(\Omega)} \left(\int_\Omega (4\pi t)^{-n} e^{-|x-y|^2/(2t)} dy\right)^{1/2} \\
&\le \left(\int_{\mathbb{R}^n} (4\pi t)^{-n} e^{-|x-y|^2/(2t)} dy\right)^{1/2} \\
&= (4\pi t)^{-n/2} t^{n/4} \left(\int_{\mathbb{R}^n} e^{-|y|^2/2} dy\right)^{1/2} \\
&= (4\pi t)^{-n/2} t^{n/4} (2\pi)^{n/4} = \pi^{-n/4} 2^{-3n/4} t^{-n/4},
\end{aligned}$$

i.e., for all $t > 0$,

$$|\varphi(x)| \le \pi^{-n/4} 2^{-3n/4} t^{-n/4} e^{t\lambda}.$$

Taking the infimum of the right-hand side with respect to $t > 0$, we obtain (33.15). □

Chapter 34
Integral Operators with Weak Singularities. Integral Equations of the First and Second Kinds.

Let Ω be a bounded domain in $\mathbb{R}^n$. Then

$$Af(x) = \int_\Omega K(x,y)f(y)\mathrm{d}y$$

is an *integral operator* in $L^2(\Omega)$ with *kernel K*.

Definition 34.1. An integral operator A is said to be an *operator with weak singularity* if its kernel $K(x,y)$ is continuous for all $x,y \in \Omega$, $x \neq y$, and there are positive constants M and $\alpha \in (0,n]$ such that

$$|K(x,y)| \leq M|x-y|^{\alpha-n}, \quad x \neq y.$$

Remark 34.2. If $K(x,y)$ is continuous for all $x,y \in \Omega$ and bounded, then this integral operator is considered also an operator with weak singularity.

If we have two integral operators A_1 and A_2 with kernels K_1 and K_2, respectively, then we can consider their composition as follows:

$$\begin{aligned}(A_1 \circ A_2)f(x) &= \int_\Omega K_1(x,y)A_2 f(y)\mathrm{d}y = \int_\Omega K_1(x,y)\left(\int_\Omega K_2(y,z)f(z)\mathrm{d}z\right)\mathrm{d}y \\ &= \int_\Omega \left(\int_\Omega K_1(x,y)K_2(y,z)\mathrm{d}y\right) f(z)\mathrm{d}z,\end{aligned}$$

and analogously

$$(A_2 \circ A_1)f(x) = \int_\Omega \left(\int_\Omega K_2(x,y)K_1(y,z)\mathrm{d}y\right) f(z)\mathrm{d}z,$$

assuming that the conditions of Fubini's theorem are fulfilled.

V. Serov, *Fourier Series, Fourier Transform and Their Applications to Mathematical Physics*, Applied Mathematical Sciences 197,
DOI 10.1007/978-3-319-65262-7_34

So, we may conclude that the compositions $A_1 \circ A_2$ and $A_2 \circ A_1$ are again integral operators with kernels

$$\begin{aligned} K(x,y) &= \int_\Omega K_1(x,z)K_2(z,y)\mathrm{d}z, \\ \widetilde{K}(x,y) &= \int_\Omega K_2(x,z)K_1(z,y)\mathrm{d}z, \end{aligned} \tag{34.1}$$

respectively. In general, $K(x,y) \neq \widetilde{K}(x,y)$, that is, $A_1 \circ A_2 \neq A_2 \circ A_1$.

Returning to integral operators with weak singularities, we obtain a very important property of them.

Lemma 34.3. *If A_1 and A_2 are integral operators with weak singularities, then $A_1 \circ A_2$ and $A_2 \circ A_1$ are also integral operators with weak singularities. Moreover, if*

$$|K_1(x,y)| \le M_1|x-y|^{\alpha_1-n} \quad \text{and} \quad |K_2(x,y)| \le M_2|x-y|^{\alpha_2-n}, \tag{34.2}$$

then there is $M > 0$ such that

$$|K(x,y)| \le M \begin{cases} |x-y|^{\alpha_1+\alpha_2-n}, & \alpha_1+\alpha_2 < n, \\ 1+|\log|x-y||, & \alpha_1+\alpha_2 = n, \\ 1, & \alpha_1+\alpha_2 > n. \end{cases} \tag{34.3}$$

The same estimates hold for the kernel $\widetilde{K}(x,y)$.

Proof. Using (34.1) and (34.2), we obtain

$$|K(x,y)| \le M_1M_2 \int_\Omega |x-z|^{\alpha_1-n}|z-y|^{\alpha_2-n}\mathrm{d}z.$$

If $\alpha_1+\alpha_2 < n$, then changing the variable $z = y+u|x-y|$, we have

$$x-z = |x-y|(e_0-u), \quad |e_0| = 1,$$

and

$$|K(x,y)| \le M_1M_2|x-y|^{\alpha_1+\alpha_2-n} \int_{\mathbb{R}^n} |u-e_0|^{\alpha_1-n}|u|^{\alpha_2-n}\mathrm{d}u. \tag{34.4}$$

In order to estimate the latter integral we consider three different cases:

$$|u| \le 1/2, \quad 1/2 \le |u| \le 3/2, \quad |u| \ge 3/2.$$

In the first case,

$$|u-e_0| \ge |e_0|-|u| \ge 1-1/2 = 1/2,$$

and therefore

$$\int_{|u|\le 1/2} |u-e_0|^{\alpha_1-n}|u|^{\alpha_2-n}\mathrm{d}u \le 2^{n-\alpha_1}\int_{|u|\le 1/2} |u|^{\alpha_2-n}\mathrm{d}u$$
$$= 2^{n-\alpha_1}\int_0^{1/2} r^{\alpha_2-1}\mathrm{d}r\int_{\mathbb{S}^{n-1}}\mathrm{d}\theta = \frac{2^{n-\alpha_1-\alpha_2}}{\alpha_2}|\mathbb{S}^{n-1}|,$$

where $|\mathbb{S}^{n-1}|$ denotes the area of the unit sphere in $\mathbb{R}^n$.

In the third case,

$$|u-e_0| \ge |u|-|e_0| \ge |u|-1 \ge |u|-\frac{2}{3}|u| = \frac{|u|}{3},$$

and we have analogously

$$\int_{|u|\ge 3/2} |u-e_0|^{\alpha_1-n}|u|^{\alpha_2-n}\mathrm{d}u$$
$$\le 3^{n-\alpha_1}|\mathbb{S}^{n-1}|\int_{3/2}^{\infty} r^{\alpha_1+\alpha_2-n-1}\mathrm{d}r = \frac{2^{n-\alpha_1-\alpha_2}3^{\alpha_2}}{n-\alpha_1-\alpha_2}|\mathbb{S}^{n-1}|.$$

In the case $1/2 \le |u| \le 3/2$ we have that $|u-e_0| \le 5/2$, and so

$$\int_{1/2\le|u|\le 3/2} |u-e_0|^{\alpha_1-n}|u|^{\alpha_2-n}\mathrm{d}u$$
$$\le 2^{n-\alpha_2}\int_{|u-e_0|\le 5/2} |u-e_0|^{\alpha_1-n}\mathrm{d}u = \frac{2^{n-\alpha_1-\alpha_2}5^{\alpha_1}}{\alpha_1}|\mathbb{S}^{n-1}|. \tag{34.5}$$

Combining (34.4)–(34.5), we obtain (34.3) for the case $\alpha_1+\alpha_2<n$. It can be mentioned here that the estimate (34.3) in this case holds also in the case of an arbitrary (not necessarily bounded) domain Ω.

If now $\alpha_1+\alpha_2=n$, then the proof of (34.3) will be a little bit different, and it holds only for a bounded domain Ω. Indeed, for every $z\in\Omega$ and

$$|x-z| \le \frac{|x-y|}{2} \quad \text{or} \quad |z-y| \le \frac{|x-y|}{2},$$

we have in both cases that

$$|K(x,y)| \le M_1M_2 2^{n-\alpha_2}|x-y|^{\alpha_2-n}\int_{\Omega'}|x-z|^{\alpha_1-n}\mathrm{d}z$$
$$\le M_1M_2 2^{n-\alpha_2}|\mathbb{S}^{n-1}||x-y|^{\alpha_2-n}\int_0^{|x-y|/2} r^{\alpha_1-1}\mathrm{d}r$$
$$= \frac{M_1M_2}{\alpha_1}|\mathbb{S}^{n-1}| \quad \text{or} \quad \frac{M_1M_2}{\alpha_2}|\mathbb{S}^{n-1}|. \tag{34.6}$$

If $z \in \Omega$ does not belong to these balls with radius $|x-y|/2$, then we consider two cases: $|z-x| \geq |z-y|$ and $|z-x| \leq |z-y|$. In both cases we have

$$\begin{aligned} |K(x,y)| &\leq M_1 M_2 \int_{\Omega \setminus \Omega'} \frac{dz}{|x-z|^n} \leq M_1 M_2 |\mathbb{S}^{n-1}| \int_{|x-y|/2}^{d} \frac{dr}{r} \\ &= M_1 M_2 |\mathbb{S}^{n-1}| \log \frac{2d}{|x-y|}, \end{aligned} \tag{34.7}$$

where $d = \operatorname{diam} \Omega$. The estimates (34.6) and (34.7) give us (34.3) in the case $\alpha_1 + \alpha_2 = n$.

If finally $\alpha_1 + \alpha_2 > n$, then since Ω is bounded, we can analogously obtain (34.3) in this case. This finishes the proof. □

Remark 34.4. In the case $\alpha_1 + \alpha_2 = n$, since for all $0 < t < 1$,

$$|\log t| \leq C_\varepsilon t^{-\varepsilon}, \quad \varepsilon > 0,$$

instead of a logarithmic singularity in (34.3) we may consider a weak singularity for the kernel $K(x,y)$ as

$$|K(x,y)| \leq M_\varepsilon |x-y|^{-\varepsilon},$$

where $\varepsilon > 0$ can be chosen appropriately.

Let A be an integral operator in $L^2(\Omega)$ with weak singularity. Then since $0 < \alpha \leq n$, we have

$$\int_\Omega |x-y|^{\alpha-n} dy \leq \beta \quad \text{and} \quad \int_\Omega |x-y|^{\alpha-n} dx \leq \beta,$$

where

$$\beta = \sup_{x \in \Omega} \int_\Omega |x-y|^{\alpha-n} dy < \infty.$$

Schur's test (see Example 26.2) shows that A is bounded in $L^2(\Omega)$ and

$$\|A\|_{L^2(\Omega) \to L^2(\Omega)} \leq M\beta.$$

We can prove even more.

Theorem 34.5. *An integral operator with weak singularity is compact in $L^2(\Omega)$.*

Proof. Let us introduce the function

$$\chi_\sigma(t) = \begin{cases} 1, & 0 \leq t \leq \sigma, \\ 0, & t > \sigma. \end{cases}$$

Then for all $\sigma > 0$ we may write

$$K(x,y) = \chi_\sigma(|x-y|)K(x,y) + (1-\chi_\sigma(|x-y|))K(x,y) =: K_1(x,y) + K_2(x,y).$$

The integral operator with kernel $K_2(x,y)$ is a Hilbert–Schmidt operator for all $\sigma > 0$, since

$$\int_\Omega \int_\Omega |K_2(x,y)|^2 \mathrm{d}x\mathrm{d}y \leq M^2 \iint_{\sigma \leq |x-y| \leq d} |x-y|^{2\alpha-2n} \mathrm{d}x\mathrm{d}y$$

is finite. It is therefore compact in $L^2(\Omega)$ (see Exercise 28.10). For the integral operator A_1 with kernel $K_1(x,y)$ we proceed as follows:

$$\|A_1 f\|^2_{L^2(\Omega)} = (A_1 f, A_1 f)_{L^2(\Omega)} = (f, A_1^* \circ A_1 f)_{L^2(\Omega)}, \tag{34.8}$$

where A_1^* is the adjoint operator with kernel

$$K_1^*(x,y) = \chi_\sigma(|x-y|)\overline{K(y,x)},$$

which is also an operator with weak singularity. Using Lemma 34.3, we can estimate the right-hand side of (34.8) from above as

$$\begin{aligned}\int_\Omega \int_\Omega |K_\sigma(x,y)||f(x)||f(y)|\mathrm{d}x\mathrm{d}y &\leq \frac{1}{2}\int_\Omega \int_\Omega |K_\sigma(x,y)||f(x)|^2 \mathrm{d}x\mathrm{d}y \\ &\quad + \frac{1}{2}\int_\Omega \int_\Omega |K_\sigma(x,y)||f(y)|^2 \mathrm{d}x\mathrm{d}y,\end{aligned} \tag{34.9}$$

where $K_\sigma(x,y)$ is the kernel of the operator with weak singularity, i.e.,

$$|K_\sigma(x,y)| \leq M \begin{cases} |x-y|^{2\alpha-n}, & \alpha < n/2, \\ |x-y|^{-\varepsilon}, & \alpha = n/2, \\ 1, & \alpha > n/2, \end{cases}$$

where $\varepsilon > 0$ can be chosen as small as we want.

Let us note also that the definition of $\chi_\sigma(t)$ implies that $K_\sigma(x,y) = 0$ for $|x-y| > 2\sigma$. Thus (see (34.8) and (34.9)) we have $(\alpha < 2n)$

$$\begin{aligned}\|A_1 f\|^2_{L^2(\Omega)} &\leq M \iint_{|x-y| \leq 2\sigma} |x-y|^{2\alpha-n} |f(x)|^2 \mathrm{d}x\mathrm{d}y \\ &\leq M \int_\Omega |f(x)|^2 \int_{|x-y| \leq 2\sigma} |x-y|^{2\alpha-n} \mathrm{d}y\mathrm{d}x \\ &= M \|f\|^2_{L^2(\Omega)} \frac{(2\sigma)^{2\alpha}}{2\alpha} |\mathbb{S}^{n-1}| \to 0,\end{aligned}$$

as $\sigma \to 0$. This means that

$$\|A_1\|_{L^2(\Omega)\to L^2(\Omega)} \to 0, \quad \sigma \to 0.$$

The same fact is valid for the cases $\alpha \geq n/2$. Thus,

$$\|A - A_2\|_{L^2(\Omega)\to L^2(\Omega)} \leq \|A_1\|_{L^2(\Omega)\to L^2(\Omega)} \to 0$$

as $\sigma \to 0$. But A_2 is compact for every $\sigma > 0$, and therefore, A is also compact as the limit of compact operators. This completes the proof. □

We want now to expand the analysis of integral operators with weak singularity defined on domains in $\mathbb{R}^n$ to integral operators with weak singularity defined on surfaces of dimension $n-1$.

Assume that $\partial\Omega$ is the boundary of a bounded domain of class C^1. This means, roughly speaking, that at every point $x \in \partial\Omega$ there is a tangent plane with normal vector $\nu(x)$ that is continuous function on $\partial\Omega$, and the surface differential $\mathrm{d}\sigma(y)$ in a neighborhood of each point $x \in \partial\Omega$ satisfies the inequality (see [22])

$$\mathrm{d}\sigma(y) \leq c_0 \rho^{n-2} \mathrm{d}\rho \mathrm{d}\theta,$$

where (ρ, θ) are the polar coordinates in the tangent plane with origin x, and c_0 is independent of x. According to the dimension $n-1$ of the surface $\partial\Omega$, an integral operator in $L^2(\partial\Omega)$, i.e.,

$$Af(x) = \int_{\partial\Omega} K(x,y) f(y) \mathrm{d}\sigma(y),$$

is said to be with weak singularity if its kernel $K(x,y)$ is continuous for all $x, y \in \partial\Omega$, $x \neq y$, and there are constants $M > 0$ and $\alpha \in (0, n-1]$ such that

$$|K(x,y)| \leq M|x-y|^{\alpha-(n-1)}, \quad x \neq y.$$

If $K(x,y)$ is continuous everywhere, we require that K is bounded on $\partial\Omega \times \partial\Omega$. We can provide now the following theorem.

Theorem 34.6. *An integral operator with weak singularity is compact in $L^2(\partial\Omega)$.*

Proof. The proof is the same as that for Theorem 34.5. □

For Banach spaces (i.e., complete normed spaces) the same definition of compact operator holds (see Definition 28.8). We will need the compactness of these integral operators in the Banach space $C(\overline{\Omega})$. Let $\overline{\Omega}$ be a compact set in $\mathbb{R}^n$. The Banach space $C(\overline{\Omega})$ is defined as the set of all complex-valued functions $\varphi(x)$ that are continuous on $\overline{\Omega}$ with norm

$$\|\varphi\|_{C(\overline{\Omega})} := \max_{\overline{\Omega}} |\varphi(x)| \equiv \|\varphi\|_{L^\infty(\overline{\Omega})} \cdot$$

We will need also the famous *Ascoli–Arzelà theorem* (for a proof, see [22]).

Theorem 34.7. *A set $U \subset C(\overline{\Omega})$ is relatively compact if and only if*

(1) there is a constant $M > 0$ such that for all $\varphi \in U$ we have $\|\varphi\|_{L^\infty(\overline{\Omega})} \le M$ (uniform boundedness);
(2) for every $\varepsilon > 0$ there is $\delta > 0$ such that

$$|\varphi(x) - \varphi(y)| < \varepsilon$$

for all $x, y \in \overline{\Omega}$ with $|x-y| < \delta$ and all $\varphi \in U$ (equicontinuity).

Theorem 34.8. *An integral operator with continuous kernel is compact on $C(\overline{\Omega})$.*

Proof. The result follows straightforwardly from the Ascoli–Arzelà theorem. □

Theorem 34.9. *An integral operator with weak singularity is compact on $C(\overline{\Omega})$.*

Proof. The proof follows from Theorem 34.8. Indeed, let us choose a continuous function h as

$$h(t) = \begin{cases} 0, & 0 \le t \le 1/2, \\ 2t-1, & 1/2 \le t \le 1, \\ 1, & t \ge 1, \end{cases}$$

and the integral operator A_k with kernel $K_k(x,y)$ given by

$$K_k(x,y) = \begin{cases} h(k|x-y|)K(x,y), & x \ne y, \\ 0, & x = y. \end{cases}$$

The kernel $K_k(x,y)$ is continuous for every $k = 1,2,\ldots$, and therefore A_k is compact on $C(\overline{\Omega})$. Moreover,

$$\begin{aligned} |A\varphi(x) - A_k\varphi(x)| &= \left| \int_{\overline{\Omega}} (1 - h(k|x-y|))K(x,y)\varphi(y)\mathrm{d}y \right| \\ &\le \|\varphi\|_{L^\infty(\overline{\Omega})} \int_{|x-y|\le 1/k} |K(x,y)|\mathrm{d}y \\ &\le M\|\varphi\|_{L^\infty(\overline{\Omega})} \int_{|x-y|\le 1/k} |x-y|^{\alpha-n}\mathrm{d}y \to 0 \end{aligned}$$

as $k \to \infty$ uniformly in $x \in \overline{\Omega}$. Thus, A is compact as the norm limit of compact operators. □

There is a very useful and quite general result for integral operators with weak singularity for both domains and surfaces in $\mathbb{R}^n$.

Theorem 34.10. *An integral operator with weak singularity transforms bounded functions into continuous functions.*

Proof. We give the proof for domains in $\mathbb{R}^n$. The proof for surfaces in $\mathbb{R}^n$ is the same. Let $x, y \in \Omega$ and $|x-y| < \delta$. Then

$$\begin{aligned}|Af(x) - Af(y)| &\le \int_{|x-z|<2\delta} (|K(x,z)| + |K(y,z)|)|f(z)|\mathrm{d}z \\ &+ \int_{\Omega \setminus \{|x-z|<2\delta\}} |K(x,z) - K(y,z)||f(z)|\mathrm{d}z \\ &\le M\|f\|_{L^\infty(\Omega)} \int_{|x-z|<2\delta} (|x-z|^{\alpha-n} + |y-z|^{\alpha-n})\mathrm{d}z \\ &+ \|f\|_{L^\infty(\Omega)} \int_{\Omega \setminus \{|x-z|<2\delta\}} |K(x,z) - K(y,z)|\mathrm{d}z =: I_1 + I_2.\end{aligned}$$

Since $|z-y| \le |x-z| + |x-y|$, we have

$$I_1 \le 2M\|f\|_{L^\infty(\Omega)} |\mathbb{S}^{n-1}| \int_0^{3\delta} r^{\alpha-1}\mathrm{d}r = 2M|\mathbb{S}^{n-1}| \, \|f\|_{L^\infty(\Omega)} \frac{(3\delta)^\alpha}{\alpha} \to 0$$

as $\delta \to 0$. On the other hand, for $|x-y| < \delta$ and $|x-z| \ge 2\delta$ we have that

$$|y-z| \ge |x-z| - |x-y| > 2\delta - \delta = \delta.$$

So the continuity of the kernel K outside of the diagonal implies that

$$K(x,z) - K(y,z) \to 0, \quad \delta \to 0,$$

uniformly in $z \in \Omega \setminus \{|x-z| < 2\delta\}$. Since Ω is bounded, we obtain that $I_2 \to 0$ as $\delta \to 0$. This completes the proof. □

Exercise 34.1. Prove that if A is as in Theorem 34.10, then $f(x) + Af(x) \in C(\overline{\Omega})$ for $f \in L^2(\Omega)$ implies $f \in C(\overline{\Omega})$.

We are now in a position to extend the solvability conditions (Fredholm alternative; see Theorem 28.16) to equations in Hilbert space with compact but not necessarily self-adjoint operators.

Theorem 34.11 (Fredholm alternative II). *Suppose $A : H \to H$ is compact. For all $\mu \in \mathbb{C}$ either the equations*

$$(I - \mu A)f = g, \quad (I - \overline{\mu}A^*)f' = g',$$

have the unique solutions f and f' for any given g and g' from H or the corresponding homogeneous equations

$$(I - \mu A)f = 0, \quad (I - \overline{\mu}A^*)f' = 0 \tag{34.10}$$

have nontrivial solutions such that

$$\dim N(I - \mu A) = \dim N(I - \overline{\mu} A^*) < \infty,$$

and in this case equations (34.10) *have solutions if and only if*

$$\begin{aligned} g \perp N(I - \overline{\mu} A^*) \quad &\Leftrightarrow \quad g \in R(I - \mu A), \\ g' \perp N(I - \mu A) \quad &\Leftrightarrow \quad g' \in R(I - \overline{\mu} A^*), \end{aligned}$$

respectively.

Proof. Riesz's lemma (see Theorem 28.14) and Exercise 26.7 give

$$\begin{aligned} R(I - \mu A) &= N(I - \overline{\mu} A^*)^{\perp}, \\ R(I - \overline{\mu} A^*) &= N(I - \mu A)^{\perp}. \end{aligned}$$

Let us first prove that one always has

$$\dim N(I - \mu A) = \dim N(I - \overline{\mu} A^*).$$

These two dimensions are finite due to Riesz (see Proposition 28.13). Since every compact operator is a norm limit of a sequence of operators of finite rank (see Chapter 28 for details), for every $\mu \in \mathbb{C}$, $\mu \neq 0$, we have

$$I - \mu A = -\mu A_0 + (I - \mu A_1),$$

where A_0 is of finite rank and $\|\mu A_1\| < 1$. Then $(I - \mu A_1)^{-1}$ exists and

$$(I - \mu A_1)^{-1}(I - \mu A) = I - \mu (I - \mu A_1)^{-1} A_0 =: I - A_2,$$

where A_2 is of finite rank too. Analogously, since $(I - \overline{\mu} A_1^*)^{-1}$ exists, we must have

$$(I - \overline{\mu} A^*)(I - \overline{\mu} A_1^*)^{-1} = I - \overline{\mu} A_0^* (I - \overline{\mu} A_1^*)^{-1} =: I - A_2^*,$$

where A_2^* is adjoint to A_2 and is of finite rank too. These representations allow us to conclude that

$$\begin{aligned} &g \in N(I - \mu A) \Leftrightarrow g \in N(I - A_2), \\ &g' \in N(I - \mu A_2^*) \Leftrightarrow (I - \overline{\mu} A_1^*)^{-1} g' \in N(I - \overline{\mu} A^*). \end{aligned}$$

Thus, it suffices to show that the numbers of independent solutions of the equations

$$g = A_2 g, \quad g' = A_2^* g'$$

are equal.

Since we know that the ranks of A_2 and A_2^* are finite, we may represent the mappings of the operators $I - A_2$ and $I - A_2^*$ as the mappings of matrices $I - M_2$ and $I - M_2^*$ with adjoint matrices M_2 and M_2^*. But the ranks of the adjoint matrices are equal, and therefore the numbers of independent solutions of the equations $g = A_2 g$ and $g' = A_2^* g'$ are equal.

The next step is the following: if $R(I - \mu A) = H$, then $N(I - \overline{\mu}A^*) = \{0\}$, and consequently $N(I - \mu A) = \{0\}$ and $R(I - \overline{\mu}A^*) = H$ (see Exercise 26.7). This means that both $(I - \mu A)^{-1}$ and $(I - \overline{\mu}A^*)^{-1}$ exist, and the unique solutions of (34.10) are given by

$$f = (I - \mu A)^{-1} g, \quad f' = (I - \overline{\mu}A^*)^{-1} g'.$$

If $N(I - \mu A)$ and $N(I - \overline{\mu}A^*)$ are not zero, then $R(I - \mu A)$ and $R(I - \overline{\mu}A^*)$ are proper subspaces of H, and equations (34.10) have solutions if and only if

$$g \in R(I - \mu A), \quad g' \in R(I - \overline{\mu}A^*).$$

This is equivalent (see Exercise 26.7) to

$$g \perp N(I - \overline{\mu}A^*), \quad g' \perp N(I - \mu A).$$

This completes the proof. □

We will now demonstrate this Fredholm alternative for integral operators. Let $\Omega \subset \mathbb{R}^n$ be a domain, and let

$$Af(x) = \int_\Omega K(x,y) f(y) dy$$

be a compact integral operator in $L^2(\Omega)$. Then its adjoint is defined as

$$A^* f(x) = \int_\Omega \overline{K(y,x)} f(y) dy.$$

Hence, the Fredholm alternative for these operators reads as follows: either the equations

$$\begin{aligned} f(x) - \mu \int_\Omega K(x,y) f(y) dy &= g(x), \\ f'(x) - \overline{\mu} \int_\Omega \overline{K(y,x)} f'(y) dy &= g'(x), \end{aligned} \tag{34.11}$$

are uniquely solvable for all g and g' from $L^2(\Omega)$ or the equations

$$\begin{aligned} f(x) &= \mu \int_\Omega K(x,y) f(y) dy, \\ f'(x) &= \overline{\mu} \int_\Omega \overline{K(y,x)} f'(y) dy, \end{aligned} \tag{34.12}$$

have the same (finite) number of linearly independent solutions. And in this case, equations (34.11) are solvable if and only if g and g' are orthogonal to every solution f and f' of the equations (34.12), respectively.

Definition 34.12. Equations (34.11) and (34.12) are called *integral equations of the second and first kinds*, respectively.

Exercise 34.2. Consider in $L^2(a,b)$ the integral equation

$$\varphi(x) - \int_a^b e^{x-y}\varphi(y)dy = f(x), \quad x \in [a,b],$$

where $f \in L^2(a,b)$. Solve this equation and formulate the Fredholm alternative for it.

Example 34.13. (Boundary value problems) Consider the second-order ordinary differential equation

$$a_0(x)u''(x) + a_1(x)u'(x) + a_2(x)u(x) = f(x)$$

on the interval $[0,1]$ with coefficients $f, a_2 \in L^2(0,1)$, $a_1 \in W_2^1(0,1)$ and with smooth $a_0(x) \geq c_0 > 0$ subject to the boundary conditions

$$u(0) = u_0, \quad u(1) = u_1.$$

Dividing this equation by $a_0(x)$, we may consider the boundary value problem in the form

$$u'' + a_1(x)u' + a_2(x)u = f, \quad u(0) = u_0, u(1) = u_1.$$

Using Green's function $G(x,y)$ of the form

$$G(x,y) = \begin{cases} y(1-x), & 0 \leq y \leq x \leq 1, \\ x(1-y), & 0 \leq x \leq y \leq 1, \end{cases}$$

we can rewrite this boundary value problem as

$$u(x) = \varphi_0(x) + \int_0^1 G(x,y)(a_1(y)u' + a_2(y)u - f(y))dy,$$

where $\varphi_0(x) = u_0(1-x) + u_1 x$. Integration by parts implies

$$\begin{aligned} u(x) = \varphi_0(x) &- \int_0^1 G(x,y)f(y)dy + G(x,y)a_1(y)u(y)|_0^1 \\ &- \int_0^1 \left[\partial_y G(x,y)a_1(y) + G(x,y)a_1'(y)\right] u(y)dy + \int_0^1 G(x,y)a_2(y)u(y)dy. \end{aligned}$$

Since $G(x,1) = G(x,0) = 0$, this equation can be rewritten as

$$u(x) = \widetilde{\varphi_0}(x) - \int_0^1 K(x,y)u(y)\mathrm{d}y,$$

where

$$\widetilde{\varphi_0}(x) = \varphi_0(x) - \int_0^1 G(x,y)f(y)\mathrm{d}y$$

and

$$K(x,y) = \partial_y G(x,y)a_1(y) + G(x,y)a_1'(y) - G(x,y)a_2(y).$$

Exercise 34.3. (1) Prove that $K(x,y)$ is a Hilbert–Schmidt kernel on $[0,1] \times [0,1]$.
(2) Prove that the boundary value problem and this integral equation of the second kind are equivalent.
(3) Formulate the solvability condition for the boundary value problem using the Fredholm alternative for this integral operator.

Chapter 35
Volterra and Singular Integral Equations

In this chapter we consider integral equations of special types on a finite interval $[a,b]$. We consider the Lebesgue space $L^\infty(a,b)$ and the Hölder space $C^\alpha(a,b)$ (which are not Hilbert spaces but normed spaces) instead of the Hilbert space L^2: The norms of the spaces $L^\infty(a,b)$ and $C^\alpha[a,b]$ are defined as follows:

$$\|f\|_{L^\infty(a,b)} = \inf\{M : |f(x)| \le M \text{ a.e. on } (a,b)\},$$

$$\|f\|_{C^\alpha[a,b]} = \|f\|_{L^\infty(a,b)} + \sup_{x,y\in[a,b]} \frac{|f(x)-f(y)|}{|x-y|^\alpha},$$

where $0 < \alpha \le 1$.

The fact that f belongs to the Hölder space $C^\alpha[a,b]$ is equivalent to the fact that $f \in L^\infty(a,b)$ and there is a constant $c_0 > 0$ such that for all h (sufficiently small),

$$|f(x+h) - f(x)| \le c_0|h|^\alpha,$$

where $x, x+h \in [a,b]$.

Definition 35.1. Integral equations in $L^\infty(a,b)$ of the form

$$f(x) = \int_a^x K(x,y)\varphi(y)\mathrm{d}y$$

and

$$\varphi(x) = f(x) + \int_a^x K(x,y)\varphi(y)\mathrm{d}y, \tag{35.1}$$

where $x \subset [a,b]$ and $\sup_{x,y\in[a,b]} |K(x,y)| < \infty$, are called *Volterra integral equations of the first and second kinds*, respectively.

V. Serov, *Fourier Series, Fourier Transform and Their Applications to Mathematical Physics*, Applied Mathematical Sciences 197,
DOI 10.1007/978-3-319-65262-7_35

Theorem 35.2. *For each $f \in L^\infty(a,b)$ the Volterra integral equation of the second kind has a unique solution $\varphi \in L^\infty(a,b)$ such that*

$$|\varphi(x)| \le \mathrm{e}^{M(x-a)} \|f\|_{L^\infty(a,b)} \tag{35.2}$$

for all $x \in [a,b]$ and

$$\|\varphi\|_{L^\infty(a,b)} \le \|f\|_{L^\infty(a,b)} \mathrm{e}^{M(b-a)}, \tag{35.3}$$

where $M = \sup_{x,y\in[a,b]} |K(x,y)|$.

Proof. We introduce the iterations of the equation (35.1) by

$$\varphi_{j+1}(x) := \int_a^x K(x,y)\varphi_j(y)\mathrm{d}y, \quad j = 0,1,2,\ldots,$$

with $\varphi_0 = f$. Let us prove by induction that

$$|\varphi_j(x)| \le \frac{(M(x-a))^j}{j!} \|f\|_{L^\infty(a,b)}, \quad j = 0,1,\ldots. \tag{35.4}$$

Indeed, this estimate clearly holds for $j = 0$. Assume that (35.4) has been proved for some $j \ge 0$. Then

$$\begin{aligned} |\varphi_{j+1}(x)| &\le \int_a^x |K(x,y)||\varphi_j(y)|\mathrm{d}y \le M \int_a^x \frac{(M(y-a))^j}{j!} \|f\|_{L^\infty(a,b)} \mathrm{d}y \\ &= M^{j+1} \|f\|_{L^\infty(a,b)} \int_a^x \frac{(y-a)^j}{j!}\mathrm{d}y = M^{j+1} \|f\|_{L^\infty(a,b)} \frac{(x-a)^{j+1}}{(j+1)!}. \end{aligned}$$

This proves (35.4).

Let us introduce the function

$$\varphi(x) := \sum_{j=0}^{\infty} \varphi_j(x). \tag{35.5}$$

Then from (35.4) we obtain for all $x \in [a,b]$ that

$$|\varphi(x)| \le \|f\|_{L^\infty(a,b)} \sum_{j=0}^{\infty} \frac{(M(x-a))^j}{j!} = \|f\|_{L^\infty(a,b)} \mathrm{e}^{M(x-a)}.$$

Thus, the function $\varphi(x)$ is well defined by the series (35.5), since this series is uniformly convergent with respect to $x \in [a,b]$.

It remains now to show that this $\varphi(x)$ solves (35.1). Since the series (35.5) converges uniformly, we may integrate it term by term and obtain

$$\int_a^x K(x,y)\varphi(y)\mathrm{d}y = \sum_{j=0}^{\infty} \int_a^x K(x,y)\varphi_j(y)\mathrm{d}y = \sum_{j=0}^{\infty} \varphi_{j+1}(x)$$
$$= \sum_{j=1}^{\infty} \varphi_j(x) + \varphi_0(x) - f(x) = \varphi(x) - f(x).$$

So (35.1) holds with this φ. The estimate (35.3) then follows immediately from (35.2). Finally, the uniqueness of this solution follows from (35.3) too. □

Corollary 35.3. *The homogeneous equation*

$$\varphi(x) = \int_a^x K(x,y)\varphi(y)\mathrm{d}y$$

has only the trivial solution in $L^\infty(a,b)$.

Proof. The result follows from (35.3).

In general, integral equations of the first kind are more delicate with respect to solvability than equations of the second kind. However, in some cases, Volterra integral equations of the first kind can be treated by reducing them to equations of the second kind. Indeed, consider for $x \in [a,b]$,

$$\int_a^x K(x,y)\varphi(y)\mathrm{d}y = f(x), \tag{35.6}$$

and assume that the derivatives $\frac{\partial K}{\partial x}(x,y)$ and $f'(x)$ exist and are bounded and that $K(x,x) \neq 0$ for all $x \in [a,b]$. Then, differentiating with respect to x reduces (35.6) to

$$\varphi(x)K(x,x) + \int_a^x \frac{\partial K}{\partial x}(x,y)\varphi(y)\mathrm{d}y = f'(x),$$

or

$$\varphi(x) = \frac{f'(x)}{K(x,x)} - \int_a^x \frac{\frac{\partial K}{\partial x}(x,y)}{K(x,x)}\varphi(y)\mathrm{d}y. \tag{35.7}$$

Exercise 35.1. Show that (35.6) and (35.7) are equivalent if $f(a) = 0$.

The second possibility occurs if we assume that

$$\frac{\partial K}{\partial y}(x,y)$$

exists and is bounded and that $K(x,x) \neq 0$ for all $x \in [a,b]$. In this case, setting

$$\psi(x) := \int_a^x \varphi(y)\mathrm{d}y, \quad \psi' = \varphi$$

and performing integration by parts in (35.6) yields

$$f(x) = \int_a^x K(x,y)\psi'(y)\mathrm{d}y = K(x,y)\psi(y)|_a^x - \int_a^x \frac{\partial K}{\partial y}(x,y)\psi(y)\mathrm{d}y$$

$$= K(x,x)\psi(x) - \int_a^x \frac{\partial K}{\partial y}(x,y)\psi(y)\mathrm{d}y,$$

or

$$\psi(x) = \frac{f(x)}{K(x,x)} + \int_a^x \frac{\frac{\partial K}{\partial y}(x,y)}{K(x,x)}\psi(y)\mathrm{d}y.$$

There is an interesting generalization of equation (35.1) when the kernel has weak singularities. More precisely, we consider (35.1) in the space $L^\infty(a,b)$ and assume that the kernel $K(x,y)$ satisfies the estimate

$$|K(x,y)| \le M|x-y|^{-\alpha}, \quad x,y \in [a,b],\, x \ne y,$$

with some $0 < \alpha < 1$. If we consider again the iterations

$$\varphi_j(x) := \int_a^x K(x,y)\varphi_{j-1}(y)\mathrm{d}y, \quad j = 1,2,\ldots,$$

with $\varphi_0 = f$, then it can be proved by induction that for all $x \in [a,b]$ we have

$$|\varphi_j(x)| \le \left(\frac{M(x-a)^{1-\alpha}}{1-\alpha}\right)^j \|f\|_{L^\infty(a,b)}, \quad j = 0,1,\ldots.$$

Indeed, since this clearly holds for $j = 0$, assume that it has been proved for some $j \ge 0$. Then

$$\begin{aligned}
|\varphi_{j+1}(x)| &\le \int_a^x |K(x,y)||\varphi_j(y)|\mathrm{d}y \\
&\le M\frac{M^j}{(1-\alpha)^j}\int_a^x |x-y|^{-\alpha}((y-a)^{1-\alpha})^j \|f\|_{L^\infty(a,b)}\mathrm{d}y \\
&\le \frac{M^{j+1}}{(1-\alpha)^j}((x-a)^{1-\alpha})^j \|f\|_{L^\infty(a,b)} \int_a^x (x-y)^{-\alpha}\mathrm{d}y \\
&\le \frac{M^{j+1}}{(1-\alpha)^j}((x-a)^{1-\alpha})^j \|f\|_{L^\infty(a,b)} \frac{(x-a)^{1-\alpha}}{1-\alpha} \\
&= \left(\frac{M(x-a)^{1-\alpha}}{1-\alpha}\right)^{j+1} \|f\|_{L^\infty(a,b)}.
\end{aligned}$$

If we assume now that

$$\frac{M(b-a)^{1-\alpha}}{1-\alpha} < 1,$$

then the series

$$\sum_{j=0}^{\infty} \varphi_j(x)$$

converges uniformly on the interval $[a,b]$, and the function φ defined by

$$\varphi(x) := \sum_{j=0}^{\infty} \varphi_j(x)$$

solves therefore the inhomogeneous integral equation (35.1). Moreover, the following estimates hold:

$$|\varphi(x)| \le \frac{\|f\|_{L^\infty(a,b)}}{1 - \frac{M(x-a)^{1-\alpha}}{1-\alpha}}, \quad x \in [a,b],$$

and

$$\|\varphi\|_{L^\infty(a,b)} \le \frac{\|f\|_{L^\infty(a,b)}}{1 - \frac{M(b-a)^{1-\alpha}}{1-\alpha}}.$$

Exercise 35.2. Show that the Volterra integral equation of the first kind

$$\varphi(x) = \lambda \int_a^x \mathrm{e}^{-(x-y)} \varphi(y) \mathrm{d}y$$

has, for all λ, only the trivial solution in $L^\infty(a,b)$.

Definition 35.4. Let $0 < \alpha < 1$, $\varphi \in C^\alpha[-a,a]$, and suppose that φ is periodic, i.e., $\varphi(-a) = \varphi(a)$. In this space an integral equation of the form

$$\varphi(x) = f(x) + \lambda \,\text{p.v.} \int_{-a}^{a} \frac{\varphi(x+y)\mathrm{d}y}{y}, \quad \lambda \in \mathbb{C}, \tag{35.8}$$

is understood in the sense that

$$\text{p.v.} \int_{-a}^{a} \frac{\varphi(x+y)\mathrm{d}y}{y} = \lim_{\varepsilon \to 0} \int_{|y| \ge \varepsilon, y \in [-a,a]} \frac{\varphi(x+y)\mathrm{d}y}{y} \tag{35.9}$$

and the function φ is extended periodically (with period $2a$) to the whole line.

Due to (35.9), we have that

$$\text{p.v.} \int_{-a}^{a} \frac{\mathrm{d}y}{y} = 0.$$

Thus

$$\text{p.v.}\int_{-a}^{a}\frac{\varphi(x+y)\mathrm{d}y}{y}=\text{p.v.}\int_{-a}^{a}\frac{\varphi(x+y)-\varphi(x)}{y}\mathrm{d}y=\int_{-a}^{a}\frac{\varphi(x+y)-\varphi(x)}{y}\mathrm{d}y,$$

and the latter integral can be understood in the usual sense for periodic $\varphi\in C^{\alpha}[-a,a]$, since

$$\left|\int_{-a}^{a}\frac{\varphi(x+y)-\varphi(x)}{y}\mathrm{d}y\right|\le c_0\int_{-a}^{a}\frac{|y|^{\alpha}}{|y|}\mathrm{d}y=2c_0\int_{0}^{a}\xi^{\alpha-1}\mathrm{d}\xi=2c_0\frac{a^{\alpha}}{\alpha}. \quad (35.10)$$

Inequality (35.10) shows us that for every $\varphi\in C^{\alpha}[-a,a]$ the integral in (35.8) is uniformly bounded and also periodic with period $2a$. But even more is true.

Proposition 35.5. *For every $2a$-periodic $\varphi\in C^{\alpha}[-a,a]$ with $0<\alpha<1$ the integral in* (35.8) *defines a $2a$-periodic function of x that belongs to the same Hölder space $C^{\alpha}[-a,a]$.*

Proof. Let us denote by $g(x)$ the integral in (35.8). For $|h|>0$ sufficiently small we have

$$\begin{aligned}
g(x+h)-g(x)&=\int_{-a}^{a}\frac{\varphi(x+h+y)-\varphi(x+h)}{y}\mathrm{d}y-\int_{-a}^{a}\frac{\varphi(x+y)-\varphi(x)}{y}\mathrm{d}y\\
&=\int_{|y|\le 3|h|}\frac{\varphi(x+h+y)-\varphi(x+h)}{y}\mathrm{d}y-\int_{|y|\le 3|h|}\frac{\varphi(x+y)-\varphi(x)}{y}\mathrm{d}y\\
&\quad+\int_{|y|\ge 3|h|}\frac{\varphi(x+h+y)-\varphi(x)}{y}\mathrm{d}y-\int_{|y|\ge 3|h|}\frac{\varphi(x+y)-\varphi(x)}{y}\mathrm{d}y\\
&=:I_1+I_2.
\end{aligned}$$

For the first integral I_1 we have

$$\begin{aligned}
|I_1|&\le\int_{|y|\le 3|h|}\frac{|\varphi(x+y+h)-\varphi(x+h)|}{|y|}\mathrm{d}y+\int_{|y|\le 3|h|}\frac{|\varphi(x+y)-\varphi(x)|}{|y|}\mathrm{d}y\\
&\le c_0\int_{|y|\le 3|h|}\frac{|y|^{\alpha}}{|y|}\mathrm{d}y+c_0\int_{|y|\le 3|h|}\frac{|y|^{\alpha}}{|y|}\mathrm{d}y\\
&\le 4c_0\int_{0}^{3|h|}\xi^{\alpha-1}\mathrm{d}\xi=4c_0\frac{(3|h|)^{\alpha}}{\alpha}=\frac{4c_0 3^{\alpha}}{\alpha}|h|^{\alpha}. \quad (35.11)
\end{aligned}$$

For the estimation of I_2 we first rewrite it as (we change variables in the first integral)

$$
\begin{aligned}
I_2 &= \int_{|z-h|\geq 3|h|} \frac{\varphi(z+x)-\varphi(x)}{z-h}\mathrm{d}z - \int_{|z|\geq 3|h|} \frac{\varphi(z+x)-\varphi(x)}{z}\mathrm{d}z \\
&= \int_{|z|\geq 3|h|} (\varphi(z+x)-\varphi(x)) \left[\frac{1}{z-h} - \frac{1}{z}\right] \mathrm{d}z \\
&- \int_{\{|z-h|\geq 3|h|\}\setminus\{|z|\geq 3|h|\}} \frac{\varphi(z+x)-\varphi(x)}{z-h}\mathrm{d}z.
\end{aligned}
$$

Then we have

$$
\begin{aligned}
|I_2| &\leq \int_{\substack{|z|\geq 3|h| \\ z\in[-a,a]}} \frac{|\varphi(z+x)-\varphi(x)||h|\mathrm{d}z}{|z|\cdot|z-h|} + \int_{2|h|\leq|z|\leq 3|h|} \frac{|\varphi(z+x)-\varphi(x)|}{|z-h|}\mathrm{d}z \\
&\leq c_0|h| \int_{a\geq|z|\geq 3|h|} \frac{|z|^\alpha}{|z|\cdot 2|z|/3}\mathrm{d}z + c_0 \int_{2|h|\leq|z|\leq 3|h|} \frac{|z|^\alpha}{|z|/2}\mathrm{d}z \\
&= 2\cdot\frac{3c_0}{2}|h| \int_{3|h|}^{a} \xi^{\alpha-2}\mathrm{d}\xi + 4c_0 \int_0^{3|h|} \xi^{\alpha-1}\mathrm{d}\xi \\
&= 3c_0|h| \left.\frac{\xi^{\alpha-1}}{\alpha-1}\right|_{3|h|}^{a} + 4c_0\frac{(3|h|)^\alpha}{\alpha} = 3c_0|h|\left(\frac{(3|h|)^{\alpha-1}}{1-\alpha} - \frac{a^{\alpha-1}}{1-\alpha}\right) + \frac{4c_0 3^\alpha}{\alpha}|h|^\alpha \\
&< \frac{3^\alpha c_0}{1-\alpha}|h|^\alpha + \frac{4c_0 3^\alpha}{\alpha}|h|^\alpha = c_0 3^\alpha\left(\frac{1}{1-\alpha} + \frac{4}{\alpha}\right)|h|^\alpha,
\end{aligned}
\tag{35.12}
$$

since $0 < \alpha < 1$. Estimates (35.11)–(35.12) show that this proposition is completely proved. □

If we denote by

$$
A\varphi(x) := \text{p. v.} \int_{-a}^{a} \frac{\varphi(x+y)\mathrm{d}y}{y} \tag{35.13}
$$

a periodic linear operator on $C^\alpha[-a,a]$, $0 < \alpha < 1$, then Proposition 35.5 gives that A is bounded in this space. But this operator is not compact there. Nevertheless, the following holds.

Corollary 35.6. *There is $\lambda_0 > 0$ such that for all $\lambda \in \mathbb{C}$, $|\lambda| < \lambda_0$, and periodic $f \in C^\alpha[-a,a]$, $0 < \alpha < 1$, the integral equation* (35.8) *has a unique solution in $C^\alpha[-a,a]$, $0 < \alpha < 1$.*

Proof. Since the operator A from (35.13) is a bounded linear operator in the space $C^\alpha[-a,a]$, it follows that

$$
\|A\|_{C^\alpha\to C^\alpha} \leq c_0
$$

with some constant $c_0 > 0$. If we choose now $\lambda_0 = 1/c_0$, then for all $\lambda \in \mathbb{C}$, $|\lambda| < \lambda_0$, the operator $I - \lambda A$ will be invertible in the space $C^\alpha[-a,a]$, since

$$
\|\lambda A\|_{C^\alpha\to C^\alpha} < 1.
$$

This fact implies that the integral equation (35.8) can be solved uniquely in this space, and the unique solution φ can be obtained as

$$\varphi = (I - \lambda A)^{-1} f.$$

This is equivalent to the fact that (35.8) can be solved by iterations. □

Chapter 36
Approximate Methods

In this chapter we will study approximate solution methods for equations in a Hilbert space H of the form

$$A\varphi = f, \quad (I-A)\varphi = f \tag{36.1}$$

with a bounded or compact operator A. The fundamental concept for solving equations (36.1) approximately is to replace them by the equations

$$A_n\varphi_n = f_n, \quad (I-A_n)\varphi_n = f_n, \tag{36.2}$$

respectively. For practical purposes, the approximating equations (36.2) will be chosen so that they can be reduced to a finite-dimensional linear system.

We will begin with some general results that are the basis of our considerations.

Theorem 36.1. *Let $A : H \to H$ be a bounded linear operator with bounded inverse A^{-1}. Assume that the sequence $A_n : H \to H$ of bounded linear operators is norm convergent to A, i.e.,*

$$\|A_n - A\| \to 0, \quad n \to \infty.$$

Then for all n such that

$$\left\|A^{-1}(A_n - A)\right\| < 1,$$

the inverse operators A_n^{-1} exist and

$$\left\|A_n^{-1}\right\| \le \frac{\left\|A^{-1}\right\|}{1 - \|A^{-1}(A_n - A)\|}.$$

Moreover, the solutions of (36.1) *and* (36.2) *satisfy the error estimate*

$$\|\varphi_n - \varphi\| \le \frac{\left\|A^{-1}\right\|}{1 - \|A^{-1}(A_n - A)\|}\Big(\|(A_n - A)\varphi\| + \|f_n - f\|\Big).$$

V. Serov, *Fourier Series, Fourier Transform and Their Applications to Mathematical Physics*, Applied Mathematical Sciences 197,
DOI 10.1007/978-3-319-65262-7_36

Proof. Since A^{-1} exists, we may write

$$A^{-1}A_n = I - A^{-1}(A - A_n).$$

Since

$$\left\|A^{-1}(A_n - A)\right\| < 1$$

for n sufficiently large, for these values of n we have that

$$\left(I - A^{-1}(A - A_n)\right)^{-1}$$

exists by the Neumann series. Thus,

$$\left(A^{-1}A_n\right)^{-1} = \left(I - A^{-1}(A - A_n)\right)^{-1},$$

or

$$A_n^{-1}A = \left(I - A^{-1}(A - A_n)\right)^{-1},$$

or

$$A_n^{-1} = \left(I - A^{-1}(A - A_n)\right)^{-1} A^{-1}.$$

The error estimate follows immediately from the representation

$$\varphi_n - \varphi = A_n^{-1}(A - A_n)\varphi + A_n^{-1}(f_n - f).$$

This completes the proof. □

Theorem 36.2. *Assume that $A_n^{-1} : H \to H$ exist for all $n \geq n_0$ and that their norms are uniformly bounded for such n. Let $\|A_n - A\| \to 0$ as $n \to \infty$. Then the inverse operator A^{-1} exists and*

$$\left\|A^{-1}\right\| \leq \frac{\left\|A_n^{-1}\right\|}{1 - \left\|A_n^{-1}(A_n - A)\right\|}$$

for all $n \geq n_0$ with $\left\|A_n^{-1}(A_n - A)\right\| < 1$.

Exercise 36.1. Prove Theorem 36.2 and obtain the error estimate in this case.

Definition 36.3. A sequence $\{A_n\}_{n=1}^{\infty}$ of compact operators in a Hilbert space H is said to be *collectively compact* if for every bounded set $U \subset H$, the image

$$J = \{A_n\varphi : \varphi \in U,\ n = 1, 2, \ldots\}$$

is *relatively compact*, i.e., if every sequence from J contains a convergent subsequence.

Exercise 36.2. Assume that a sequence of compact operators $\{A_n\}_{n=1}^{\infty}$ is collectively compact and converges pointwise to A in H, i.e.,

$$\lim_{n\to\infty} A_n\varphi = A\varphi, \quad \varphi \in H.$$

Prove that the limit operator A is compact.

Exercise 36.3. Under the same assumptions for $\{A_n\}_{n=1}^{\infty}$ as in Exercise 36.2, prove that

$$\|(A_n - A)A\| \to 0, \quad \|(A_n - A)A_n\| \to 0,$$

as $n \to \infty$.

Theorem 36.4. *Let $A : H \to H$ be a compact operator and let $I - A$ be injective. Assume that the sequence $A_n : H \to H$ is collectively compact and pointwise convergent to A. Then for all n such that*

$$\left\|(I-A)^{-1}(A_n - A)A_n\right\| < 1,$$

the inverse operators $(I - A_n)^{-1}$ exist and the solutions of (36.1) *and* (36.2) *satisfy the error estimate*

$$\|\varphi_n - \varphi\| \le \frac{1 + \left\|(I-A)^{-1}A_n\right\|}{1 - \|(I-A)^{-1}(A_n - A)A_n\|}\Big(\|(A_n - A)\varphi\| + \|f_n - f\|\Big).$$

Proof. By Riesz's theorem (see Theorem 28.15), the inverse operator $(I - A)^{-1}$ exists and is bounded. Due to Exercise 36.3,

$$\|(A_n - A)A_n\| \to 0, \quad n \to \infty.$$

Therefore, for n sufficiently large we have

$$\left\|(I-A)^{-1}(A_n - A)A_n\right\| < 1.$$

This fact allows us to conclude (as in Theorem 36.1) that $(I - A_n)^{-1}$ exists and

$$\left\|(I - A_n)^{-1}\right\| \le \frac{1 + \left\|(I-A)^{-1}A_n\right\|}{1 - \|(I-A)^{-1}(A_n - A)A_n\|}.$$

The error estimate follows from this inequality and the representation

$$\varphi_n - \varphi = (I - A_n)^{-1}\big((A_n - A)\varphi + f_n - f\big).$$

This yields the claim. □

Corollary 36.5. *Let A_n be as in Theorem 36.4. Assume that the inverse operators $(I-A_n)^{-1}$ exist and are uniformly bounded for all $n \geq n_0$. Then the inverse $(I-A)^{-1}$ exists if*

$$\left\|(I-A_n)^{-1}(A_n-A)A\right\| < 1.$$

The solutions of (36.1) *and* (36.2) *satisfy the error estimate*

$$\|\varphi_n-\varphi\| \leq \frac{1+\left\|(I-A_n)^{-1}A\right\|}{1-\|(I-A_n)^{-1}(A_n-A)A\|}\left(\|(A_n-A)\varphi\|+\|f_n-f\|\right).$$

Theorem 36.6. *Let $A: H \to H$ be a bounded linear operator with $\|A\| < 1$. Then the* successive approximations

$$\varphi_{n+1} := A\varphi_n + f, \quad n = 0, 1, \ldots, \tag{36.3}$$

converge for each $f \in H$ and each $\varphi_0 \in H$ to the unique solution of (36.1).

Proof. The condition $\|A\| < 1$ implies the existence and boundedness of the inverse operator $(I-A)^{-1}$ and the existence of the unique solution of (36.1) as

$$\varphi = (I-A)^{-1}f.$$

It remains only to show that the successive approximations converge to φ for all $\varphi_0 \in H$. The definition (36.3) implies

$$\|\varphi_{n+1}-\varphi_n\| \leq \|A\|\,\|\varphi_n-\varphi_{n-1}\| \leq \cdots \leq \|A\|^n\,\|\varphi_1-\varphi_0\|.$$

Hence for each $p \in \mathbb{N}$ we have

$$\begin{aligned}
\left\|\varphi_{n+p}-\varphi_n\right\| &\leq \left\|\varphi_{n+p}-\varphi_{n+p-1}\right\| + \cdots + \|\varphi_{n+1}-\varphi_n\| \\
&\leq \left(\|A\|^{n+p-1}+\|A\|^{n+p-2}+\cdots+\|A\|^n\right)\|\varphi_1-\varphi_0\| \\
&\leq \frac{\|A\|^n}{1-\|A\|}\|\varphi_1-\varphi_0\| \to 0
\end{aligned}$$

as $n \to \infty$ uniformly in $p \in \mathbb{N}$. This means that $\{\varphi_n\}$ is a Cauchy sequence in the Hilbert space H. Therefore, there exists a unique limit

$$\varphi = \lim_{n\to\infty} \varphi_n.$$

It is clear that this φ solves (36.1) uniquely. □

We will return to the integral operators

$$Af(x) = \int_\Omega K(x,y)f(y)\mathrm{d}y, \tag{36.4}$$

where $K(x,y)$ is assumed to be in $L^2(\Omega \times \Omega)$. In that case, as we know, A is compact in $L^2(\Omega)$.

Definition 36.7. A function $K_n(x,y) \in L^2(\Omega \times \Omega)$ is said to be a *degenerate kernel* if

$$K_n(x,y) = \sum_{j=1}^{n} a_j(x)b_j(y),$$

with some functions $a_j, b_j \in L^2(\Omega)$.

We consider integral equations of the second kind with a degenerate kernel $K_n(x,y)$, i.e.,

$$\varphi_n(x) - \int_{\Omega} \sum_{j=1}^{n} a_j(x)b_j(y)\varphi_n(y)\mathrm{d}y = f(x), \tag{36.5}$$

in the form

$$\varphi_n(x) - \sum_{j=1}^{n} \gamma_j a_j(x) = f(x),$$

where $\gamma_j = (\varphi_n, \overline{b_j})_{L^2(\Omega)}$. This means that the solution φ_n of (36.5) is necessarily represented as

$$\varphi_n(x) = f(x) + \sum_{j=1}^{n} \gamma_j a_j(x) \tag{36.6}$$

such that the coefficients γ_j (which are to be determined) satisfy the linear system

$$\gamma_j - \sum_{k=1}^{n} \gamma_k (a_k, \overline{b_j})_{L^2(\Omega)} = (f, \overline{b_j})_{L^2(\Omega)} = f_j, \quad j = 1, 2, \ldots, n. \tag{36.7}$$

Hence, the solution φ_n of (36.5) (see also (36.6)) can be obtained whenever we can solve the linear system (36.7) uniquely with respect to γ_j.

Let us consider now integral equations of the second kind with compact self-adjoint operator (36.4), i.e.,

$$\varphi(x) - A\varphi(x) = f(x). \tag{36.8}$$

The main idea is to approximate the kernel $K(x,y)$ from (36.8) by the degenerate kernel $K_n(x,y)$ from (36.5) such that

$$\|K(x,y) - K_n(x,y)\|_{L^2(\Omega\times\Omega)} \to 0 \tag{36.9}$$

as $n \to \infty$ and such that in addition, the inverse operators $(I - A_n)^{-1}$ exist and are uniformly bounded in n.

In that case the system (36.7) is uniquely solvable, and we obtain an approximate solution φ_n such that

$$\|\varphi - \varphi_n\|_{L^2(\Omega)} \to 0, \quad n \to \infty.$$

Indeed, equations (36.5) and (36.8) imply

$$(\varphi - \varphi_n) - A_n(\varphi - \varphi_n) = (A - A_n)\varphi.$$

Since $(I - A_n)^{-1}$ exist and are uniformly bounded, we have

$$\|\varphi - \varphi_n\| \le \left\|(I - A_n)^{-1}\right\| \|A - A_n\| \|\varphi\| \to 0$$

as $n \to \infty$ by (36.9). The unique solvability of (36.7) (or the uniqueness of φ_n) follows from the existence of the inverse operators $(I - A_n)^{-1}$.

We may justify this choice of the degenerate kernel $K_n(x,y)$ by the following considerations. Let $\{e_j\}_{j=1}^{\infty}$ be an orthonormal basis in $L^2(\Omega)$. Then $K(x,y) \in L^2(\Omega \times \Omega)$ as a function of $x \in \Omega$ (with parameter $y \in \Omega$) can be represented by

$$K(x,y) = \sum_{j=1}^{\infty} (K(\cdot,y), e_j)_{L^2} e_j(x).$$

Then

$$\left\| K(x,y) - \sum_{j=1}^{n} (K(\cdot,y), e_j)_{L^2} e_j \right\|_{L^2(\Omega\times\Omega)} \to 0$$

as $n \to \infty$, and we may consider the degenerate kernel $K_n(x,y)$ in the form

$$K_n(x,y) = \sum_{j=1}^{n} e_j(x) b_j(y),$$

where $b_j(y) = (K(\cdot,y), e_j)_{L^2}$. The system (36.7) transforms in this case to

$$\gamma_j - \sum_{k=1}^{n} \gamma_k (e_k, (e_j, K(\cdot,y))_{L^2})_{L^2(\Omega)} = f_j.$$

If, for example, e_j are the normalized eigenfunctions of the operator A with corresponding eigenvalues λ_j, then the latter system can be rewritten as

$$\gamma_j - \lambda_j \gamma_j = f_j, \quad j = 1, 2, \ldots, n.$$

We assume that $\lambda_j \ne 1$, so that γ_j can be uniquely determined as

$$\gamma_j = \frac{f_j}{1 - \lambda_j},$$

and therefore φ_n is equal to

$$\varphi(x) = f(x) + \sum_{j=1}^{n} \frac{f_j}{1-\lambda_j} e_j(x).$$

A different method goes back to Nyström. Let us consider instead of an integral operator A with kernel $K(x,y)$ a sequence of numerical integration operators

$$A_n\varphi(x) = \sum_{j=1}^{n} \alpha_j^{(n)} K(x, x_j^{(n)})\varphi(x_j^{(n)}).$$

We assume that the points $x_j^{(n)}$ and the weights $\alpha_j^{(n)}$ are chosen so that

$$\|A\varphi - A_n\varphi\|_{L^2}^2$$
$$= \int_\Omega \left| \int_\Omega K(x,y)\varphi(y) - \frac{1}{|\Omega|} \sum_{j=1}^{n} \alpha_j^{(n)} K(x, x_j^{(n)})\varphi(x_j^{(n)}) \mathrm{d}y \right|^2 \mathrm{d}x \to 0$$

as $n \to \infty$. The main problem here is to choose the weights $\alpha_j^{(n)}$ and the points $x_j^{(n)}$ with this approximation property. The original Nyström method was constructed for continuous kernels $K(x,y)$.

In Hilbert spaces it is more natural to consider projection methods.

Definition 36.8. Let $A : H \to H$ be an injective bounded linear operator. Let $P_n : H \to H_n$ be projection operators such that $\dim H_n = n$. For given $f \in H$, the *projection method* generated by H_n and P_n approximates the equation $A\varphi = f$ by the projection equation

$$P_n A \varphi_n = P_n f, \quad \varphi_n \in H. \tag{36.10}$$

This projection method is said to be convergent if there is $n_0 \in \mathbb{N}$ such that for each $f \in H$ the approximating equation (36.10) has a unique solution $\varphi_n \in H_n$ for all $n \geq n_0$ and

$$\varphi_n \to \varphi, \quad n \to \infty,$$

where φ is the unique solution of the equation $A\varphi = f$.

Theorem 36.9. *A projection method converges if and only if there exist $n_0 \in \mathbb{N}$ and $M > 0$ such that for all $n \geq n_0$ the operators*

$$P_n A : H \to H$$

are invertible and the operators $(P_n A)^{-1} P_n A : H \to H$ are uniformly bounded, i.e.,

$$\left\|(P_n A)^{-1} P_n A\right\| \leq M, \quad n \geq n_0.$$

In case of convergence we have the error estimate

$$\|\varphi_n - \varphi\| \le (1+M) \inf_{\psi \in H_n} \|\psi - \varphi\|.$$

Proof. If a projection method converges then, by definition, the P_nA are invertible, and the uniform boundedness follows from the Banach–Steinhaus theorem.

Conversely, under the assumptions of the theorem,

$$\varphi_n - \varphi = ((P_nA)^{-1}P_nA - I)\varphi.$$

Since for all $\psi \in H_n$ we have trivially $(P_nA)^{-1}P_nA\psi = \psi$, it follows that

$$\varphi_n - \varphi = ((P_nA)^{-1}P_nA - I)(\varphi - \psi),$$

and the error estimate follows. □

Remark 36.10. Projection methods make sense, and we can expect convergence only if the subspaces H_n possess the *denseness property*

$$\inf_{\psi \in H_n} \|\psi - \varphi\| \to 0, \quad n \to \infty.$$

Theorem 36.11. *Assume that $A : H \to H$ is compact, $I - A$ is injective, and the projection operators $P_n : H \to H_n$ converge pointwise, i.e., $P_n\varphi \to \varphi, n \to \infty$ for each $\varphi \in H$. Then the projection method for $I - A$ converges.*

Proof. By Riesz's theorem (see Theorem 28.15), the operator $I - A$ has a bounded inverse. Since $P_n\varphi \to \varphi$ as $n \to \infty$, we have $P_nA\varphi \to A\varphi$ as $n \to \infty$, too. At the same time, the sequence P_nA is collectively compact, since A is compact and P_n is of finite rank. Thus, due to Exercise 36.3 we have

$$\|(P_nA - A)P_nA\| \to 0, \quad n \to \infty. \tag{36.11}$$

Then the operators $(I - P_nA)^{-1}$ exist and are uniformly bounded. Indeed, writing

$$B_n := I + (I - A)^{-1}P_nA,$$

we obtain

$$\begin{aligned} B_n(I - P_nA) &= (I - P_nA) + (I - A)^{-1}P_nA(I - P_nA) \\ &= I - (I - A)^{-1}(P_nA - A)P_nA =: I - S_n. \end{aligned}$$

But it is easy to see from (36.11) that

$$\|S_n\| \to 0, \quad n \to \infty.$$

Hence, both $I-P_nA$ and B_n are injective. Since P_nA is compact, we have that $(I-P_nA)^{-1}$ is bounded. As a consequence of this fact we have that

$$(I-P_nA)^{-1}=(I-S_n)^{-1}B_n.$$

The definition of B_n implies

$$\|B_n\|\le 1+\left\|(I-A)^{-1}\right\|\|A\|.$$

Therefore $\left\|(I-P_nA)^{-1}\right\|$ is uniformly bounded in n. The exact equation $\varphi-A\varphi=f$ and (36.10) with operator $I-A$ lead to

$$(I-P_nA)(\varphi_n-\varphi)=P_nA\varphi-A\varphi+P_nf-f,$$

which implies also the error estimate

$$\|\varphi_n-\varphi\|\le\left\|(I-P_nA)^{-1}\right\|\left(\|P_nA\varphi-A\varphi\|+\|P_nf-f\|\right).$$

The proof is concluded. □

Corollary 36.12. *Under the assumptions of Theorem 36.11 and provided additionally that*

$$\|P_nA-A\|\longrightarrow 0,\quad n\to\infty,$$

the approximate equation (36.10) *with $I-A$ is uniquely solvable for each $f\in H$, and we have the error estimate*

$$\|\varphi_n-\varphi\|\le M\|P_n\varphi-\varphi\|,$$

where M is an upper bound for the norm $\left\|(I-P_nA)^{-1}\right\|$.

Proof. The existence of the inverse operators $(I-P_nA)^{-1}$ and their uniform boundedness follow from

$$I-P_nA=(I-A)-(P_nA-A)=(I-A)\left[I-(I-A)^{-1}(P_nA-A)\right],$$

$$(I-P_nA)^{-1}=\left[I-(I-A)^{-1}(P_nA-A)\right]^{-1}(I-A)^{-1},$$

and

$$\left\|(I-P_nA)^{-1}\right\|\le\frac{\left\|(I-A)^{-1}\right\|}{1-\|(I-A)^{-1}\|\,\|(P_nA-A)\|}.$$

From

$$(\varphi-\varphi_n)-P_nA(\varphi-\varphi_n)=\varphi-P_n\varphi$$

and
$$\|\varphi_n - \varphi\| \le \left\|(I - P_n A)^{-1}\right\| \|P_n\varphi - \varphi\|$$

we obtain the error estimate. □

Let us return to the projection equation (36.10). It can be rewritten equivalently as

$$(A\varphi_n - f, g) = 0, \quad g \in H_n. \tag{36.12}$$

Indeed, if $g \in H_n$, then $g = P_n g$, $P_n^* = P_n$, and hence

$$0 = (A\varphi_n - f, g) = (A\varphi_n - f, P_n g) = (P_n(A\varphi_n - f), g),$$

or

$$P_n(A\varphi_n - f) = 0,$$

since H_n is considered here to be a Hilbert space. This is the basis for the following *Galerkin projection method*.

Assume that $\{e_j\}_{j=1}^{\infty}$ is an orthonormal basis in a Hilbert space H. Considering

$$H_n = \overline{\text{span}(e_1, \ldots, e_n)},$$

we have for the solution φ_n of the projection equation (36.10) the representation

$$\varphi_n(x) = \sum_{j=1}^{n} \gamma_j e_j. \tag{36.13}$$

The task here is to find (if possible uniquely) the coefficients γ_j such that φ_n from (36.13) solves (36.10). Since (36.12) is equivalent to (36.10), we have from (36.13) that

$$(A\varphi_n, g) = (f, g), \quad g \in H_n,$$

or

$$(A\varphi_n, e_k) = (f, e_k) = f_k, \quad k = 1, 2, \ldots, n,$$

or

$$\sum_{j=1}^{n} \gamma_j (Ae_j, e_k) = f_k,$$

or

$$M\vec{\gamma} = \vec{f}, \tag{36.14}$$

where $\vec{\gamma} = (\gamma_1, \ldots, \gamma_n)$, $\vec{f} = (f_1, \ldots, f_n)$, and $M = \{a_{jk}\}_{n\times n}$ with $a_{jk} = (Ae_j, e_k)$. If the operator A is invertible, then the matrix M is invertible too, and $\vec{\gamma}$ can be obtained uniquely as

$$\vec{\gamma} = M^{-1}\vec{f}.$$

As a result of this consideration we obtain $\varphi_n(x)$ uniquely from (36.13). It remains only to check that this φ_n converges to the solution of the exact equation $A\varphi = f$. In order to verify this fact it is enough to apply Theorem 36.9.

We apply now this projection method to the equation (36.8) with compact operator A.

Theorem 36.13. *Let $A : H \to H$ be compact and let $I - A$ be injective. Then the Galerkin projection method converges.*

Proof. By Riesz's theorem, the operator $I - A$ has bounded inverse. Therefore, φ_n from (36.13) is uniquely defined with γ_j that satisfies equation (36.14) with matrix $M = \{a_{jk}\}_{n\times n}$, $a_{jk} = ((I-A)e_j, e_k)$. Since

$$\|P_n\varphi - \varphi\|_H^2 = \sum_{j=n+1}^{\infty} |(\varphi, e_j)|^2 \to 0, \quad n \to \infty,$$

we may apply Theorem 36.11 and conclude the proof of this theorem. □

Part IV
Partial Differential Equations

Chapter 37
Introduction

We consider the Euclidean space $\mathbb{R}^n$, $n \geq 1$, with elements $x = (x_1, \ldots, x_n)$. The Euclidean length of x is defined by

$$|x| = \sqrt{x_1^2 + \cdots + x_n^2}$$

and the standard inner product by

$$(x, y) = x_1 y_1 + \cdots + x_n y_n.$$

We use the Cauchy–Bunyakovsky–Schwarz inequality in $\mathbb{R}^n$:

$$|(x, y)| \leq |x| \cdot |y|.$$

Equality here occurs if and only if $x = \lambda y$ for some $\lambda \in \mathbb{R}$. By $B_R(x)$ and $S_R(x)$ we denote the ball and sphere of radius $R > 0$ with center x:

$$B_R(x) := \{y \in \mathbb{R}^n : |x - y| < R\}, \quad S_R(x) := \{y \in \mathbb{R}^n : |x - y| = R\}.$$

We say that $\Omega \subset \mathbb{R}^n$, $n \geq 2$, is an open set if for every $x \in \Omega$ there is $R > 0$ such that $B_R(x) \subset \Omega$. If $n = 1$, by an open set we mean an open interval (a, b), $a < b$.

We say that $\Omega \subset \mathbb{R}^n$, $n \geq 2$, is a closed set if $\mathbb{R}^n \setminus \Omega$ is open. This is equivalent to the fact that $\Omega' \subset \Omega$, where Ω' denotes the set of limit points of Ω, i.e.,

$$\Omega' = \{y \in \mathbb{R}^n : \exists \{x^{(k)}\}_{k=1}^{\infty} \subset \Omega, |x^{(k)} - y| \to 0, k \to \infty\}.$$

The closure $\overline{\Omega}$ of the set Ω is defined as $\overline{\Omega} = \Omega \cup \Omega'$. We say that Ω is bounded if there is $R > 0$ such that $\Omega \subset B_R(0)$. A closed and bounded set in $\mathbb{R}^n$ is called *compact*. The boundary $\partial\Omega$ of the set $\Omega \subset \mathbb{R}^n$ is defined as

V. Serov, *Fourier Series, Fourier Transform and Their Applications to Mathematical Physics*, Applied Mathematical Sciences 197,
DOI 10.1007/978-3-319-65262-7_37

$$\partial\Omega = \overline{\Omega} \cap \left(\overline{\mathbb{R}^n \setminus \Omega}\right).$$

An n-tuple $\alpha = (\alpha_1, \ldots, \alpha_n)$ of nonnegative integers will be called a *multi-index*. We define

(1) $|\alpha| = \sum_{j=1}^{n} \alpha_j$.
(2) $\alpha + \beta = (\alpha_1 + \beta_1, \ldots, \alpha_n + \beta_n)$ with $|\alpha + \beta| = |\alpha| + |\beta|$.
(3) $\alpha! = \alpha_1! \cdots \alpha_n!$ with $0! = 1$.
(4) $\alpha \geq \beta$ if and only if $\alpha_j \geq \beta_j$ for each $j = 1, 2, \ldots, n$. Moreover, $\alpha > \beta$ if and only if $\alpha \geq \beta$ and there exists j_0 such that $\alpha_{j_0} > \beta_{j_0}$.
(5) If $\alpha \geq \beta$, then $\alpha - \beta = (\alpha_1 - \beta_1, \ldots, \alpha_n - \beta_n)$ and $|\alpha - \beta| = |\alpha| - |\beta|$.
(6) For $x \in \mathbb{R}^n$ we define

$$x^\alpha = x_1^{\alpha_1} \cdots x_n^{\alpha_n}$$

with $0^0 = 1$.

We will use the shorthand notation

$$\partial_j = \frac{\partial}{\partial x_j}, \quad \partial^\alpha = \partial_1^{\alpha_1} \cdots \partial_n^{\alpha_n} \equiv \frac{\partial^{|\alpha|}}{\partial x_1^{\alpha_1} \cdots \partial x_n^{\alpha_n}}.$$

This part assumes that the reader is familiar also with the following concepts:

(1) The Lebesgue integral in a bounded domain $\Omega \subset \mathbb{R}^n$ and in $\mathbb{R}^n$.
(2) The Banach spaces ($L^p, 1 \leq p \leq \infty$, C^k) and Hilbert spaces (L^2). If $1 \leq p < \infty$, then we set

$$L^p(\Omega) := \{f : \Omega \to \mathbb{C} \text{ measurable} : \|f\|_{L^p(\Omega)} := \left(\int_\Omega |f(x)|^p \mathrm{d}x\right)^{1/p} < \infty\},$$

while

$$L^\infty(\Omega) := \{f : \Omega \to \mathbb{C} \text{ measurable} : \|f\|_{L^\infty(\Omega)} := \operatorname*{ess\,sup}_{x \in \Omega} |f(x)| < \infty\}.$$

Moreover,

$$C^k(\overline{\Omega}) := \{f : \overline{\Omega} \to \mathbb{C} : \|f\|_{C^k(\overline{\Omega})} := \max_{x \in \overline{\Omega}} \sum_{|\alpha| \leq k} |\partial^\alpha f(x)| < \infty\},$$

where $\overline{\Omega}$ is the closure of Ω. We say that $f \in C^\infty(\Omega)$ if $f \in C^k(\overline{\Omega_1})$ for all $k \in \mathbb{N}$ and for all bounded subsets $\Omega_1 \subset \Omega$. The space $C^\infty(\Omega)$ is not a normed space. The inner product in $L^2(\Omega)$ is denoted by

$$(f, g)_{L^2(\Omega)} = \int_\Omega f(x)\overline{g(x)}\mathrm{d}x.$$

Also in $L^2(\Omega)$, the duality pairing is given by

$$\langle f, g\rangle_{L^2(\Omega)} = \int_\Omega f(x)g(x)\mathrm{d}x.$$

(3) Hölder's inequality: Let $1 \le p \le \infty, u \in L^p$ and $v \in L^{p'}$ with

$$\frac{1}{p} + \frac{1}{p'} = 1.$$

Then $uv \in L^1$ and

$$\int_\Omega |u(x)v(x)|\mathrm{d}x \le \left(\int_\Omega |u(x)|^p \mathrm{d}x\right)^{\frac{1}{p}} \left(\int_\Omega |v(x)|^{p'} \mathrm{d}x\right)^{\frac{1}{p'}},$$

where the Hölder conjugate exponent p' of p is obtained via

$$p' = \frac{p}{p-1},$$

with the understanding that $p' = \infty$ if $p = 1$ and $p' = 1$ if $p = \infty$.

(4) Lebesgue's dominated convergence theorem:
Let $A \subset \mathbb{R}^n$ be measurable and let $\{f_k\}_{k=1}^\infty$ be a sequence of measurable functions converging to $f(x)$ pointwise in A. If there exists a function $g \in L^1(A)$ such that $|f_k(x)| \le g(x)$ in A, then $f \in L^1(A)$ and

$$\lim_{k\to\infty} \int_A f_k(x)\mathrm{d}x = \int_A f(x)\mathrm{d}x.$$

(5) Fubini's theorem on the interchange of the order of integration:

$$\int_{X\times Y} f(x,y)\mathrm{d}x\mathrm{d}y = \int_X \mathrm{d}x \left(\int_Y f(x,y)\mathrm{d}y\right) = \int_Y \mathrm{d}y \left(\int_X f(x,y)\mathrm{d}x\right),$$

if $f \in L^1(X \times Y)$.

Exercise 37.1. Prove the generalized *Leibniz formula*

$$\partial^\alpha(fg) = \sum_{\beta\le\alpha} C_\alpha^\beta \partial^\beta f \partial^{\alpha-\beta} g,$$

where the generalized binomial coefficients are defined as

$$C_\alpha^\beta = \frac{\alpha!}{\beta!(\alpha-\beta)!} = C_\alpha^{\alpha-\beta}.$$

Hypersurfaces

A set $S \subset \mathbb{R}^n$ is called a *hypersurface* of class C^k, $k = 1,2,\ldots,\infty$, if for every $x_0 \in S$ there exist an open set $V \subset \mathbb{R}^n$ containing x_0 and a real-valued function $\varphi \in C^k(V)$ such that

$$\nabla\varphi \equiv (\partial_1\varphi, \ldots \partial_n\varphi) \neq 0 \quad \text{on} \quad S \cap V,$$

$$S \cap V = \{x \in V : \varphi(x) = 0\}.$$

By the implicit function theorem we can solve the equation $\varphi(x) = 0$ near x_0 to obtain

$$x_n = \psi(x_1, \ldots, x_{n-1})$$

for some C^k function ψ. A neighborhood of x_0 in S can then be mapped to a piece of the *hyperplane* $\widetilde{x_n} = 0$ by

$$x \mapsto (x', x_n - \psi(x')),$$

where $x' = (x_1, \ldots, x_{n-1})$. The vector $\nabla\varphi$ is perpendicular to S at $x \in S \cap V$. The vector $\nu(x)$, which is defined as

$$\nu(x) := \pm\frac{\nabla\varphi}{|\nabla\varphi|},$$

is called the *normal* to S at x. It can be proved that

$$\nu(x) = \pm\frac{(\nabla\psi, -1)}{\sqrt{|\nabla\psi|^2 + 1}}.$$

If S is the boundary of a domain $\Omega \subset \mathbb{R}^n$, $n \geq 2$, we always choose the orientation so that $\nu(x)$ *points out* of Ω, and we define the normal derivative of u on S by

$$\partial_\nu u := \nu \cdot \nabla u \equiv \nu_1\frac{\partial u}{\partial x_1} + \cdots + \nu_n\frac{\partial u}{\partial x_n}.$$

Thus ν and $\partial_\nu u$ are C^{k-1} functions.

Example 37.1. Let $S_r(y) = \{x \in \mathbb{R}^n : |x - y| = r\}$. Then

$$\nu(x) = \frac{x - y}{r} \quad \text{and} \quad \partial_\nu = \frac{1}{r}\sum_{j=1}^{n}(x_j - y_j)\frac{\partial}{\partial x_j} = \frac{\partial}{\partial r}.$$

The divergence theorem

Let $\Omega \subset \mathbb{R}^n$ be a bounded domain with C^1 boundary $S = \partial\Omega$ and let F be a C^1 vector field on $\overline{\Omega}$. Then

$$\int_\Omega \nabla \cdot F \mathrm{d}x = \int_S F \cdot \nu \mathrm{d}\sigma(x).$$

Corollary 37.2 (Integration by parts). *Let f and g be C^1 functions on $\overline{\Omega}$. Then*

$$\int_\Omega \partial_j f \cdot g \mathrm{d}x = -\int_\Omega f \cdot \partial_j g \mathrm{d}x + \int_S f \cdot g \nu_j \mathrm{d}\sigma(x).$$

Let f and g be locally integrable functions on $\mathbb{R}^n$, i.e., integrable on every bounded set from $\mathbb{R}^n$. The *convolution* $f * g$ of f and g is defined by

$$(f * g)(x) = \int_{\mathbb{R}^n} f(x-y)g(y)\mathrm{d}y = (g * f)(x),$$

provided that the integral in question exists. The basic theorem on the existence of convolutions is the following (*Young's inequality for convolution*):

Proposition 37.3 (Young's inequality). *Let $f \in L^1(\mathbb{R}^n)$ and $g \in L^p(\mathbb{R}^n)$, $1 \le p \le \infty$. Then $f * g \in L^p(\mathbb{R}^n)$ and*

$$\|f * g\|_{L^p} \le \|f\|_{L^1} \|g\|_{L^p}.$$

Proof. Let $p = \infty$. Then

$$|(f * g)(x)| \le \int_{\mathbb{R}^n} |f(x-y)||g(y)|\mathrm{d}y \le \|g\|_{L^\infty} \int_{\mathbb{R}^n} |f(x-y)|\mathrm{d}y = \|g\|_{L^\infty} \|f\|_{L^1}.$$

Now let $1 \le p < \infty$. Then it follows from Hölder's inequality and Fubini's theorem that

$$\begin{aligned}
\int_{\mathbb{R}^n} |(f * g)(x)|^p \mathrm{d}x &\le \int_{\mathbb{R}^n} \left(\int_{\mathbb{R}^n} |f(x-y)||g(y)|\mathrm{d}y \right)^p \mathrm{d}x \\
&\le \int_{\mathbb{R}^n} \left(\int_{\mathbb{R}^n} |f(x-y)|\mathrm{d}y \right)^{p/p'} \int_{\mathbb{R}^n} |f(x-y)||g(y)|^p \mathrm{d}y\mathrm{d}x \\
&\le \|f\|_{L^1}^{p/p'} \int_{\mathbb{R}^n} \int_{\mathbb{R}^n} |f(x-y)||g(y)|^p \mathrm{d}y\mathrm{d}x \\
&\le \|f\|_{L^1}^{p/p'} \int_{\mathbb{R}^n} |g(y)|^p \mathrm{d}y \int_{\mathbb{R}^n} |f(x-y)|\mathrm{d}x \\
&= \|f\|_{L^1}^{p/p'} \|g\|_{L^p}^p \|f\|_{L^1} = \|f\|_{L^1}^{p/p'+1} \|g\|_{L^p}^p.
\end{aligned}$$

Thus, we have finally

$$\|f*g\|_{L^p} \le \|f\|_{L^1}^{1/p'+1/p} \|g\|_{L^p} = \|f\|_{L^1} \|g\|_{L^p},$$

and the proof is complete. $\square$

Exercise 37.2. Suppose $1 \le p,q,r \le \infty$ and $\frac{1}{p}+\frac{1}{q}=\frac{1}{r}+1$. Prove that if $f \in L^p(\mathbb{R}^n)$ and $g \in L^q(\mathbb{R}^n)$, then $f*g \in L^r(\mathbb{R}^n)$ and

$$\|f*g\|_r \le \|f\|_p \|g\|_q.$$

In particular,

$$\|f*g\|_{L^\infty} \le \|f\|_{L^p} \|g\|_{L^{p'}}.$$

Definition 37.4. Let $u \in L^1(\mathbb{R}^n)$ with

$$\int_{\mathbb{R}^n} u(x)\mathrm{d}x = 1.$$

Then $u_\varepsilon(x) := \varepsilon^{-n}u(x/\varepsilon)$, $\varepsilon > 0$, is called an *approximation to the identity*.

Proposition 37.5. *Let $u_\varepsilon(x)$ be an approximation to the identity. Then for every function $\varphi \in L^\infty(\mathbb{R}^n)$ that is continuous at $\{0\}$ we have*

$$\lim_{\varepsilon\to 0+} \int_{\mathbb{R}^n} u_\varepsilon(x)\varphi(x)\mathrm{d}x = \varphi(0).$$

Proof. Since $u_\varepsilon(x)$ is an approximation to the identity, we have

$$\int_{\mathbb{R}^n} u_\varepsilon(x)\varphi(x)\mathrm{d}x - \varphi(0) = \int_{\mathbb{R}^n} u_\varepsilon(x)(\varphi(x)-\varphi(0))\mathrm{d}x,$$

and thus

$$\begin{aligned}\left|\int_{\mathbb{R}^n} u_\varepsilon(x)\varphi(x)\mathrm{d}x - \varphi(0)\right| &\le \int_{|x|\le\sqrt{\varepsilon}} |u_\varepsilon(x)||\varphi(x)-\varphi(0)|\mathrm{d}x \\ &\quad + \int_{|x|>\sqrt{\varepsilon}} |u_\varepsilon(x)||\varphi(x)-\varphi(0)|\mathrm{d}x \\ \le \sup_{|x|\le\sqrt{\varepsilon}} |\varphi(x)-\varphi(0)| &\int_{\mathbb{R}^n} |u_\varepsilon(x)|\mathrm{d}x + 2\|\varphi\|_{L^\infty} \int_{|x|>\sqrt{\varepsilon}} |u_\varepsilon(x)|\mathrm{d}x \\ \le \sup_{|x|\le\sqrt{\varepsilon}} |\varphi(x)-\varphi(0)| &\cdot \|u\|_{L^1} + 2\|\varphi\|_{L^\infty} \int_{|y|>1/\sqrt{\varepsilon}} |u(y)|\mathrm{d}y \to 0\end{aligned}$$

as $\varepsilon \to 0$.

$\square$

Example 37.6. Let $u(x)$ be defined as

$$u(x) = \begin{cases} \frac{\sin x_1}{2} \cdots \frac{\sin x_n}{2}, & x \in [0,\pi]^n, \\ 0, & x \notin [0,\pi]^n. \end{cases}$$

Then $u_\varepsilon(x)$ is an approximation to the identity and

$$\lim_{\varepsilon \to 0} (2\varepsilon)^{-n} \int_0^{\varepsilon\pi} \cdots \int_0^{\varepsilon\pi} \prod_{j=1}^n \sin \frac{x_j}{\varepsilon} \varphi(x) \mathrm{d}x = \varphi(0).$$

Fourier transform

If $f \in L^1(\mathbb{R}^n)$, its *Fourier transform* $\widehat{f}$ or $\mathscr{F}(f)$ is the bounded function on $\mathbb{R}^n$ defined by

$$\widehat{f}(\xi) = (2\pi)^{-n/2} \int_{\mathbb{R}^n} \mathrm{e}^{-\mathrm{i}x\cdot\xi} f(x) \mathrm{d}x.$$

Clearly $\widehat{f}(\xi)$ is well defined for all ξ and $\left\|\widehat{f}\right\|_\infty \le (2\pi)^{-n/2} \|f\|_1$.

The Riemann–Lebesgue lemma

If $f \in L^1(\mathbb{R}^n)$, then $\widehat{f}$ is continuous and tends to zero at infinity.

Proof. Let us first prove that $\mathscr{F}f(\xi)$ is continuous (even uniformly continuous) in $\mathbb{R}^n$. Indeed,

$$\begin{aligned} |\mathscr{F}f(\xi+h) - \mathscr{F}f(\xi)| &\le (2\pi)^{-n/2} \int_{\mathbb{R}^n} |f(x)| \cdot |\mathrm{e}^{-\mathrm{i}(x,h)} - 1| \mathrm{d}x \\ &\le \int_{|x||h| \le \sqrt{|h|}} |f(x)||x||h| \mathrm{d}x + 2 \int_{|x||h| > \sqrt{|h|}} |f(x)| \mathrm{d}x \\ &\le \sqrt{|h|}\, \|f\|_{L^1} + 2 \int_{|x| > 1/\sqrt{|h|}} |f(x)| \mathrm{d}x \to 0 \end{aligned}$$

as $|h| \to 0$, since $f \in L^1(\mathbb{R}^n)$.

To prove that $\mathscr{F}f(\xi) \to 0$ as $|\xi| \to 0$ we proceed as follows. Since $\mathrm{e}^{\mathrm{i}\pi} = -1$, we have

$$\begin{aligned} 2\mathscr{F}f(\xi) &= (2\pi)^{-n/2} \int_{\mathbb{R}^n} f(x) \mathrm{e}^{-\mathrm{i}(x,\xi)} \mathrm{d}x - (2\pi)^{-n/2} \int_{\mathbb{R}^n} f(x) \mathrm{e}^{-\mathrm{i}(x-\pi\xi/|\xi|^2,\xi)} \mathrm{d}x \\ &= (2\pi)^{-n/2} \int_{\mathbb{R}^n} f(x) \mathrm{e}^{-\mathrm{i}(x,\xi)} \mathrm{d}x - (2\pi)^{-n/2} \int_{\mathbb{R}^n} f(y + \pi\xi/|\xi|^2) \mathrm{e}^{-\mathrm{i}(y,\xi)} \mathrm{d}y \\ &= -(2\pi)^{-n/2} \int_{\mathbb{R}^n} (f(x+\pi\xi/|\xi|^2) - f(x)) \mathrm{e}^{-\mathrm{i}(x,\xi)} \mathrm{d}x. \end{aligned}$$

Hence

$$\begin{aligned} 2|\mathscr{F}f(\xi)| &\le (2\pi)^{-n/2}\int_{\mathbb{R}^n}|f(x+\pi\xi/|\xi|^2)-f(x)|\mathrm{d}x \\ &= (2\pi)^{-n/2}\left\|f(\cdot+\pi\xi/|\xi|^2)-f(\cdot)\right\|_{L^1}\to 0 \end{aligned}$$

as $|\xi|\to\infty$, since $f\in L^1(\mathbb{R}^n)$. □

Exercise 37.3. Prove that if $f,g\in L^1(\mathbb{R}^n)$, then $\widehat{f*g}=(2\pi)^{n/2}\widehat{f}\widehat{g}$.

Exercise 37.4. Suppose $f\in L^1(\mathbb{R}^n)$. Prove the following:

(1) If $f_h(x)=f(x+h)$, then $\widehat{f_h}=\mathrm{e}^{ih\cdot\xi}\widehat{f}$.
(2) If $T:\mathbb{R}^n\to\mathbb{R}^n$ is linear and invertible, then $\widehat{f\circ T}=|\det T|^{-1}\widehat{f}\left((T^{-1})'\xi\right)$, where T' is the adjoint matrix.
(3) If T is a rotation, that is, $T'=T^{-1}$ (and $|\det T|=1$), then $\widehat{f\circ T}=\widehat{f}\circ T$.

Exercise 37.5. Prove that

$$\partial^\alpha\widehat{f}=\widehat{(-\mathrm{i}x)^\alpha f},\quad \widehat{\partial^\alpha f}=(\mathrm{i}\xi)^\alpha\widehat{f}.$$

Exercise 37.6. Prove that if $f,g\in L^1(\mathbb{R}^n)$, then

$$\int_{\mathbb{R}^n}f(\xi)\widehat{g}(\xi)\mathrm{d}\xi=\int_{\mathbb{R}^n}\widehat{f}(\xi)g(\xi)\mathrm{d}\xi.$$

For $f\in L^1(\mathbb{R}^n)$ we define the inverse Fourier transform of f by

$$\mathscr{F}^{-1}f(x)=(2\pi)^{-n/2}\int_{\mathbb{R}^n}\mathrm{e}^{\mathrm{i}x\cdot\xi}f(\xi)\mathrm{d}\xi.$$

It is clear that

$$\mathscr{F}^{-1}f(x)=\mathscr{F}f(-x),\quad \mathscr{F}^{-1}f=\mathscr{F}(\overline{f}),$$

and for $f,g\in L^1(\mathbb{R}^n)$,

$$(\mathscr{F}f,g)_{L^2}=(f,\mathscr{F}^{-1}g)_{L^2}.$$

The *Schwartz space* $S(\mathbb{R}^n)$ is defined as

$$S(\mathbb{R}^n)=\left\{f\in C^\infty(\mathbb{R}^n):\sup_{x\in\mathbb{R}^n}|x^\alpha\partial^\beta f(x)|<\infty,\text{ for all multi-indices }\alpha\text{ and }\beta\right\}.$$

The Fourier inversion formula

If $f\in S(\mathbb{R}^n)$, then $(\mathscr{F}^{-1}\mathscr{F})f=f$.

Exercise 37.7. Prove the Fourier inversion formula for $f\in S(\mathbb{R}^n)$.

Plancherel's theorem

The Fourier transform on S extends uniquely to a unitary isomorphism of $L^2(\mathbb{R}^n)$ onto itself, i.e.,

$$\left\|\widehat{f}\right\|_2 = \|f\|_2.$$

This formula is called *Parseval's equality*.

The *support* of a function $f : \mathbb{R}^n \to \mathbb{C}$, denoted by $\operatorname{supp} f$, is the set

$$\operatorname{supp} f = \overline{\{x \in \mathbb{R}^n : f(x) \neq 0\}}.$$

Exercise 37.8. Prove that if $f \in L^1(\mathbb{R}^n)$ has compact support, then $\widehat{f}$ extends to an entire holomorphic function on $\mathbb{C}^n$.

Exercise 37.9. Prove that if $f \in C_0^\infty(\mathbb{R}^n)$, i.e., $f \in C^\infty(\mathbb{R}^n)$ with compact support, is supported in $\{x \in \mathbb{R}^n : |x| \leq R\}$, then for every multi-index α we have

$$|(\mathrm{i}\xi)^\alpha \hat{f}(\xi)| \leq (2\pi)^{-n/2} \mathrm{e}^{R|\operatorname{Im}\xi|} \|\partial^\alpha f\|_1,$$

that is, $\widehat{f}(\xi)$ is decays rapidly $|\operatorname{Re}\xi| \to \infty$ when $|\operatorname{Im}\xi|$ remains bounded.

Distributions

We say that $\varphi_j \to \varphi$ in $C_0^\infty(\Omega)$, $\Omega \subset \mathbb{R}^n$ open, if φ_j are all supported in a common compact set $K \subset \Omega$ and

$$\sup_{x \in K} |\partial^\alpha \varphi_j(x) - \partial^\alpha \varphi(x)| \to 0, \quad j \to \infty$$

for all α. A *distribution* on Ω is a linear functional u on $C_0^\infty(\Omega)$ that is continuous, i.e.,

(1) $u : C_0^\infty(\Omega) \to \mathbb{C}$. The action of u on $\varphi \in C_0^\infty(\Omega)$ is denoted by $\langle u, \varphi \rangle$. The set of all distributions is denoted by $\mathscr{D}'(\Omega)$.
(2) $\langle u, c_1\varphi_1 + c_2\varphi_2 \rangle = c_1\langle u, \varphi_1 \rangle + c_2\langle u, \varphi_2 \rangle$.
(3) If $\varphi_j \to \varphi$ in $C_0^\infty(\Omega)$, then $\langle u, \varphi_j \rangle \to \langle u, \varphi \rangle$ in $\mathbb{C}$ as $j \to \infty$. This is equivalent to the following condition: for all $K \subset \Omega$ there exist a constant C_K and an integer N_K such that for all $\varphi \in C_0^\infty(K)$,

$$|\langle u, \varphi \rangle| \leq C_K \sum_{|\alpha| \leq N_K} \|\partial^\alpha \varphi\|_\infty.$$

Remark 37.7. If $u \in L^1_{\mathrm{loc}}(\Omega)$, $\Omega \subset \mathbb{R}^n$ open, then u can be regarded as a distribution (in that case, a *regular distribution*) as follows:

$$\langle u, \varphi \rangle := \int_\Omega u(x)\varphi(x)\mathrm{d}x, \quad \varphi \in C_0^\infty(\Omega).$$

The Dirac δ-function

The Dirac δ-function is defined as

$$\langle \delta, \varphi \rangle = \varphi(0), \quad \varphi \in C_0^\infty(\Omega).$$

It is not a regular distribution.

Example 37.8. Let $u_\varepsilon(x)$ be an approximation to the identity. Then

$$\widehat{u_\varepsilon}(\xi) = (2\pi)^{-n/2} \int_{\mathbb{R}^n} \varepsilon^{-n} u(\frac{x}{\varepsilon}) \mathrm{e}^{-\mathrm{i}(x,\xi)}\mathrm{d}x = (2\pi)^{-n/2} \int_{\mathbb{R}^n} u(y)\mathrm{e}^{-\mathrm{i}(y,\varepsilon\xi)}\mathrm{d}y = \widehat{u}(\varepsilon\xi).$$

In particular,

$$\lim_{\varepsilon\to 0+} \widehat{u_\varepsilon}(\xi) = \lim_{\varepsilon\to 0+} \widehat{u}(\varepsilon\xi) = (2\pi)^{-n/2}.$$

Applying Proposition 37.5, we may conclude that

(1) $\lim_{\varepsilon\to 0+}\langle u_\varepsilon, \varphi\rangle = \varphi(0)$ i.e. $\lim_{\varepsilon\to 0+} u_\varepsilon = \delta$ in the sense of distributions, and
(2) $\widehat{\delta} = (2\pi)^{-n/2}\cdot 1$.

We can extend the operations from functions to distributions as follows:

$$\langle \partial^\alpha u, \varphi \rangle = \langle u, (-1)^{|\alpha|}\partial^\alpha \varphi \rangle,$$

$$\langle fu, \varphi \rangle = \langle u, f\varphi \rangle, \quad f \in C^\infty(\Omega),$$

$$\langle u * \psi, \varphi \rangle = \langle u, \varphi * \widetilde{\psi} \rangle, \quad \psi \in C_0^\infty(\Omega),$$

where $\widetilde{\psi}(x) = \psi(-x)$. It is possible to show that $u * \psi$ is actually a C^∞ function and

$$\partial^\alpha(u * \psi) = u * \partial^\alpha \psi.$$

A *tempered distribution* is a continuous linear functional on $S(\mathbb{R}^n)$. In addition to the preceding operations for tempered distributions we can define the Fourier transform by

$$\langle \widehat{u}, \varphi \rangle = \langle u, \widehat{\varphi} \rangle, \quad \varphi \in S.$$

Exercise 37.10. Prove that if u is a tempered distribution and $\psi \in S$, then

$$\widehat{u * \psi} = (2\pi)^{n/2}\widehat{\psi}\widehat{u}.$$

Exercise 37.11. Prove that

(1) $\widehat{\delta} = (2\pi)^{-n/2} \cdot 1, \quad \widehat{1} = (2\pi)^{n/2}\delta.$
(2) $\widehat{\partial^\alpha \delta} = (\mathrm{i}\xi)^\alpha (2\pi)^{-n/2}.$
(3) $\widehat{x^\alpha} = \mathrm{i}^{|\alpha|}\partial^\alpha(\widehat{1}) = \mathrm{i}^{|\alpha|}(2\pi)^{n/2}\partial^\alpha \delta.$

Chapter 38
Local Existence Theory

A *partial differential equation of order* $k \in \mathbb{N}$ is an equation of the form

$$F\left(x, (\partial^\alpha u)_{|\alpha|\le k}\right) = 0, \tag{38.1}$$

where F is a function of the variables $x \in \Omega \subset \mathbb{R}^n$, $n \ge 2$, and $(\partial^\alpha u)_{|\alpha|\le k}$.

A complex-valued function $u(x)$ on Ω is a *classical solution* of (38.1) if the derivatives $\partial^\alpha u$ occurring in F exist on Ω and

$$F\left(x, (\partial^\alpha u(x))_{|\alpha|\le k}\right) = 0$$

pointwise for all $x \in \Omega$. The equation (38.1) is said to be *linear* if it can be written as

$$\sum_{|\alpha|\le k} a_\alpha(x)\partial^\alpha u(x) = f(x) \tag{38.2}$$

for some known functions a_α and f. In this case we speak of the (linear) *differential operator*

$$L(x,\partial) \equiv \sum_{|\alpha|\le k} a_\alpha(x)\partial^\alpha$$

and write (38.2) simply as $Lu = f$. If the coefficients $a_\alpha(x)$ belong to $C^\infty(\Omega)$, we can apply the operator L to any distribution $u \in \mathscr{D}'(\Omega)$, and u is called a *distributional solution* (or *weak solution*) of (38.2) if equation (38.2) holds in the sense of distributions, i.e.,

$$\sum_{|\alpha|\le k} (-1)^{|\alpha|}\langle u, \partial^\alpha(a_\alpha\varphi)\rangle = \langle f, \varphi\rangle,$$

where $\varphi \in C_0^\infty(\Omega)$. Let us list some examples. Here and throughout we set $u_t = \frac{\partial u}{\partial t}$, $u_{tt} = \frac{\partial^2 u}{\partial t^2}$, and so forth.

V. Serov, *Fourier Series, Fourier Transform and Their Applications to Mathematical Physics*, Applied Mathematical Sciences 197,
DOI 10.1007/978-3-319-65262-7_38

(1) The *eikonal equation*

$$|\nabla u|^2 = c^2,$$

where $\nabla u = (\partial_1 u, \ldots, \partial_n u)$ is the *gradient* of u.

(2) (a) The *heat (or evolution) equation*

$$u_t = k\Delta u.$$

(b) The *wave equation*

$$u_{tt} = c^2 \Delta u.$$

(c) The *Poisson equation*

$$\Delta u = f,$$

where $\Delta \equiv \nabla \cdot \nabla = \partial_1^2 + \cdots + \partial_n^2$ is the *Laplacian* (or the *Laplace operator*).

(3) The *telegrapher's equation*

$$u_{tt} = c^2 \Delta u - \alpha u_t - m^2 u.$$

(4) The *Sine–Gordon equation*

$$u_{tt} = c^2 \Delta u - \sin u.$$

(5) The *biharmonic equation*

$$\Delta^2 u \equiv \Delta(\Delta u) = 0.$$

(6) The *Korteweg–de Vries equation*

$$u_t + cu \cdot u_x + u_{xxx} = 0.$$

In the linear case, a simple measure of the "strength" of a differential operator is provided by the notion of *characteristics*. If $L(x, \partial) = \sum_{|\alpha| \le k} a_\alpha(x) \partial^\alpha$, then its *characteristic form* (or *principal symbol*) at $x \in \Omega$ is the homogeneous polynomial of degree k defined by

$$\chi_L(x, \xi) = \sum_{|\alpha| = k} a_\alpha(x) \xi^\alpha, \quad \xi \in \mathbb{R}^n.$$

A nonzero ξ is said to be *characteristic* for L at x if $\chi_L(x, \xi) = 0$, and the set of all such ξ is called the *characteristic variety* of L at x, denoted by $\mathrm{char}_x(L)$. In other words,

$$\mathrm{char}_x(L) = \{\xi \neq 0 : \chi_L(x, \xi) = 0\}.$$

In particular, L is said to be *elliptic* at x if $\mathrm{char}_x(L) = \emptyset$ and elliptic in Ω if it is elliptic at every $x \in \Omega$.

Example 38.1. (1) $L = \partial_1\partial_2$ with

$$\mathrm{char}_x(L) = \{\xi \in \mathbb{R}^2 : \xi_1 = 0 \text{ or } \xi_2 = 0, \xi_1^2 + \xi_2^2 > 0\}.$$

(2) $L = \frac{1}{2}(\partial_1 + i\partial_2)$ is the *Cauchy–Riemann operator* on $\mathbb{R}^2$. It is elliptic in $\mathbb{R}^2$.
(3) $L = \Delta$ is elliptic in $\mathbb{R}^n$.
(4) $L = \partial_1 - \sum_{j=2}^n \partial_j^2$, $\mathrm{char}_x(L) = \{\xi \in \mathbb{R}^n \setminus \{0\} : \xi_j = 0, j = 2,3,\dots,n\}$.
(5) $L = \partial_1^2 - \sum_{j=2}^n \partial_j^2$, $\mathrm{char}_x(L) = \left\{\xi \in \mathbb{R}^n \setminus \{0\} : \xi_1^2 = \sum_{j=2}^n \xi_j^2\right\}$.

Let $\nu(x)$ be the normal to S at x. A hypersurface S is said to be *characteristic* for L at $x \in S$ if $\nu(x) \in \mathrm{char}_x(L)$, i.e.,

$$\chi_L(x, \nu(x)) = 0,$$

and S is said to be *non-characteristic* if it is not characteristic at every point, that is, if for all $x \in S$,

$$\chi_L(x, \nu(x)) \neq 0.$$

It is clear that every S is noncharacteristic for elliptic operators. The lines

$$S = \{x \in \mathbb{R}^n : x_1 \neq 0, x_2 = \cdots = x_n = 0\}$$

are characteristic for the heat operator, and the cones

$$S_\pm = \{x \in \mathbb{R}^n : x_1 = \pm\sqrt{x_2^2 + \cdots x_n^2}\}$$

are characteristic for the wave operator.

Let us consider the first-order linear equation

$$Lu \equiv \sum_{j=1}^n a_j(x)\partial_j u + b(x)u = f(x), \tag{38.3}$$

where a_j, b, and f are assumed to be C^1 functions of x. We assume also that a_j, b, and f are real-valued. Suppose we wish to find a solution u of (38.3) with given initial values $u = g$ on the hypersurface S (g is also real-valued). It is clear that

$$\mathrm{char}_x(L) = \left\{\xi \neq 0 : \vec{A} \cdot \xi = 0\right\},$$

where $\vec{A} = (a_1, \dots, a_n)$. This implies that $\mathrm{char}_x(L) \cup \{0\}$ is the hyperplane orthogonal to $\vec{A}$, and therefore, S is characteristic at x if and only if $\vec{A}$ is tangent to S at x ($\vec{A} \cdot \nu = 0$). Then

$$\sum_{j=1}^{n} a_j(x)\partial_j u(x) = \sum_{j=1}^{n} a_j(x)\partial_j g(x), \quad x \in S,$$

is completely determined as a set of certain directional derivatives of φ (see the definition of S) along S at x, and it may be impossible to make this sum equal to $f(x) - b(x)u(x)$ (in order to satisfy (38.3)). Indeed, let us assume that u_1 and u_2 have the same value g on S. This means that $u_1 - u_2 = 0$ on S, or (more or less equivalently)

$$u_1 - u_2 = \varphi \cdot \gamma,$$

where $\varphi = 0$ on S (φ defines this surface) and $\gamma \neq 0$ on S. Next,

$$(\vec{A}\cdot\nabla)u_1 - (\vec{A}\cdot\nabla)u_2 = (\vec{A}\cdot\nabla)(\varphi\gamma) = \gamma(\vec{A}\cdot\nabla)\varphi + \varphi(\vec{A}\cdot\nabla)\gamma = 0,$$

since S is characteristic for L ($(\vec{A}\cdot\nabla)\varphi = 0 \Leftrightarrow (\vec{A}\cdot\frac{\nabla}{|\nabla|})\varphi = 0 \Leftrightarrow \vec{A}\cdot\nu = 0$). Therefore, to make the initial value problem well defined we must assume that S is noncharacteristic for this problem.

Let us assume that S is noncharacteristic for L and $u = g$ on S. We define the *integral curves* for (38.3) as the parametrized curves $x(t)$ that satisfy the system

$$\dot{x} = \vec{A}(x), \quad x = x(t) = (x_1(t), \ldots, x_n(t)), \tag{38.4}$$

of ordinary differential equations, where

$$\dot{x} = (x_1'(t), \ldots, x_n'(t)).$$

Along one of those curves a solution u of (38.3) must satisfy

$$\frac{\mathrm{d}u}{\mathrm{d}t} = \frac{\mathrm{d}}{\mathrm{d}t}(u(x(t))) = \sum_{j=1}^{n} \dot{x}_j \frac{\partial u}{\partial x_j} = (\vec{A}\cdot\nabla)u = f - bu \equiv f(x(t)) - bu(x(t)),$$

or

$$\frac{\mathrm{d}u}{\mathrm{d}t} = f - bu. \tag{38.5}$$

By the existence and uniqueness theorem for ordinary differential equations there is a unique solution (unique curve) of (38.4) with $x(0) = x_0$. Along this curve the solution $u(x)$ of (38.3) must be the solution of (38.5) with $u(0) = u(x(0)) = u(x_0) = g(x_0)$. Moreover, since S is noncharacteristic, $x(t) \notin S$ for $t \neq 0$, at least for small t, and the curves $x(t)$ fill out a neighborhood of S. Thus we have proved the following theorem.

Theorem 38.2. *Assume that S is a surface of class C^1 that is noncharacteristic for (38.3), and that a_j, b, f, and g are real-valued C^1 functions. Then for every sufficiently small neighborhood U of S in $\mathbb{R}^n$ there is a unique solution $u \in C^1$ of (38.3) on U that satisfies $u = g$ on S.*

Remark 38.3. The method that was presented above is called the *method of characteristics*.

The following two examples demonstrate the necessity of noncharacteristic surfaces for boundary value problems.

Example 38.4. (1) In $\mathbb{R}^2$, solve $x_2\partial_1 u + x_1\partial_2 u = u$ with $u(x_1,0) = g(x_1)$ on the line $x_2 = 0$.

Since $\nu(x) = (0,1)$ and since $\chi_L(x,\xi) = x_2\xi_1 + x_1\xi_2$, we have

$$\chi_L(x,\nu(x)) = x_2 \cdot 0 + x_1 \cdot 1 = x_1 \neq 0,$$

so that the lines $x_1 > 0$ and $x_1 < 0$ are noncharacteristic. The system (38.4)–(38.5) to be solved is

$$\dot{x}_1 = x_2, \quad \dot{x}_2 = x_1, \quad \dot{u} = u,$$

with initial conditions

$$(x_1,x_2)|_{t=0} = (x_1^0,0), \quad u(0) = g(x_1^0),$$

on S. We obtain

$$x_1 = \frac{x_1^0}{2}(\mathrm{e}^t + \mathrm{e}^{-t}), \quad x_2 = \frac{x_1^0}{2}(\mathrm{e}^t - \mathrm{e}^{-t}), \quad u = g(x_1^0)\mathrm{e}^t.$$

These equations imply

$$x_1 + x_2 = x_1^0\mathrm{e}^t, \quad x_1 - x_2 = x_1^0\mathrm{e}^{-t}, \quad x_1^2 - x_2^2 = (x_1^0)^2.$$

So

$$\mathrm{e}^t = \pm\frac{x_1 + x_2}{\sqrt{x_1^2 - x_2^2}}, \quad x_1^2 > x_2^2,$$

and thus

$$u(x_1,x_2) = \pm g\left(\pm\sqrt{x_1^2 - x_2^2}\right)\frac{x_1 + x_2}{\sqrt{x_1^2 - x_2^2}},$$

where we have a plus sign for $x_1 + x_2 > 0$ and a minus sign for $x_1 + x_2 < 0$.

(2) In $\mathbb{R}^2$, solve $x_1\partial_1 u + x_2\partial_2 u = u$ with $u(x_1,0) = g(x_1)$ on the line $x_2 = 0$. Compared to previous example, in this case the line $x_2 = 0$ is characteristic, since $x_1 \cdot 0 + x_2 \cdot 1 = 0$ on S. The system (38.4)–(38.5) gives in this case that

$$x_1 = x_1^0\mathrm{e}^t, \quad x_2 \equiv 0, \quad u = g(x_1^0)\mathrm{e}^t.$$

This means that $u(x_1,x_2)$ is a function of only one variable x_1, and the original equation transforms to

$$x_1\partial_1 u = u_1,$$

which has only the solution $u_1(x_1) = cx_1$, where c is a constant. But then we have a contradiction, since the equality $cx_1 = g(x_1)$ is impossible for an arbitrary C^1 function g.

Let us consider more examples in which we apply the method of characteristics.

Example 38.5. In $\mathbb{R}^3$, solve $x_1\partial_1 u + 2x_2\partial_2 u + \partial_3 u = 3u$ with $u = g(x_1,x_2)$ in the plane $x_3 = 0$.

Since $S = \{x \in \mathbb{R}^3 : x_3 = 0\}$, we have $\nu(x) = (0,0,1)$, and since $\chi_L(x,\xi) = x_1\xi_1 + 2x_2\xi_2 + \xi_3$, we must have

$$\chi_L(x,\nu(x)) = x_1 \cdot 0 + 2x_2 \cdot 0 + 1 \cdot 1 = 1 \neq 0,$$

so that S is noncharacteristic. The system (38.4)–(38.5) to be solved is

$$\dot{x}_1 = x_1, \quad \dot{x}_2 = 2x_2, \quad \dot{x}_3 = 1, \quad \dot{u} = 3u,$$

with initial conditions

$$(x_1,x_2,x_3)|_{t=0} = (x_1^0,x_2^0,0), \quad u(0) = g(x_1^0,x_2^0),$$

on S. We obtain

$$x_1 = x_1^0 \mathrm{e}^t, \quad x_2 = x_2^0 \mathrm{e}^{2t}, \quad x_3 = t, \quad u = g(x_1^0,x_2^0)\mathrm{e}^{3t}.$$

These equations imply

$$x_1^0 = x_1\mathrm{e}^{-t} = x_1\mathrm{e}^{-x_3}, \quad x_2^0 = x_2\mathrm{e}^{-2t} = x_2\mathrm{e}^{-2x_3}.$$

Therefore,

$$u(x) = u(x_1,x_2,x_3) = g(x_1\mathrm{e}^{-x_3}, x_2\mathrm{e}^{-2x_3})\mathrm{e}^{3x_3}.$$

Example 38.6. In $\mathbb{R}^3$, solve $\partial_1 u + x_1\partial_2 u - \partial_3 u = u$ with $u(x_1,x_2,1) = x_1 + x_2$.

Since $S = \{x \in \mathbb{R}^3 : x_3 = 1\}$, we have $\nu(x) = (0,0,1)$, and therefore,

$$\chi_L(x,\nu(x)) = 1 \cdot 0 + x_1 \cdot 0 - 1 \cdot 1 = -1 \neq 0,$$

and S is noncharacteristic. The system (38.4)–(38.5) for this problem becomes

$$\dot{x}_1 = 1, \quad \dot{x}_2 = x_1, \quad \dot{x}_3 = -1, \quad \dot{u} = u,$$

with

$$(x_1,x_2,x_3)|_{t=0} = (x_1^0,x_2^0,1), \quad u(0) = x_1^0 + x_2^0.$$

We obtain

$$x_1 = t + x_1^0, \quad x_2 = \frac{t^2}{2} + tx_1^0 + x_2^0, \quad x_3 = -t+1, \quad u = (x_1^0 + x_2^0)e^t.$$

Then,

$$t = 1 - x_3, \quad x_1^0 = x_1 - t = x_1 + x_3 - 1,$$

$$x_2^0 = x_2 - \frac{(1-x_3)^2}{2} - (1-x_3)(x_1+x_3-1) = \frac{1}{2} - x_1 + x_2 - x_3 + x_1x_3 + \frac{x_3^2}{2},$$

and finally,

$$u = \left(\frac{x_3^2}{2} + x_1x_3 + x_2 - \frac{1}{2}\right)e^{1-x_3}.$$

Now let us generalize this technique to *quasilinear equations*, or to the equations of the form

$$\sum_{j=1}^{n} a_j(x,u)\partial_j u = b(x,u), \tag{38.6}$$

where a_j, b, and u are real-valued. If u is a function of x, the normal to the graph of u in $\mathbb{R}^{n+1}$ is proportional to $(\nabla u, -1)$, so (38.6) just says that the vector field

$$\vec{A}(x,y) := (a_1,\dots,a_n,b) \in \mathbb{R}^{n+1}$$

is tangent to the graph $y = u(x)$ at every point. This suggests that we look at the integral curves of $\vec{A}$ in $\mathbb{R}^{n+1}$ given by solving the ordinary differential equations

$$\dot{x}_j = a_j(x,y), \quad j = 1,2,\dots,n, \quad \dot{y} = b(x,y).$$

Suppose we are given initial data $u = g$ on S. If we form the submanifold

$$S^* := \{(x,g(x)) : x \in S\}$$

in $\mathbb{R}^{n+1}$, then the graph of the solution should be the hypersurface generated by the integral curves of $\vec{A}$ passing through S^*. Again, we need to assume that S is noncharacteristic in the sense that the vector

$$(u_1(x,g(x)),\dots,a_n(x,g(x))), \quad x \in S,$$

should not be tangent to S at x, or

$$\sum_{j=1}^{n} a_j(x, g(x))\nu_j(x) \neq 0.$$

Suppose u is a solution of (38.6). If we solve

$$\dot{x}_j = a_j(x, u(x)), \quad j = 1, 2, \ldots, n,$$

with $x_j(0) = x_j^0$, then writing the solution u via integral curves as $y(t) = u(x(t))$, we obtain that

$$\dot{y} = \sum_{j=1}^{n} \partial_j u \cdot \dot{x}_j = \sum_{j=1}^{n} a_j(x, u)\partial_j u = b(x, u) = b(x, y).$$

Thus, as in the linear case, u solves (38.6) with given initial data g on S.

Example 38.7. In $\mathbb{R}^2$, solve $u\partial_1 u + \partial_2 u = 1$ with $u = s/2$ on the segment $x_1 = x_2 = s$, where $s > 0$, $s \neq 2$, is a parameter.

Since $\vec{\varphi}(s) = (s, s)$, it follows that ($x' = x_1 = s$)

$$\det \begin{pmatrix} \frac{\partial x_1}{\partial s} & a_1(s, s, s/2) \\ \frac{\partial x_2}{\partial s} & a_2(s, s, s/2) \end{pmatrix} = \det \begin{pmatrix} 1 & s/2 \\ 1 & 1 \end{pmatrix} = 1 - s/2 \neq 0,$$

for $s > 0$, $s \neq 2$. The system (38.4)–(38.5) for this problem is

$$\dot{x}_1 = u, \quad \dot{x}_2 = 1, \quad \dot{u} = 1,$$

with

$$(x_1, x_2, u)|_{t=0} = (x_1^0, x_2^0, \frac{x_1^0}{2}) = (s, s, s/2).$$

Then

$$u = t + s/2, \quad x_2 = t + s, \quad \dot{x}_1 = t + s/2,$$

so that $x_1 = \frac{t^2}{2} + \frac{st}{2} + s$. This implies

$$x_1 - x_2 = t^2/2 + t(s/2 - 1).$$

For s and t in terms of x_1 and x_2 we obtain

$$\frac{s}{2} = 1 + \frac{1}{t}\left(x_1 - x_2 - \frac{t^2}{2}\right), \quad t = \frac{2(x_1 - x_2)}{x_2 - 2}.$$

Hence

$$\begin{aligned}
u &= \frac{2(x_1-x_2)}{x_2-2}+1+\frac{x_1-x_2}{t}-\frac{t}{2}\\
&= \frac{2(x_1-x_2)}{x_2-2}+1+\frac{x_2-2}{2}-\frac{x_1-x_2}{x_2-2}\\
&= \frac{x_1-x_2}{x_2-2}+1+\frac{x_2-2}{2}=\frac{x_1-x_2}{x_2-2}+\frac{x_2}{2}=\frac{2x_1-4x_2+x_2^2}{2(x_2-2)}.
\end{aligned}$$

Exercise 38.1. In $\mathbb{R}^2$, solve $x_1^2\partial_1 u+x_2^2\partial_2 u=u^2$ with $u\equiv 1$ when $x_2=2x_1$.

Exercise 38.2. In $\mathbb{R}^2$, solve $u\partial_1 u+x_2\partial_2 u=x_1$ with $u(x_1,1)=2x_1$.

Example 38.8. Consider the *Burgers equation*

$$u\partial_1 u+\partial_2 u=0$$

in $\mathbb{R}^2$ with $u(x_1,0)=h(x_1)$, where h is a known C^1 function. It is clear that $S:=\{x\in\mathbb{R}^2 : x_2=0\}$ is noncharacteristic for this quasilinear equation, since

$$\det\begin{pmatrix}1 & h(x_1)\\ 0 & 1\end{pmatrix}=1\neq 0,$$

and $\nu(x)=(0,1)$. Now we have to solve the ordinary differential equations

$$\dot{x}_1=u,\quad \dot{x}_2=1,\quad \dot{u}=0,$$

with

$$(x_1,x_2,u)|_{t=0}=(x_1^0,0,h(x_1^0)).$$

We obtain

$$x_2=t,\quad u\equiv h(x_1^0),\quad x_1=h(x_1^0)t+x_1^0,$$

so that

$$x_1-x_2h(x_1^0)-x_1^0=0.$$

Let us assume that

$$-x_2h_1'(x_1^0)-1\neq 0.$$

By this condition, the last equation defines an implicit function $x_1^0=g(x_1,x_2)$. Therefore, the solution u of the Burgers equation has the form

$$u(x_1,x_2)=h(g(x_1,x_2)).$$

Let us consider two particular cases:

(1) If $h(x_1^0) = ax_1^0 + b, a \neq 0$, then

$$u(x_1, x_2) = \frac{ax_1 + b}{ax_2 + 1}, \quad x_2 \neq -\frac{1}{a}.$$

(2) If $h(x_1^0) = a(x_1^0)^2 + bx_1^0 + c, a \neq 0$, then

$$u(x_1, x_2) = a\left(\frac{-x_2 b - 1 + \sqrt{(x_2 b + 1)^2 - 4ax_2(cx_2 - x_1)}}{2ax_2}\right)^2 + b\left(\frac{-x_2 b - 1 + \sqrt{(x_2 b + 1)^2 - 4ax_2(cx_2 - x_1)}}{2ax_2}\right) + c,$$

with $D = (x_2 b + 1)^2 - 4ax_2(cx_2 - x_1) > 0.$

Let us consider again the linear equation (38.2) of order k, i.e.,

$$\sum_{|\alpha| \le k} a_\alpha(x) \partial^\alpha u(x) = f(x).$$

Let S be a hypersurface of class C^k. If u is a C^k function defined near S, the quantities

$$u, \partial_\nu u, \ldots, \partial_\nu^{k-1} u \tag{38.7}$$

on S are called the *Cauchy data* of u on S. And the *Cauchy problem* is to solve (38.2) with the Cauchy data (38.7). We shall consider $\mathbb{R}^n$, $n \ge 2$, to be $\mathbb{R}^{n-1} \times \mathbb{R}$ and denote the coordinates by (x, t), where $x = (x_1, \ldots, x_{n-1})$. We can make a change of coordinates from $\mathbb{R}^n$ to $\mathbb{R}^{n-1} \times \mathbb{R}$ so that $x_0 \in S$ is mapped to $(0,0)$ and a neighborhood of x_0 in S is mapped into the hyperplane $t = 0$. In that case $\partial_\nu = \frac{\partial}{\partial t}$ on $S = \{(x,t) : t = 0\}$, and equation (38.2) can be written in the new coordinates as

$$\sum_{|\alpha| + j \le k} a_{\alpha,j}(x,t) \partial_x^\alpha \partial_t^j u = f(x,t) \tag{38.8}$$

with the Cauchy data

$$\partial_t^j u(x, 0) = \varphi_j(x), \quad j = 0, 1, \ldots, k-1. \tag{38.9}$$

Since $\nu = (0, 0, \ldots, 0, 1)$, the assumption that S is noncharacteristic means that

$$\chi_L(x, 0, \nu(x, 0)) \equiv a_{0,k}(x, 0) \neq 0.$$

Hence by continuity, $a_{0,k}(x,t) \neq 0$ for small t, and we can solve (38.8) for $\partial_t^k u$:

$$\partial_t^k u(x,t) = (a_{0,k}(x,t))^{-1} \left(f - \sum_{|\alpha|+j \le k, j<k} a_{\alpha,j} \partial_x^\alpha \partial_t^j u \right) \tag{38.10}$$

with the Cauchy data (38.9).

Example 38.9. The line $t = 0$ is noncharacteristic for $\partial_t^2 u = \partial_x^2 u$ in $\mathbb{R}^2$. The Cauchy problem $u(x,0) = g_0(x)$, $\partial_t u(x,0) = g_1(x)$, has a unique solution in appropriate classes for g_0 and g_1. This can be proved by the method of separation of variables (see Section 14.2).

Example 38.10. The line $t = 0$ is characteristic for $\partial_x \partial_t u = 0$ in $\mathbb{R}^2$, and we will therefore have some problems with the solutions. Indeed, if u is a solution of this equation with Cauchy data $u(x,0) = g_0(x)$ and $\partial_t u(x,0) = g_1(x)$, then $\partial_x g_1 = 0$, that is, $g_1 \equiv$ constant. Thus the Cauchy problem is not solvable in general. On the other hand, if g_1 is constant, then there is no uniqueness, because we can take $u(x,t) = g_0(x) + f(t)$ with any $f(t)$ such that $f(0) = 0$ and $f'(0) = g_1$.

Example 38.11. The line $t = 0$ is characteristic for $\partial_x^2 u - \partial_t u = 0$ in $\mathbb{R}^2$. Here if we are given $u(x,0) = g_0(x)$, then $\partial_t u(x,0)$ is already completely determined by $\partial_t u(x,0) = g_0''(x)$. So, again the Cauchy problem has "bad" behavior.

Let us now formulate and give a sketch of the proof of the famous *Cauchy–Kowalevski theorem* for the linear case.

Theorem 38.12. *If $a_{\alpha,j}(x,t), \varphi_0(x), \dots, \varphi_{k-1}(x)$ are real-analytic near the origin in $\mathbb{R}^n$, then there is a neighborhood of the origin on which the Cauchy problem* (38.10)–(38.9) *has a unique real-analytic solution.*

Proof. The uniqueness of the analytic solution follows from the fact that an analytic function is completely determined by the values of its derivatives at one point (see the Taylor formula or the Taylor series). Indeed, for all α and $j = 0, 1, \dots, k-1$,

$$\partial_x^\alpha \partial_t^j u(x,0) = \partial_x^\alpha \varphi_j(x).$$

Therefore,

$$\partial_t^k u|_{t=0} = (a_{0,k})^{-1} \left(f(x,0) - \sum_{|\alpha|+j \le k, j<k} a_{\alpha,j}(x,0) \partial_x^\alpha \varphi_j(x) \right),$$

and moreover,

$$\partial_t^k u(x,t) = (a_{0,k})^{-1} \left(f(x,t) - \sum_{|\alpha|+j \le k, j<k} a_{\alpha,j}(x,t) \partial_x^\alpha \partial_t^j u \right).$$

Then all derivatives of u can be defined from this equation by

$$\partial_t^{k+1} u = \partial_t \left(\partial_t^k u \right).$$

Next, let us denote by $y_{\alpha,j} = \partial_x^\alpha \partial_t^j u$ and by $Y = (y_{\alpha,j})$ this vector. Then equation (38.10) can be rewritten as

$$y_{0,k} = \left(a_{0,k}\right)^{-1} \left(f - \sum_{|\alpha|+j \le k, j<k} a_{\alpha,j} y_{\alpha,j} \right),$$

or

$$\partial_t \left(y_{0,k-1}\right) = \left(a_{0,k}\right)^{-1} \left(f - \sum_{|\alpha|+j \le k, j<k} a_{\alpha,j} \partial_{x_j} y_{(\alpha-\vec{j}),j} \right),$$

and therefore, the Cauchy problem (38.10)–(38.9) becomes

$$\begin{cases} \partial_t Y = \sum_{j=1}^{n-1} A_j \partial_{x_j} Y + B \\ Y(x,0) = \Phi(x), \quad x \in \mathbb{R}^{n-1}, \end{cases} \tag{38.11}$$

where Y, B, and Φ are analytic vector-valued functions and the A_j are analytic matrix-valued functions. Without loss of generality we can assume that $\Phi \equiv 0$. Let $Y = (y_1, \ldots, y_N)$, $B = (b_1, \ldots, b_N)$, $A_j = (a_{ml}^{(j)})_{m,l=1}^N$. We seek a solution $Y = (y_1, \ldots, y_N)$ of the form

$$y_m = \sum C_{\alpha,j}^{(m)} x^\alpha t^j, \quad m = 1, 2, \ldots, N.$$

The Cauchy data tell us that $C_{\alpha,0}^{(m)} = 0$ for all α and m, since we assumed $\Phi \equiv 0$. To determine $C_{\alpha,j}^{(m)}$ for $j > 0$, we substitute y_m into (38.11) and get for $m = 1, 2, \ldots, N$, that

$$\partial_t y_m = \sum a_{ml}^{(j)} \partial_{x_j} y_l + b_m(x,y),$$

or

$$\sum C_{\alpha,j}^{(m)} j x^\alpha t^{j-1} = \sum_{j,l} \sum_{\beta,r} \left(a_{ml}^{(j)}\right)_{\beta r} x^\beta t^r \sum C_{\alpha,j}^{(m)} \alpha_j x^{\alpha-\vec{j}} t^j + \sum b_{\alpha_j}^{(m)} x^\alpha t^j.$$

It can be proved that this equation determines uniquely the coefficients $C_{\alpha,j}^{(m)}$ and therefore the solution $Y = (y_1, \ldots, y_N)$. □

Remark 38.13. Consider the following example in $\mathbb{R}^2$, due to Hadamard, which sheds light on the Cauchy problem:

$$\Delta u = 0, \quad u(x_1,0) = 0, \quad \partial_2 u(x_1,0) = k e^{-\sqrt{k}} \sin(x_1 k), \quad k \in \mathbb{N}.$$

This problem is noncharacteristic on $\mathbb{R}^2$, since Δ is elliptic in $\mathbb{R}^2$. We look for $u(x_1,x_2) = u_1(x_1)u_2(x_2)$. Then

$$u_1''u_2 + u_2''u_1 = 0,$$

which implies that

$$\frac{u_1''}{u_1} = -\frac{u_2''}{u_2} = -\lambda = \text{constant}.$$

Next, the general solutions of

$$u_1'' = -\lambda u_1$$

and

$$u_2'' = \lambda u_2$$

are

$$u_1 = A\sin(\sqrt{\lambda}x_1) + B\cos(\sqrt{\lambda}x_1)$$

and

$$u_2 = C\sinh(\sqrt{\lambda}x_2) + D\cosh(\sqrt{\lambda}x_2),$$

respectively. But $u_2(0) = 0, u_2'(0) = 1$ and $u_1(x_1) = ke^{-\sqrt{k}}\sin(kx_1)$. Thus $D = 0$, $B = 0$, $k = \sqrt{\lambda}$, $A = ke^{-\sqrt{k}}$, and $C = \frac{1}{k} = \frac{1}{\sqrt{\lambda}}$. So we finally have

$$u(x_1,x_2) = ke^{-\sqrt{k}}\sin(kx_1)\frac{1}{k}\sinh(kx_2) = e^{-\sqrt{k}}\sin(kx_1)\sinh(kx_2).$$

As $k \to +\infty$, the Cauchy data and their derivatives (for $x_2 = 0$) of all orders tend uniformly to zero, since $e^{-\sqrt{k}}$ decays faster than polynomially. But if $x_2 \neq 0$ (more precisely, $x_2 > 0$), then

$$\lim_{k\to+\infty} e^{-\sqrt{k}}\sin(kx_1)\sinh(kx_2) = \infty,$$

if we choose, for example, $x_2 = 1$ and $x_1^{(k)} = \pi/(2k) + 2\pi$. Hence $u(x_1,x_2)$ is not bounded. But the solution of the original problem that corresponds to the limiting case $k = \infty$ is of course $u \equiv 0$, since $u(x_1,0) = 0$ and $\partial_2 u(x_1,0) = 0$ in the limiting case. Hence the solution of the Cauchy problem may not depend continuously on the Cauchy data. This means by Hadamard that the Cauchy problem for elliptic operators is "ill-posed," even when this problem is noncharacteristic.

Remark 38.14. This example of Hadamard's shows that the solution of the Cauchy problem may not depend continuously on the Cauchy data. By the terminology of Hadamard, "the Cauchy problem for the Laplacian is not well posed, but it is ill posed." Due to Hadamard and Tikhonov, a problem is called *well posed* if the following conditions are satisfied:

(1) existence;
(2) uniqueness;
(3) stability or continuous dependence on data.

Otherwise, it is called *ill posed*.

Let us consider one more important example due to H. Lewy. Let L be the first-order differential operator in $\mathbb{R}^3$ ($(x,y,t) \in \mathbb{R}^3$) given by

$$L \equiv \frac{\partial}{\partial x} + \mathrm{i}\frac{\partial}{\partial y} - 2\mathrm{i}(x+\mathrm{i}y)\frac{\partial}{\partial t}. \tag{38.12}$$

Theorem 38.15 (The *Hans Lewy example*). *Let f be a continuous real-valued function depending only on t. If there is a C^1 function u satisfying $Lu = f$, with the operator L from* (38.12), *in some neighborhood of the origin, then $f(t)$ necessarily is analytic at $t = 0$.*

Remark 38.16. This example shows that the assumption of analyticity of f in Theorem 38.12 in the linear equation cannot be omitted (it is essential). It appears necessarily, since $Lu = f$ with L from (38.12) has no C^1 solution unless f is analytic.

Proof. Suppose $x^2 + y^2 < R^2$, $|t| < R$, and set $z = x + \mathrm{i}y = r\mathrm{e}^{\mathrm{i}\theta}$. Let us denote by $V(t)$ the function

$$V(t) := \int_{|z|=r} u(x,y,t)\mathrm{d}\sigma(z) = \mathrm{i}r\int_0^{2\pi} u(r,\theta,t)\mathrm{e}^{\mathrm{i}\theta}\mathrm{d}\theta,$$

where $u(x,y,t)$ is the C^1 solution of the equation $Lu = f$ with L from (38.12). We continue to denote u in polar coordinates also by u. By the divergence theorem for $F := (u, \mathrm{i}u)$ we get

$$\begin{aligned}
\mathrm{i}\int_{|z|<r} \nabla \cdot F\mathrm{d}x\mathrm{d}y &\equiv \mathrm{i}\int_{|z|<r}\left(\frac{\partial u}{\partial x} + \mathrm{i}\frac{\partial u}{\partial y}\right)\mathrm{d}x\mathrm{d}y = \mathrm{i}\int_{|z|=r}(u,\mathrm{i}u)\cdot \nu\mathrm{d}\sigma(z)\\
&= \mathrm{i}\int_{|z|=r}\left(u\frac{x}{r} + \mathrm{i}u\frac{y}{r}\right)\mathrm{d}\sigma(z) = \mathrm{i}\int_{|z|=r} u\mathrm{e}^{\mathrm{i}\theta}\mathrm{d}\sigma(z)\\
&= \mathrm{i}r\int_0^{2\pi} u\mathrm{e}^{\mathrm{i}\theta}\mathrm{d}\theta \equiv V(t).
\end{aligned}$$

But on the other hand, in polar coordinates,

$$V(t) \equiv \mathrm{i}\int_{|z|<r}\left(\frac{\partial u}{\partial x} + \mathrm{i}\frac{\partial u}{\partial y}\right)\mathrm{d}x\mathrm{d}y = \mathrm{i}\int_0^r\int_0^{2\pi}\left(\frac{\partial u}{\partial x} + \mathrm{i}\frac{\partial u}{\partial y}\right)(\rho,\theta,t)\rho\mathrm{d}\rho\mathrm{d}\theta.$$

This implies that

$$\begin{aligned}\frac{\partial V}{\partial r} &= \mathrm{i}r\int_0^{2\pi}\left(\frac{\partial u}{\partial x}+\mathrm{i}\frac{\partial u}{\partial y}\right)(r,\theta,t)\mathrm{d}\theta = \int_{|z|=r}\left(\frac{\partial u}{\partial x}+\mathrm{i}\frac{\partial u}{\partial y}\right)(x,y,t)2r\frac{\mathrm{d}\sigma(z)}{2z}\\ &= 2r\int_{|z|=r}\left(\mathrm{i}\frac{\partial u}{\partial t}+\frac{f(t)}{2z}\right)\mathrm{d}\sigma(z) = 2r\left(\mathrm{i}\frac{\partial V}{\partial t}+f(t)\int_{|z|=r}\frac{\mathrm{d}\sigma(z)}{2z}\right)\\ &= 2r\left(\mathrm{i}\frac{\partial V}{\partial t}+\mathrm{i}\pi f(t)\right).\end{aligned}$$

We therefore have the following equation for V:

$$\frac{1}{2r}\frac{\partial V}{\partial r} = \mathrm{i}\left(\frac{\partial V}{\partial t}+\pi f(t)\right). \tag{38.13}$$

Let us introduce now a new function $U(s,t) = V(s)+\pi F(t)$, where $s = r^2$ and $F' = f$. The function F exists because f is continuous. It follows from (38.13) that

$$\frac{1}{2r}\frac{\partial V}{\partial r} \equiv \frac{\partial V}{\partial s}, \quad \frac{\partial U}{\partial s} = \frac{\partial V}{\partial s}, \quad \frac{\partial U}{\partial s} = \mathrm{i}\frac{\partial U}{\partial t}.$$

Hence

$$\frac{\partial U}{\partial t}+\mathrm{i}\frac{\partial U}{\partial s} = 0. \tag{38.14}$$

Since (38.14) is the Cauchy–Riemann equation, we have that U is a holomorphic (analytic) function of the variable $w = t+\mathrm{i}s$, in the region $0 < s < R^2$, $|t| < R$, and U is continuous up to $s = 0$. Next, since $U(0,t) = \pi F(t)$ ($V = 0$ when $s = 0$, i.e., $r = 0$) and $f(t)$ is real-valued, it follows that $U(0,t)$ is also real-valued. Therefore, by the Schwarz reflection principle (see complex analysis), the formula

$$U(-s,t) := \overline{U(s,t)}$$

gives a holomorphic continuation of U to a full neighborhood of the origin. In particular, $U(0,t) = \pi F(t)$ is analytic in t, hence so is $f(t) \equiv F'(t)$. □

Chapter 39
The Laplace Operator

We consider what is perhaps the most important of all partial differential operators, the*Laplace operator* (*Laplacian*) on $\mathbb{R}^n$, defined by

$$\Delta = \sum_{j=1}^{n} \partial_j^2 \equiv \nabla \cdot \nabla.$$

We will begin with a quite general fact about partial differential operators.

Definition 39.1. (1) A linear transformation T on $\mathbb{R}^n$ is called a*rotation* if $T' = T^{-1}$.
(2) Let h be a fixed vector in $\mathbb{R}^n$. The*translation* transformation $T_h f(x) := f(x+h)$ is called a.

Theorem 39.2. *Suppose that L is a linear partial differential operator on $\mathbb{R}^n$. Then L commutes with translations and rotations if and only if L is a polynomial in Δ, that is, $L \equiv \sum_{j=0}^{m} a_j \Delta^j$.*

Proof. Let

$$L(x,\partial) \equiv \sum_{|\alpha|\le k} a_\alpha(x) \partial^\alpha$$

commute with a translation T_h. Then

$$\sum_{|\alpha|\le k} a_\alpha(x) \partial^\alpha f(x+h) = \sum_{|\alpha|\le k} a_\alpha(x+h) \partial^\alpha f(x+h).$$

This implies that the $a_\alpha(x)$ must be constants (because $a_\alpha(x) \equiv a_\alpha(x+h)$ for all h), say a_α. Next, since L now has constant coefficients, we have (see Exercise 37.5)

$$\widehat{Lu}(\xi) = P(\xi)\widehat{u}(\xi),$$

V. Serov, *Fourier Series, Fourier Transform and Their Applications to Mathematical Physics*, Applied Mathematical Sciences 197,
DOI 10.1007/978-3-319-65262-7_39

where the polynomial $P(\xi)$ is defined by

$$P(\xi) = \sum_{|\alpha|\le k} a_\alpha (\mathrm{i}\xi)^\alpha.$$

Recall from Exercise 37.4 that if T is a rotation, then

$$\widehat{u\circ T}(\xi) = (\widehat{u}\circ T)(\xi).$$

Therefore,

$$(\widehat{Lu)(Tx})(\xi) = \widehat{Lu}(T\xi),$$

or

$$P(\xi)\widehat{u(Tx)}(\xi) = P(T\xi)\widehat{u}(T\xi).$$

This forces

$$P(\xi) = P(T\xi).$$

Write $\xi = |\xi|\theta$, where $\theta \in \mathbb{S}^{n-1} = \{x \in \mathbb{R}^n : |x| = 1\}$ is the direction of ξ. Then $T\xi = |\xi|\theta'$ with some $\theta' \in \mathbb{S}^{n-1}$. But

$$0 = P(\xi) - P(T\xi) = P(|\xi|\theta) - P(|\xi|\theta')$$

shows that $P(\xi)$ does not depend on the angle θ of ξ. Therefore, $P(\xi)$ is radial, that is,

$$P(\xi) = P_1(|\xi|) = \sum_{|\alpha|\le k} a'_\alpha |\xi|^{|\alpha|}.$$

But since we know that $P(\xi)$ is a polynomial, $|\alpha|$ must be even:

$$P(\xi) = \sum_j a_j |\xi|^{2j}.$$

By Exercise 37.5 we have that

$$\widehat{\Delta u}(\xi) = -|\xi|^2\widehat{u}(\xi).$$

It follows by induction that

$$\widehat{\Delta^j u}(\xi) = (-1)^j|\xi|^{2j}\widehat{u}(\xi), \quad j = 0, 1, \ldots.$$

Taking the inverse Fourier transform, we obtain

$$Lu = \mathscr{F}^{-1}(P(\xi)\widehat{u}(\xi)) = \mathscr{F}^{-1}\sum_j a_j|\xi|^{2j}\widehat{u}(\xi) = \mathscr{F}^{-1}\sum_j a'_j\widehat{\Delta^j u}(\xi) = \sum_j a'_j\Delta^j u.$$

Conversely, let

$$Lu = \sum_j a_j \Delta^j u.$$

It is clear by the chain rule that the Laplacian commutes with translations T_h and rotations T. By induction, the same is true for any power of Δ, and so for L as well. □

Lemma 39.3. *If $f(x) = \varphi(r)$, $r = |x|$, that is, f is radial, then $\Delta f = \varphi''(r) + \frac{n-1}{r}\varphi'(r)$.*

Proof. Since $\frac{\partial r}{\partial x_j} = \frac{x_j}{r}$, it follows that

$$\begin{aligned}
\Delta f &= \sum_{j=1}^{n} \partial_j(\partial_j \varphi(r)) = \sum_{j=1}^{n} \partial_j \left(\frac{x_j}{r}\varphi'(r)\right) \\
&= \sum_{j=1}^{n} \varphi'(r)\partial_j\left(\frac{x_j}{r}\right) + \sum_{j=1}^{n} \frac{x_j^2}{r^2}\varphi''(r) \\
&= \sum_{j=1}^{n} \left(\frac{1}{r} - \frac{x_j^2}{r^3}\right)\varphi'(r) + \sum_{j=1}^{n} \frac{x_j^2}{r^2}\varphi''(r) \\
&= \frac{n}{r}\varphi'(r) - \frac{1}{r^3}\sum_{j=1}^{n} x_j^2 \varphi'(r) + \varphi''(r) = \varphi''(r) + \frac{n-1}{r}\varphi'(r).
\end{aligned}$$

This completes the proof. □

Corollary 39.4. *If $f(x) = \varphi(r)$, then $\Delta f = 0$ on $\mathbb{R}^n \setminus \{0\}$ if and only if*

$$\varphi(r) = \begin{cases} a + br^{2-n}, & n \neq 2, \\ a + b\log r, & n = 2, \end{cases}$$

where a and b are arbitrary constants.

Proof. If $\Delta f = 0$, then by Lemma 39.3, we have

$$\varphi''(r) + \frac{n-1}{r}\varphi'(r) = 0.$$

Define $\psi(r) := \varphi'(r)$. Since ψ solves the first-order differential equation

$$\psi'(r) + \frac{n-1}{r}\psi(r) = 0,$$

$\psi(r)$ can be found by the use of an integrating factor. Indeed, multiply by $e^{(n-1)\log r} = r^{n-1}$ to get

$$r^{n-1}\psi'(r)+(n-1)r^{n-2}\psi(r)=0,$$

or

$$(r^{n-1}\psi(r))'=0.$$

It follows that

$$\varphi'(r)=\psi(r)=cr^{1-n}.$$

Integrate once more to arrive at

$$\varphi(r)=\begin{cases}\frac{cr^{2-n}}{2-n}+c_1, & n\neq 2,\\ c\log r+c_1, & n=2,\end{cases}=\begin{cases}ar+b, & n=1,\\ a\log r+b, & n=2,\\ ar^{2-n}+b, & n\geq 3.\end{cases}$$

In the opposite direction the result follows from elementary differentiation. □

Definition 39.5. A C^2 function u on an open set $\Omega\subset\mathbb{R}^n$ is said to be *harmonic on* Ω if $\Delta u=0$ on Ω.

Exercise 39.1. For $u,v\in C^2(\Omega)\cap C^1(\overline{\Omega})$ and for $S=\partial\Omega$, which is a surface*Green's identities* of class C^1, prove the following:

(1)

$$\int_\Omega(v\Delta u-u\Delta v)\,\mathrm{d}x=\int_S(v\partial_\nu u-u\partial_\nu v)\,\mathrm{d}\sigma;$$

(2)

$$\int_\Omega(v\Delta u+\nabla v\cdot\nabla u)\,\mathrm{d}x=\int_S v\partial_\nu u\mathrm{d}\sigma.$$

Exercise 39.2. Prove that if u is harmonic on Ω and $u\in C^1(\overline{\Omega})$, then

$$\int_S\partial_\nu u\mathrm{d}\sigma=0.$$

Corollary 39.6 (From Green's identities). *Let $u\in C^1(\overline{\Omega})$ be harmonic on Ω.*

(1) if $u=0$ on S, then $u\equiv 0$;
(2) if $\partial_\nu u=0$ on S, then $u\equiv$ constant.

Proof. By taking real and imaginary parts, it suffices to consider real-valued functions. If we let $u=v$ in part (2) of Exercise 39.1, we obtain

$$\int_\Omega|\nabla u|^2\mathrm{d}x=\int_S u\partial_\nu u\mathrm{d}\sigma(x).$$

In the case (1) we get $\nabla u \equiv 0$, or $u \equiv$ constant. But $u \equiv 0$ on S implies that $u \equiv 0$. In the case (2) we can conclude only that $u \equiv$ constant. □

Theorem 39.7 (The *mean value theorem*). *Suppose u is harmonic on an open set $\Omega \subset \mathbb{R}^n$. If $x \in \Omega$ and $r > 0$ is small enough that $\overline{B_r(x)} \subset \Omega$, then*

$$u(x) = \frac{1}{r^{n-1}\omega_n} \int_{|x-y|=r} u(y) \mathrm{d}\sigma(y) \equiv \frac{1}{\omega_n} \int_{|y|=1} u(x+ry) \mathrm{d}\sigma(y),$$

where $\omega_n = \frac{2\pi^{n/2}}{\Gamma(n/2)}$ is the area of the unit sphere in $\mathbb{R}^n$.

Proof. Let us apply Green's identity (1) with u and $v = |y|^{2-n}$ if $n \neq 2$, and $v = \log|y|$ if $n = 2$ in the domain

$$B_r(x) \backslash \overline{B_\varepsilon(x)} = \{y \in \mathbb{R}^n : \varepsilon < |x-y| < r\}.$$

Then for $v(y-x)$ we obtain ($n \neq 2$)

$$\begin{aligned} 0 &= \int_{B_r(x)\backslash \overline{B_\varepsilon(x)}} (v\Delta u - u\Delta v) \mathrm{d}y \\ &= \int_{|x-y|=r} (v\partial_\nu u - u\partial_\nu v) \mathrm{d}\sigma(y) - \int_{|x-y|=\varepsilon} (v\partial_\nu u - u\partial_\nu v) \mathrm{d}\sigma(y) \\ &= r^{2-n} \int_{|x-y|=r} \partial_\nu u \mathrm{d}\sigma(y) - (2-n) r^{1-n} \int_{|x-y|=r} u \mathrm{d}\sigma(y) \\ &- \varepsilon^{2-n} \int_{|x-y|=\varepsilon} \partial_\nu u \mathrm{d}\sigma(y) + (2-n)\varepsilon^{1-n} \int_{|x-y|=\varepsilon} u \mathrm{d}\sigma(y). \end{aligned} \quad (39.1)$$

In order to get (39.1) we took into account that

$$\partial_\nu = \nu \cdot \nabla = \frac{x-y}{r}\frac{x-y}{r}\frac{\mathrm{d}}{\mathrm{d}r} = \frac{\mathrm{d}}{\mathrm{d}r}$$

for the sphere. Since u is harmonic, due to Exercise 39.2 we can get from (39.1) that for all $\varepsilon > 0$, $\varepsilon < r$,

$$\varepsilon^{1-n} \int_{|x-y|=\varepsilon} u \mathrm{d}\sigma(y) = r^{1-n} \int_{|x-y|=r} u \mathrm{d}\sigma(y).$$

Therefore,

$$\begin{aligned} \lim_{\varepsilon\to 0} \varepsilon^{1-n} \int_{|x-y|=\varepsilon} u(y) \mathrm{d}\sigma(y) &= \lim_{\varepsilon\to 0} \int_{|\theta|=1} u(x+\varepsilon\theta) \mathrm{d}\theta \\ &= \omega_n u(x) = r^{1-n} \int_{|x-y|=r} u(y) \mathrm{d}\sigma(y). \end{aligned}$$

This proves the theorem, because the latter steps hold for $n = 2$ also. □

Corollary 39.8. *If u and r are as in Theorem 39.7, then*

$$u(x) = \frac{n}{r^n \omega_n} \int_{|x-y|\le r} u(y)\mathrm{d}y \equiv \frac{n}{\omega_n} \int_{|y|\le 1} u(x+ry)\mathrm{d}y, \quad x \in \Omega. \tag{39.2}$$

Proof. Perform integration in polar coordinates and apply Theorem 39.7. □

Remark 39.9. It follows from the latter formula that

$$\mathrm{vol}\{y : |y| \le 1\} = \frac{\omega_n}{n}.$$

Exercise 39.3. Assume that u is harmonic in Ω. Let $\chi(x) \in C_0^\infty(B_1(0))$ be such that $\chi(x) = \chi_1(|x|)$ and $\int_{\mathbb{R}^n} \chi(x)\mathrm{d}x = 1$. Define an approximation to the identity by $\chi_\varepsilon(\cdot) = \varepsilon^{-n}\chi(\varepsilon^{-1}\cdot)$. Prove that

$$u(x) = \int_{B_\varepsilon(x)} \chi_\varepsilon(x-y)u(y)\mathrm{d}y$$

for $x \in \Omega_\varepsilon := \{x \in \Omega : \overline{B_\varepsilon(x)} \subset \Omega\}$.

Corollary 39.10. *If u is harmonic on* Ω*, then* $u \in C^\infty(\Omega)$.

Proof. The statement follows from Exercise 39.3, since the function χ_ε is compactly supported and we may thus differentiate under the integral sign as often as we please. □

Corollary 39.11. *If* $\{u_k\}_{k=1}^\infty$ *is a sequence of harmonic functions on an open set* $\Omega \subset \mathbb{R}^n$ *that converges uniformly on compact subsets of* Ω *to a limit u, then u is harmonic on* Ω.

Theorem 39.12 (The *maximum principle*). *Suppose* $\Omega \subset \mathbb{R}^n$ *is open and connected. If u is real-valued and harmonic on* Ω *with* $\sup_{x\in\Omega} u(x) = A < \infty$*, then either* $u < A$ *for all* $x \in \Omega$ *or* $u(x) \equiv A$ *in* Ω.

Proof. Since u is continuous on Ω, the set $\{x \in \Omega : u(x) = A\}$ is closed in Ω. On the other hand, we may conclude that if $u(x) = A$ at some point $x \in \Omega$, then $u(y) = A$ for all y in a ball about x. Indeed, if $y_0 \in B'_\sigma(x)$ and $u(y_0) < A$, then $u(y) < A$ for all y from a small neighborhood of y_0. Hence, by Corollary 39.8, for $r \le \sigma$,

$$\begin{aligned}
A = u(x) &= \frac{n}{r^n \omega_n} \int_{|x-y|\le r} u(y)\mathrm{d}y \\
&= \frac{n}{r^n \omega_n} \int_{|x-y|\le r, |y_0-y|>\varepsilon} u(y)\mathrm{d}y + \frac{n}{r^n \omega_n} \int_{|y-y_0|\le\varepsilon} u(y)\mathrm{d}y \\
&< A\left(\frac{n}{r^n \omega_n} \int_{|x-y|\le r, |y_0-y|>\varepsilon} \mathrm{d}y + \frac{n}{r^n \omega_n} \int_{|y-y_0|\le\varepsilon} \mathrm{d}y\right) \\
&= A\frac{n}{r^n \omega_n} \int_{|x-y|\le r} \mathrm{d}y = A,
\end{aligned}$$

that is, $A < A$. This contradiction proves our statement. This fact also means that the set $\{x \in \Omega : u(x) = A\}$ is also open. Hence it is either Ω (in this case $u \equiv A$ in Ω) or the empty set (in this case $u(x) < A$ in Ω). □

Corollary 39.13. *Suppose $\Omega \subset \mathbb{R}^n$ is connected, open, and bounded. If u is real-valued and harmonic on Ω and continuous on $\overline{\Omega}$, then the maximum and minimum of u on $\overline{\Omega}$ are achieved only on $\partial\Omega$.*

Corollary 39.14 (The uniqueness theorem). *Suppose Ω is as in Corollary 39.13. If u_1 and u_2 are harmonic on Ω and continuous in $\overline{\Omega}$ (possibly complex-valued) and $u_1 = u_2$ on $\partial\Omega$, then $u_1 = u_2$ on $\overline{\Omega}$.*

Proof. The real and imaginary parts of $u_1 - u_2$ and $u_2 - u_1$ are harmonic on Ω. Hence they must achieve their maxima on $\partial\Omega$. These maxima are therefore zero, so $u_1 \equiv u_2$. □

Theorem 39.15 ***(Liouville's theorem).*** *If u is bounded and harmonic on $\mathbb{R}^n$, then $u \equiv$ constant.*

Proof. For all $x \in \mathbb{R}^n$ and $|x| \le R$, by Corollary 39.8 we have

$$|u(x) - u(0)| = \frac{n}{R^n \omega_n} \left| \int_{B_R(x)} u(y)\mathrm{d}y - \int_{B_R(0)} u(y)\mathrm{d}y \right| \le \frac{n}{R^n \omega_n} \int_D |u(y)|\mathrm{d}y,$$

where

$$D = (B_R(x) \backslash B_R(0)) \cup (B_R(0) \backslash B_R(x))$$

is the symmetric difference of the balls $B_R(x)$ and $B_R(0)$. Therefore, we obtain

$$\begin{aligned} |u(x) - u(0)| &\le \frac{n\|u\|_\infty}{R^n \omega_n} \int_{R-|x| \le |y| \le R+|x|} \mathrm{d}y \le \frac{n\|u\|_\infty}{R^n \omega_n} \int_{R-|x|}^{R+|x|} r^{n-1}\mathrm{d}r \int_{|\theta|=1} \mathrm{d}\theta \\ &= \frac{(R+|x|)^n - (R-|x|)^n}{R^n} \|u\|_\infty = O\left(\frac{1}{R}\right) \|u\|_\infty. \end{aligned}$$

Hence the difference $|u(x) - u(0)|$ vanishes as $R \to \infty$, that is, $u(x) = u(0)$. □

Definition 39.16. A *fundamental solution* for a partial differential operator L is a distribution $K \in \mathscr{D}'$ such that

$$LK = \delta.$$

Remark 39.17. Note that a fundamental solution is not unique. Any two fundamental solutions differ by a solution of the homogeneous equation $Lu = 0$.

Exercise 39.4. Show that the characteristic function of the set

$$\{(x_1, x_2) \in \mathbb{R}^2 : x_1 > 0, x_2 > 0\}$$

is a fundamental solution for $L = \partial_1 \partial_2$.

Exercise 39.5. Prove that the Fourier transform of $\frac{1}{x_1+\mathrm{i}x_2}$ in $\mathbb{R}^2$ is equal to $-\frac{\mathrm{i}}{\xi_1+\mathrm{i}\xi_2}$.

Exercise 39.6. Show that the fundamental solution for the Cauchy–Riemann operator $L = \frac{1}{2}(\partial_1 + \mathrm{i}\partial_2)$ on $\mathbb{R}^2$ is equal to

$$\frac{1}{\pi}\frac{1}{x_1+\mathrm{i}x_2}.$$

Since the Laplacian commutes with rotations (Theorem 39.2), it should have a radial fundamental solution that must be a function of $|x|$ that is harmonic on $\mathbb{R}^n \setminus \{0\}$.

Theorem 39.18. *Let*

$$K(x) = \begin{cases} \frac{|x|^{2-n}}{(2-n)\omega_n}, & n \neq 2, \\ \frac{1}{2\pi}\log|x|, & n = 2. \end{cases} \tag{39.3}$$

Then K is a fundamental solution for Δ.

Proof. For $\varepsilon > 0$ we consider a smoothed-out version K_ε of K as

$$K_\varepsilon(x) = \begin{cases} \frac{(|x|^2+\varepsilon^2)^{\frac{2-n}{2}}}{(2-n)\omega_n}, & n \neq 2 \\ \frac{1}{4\pi}\log(|x|^2+\varepsilon^2), & n = 2. \end{cases} \tag{39.4}$$

Then $K_\varepsilon \to K$ pointwise ($x \neq 0$) as $\varepsilon \to 0+$, and K_ε and K are dominated by a fixed locally integrable function for $\varepsilon \leq 1$ (namely, by $|K|$ for $n > 2$, $|\log|x|| + 1$ for $n = 2$, and $(|x|^2+1)^{1/2}$ for $n = 1$). So by Lebesgue's dominated convergence theorem, $K_\varepsilon \to K$ in L^1_{loc} (or in the topology of distributions) as $\varepsilon \to 0+$. Hence we need to show only that $\Delta K_\varepsilon \to \delta$ as $\varepsilon \to 0$ in the sense of distributions, that is,

$$\langle \Delta K_\varepsilon, \varphi \rangle \to \varphi(0), \quad \varepsilon \to 0$$

for all $\varphi \in C_0^\infty(\mathbb{R}^n)$.

Exercise 39.7. Prove that

$$\Delta K_\varepsilon(x) = n\omega_n^{-1}\varepsilon^2(|x|^2+\varepsilon^2)^{-(\frac{n}{2}+1)} \equiv \varepsilon^{-n}\psi(\varepsilon^{-1}x)$$

for $\psi(y) = n\omega_n^{-1}(|y|^2+1)^{-(\frac{n}{2}+1)}$.

Exercise 39.7 allows us to write

$$\langle \Delta K_\varepsilon, \varphi \rangle = \int_{\mathbb{R}^n} \varphi(x)\varepsilon^{-n}\psi(\varepsilon^{-1}x)\mathrm{d}x = \int_{\mathbb{R}^n} \varphi(\varepsilon z)\psi(z)\mathrm{d}z \to \varphi(0)\int_{\mathbb{R}^n} \psi(z)\mathrm{d}z$$

as $\varepsilon \to 0+$. So it remains to show that

$$\int_{\mathbb{R}^n} \psi(z)\mathrm{d}z = 1.$$

Using Exercise 39.7, we have

$$\begin{aligned}
\int_{\mathbb{R}^n} \psi(x)\mathrm{d}x &= \frac{n}{\omega_n}\int_{\mathbb{R}^n} (|x|^2+1)^{-(\frac{n}{2}+1)}\mathrm{d}x \\
&= \frac{n}{\omega_n}\int_0^\infty r^{n-1}(r^2+1)^{-(\frac{n}{2}+1)}\mathrm{d}r \int_{|\theta|=1} \mathrm{d}\theta \\
&= n\int_0^\infty r^{n-1}(r^2+1)^{-(\frac{n}{2}+1)}\mathrm{d}r = \frac{n}{2}\int_0^\infty t^{(n-1)/2}(1+t)^{-\frac{n}{2}-1}\frac{1}{\sqrt{t}}\mathrm{d}t \\
&= \frac{n}{2}\int_0^\infty t^{n/2-1}(1+t)^{-\frac{n}{2}-1}\mathrm{d}t = \frac{n}{2}\int_0^1 \left(\frac{1}{s}-1\right)^{n/2-1} s^{\frac{n}{2}+1}\frac{\mathrm{d}s}{s^2} \\
&= \frac{n}{2}\int_0^1 (1-s)^{n/2-1}\mathrm{d}s = \frac{n}{2}\int_0^1 \tau^{n/2-1}\mathrm{d}\tau = 1.
\end{aligned}$$

This means that $\varepsilon^{-1}\psi(\varepsilon^{-1}x)$ is an approximation to the identity and

$$\Delta K_\varepsilon \to \delta.$$

But $K_\varepsilon \to K$, and so $\Delta K = \delta$ also. □

Theorem 39.19. *Suppose that*
(1) $f \in L^1(\mathbb{R}^n)$ *if* $n \geq 3$,
(2) $\int_{\mathbb{R}^2} |f(y)|(|\log|y||+1)\,\mathrm{d}y < \infty$ *if* $n = 2$,
(3) $\int_{\mathbb{R}} |f(y)|(1+|y|)\,\mathrm{d}y < \infty$ *if* $n = 1$.

Let K be given by (39.3)*. Then $f * K$ is well defined as a locally integrable function, and $\Delta(f * K) = f$ in the sense of distributions.*

Proof. Let $n \geq 3$ and set

$$\chi_1(x) = \begin{cases} 1, & x \in B_1(0), \\ 0, & x \notin B_1(0). \end{cases}$$

Then $\chi_1 K \in L^1(\mathbb{R}^n)$ and $(1-\chi_1)K \in L^\infty(\mathbb{R}^n)$. So, for $f \in L^1(\mathbb{R}^n)$ we have that $f * (\chi_1 K) \in L^1(\mathbb{R}^n)$ and $f * (1-\chi_1)K \in L^\infty(\mathbb{R}^n)$ (see Proposition 37.3). Hence $f * K \in L^1_{\mathrm{loc}}(\mathbb{R}^n)$ by addition, and we may calculate

$$\begin{aligned}
\langle \Delta(f*K), \varphi\rangle &= \langle f*K, \Delta\varphi\rangle, \quad \varphi \in C_0^\infty(\mathbb{R}^n) \\
&= \int_{\mathbb{R}^n} (f*K)(x)\Delta\varphi(x)\mathrm{d}x = \int_{\mathbb{R}^n}\int_{\mathbb{R}^n} f(y)K(x-y)\mathrm{d}y\Delta\varphi(x)\mathrm{d}x
\end{aligned}$$

$$= \int_{\mathbb{R}^n} f(y) \int_{\mathbb{R}^n} K(x-y)\Delta\varphi(x)\mathrm{d}x\mathrm{d}y = \int_{\mathbb{R}^n} f(y)\langle K(x-y), \Delta\varphi(x)\rangle \mathrm{d}y$$
$$= \int_{\mathbb{R}^n} f(y)\langle \Delta K(x-y), \varphi(x)\rangle \mathrm{d}y = \int_{\mathbb{R}^n} f(y)\langle \delta(x-y), \varphi(x)\rangle \mathrm{d}y$$
$$= \int_{\mathbb{R}^n} f(y)\varphi(y)\mathrm{d}y = \langle f, \varphi\rangle.$$

Hence $\Delta(f * K) = f$. □

Exercise 39.8. Prove Theorem 39.19 for $n = 2$.

Exercise 39.9. Prove Theorem 39.19 for $n = 1$.

Theorem 39.20. *Let Ω be a bounded domain in $\mathbb{R}^n$ (for $n = 1$ assume that $\Omega = (a,b)$) with C^1 boundary $\partial\Omega = S$. If $u \in C^1(\overline{\Omega})$ is harmonic in Ω, then*

$$u(x) = \int_S \left(u(y)\partial_{\nu_y} K(x-y) - K(x-y)\partial_\nu u(y)\right) \mathrm{d}\sigma(y), \quad x \in \Omega, \tag{39.5}$$

where $K(x)$ is the fundamental solution (39.3).

Proof. Let us consider K_ε from (39.4). Then since $\Delta u = 0$ in Ω, by Green's identity (1) (see Exercise 39.1) we have

$$\int_\Omega u(y)\Delta_y K_\varepsilon(x-y)\mathrm{d}y = \int_S \left(u(y)\partial_{\nu_y} K_\varepsilon(x-y) - K_\varepsilon(x-y)\partial_\nu u(y)\right) \mathrm{d}\sigma(y).$$

As $\varepsilon \to 0$, the right-hand side of this equation tends to the right-hand side of (39.5) for each $x \in \Omega$, since for $x \in \Omega$ and $y \in S$ there are no singularities in K. On the other hand, the left-hand side is just $(u * \Delta K_\varepsilon)(x)$ if we set $u \equiv 0$ outside Ω. According to the proof of Theorem 39.18,

$$(u * \Delta K_\varepsilon)(x) \to u(x), \quad \varepsilon \to 0,$$

completing the proof. □

Remark 39.21. If we know that $u = f$ and $\partial_\nu u = g$ on S, then

$$u(x) = \int_S \left(f(y)\partial_{\nu_y} K(x-y) - K(x-y)g(y)\right) \mathrm{d}\sigma(y)$$

is the solution of $\Delta u = 0$ with Cauchy data on S. But this problem is overdetermined, because we know from Corollary 39.14 that the solution of $\Delta u = 0$ is uniquely determined by f alone.

The following theorem concerns the spaces $C^\alpha(\Omega)$ and $C^{k,\alpha}(\Omega)$ defined by

$$C^\alpha(\Omega) \equiv C^{0,\alpha}(\Omega) = \{u \in L^\infty(\Omega) : |u(x) - u(y)| \le C|x-y|^\alpha, x, y \in \Omega\},$$
$$C^{k,\alpha}(\Omega) \equiv C^{k+\alpha}(\Omega) = \left\{u : \partial^\beta u \in C^\alpha(\Omega), |\beta| \le k\right\},$$

for $0 < \alpha < 1$ and $k \in \mathbb{N}$.

Theorem 39.22 (Regularity in Hölder spaces). *Suppose $k \ge 0$ is an integer, $0 < \alpha < 1$, and $\Omega \subset \mathbb{R}^n$ is open. If $f \in C^{k+\alpha}(\Omega)$ and u is a distributional solution of $\Delta u = f$ in Ω, then $u \in C^{k+2+\alpha}_{\mathrm{loc}}(\Omega)$.*

Proof. Since $\Delta(\partial^\beta u) = \partial^\beta \Delta u = \partial^\beta f$, we can assume without loss of generality that $k = 0$. Given $\Omega_1 \subset \Omega$ such that $\overline{\Omega_1} \subset \Omega$ choose $\varphi \in C_0^\infty(\Omega)$ such that $\varphi \equiv 1$ on Ω_1 and let $g = \varphi f$.

Since $\Delta(g * K) = g$ (see Theorem 39.19) and therefore $\Delta(g * K) = f$ in Ω_1, it follows that $u - (g * K)$ is harmonic in Ω_1 and hence C^∞ there. It is therefore enough to prove that if g is a C^α function with compact support, then $g * K \in C^{2+\alpha}$. To this end we consider $K_\varepsilon(x)$ and its derivatives. Straightforward calculations lead to following formulas ($n \ge 1$):

$$\begin{aligned}
\frac{\partial}{\partial x_j} K_\varepsilon(x) &= \omega_n^{-1} x_j (|x|^2 + \varepsilon^2)^{-n/2}, \\
\frac{\partial^2}{\partial x_i \partial x_j} K_\varepsilon(x) &= \omega_n^{-1} \begin{cases} -n x_i x_j (|x|^2 + \varepsilon^2)^{-n/2-1}, & i \ne j, \\ (|x|^2 + \varepsilon^2 - n x_j^2)(|x|^2 + \varepsilon^2)^{-n/2-1}, & i = j. \end{cases}
\end{aligned} \tag{39.6}$$

Exercise 39.10. Prove formulas (39.6).

Since $K_\varepsilon \in C^\infty$, we have $g * K_\varepsilon \in C^\infty$ also. Moreover, $\partial_j(g * K_\varepsilon) = g * \partial_j K_\varepsilon$ and $\partial_i \partial_j (g * K_\varepsilon) = g * \partial_i \partial_j K_\varepsilon$. The pointwise limits in (39.6) as $\varepsilon \to 0$ imply

$$\begin{aligned}
\frac{\partial}{\partial x_j} K(x) &= \omega_n^{-1} x_j |x|^{-n}, \\
\frac{\partial^2}{\partial x_i \partial x_j} K(x) &= \begin{cases} -n \omega_n^{-1} x_i x_j |x|^{-n-2}, & i \ne j, \\ \omega_n^{-1}(|x|^2 - n x_j^2)|x|^{-n-2}, & i = j, \end{cases}
\end{aligned} \tag{39.7}$$

for $x \ne 0$. Formulas (39.7) show that $\partial_j K(x)$ is a locally integrable function, and since g is bounded with compact support, it follows that $g * \partial_j K$ is continuous. Next, $g * \partial_j K_\varepsilon \to g * \partial_j K$ uniformly as $\varepsilon \to 0+$. This is equivalent to $\partial_j K_\varepsilon \to \partial_j K$ in the topology of distributions (see the definition). Hence $\partial_j (g * K) = g * \partial_j K$.

This argument does not work for the second derivatives, because $\partial_i \partial_j K(x)$ is not integrable. But there is a different procedure for these terms.

Let $i \ne j$. Then $\partial_i \partial_j K_\varepsilon(x)$ and $\partial_i \partial_j K(x)$ are odd functions of x_i (and x_j); see (39.6) and (39.7). Due to this fact, their integrals over an annulus $0 < a < |x| < b$ vanish. For K_ε we can even take $a = 0$.

Exercise 39.11. Prove this fact.

Therefore, for all $b > 0$ we have

$$\begin{aligned} g * \partial_i\partial_j K_\varepsilon(x) &= \int_{\mathbb{R}^n} g(x-y)\partial_i\partial_j K_\varepsilon(y)\mathrm{d}y - g(x)\int_{|y|<b} \partial_i\partial_j K_\varepsilon(y)\mathrm{d}y \\ &= \int_{|y|<b} (g(x-y)-g(x))\partial_i\partial_j K_\varepsilon(y)\mathrm{d}y + \int_{|y|\geq b} g(x-y)\partial_i\partial_j K_\varepsilon(y)\mathrm{d}y. \end{aligned}$$

If we let $\varepsilon \to 0$, we obtain

$$\begin{aligned} &\lim_{\varepsilon\to 0} g * \partial_i\partial_j K_\varepsilon(x) \\ &\quad = \int_{|y|<b} (g(x-y)-g(x))\partial_i\partial_j K(y)\mathrm{d}y + \int_{|y|\geq b} g(x-y)\partial_i\partial_j K(y)\mathrm{d}y. \end{aligned}$$

This limit exists because

$$|g(x-y)-g(x)||\partial_i\partial_j K(y)| \leq c|y|^\alpha |y|^{-n}$$

(g is C^α) and because g is compactly supported. Then, since b is arbitrary, we can let $b \to +\infty$ to obtain

$$\begin{aligned} \partial_i\partial_j(g*K)(x) &= \lim_{b\to\infty}\int_{|y|<b} (g(x-y)-g(x))\partial_i\partial_j K(y)\mathrm{d}y \\ &\quad + \lim_{b\to\infty}\int_{|y|\geq b} g(x-y)\partial_i\partial_j K(y)\mathrm{d}y \\ &= \lim_{b\to\infty}\int_{|y|<b} (g(x-y)-g(x))\partial_i\partial_j K(y)\mathrm{d}y. \end{aligned} \tag{39.8}$$

A similar result holds for $i = j$. Indeed,

$$\partial_j^2 K_\varepsilon(x) = \frac{1}{n}\varepsilon^{-n}\psi(\varepsilon^{-1}x) + K_j^\varepsilon(x),$$

where $\psi(x) = n\omega_n^{-1}(|x|^2+1)^{-n/2-1}$ and $K_j^\varepsilon = \omega_n^{-1}(|x|^2 - nx_j^2)(|x|^2+\varepsilon^2)^{-n/2-1}$ (see (39.6)). The integral I_j of K_j^ε over an annulus $a < |y| < b$ vanishes. Why is that so? First of all, I_j is independent of j by symmetry in the coordinates, that is, $I_j = I_i$ for $i \neq j$. So nI_j is the integral of $\sum_{j=1}^n K_j^\varepsilon$. But $\sum_{j=1}^n K_j^\varepsilon = 0$. Hence $I_j = 0$ also. We can therefore apply the same procedure. Since

$$g * (\varepsilon^{-n}\psi(\varepsilon^{-1}x)) \to g, \quad \varepsilon \to 0$$

(because $\varepsilon^{-n}\psi(\varepsilon^{-1}x)$ is an approximation to the identity), it follows that

$$\partial_j^2(g*K)(x) = \frac{g(x)}{n} + \lim_{b\to\infty}\int_{|y|<b}(g(x-y)-g(x))\partial_j^2 K(y)\mathrm{d}y. \tag{39.9}$$

Since the convergence in (39.8) and (39.9) is uniform, at this point we have shown that $g*K \in C^2$. But we need to prove more.

Lemma 39.23 (Calderon–Zigmund). *Let N be a C^1 function on $\mathbb{R}^n \setminus \{0\}$ that is homogeneous of degree $-n$ and satisfies*

$$\int_{a<|y|<b} N(y)\mathrm{d}y = 0$$

for all $0<a<b<\infty$. Then if g is a C^α function with compact support, $0<\alpha<1$, then

$$h(x) = \lim_{b\to\infty}\int_{|z|<b}(g(x-z)-g(x))N(z)\mathrm{d}z$$

belongs to C^α.

Proof. Let us write $h = h_1 + h_2$, where

$$h_1(x) = \int_{|z|\le 3|y|}(g(x-z)-g(x))N(z)\mathrm{d}z,$$

$$h_2(x) = \lim_{b\to\infty}\int_{3|y|<|z|<b}(g(x-z)-g(x))N(z)\mathrm{d}z.$$

We wish to estimate $h(x+y)-h(x)$. Since $\alpha>0$, we have

$$|h_1(x)| \le c\int_{|z|\le 3|y|}|z|^\alpha|z|^{-n}\mathrm{d}z = c'|y|^\alpha$$

and hence

$$|h_1(x+y)-h_1(x)| \le |h_1(x+y)|+|h_1(x)| \le 2c'|y|^\alpha.$$

On the other hand,

$$\begin{aligned}
h_2(x+y)-h_2(x) &= \lim_{b\to\infty}\int_{3|y|<|z+y|<b}(g(x-z)-g(x))N(z+y)\mathrm{d}z \\
&\quad - \lim_{b\to\infty}\int_{3|y|<|z|<b}(g(x-z)-g(x))N(z)\mathrm{d}z \\
&= \lim_{b\to\infty}\int_{3|y|<|z|<b}(g(x-z)-g(x))(N(z+y)-N(z))\mathrm{d}z \\
&\quad + \lim_{b\to\infty}\int_{\{3|y|<|z+y|<b\}\setminus\{3|y|<|z|<b\}}(g(x-z)-g(x))N(z+y)\mathrm{d}z \\
&=: I_1 + I_2.
\end{aligned}$$

It is clear that

$$\{3|y|<|z+y|\}\setminus\{3|y|<|z|\}\subset\{2|y|<|z|\}\setminus\{3|y|<|z|\}$$
$$=\{2|y|<|z|\le 3|y|\}.$$

Therefore,

$$|I_2|\le\int_{2|y|<|z|\le 3|y|}|g(x-z)-g(x)||N(z+y)|\mathrm{d}z$$
$$\le c\int_{2|y|<|z|\le 3|y|}|z|^{\alpha}|z+y|^{-n}\mathrm{d}z\le c'\int_{2|y|<|z|\le 3|y|}|z|^{\alpha-n}\mathrm{d}z=c''|y|^{\alpha}.$$

Now we observe that for $|z|>3|y|$,

$$|N(z+y)-N(z)|\le|y|\sup_{0\le t\le 1}|\nabla N(z+ty)|$$
$$\le c|y|\sup_{0\le t\le 1}|z+ty|^{-n-1}\le c'|y||z|^{-n-1},$$

because ∇N is homogeneous of degree $-n-1$, since N is homogeneous of degree $-n$. Hence

$$|I_1|\le c\int_{|z|>3|y|}|z|^{\alpha}|y||z|^{-n-1}\mathrm{d}z=c'|y|\int_{3|y|}^{\infty}\rho^{\alpha-2}\mathrm{d}\rho=c''|y|^{\alpha}.$$

Note that the condition $\alpha<1$ is needed here. Collecting the estimates for I_1 and I_2, we can see that the lemma is proved. □

In order to end the proof of Theorem 39.22 it remains to note that $\partial_i\partial_j K(x)$ satisfies all the conditions of Lemma 39.23. □

Exercise 39.12. Show that a function K_1 is a fundamental solution for $\Delta^2\equiv\Delta(\Delta)$ on $\mathbb{R}^n$ if and only if K_1 satisfies the equation

$$\Delta K_1=K,$$

where K is the fundamental solution for the Laplacian.

Exercise 39.13. Show that the following functions are the fundamental solutions for Δ^2 on $\mathbb{R}^n$:

(1) $n=4$:

$$-\frac{\log|x|}{4\omega_4};$$

(2) $n=2$:

$$\frac{|x|^2\log|x|}{8\pi};$$

(3) $n \neq 2,4$:

$$\frac{|x|^{4-n}}{2(4-n)(2-n)\omega_n}.$$

Exercise 39.14. Show that $(4\pi|x|)^{-1}e^{-c|x|}$ is the fundamental solution for $-\Delta + c^2$ on $\mathbb{R}^3$ for an arbitrary constant $c \in \mathbb{C}$.

Chapter 40
The Dirichlet and Neumann Problems

The Dirichlet problem
Given functions f in Ω and g on $S = \partial\Omega$, find a function u in $\overline{\Omega} = \Omega \cup \partial\Omega$ satisfying

$$\begin{cases} \Delta u = f, & \text{in}\, \Omega \\ u = g, & \text{on}\, S. \end{cases} \tag{D}$$

The Neumann problem
Given functions f in Ω and g on S, find a function u in $\overline{\Omega}$ satisfying

$$\begin{cases} \Delta u = f, & \text{in}\, \Omega \\ \partial_\nu u = g, & \text{on}\, S. \end{cases} \tag{N}$$

We assume that Ω is bounded with C^1 boundary. But we shall not, however, assume that Ω is connected. The uniqueness theorem (see Corollary 39.14) shows that the solution of (D) will be unique (if it exists), at least if we require $u \in C(\overline{\Omega})$. For (N) uniqueness does not hold: we can add to $u(x)$ any function that is constant on each connected component of Ω. Moreover, there is an obvious necessary condition for solvability of (N). If Ω' is a connected component of Ω, then

$$\int_{\Omega'} \Delta u \mathrm{d}x = \int_{\partial\Omega'} \partial_\nu u \mathrm{d}\sigma(x) = \int_{\partial\Omega'} g(x) \mathrm{d}\sigma(x) = \int_{\Omega'} f \mathrm{d}x,$$

that is,

$$\int_{\Omega'} f(x) \mathrm{d}x = \int_{\partial\Omega'} g(x) \mathrm{d}\sigma(x).$$

It is also clear (by linearity) that (D) can be reduced to the following *homogeneous* problems:

V. Serov, *Fourier Series, Fourier Transform and Their Applications to Mathematical Physics*, Applied Mathematical Sciences 197,
DOI 10.1007/978-3-319-65262-7_40

$$\begin{cases} \Delta v = f, & \text{in } \Omega \\ v = 0, & \text{on } S \end{cases} \tag{D$_A$}$$

$$\begin{cases} \Delta w = 0, & \text{in } \Omega \\ w = g, & \text{on } S \end{cases} \tag{D$_B$}$$

and $u := v + w$ solves (D). Similar remarks apply to (N), that is,

$$\begin{cases} \Delta v = f, & \text{in } \Omega \\ \partial_\nu v = 0, & \text{on } S \end{cases}$$

$$\begin{cases} \Delta w = 0, & \text{in } \Omega \\ \partial_\nu w = g, & \text{on } S \end{cases}$$

and $u = v + w$.

Definition 40.1. The *Green's function* for (D) in Ω is the solution $G(x,y)$ of the boundary value problem

$$\begin{cases} \Delta_x G(x,y) = \delta(x-y), & x, y \in \Omega \\ G(x,y) = 0, & x \in S, y \in \Omega. \end{cases} \tag{40.1}$$

Analogously, the Green's function for (N) in Ω is the solution $G(x,y)$ of the boundary value problem

$$\begin{cases} \Delta_x G(x,y) = \delta(x-y), & x, y \in \Omega \\ \partial_{\nu_x} G(x,y) = 0, & x \in S, y \in \Omega. \end{cases} \tag{40.2}$$

This definition allows us to write

$$G(x,y) = K(x-y) + v_y(x), \tag{40.3}$$

where K is the fundamental solution of Δ in $\mathbb{R}^n$ and for all $y \in \Omega$, the function $v_y(x)$ satisfies

$$\begin{cases} \Delta v_y(x) = 0, & \text{in } \Omega \\ v_y(x) = -K(x-y), & \text{on } S \end{cases} \tag{40.4}$$

in the case of (40.1) and

$$\begin{cases} \Delta v_y(x) = 0, & \text{in } \Omega \\ \partial_{\nu_x} v_y(x) = -\partial_{\nu_x} K(x-y), & \text{on } S \end{cases}$$

in the case of (40.2). Since (40.4) guarantees that v_y is real, it follows that so is G corresponding to (40.1).

Lemma 40.2. *The Green's function* (40.1) *exists and is unique.*

Proof. The uniqueness of G follows again from Corollary 39.14, since $K(x-y)$ in (40.4) is continuous for all $x \in S$ and $y \in \Omega$ $(x \neq y)$. The existence will be proved later. □

Lemma 40.3. *For both* (40.1) *and* (40.2) *it is true that* $G(x,y) = G(y,x)$ *for all* $x, y \in \Omega$.

Proof. Let $G(x,y)$ and $G(x,z)$ be the Green's functions for Ω corresponding to sources located at fixed y and z, $y \neq z$, respectively. Let us consider the domain

$$\Omega_\varepsilon = (\Omega \setminus \{x : |x-y| < \varepsilon\}) \setminus \{x : |x-z| < \varepsilon\},$$

see Figure 40.1.

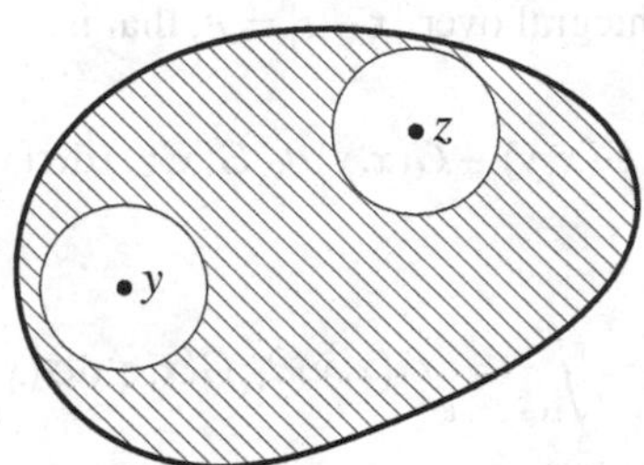

Fig. 40.1 The domain Ω_ε.

If $x \in \Omega_\varepsilon$, then $x \neq z$ and $x \neq y$, and therefore, $\Delta_x G(x,z) = 0$ and $\Delta_x G(x,y) = 0$. These facts imply

$$\begin{aligned} 0 &= \int_{\Omega_\varepsilon} (G(x,y)\Delta_x G(x,z) - G(x,z)\Delta_x G(x,y))\,\mathrm{d}x \\ &= \int_S (G(x,y)\partial_{\nu_x} G(x,z) - G(x,z)\partial_{\nu_x} G(x,y))\,\mathrm{d}\sigma(x) \\ &\quad - \int_{|x-y|=\varepsilon} (G(x,y)\partial_{\nu_x} G(x,z) - G(x,z)\partial_{\nu_x} G(x,y))\,\mathrm{d}\sigma(x) \\ &\quad - \int_{|x-z|=\varepsilon} (G(x,y)\partial_{\nu_x} G(x,z) - G(x,z)\partial_{\nu_x} G(x,y))\,\mathrm{d}\sigma(x). \end{aligned}$$

Hence by (40.1) or (40.2), for arbitrary $\varepsilon > 0$ (sufficiently small),

$$\begin{aligned} &\int_{|x-y|=\varepsilon} (G(x,y)\partial_{\nu_x} G(x,z) - G(x,z)\partial_{\nu_x} G(x,y))\,\mathrm{d}\sigma(x) \\ &= \int_{|x-z|=\varepsilon} (G(x,z)\partial_{\nu_x} G(x,y) - G(x,y)\partial_{\nu_x} G(x,z))\,\mathrm{d}\sigma(x). \end{aligned}$$

Let $n \geq 3$. Due to (40.3) for $\varepsilon \to 0$ we have

$$\begin{aligned}
&\int_{|x-y|=\varepsilon} (G(x,y)\partial_{\nu_x} G(x,z) - G(x,z)\partial_{\nu_x} G(x,y))\,\mathrm{d}\sigma(x) \\
&\approx \frac{1}{\omega_n}\int_{|x-y|=\varepsilon} \varepsilon^{2-n}\left(\frac{(x-y,x-z)}{|x-y||x-z|^n} + \partial_{\nu_x} v_z(x)\right)\mathrm{d}\sigma(x) \\
&- \int_{|x-y|=\varepsilon} G(x,z)\partial_{\nu_x} G(x,y)\mathrm{d}\sigma(x) \\
&\approx \frac{1}{\omega_n}\varepsilon^{2-n}\varepsilon^{n-1}\frac{1}{\varepsilon}\int_\theta \frac{(\varepsilon\theta,\varepsilon\theta+y-z)}{|\varepsilon\theta+y-z|^n}\mathrm{d}\theta - I_1 \approx -I_1,
\end{aligned}$$

where we have defined

$$I_1 = \int_{|x-y|=\varepsilon} G(x,z)\partial_{\nu_x} G(x,y)\mathrm{d}\sigma(x).$$

The same is true for the integral over $|x-z| = \varepsilon$, that is,

$$\int_{|x-z|=\varepsilon} (G(x,z)\partial_{\nu_x} G(x,y) - G(x,y)\partial_{\nu_x} G(x,z))\,\mathrm{d}\sigma(x) \approx -I_2, \quad \varepsilon \to 0,$$

where

$$I_2 = \int_{|x-z|=\varepsilon} G(x,y)\partial_{\nu_x} G(x,z)\mathrm{d}\sigma(x).$$

But using the previous techniques we can obtain that

$$I_1 \approx \frac{1}{\omega_n}\varepsilon^{1-n}\varepsilon^{n-1}\int_{|\theta|=1} G(\varepsilon\theta+y,z)\mathrm{d}\theta \to G(y,z), \quad \varepsilon \to 0$$

and

$$I_2 \approx \frac{1}{\omega_n}\varepsilon^{1-n}\varepsilon^{n-1}\int_{|\theta|=1} G(\varepsilon\theta+z,y)\mathrm{d}\theta \to G(z,y), \quad \varepsilon \to 0.$$

This means that $G(y,z) = G(z,y)$ for all $z \neq y$. This proof holds for $n = 2$ (and even for $n = 1$) with some simple changes. □

Lemma 40.4. *In three or more dimensions*

$$K(x-y) < G(x,y) < 0, \quad x,y \in \Omega,\, x \neq y$$

where $G(x,y)$ is the Green's function for (D).

Proof. For each fixed y, the function $v_y(x) := G(x,y) - K(x-y)$ is harmonic in Ω; see (40.4). Moreover, on $S = \partial\Omega$, $v_y(x)$ takes on the positive value

$$-K(x-y) \equiv -\frac{|x-y|^{2-n}}{\omega_n(2-n)}.$$

By the minimum principle, it follows that $v_y(x)$ is strictly positive in Ω. This proves the first inequality. □

Exercise 40.1. Prove the second inequality in Lemma 40.4.

Exercise 40.2. Show that for $n = 2$, Lemma 40.4 has the following form:

$$\frac{1}{2\pi}\log\frac{|x-y|}{h} < G(x,y) < 0, \quad x,y \in \Omega,$$

where $h \equiv \max_{x,y\in\overline{\Omega}} |x-y|$.

Exercise 40.3. Obtain the analogue of Lemma 40.4 for $n = 1$. Hint: show that the Green's function for the operator $\frac{d^2}{dx^2}$ on $\Omega = (0,1)$ is

$$G(x,y) = \begin{cases} x(y-1), & x < y \\ y(x-1), & x > y. \end{cases}$$

Remark 40.5. $G(x,y)$ may be extended naturally (because of the symmetry) to $\overline{\Omega} \times \overline{\Omega}$ by setting $G(x,y) = 0$ for $y \in S$.

Now we can solve both problems (D_A) and (D_B). Indeed, let us set $f = 0$ in (D_A) outside Ω and define

$$v(x) := \int_\Omega G(x,y)f(y)\mathrm{d}y \equiv (f * K)(x) + \int_\Omega (G(x,y) - K(x-y))\, f(y)\mathrm{d}y.$$

Then the Laplacian of the first term is f (see Theorem 39.19), and the second term is harmonic in x (since $v_y(x)$ is harmonic). Also $v(x) = 0$ on S, because the same is true for G. Thus, this $v(x)$ solves (D_A).

Consider now (D_B). We assume that g is continuous on S and we wish to find w that is continuous on $\overline{\Omega}$. Applying Green's identity (1) (together with the same limiting process as in the proof of Lemma 40.3), we obtain

$$\begin{aligned} w(x) &= \int_\Omega (w(y)\Delta_y G(x,y) - G(x,y)\Delta w(y))\,\mathrm{d}y \\ &= \int_S w(y)\partial_{\nu_y} G(x,y)\mathrm{d}\sigma(y) = \int_S g(y)\partial_{\nu_y} G(x,y)\mathrm{d}\sigma(y). \end{aligned}$$

Let us denote the last integral by (P). Since $\partial_{\nu_y} G(x,y)$ is harmonic in x and continuous in y for $x \in \Omega$ and $y \in S$, then $w(x)$ is harmonic in Ω. In order to prove that this $w(x)$ solves (D_B), it remains to prove that $w(x)$ is continuous in $\overline{\Omega}$ and $w(x)$ on S is $g(x)$. We will prove this general fact later.

Definition 40.6. The function $\partial_{\nu_y} G(x,y)$ on $\Omega \times S$ is called the *Poisson kernel* for Ω, and (P) is called the *Poisson integral.*

Now we are in a position to solve the *Dirichlet problem* in a half-space. Let

$$\Omega = \mathbb{R}^{n+1}_+ = \{(x', x_{n+1}) \in \mathbb{R}^{n+1} : x' \in \mathbb{R}^n, x_{n+1} > 0\},$$

where $n \geq 1$ now, and let $x_{n+1} = t$. Then

$$\Delta_{n+1} = \Delta_n + \partial_t^2, \quad n = 1, 2, \ldots.$$

Denote by $K(x,t)$ a fundamental solution for Δ_{n+1} in $\mathbb{R}^{n+1}$, that is,

$$K(x,t) = \begin{cases} \frac{(|x|^2+t^2)^{\frac{1-n}{2}}}{(1-n)\omega_{n+1}}, & n > 1 \\ \frac{1}{4\pi}\log(|x|^2+t^2), & n = 1. \end{cases}$$

Let us prove then that the Green's function for $\mathbb{R}^{n+1}_+$ is

$$G(x,y;t,s) = K(x-y, t-s) - K(x-y, -t-s). \tag{40.5}$$

It is clear (see (40.5)) that $G(x,y;t,0) = G(x,y;0,s) = 0$ and

$$\Delta_{n+1} G = \delta(x-y, t-s) - \delta(x-y, -t-s) = \delta(x-y)\delta(t-s),$$

because for $t, s > 0$, $-t-s < 0$, and therefore, $\delta(-t-s) = 0$. Thus G is the Dirichlet Green's function for $\mathbb{R}^{n+1}_+$. From this we immediately have the solution of (D_A) in $\mathbb{R}^{n+1}_+$ as

$$u(x,t) = \int_{\mathbb{R}^n} \int_0^\infty G(x,y;t,s) f(y,s) \mathrm{d}s \mathrm{d}y.$$

To solve (D_B) we compute the Poisson kernel for this case. Since the outward normal derivative on $\partial\mathbb{R}^{n+1}_+$ is $-\frac{\partial}{\partial t}$, the Poisson kernel becomes

$$-\frac{\partial}{\partial s} G(x,y;t,s)|_{s=0} = -\frac{\partial}{\partial s} \left(K(x-y,t-s) - K(x-y,-t-s)\right)|_{s=0}$$
$$= \frac{2t}{\omega_{n+1}(|x-y|^2+t^2)^{\frac{n+1}{2}}}. \tag{40.6}$$

Exercise 40.4. Prove (40.6).

Note that (40.6) holds for all $n \geq 1$. According to the formula for (P), the candidate for a solution to (D_B) is

$$u(x,t) = \frac{2}{\omega_{n+1}} \int_{\mathbb{R}^n} \frac{t g(y)}{(|x-y|^2+t^2)^{\frac{n+1}{2}}} \mathrm{d}y. \tag{40.7}$$

In other words, if we set

$$P_t(x) := \frac{2t}{\omega_{n+1}(|x|^2+t^2)^{\frac{n+1}{2}}}, \tag{40.8}$$

which is what is usually called the Poisson kernel for $\mathbb{R}^{n+1}_+$, the proposed solution (40.7) is simply equal to

$$u(x,t) = (g * P_t)(x). \tag{40.9}$$

Exercise 40.5. Prove that $P_t(x) = t^{-n}P_1(t^{-1}x)$ and

$$\int_{\mathbb{R}^n} P_t(y)\mathrm{d}y = 1.$$

Theorem 40.7. *Suppose $g \in L^p(\mathbb{R}^n)$, $1 \le p \le \infty$. Then $u(x,t)$ from* (40.9) *is well defined on $\mathbb{R}^{n+1}_+$ and is harmonic there. If g is bounded and uniformly continuous, then $u(x,t)$ is continuous on $\overline{\mathbb{R}^{n+1}_+}$ and $u(x,0) = g(x)$, and*

$$\|u(\cdot,t) - g(\cdot)\|_\infty \to 0$$

as $t \to 0+$.

Proof. It is clear that for all $t > 0$, $P_t(x) \in L^1(\mathbb{R}^n) \cap L^\infty(\mathbb{R}^n)$; see (40.8). Hence $P_t(x) \in L^q(\mathbb{R}^n)$ for all $q \in [1,\infty]$ with respect to x and fixed $t > 0$. Therefore, the integral in (40.9) is absolutely convergent, and the same is true if P_t is replaced by its derivative $\Delta_x P_t$ or $\partial_t^2 P_t$ (due to Young's inequality for convolution).

Since $G(x,y;t,s)$ is harmonic for $(x,t) \ne (y,s)$, it follows that $P_t(x)$ is also harmonic and

$$\Delta_x u + \partial_t^2 u = g * (\Delta_x + \partial_t^2)P_t = 0.$$

It remains to prove that if g is bounded and continuous, then

$$\|u(\cdot,t) - g(\cdot)\|_\infty \to 0$$

as $t \to 0+$, and therefore, $u(x,0) = g(x)$ and u is continuous on $\overline{\mathbb{R}^{n+1}_+}$.

We have (see Exercise 40.5)

$$\begin{aligned}
\|g * P_t - g\|_\infty &= \sup_{x\in\mathbb{R}^n}\left|\int_{\mathbb{R}^n} g(x-y)P_t(y)\mathrm{d}y - \int_{\mathbb{R}^n} g(x)P_t(y)\mathrm{d}y\right| \\
&\le \sup_{x\in\mathbb{R}^n}\int_{\mathbb{R}^n} |g(x-y)-g(x)||P_t(y)|\mathrm{d}y \\
&= \sup_{x\in\mathbb{R}^n}\int_{\mathbb{R}^n} |g(x-tz)-g(x)||P_1(z)|\mathrm{d}z
\end{aligned}$$

$$
\begin{aligned}
&= \sup_{x\in\mathbb{R}^n}\left(\int_{|z|<R} |g(x-tz)-g(x)||P_1(z)|\mathrm{d}z\right.\\
&\quad\left.+\int_{|z|\geq R} |g(x-tz)-g(x)||P_1(z)|\mathrm{d}z\right)\\
&\leq \sup_{x\in\mathbb{R}^n,|z|<R} |g(x-tz)-g(x)|+2\|g\|_\infty\int_{|z|\geq R}|P_1(z)|\mathrm{d}z<\varepsilon
\end{aligned}
$$

for t sufficiently small.

The first term in the sum on the last line can be made less than $\varepsilon/2$, since g is uniformly continuous on $\mathbb{R}^n$. The second term can be made less than $\varepsilon/2$ for R large enough, since $P_1 \in L^1(\mathbb{R}^n)$. Thus, the theorem is proved. □

Remark 40.8. The solution of this problem is not unique: if $u(x,t)$ is a solution, then so is $u(x,t)+ct$ for all $c \in \mathbb{C}$. However, we have the following theorem.

Theorem 40.9. *If $g \in C(\mathbb{R}^n)$ and $\lim_{x\to\infty} g(x) = 0$, then $u(x,t) := (g * P_t)(x) \to 0$ as $(x,t) \to \infty$ in $\mathbb{R}^{n+1}_+$, and it is the unique solution with this property.*

Proof. Assume for the moment that g has compact support, say $g = 0$ for $|x| > R$. Then $g \in L^1(\mathbb{R}^n)$ and

$$\|g * P_t\|_\infty \leq \|g\|_1 \|P_t\|_\infty \leq ct^{-n},$$

so $u(x,t) \to 0$ as $t \to \infty$ uniformly in x. On the other hand, if $0 < t \leq T$, then

$$|u(x,t)| \leq \|g\|_1 \sup_{|y|<R} |P_t(x-y)| = \|g\|_1 \sup_{|y|<R} \frac{2t}{\omega_{n+1}(|x-y|^2+t^2)^{\frac{n+1}{2}}} \leq cT|x|^{-n-1},$$

for $|x| > 2R$. Hence $u(x,t) \to 0$ as $x \to \infty$ uniformly for $t \in [0,T]$. This proves that $u(x,t)$ vanishes at infinity if $g(x)$ has compact support. For general g, choose a sequence $\{g_k\}$ of compactly supported functions that converges uniformly (in $L^\infty(\mathbb{R}^n)$) to g and let

$$u_k(x,t) = (g_k * P_t)(x).$$

Then

$$
\begin{aligned}
\|u_k - u\|_{L^\infty(\mathbb{R}^{n+1})} &= \sup_{t,x}\left|\int_{\mathbb{R}^n}(g_k-g)(y)P_t(x-y)\mathrm{d}y\right|\\
&\leq \sup_t\left(\|g_k-g\|_{L^\infty(\mathbb{R}^n)}\sup_x\int_{\mathbb{R}^n}|P_t(x-y)|\mathrm{d}y\right)\\
&= \|g_k-g\|_{L^\infty(\mathbb{R}^n)}\sup_{t>0}\int_{\mathbb{R}^n}|P_t(y)|\mathrm{d}y = \|g_k-g\|_{L^\infty(\mathbb{R}^n)} \to 0
\end{aligned}
$$

as $k \to \infty$.

Hence $u(x,t)$ vanishes at infinity. Now suppose v is another solution and let $w := v - u$. Then w vanishes at infinity and also at $t = 0$ (see Theorem 40.7). Thus $|w| < \varepsilon$ on the boundary of the cylindrical region $\{(x,t) : |x| < R, 0 < t < R\}$ for R sufficiently large, see Figure 40.2.

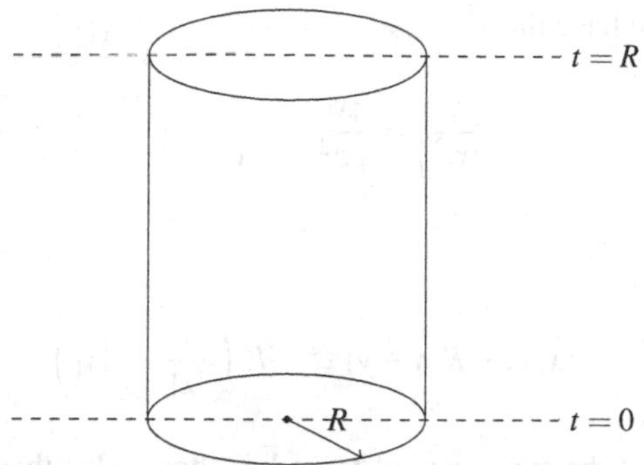

Fig. 40.2 Geometric illustration of the Poisson integral.

But since w is harmonic, it follows by the maximum principle that $|w| < \varepsilon$ in this region. Letting $\varepsilon \to 0$ and $R \to \infty$, we conclude that $w \equiv 0$. □

Let us consider now the *Dirichlet problem in a ball*. We use here the following notation:

$$B = B_1(0) = \{x \in \mathbb{R}^n : |x| < 1\}, \quad \partial B = S.$$

Exercise 40.6. Prove that

$$|x - y| = \left| \frac{x}{|x|} - y|x| \right|$$

for $x, y \in \mathbb{R}^n$, $x \neq 0$, $|y| = 1$.

Now, assuming first that $n > 2$, we define

$$\begin{aligned} G(x,y) &:= K(x-y) - K\left(\frac{x}{|x|} - y|x|\right) \\ &= \frac{1}{(2-n)\omega_n}\left(|x-y|^{2-n} - \left|\frac{x}{|x|} - y|x|\right|^{2-n}\right), \quad x \neq 0. \qquad (40.10) \end{aligned}$$

Exercise 40.6 shows that $G(x,y)$ from (40.10) satisfies $G(x,y) = 0, x \in B, y \in S$. It is also clear that $G(x,y) = G(y,x)$. This is true because

$$\left|\frac{x}{|x|} - y|x|\right|^2 = \left|\frac{x}{|x|}\right|^2 - 2(x,y) + |y|^2|x|^2 = 1 - 2(x,y) + |y|^2|x|^2$$
$$= \left|\frac{y}{|y|}\right|^2 - 2(y,x) + |x|^2|y|^2 = \left|\frac{y}{|y|} - x|y|\right|^2.$$

Next, for $x, y \in B$ we have that

$$\left|\frac{x}{|x|^2}\right| = \frac{|x|}{|x|^2} = \frac{1}{|x|} > 1$$

and $y \neq \frac{x}{|x|^2}$. Hence,

$$G(x,y) - K(x-y) \equiv -K\left(\frac{x}{|x|} - y|x|\right)$$

is harmonic in y. But the symmetry of G and K shows also that $G(x,y) - K(x-y)$ is harmonic in x. Thus, $G(x,y)$ is the Green's function for B. This also makes clear how to define G at $x = 0$ (and at $y = 0$):

$$G(0,y) = \frac{1}{(2-n)\omega_n}(|y|^{2-n} - 1),$$

since

$$\left|\frac{x}{|x|} - y|x|\right| \to 1$$

as $x \to 0$.

For $n = 2$ the analogous formulae are

$$G(x,y) = \frac{1}{2\pi}\left(\log|x-y| - \log\left|\frac{x}{|x|} - y|x|\right|\right), \quad G(0,y) = \frac{1}{2\pi}\log|y|.$$

Now we can compute the Poisson kernel $P(x,y) := \partial_{\nu_y} G(x,y)$, $x \in B, y \in S$. Since $\partial_{\nu_y} = y \cdot \nabla_y$ on S, it follows that

$$P(x,y) = -\frac{1}{\omega_n}\left(\frac{(y,x-y)}{|x-y|^n} - \frac{\left(\frac{x}{|x|} - y|x|, y|x|\right)}{\left|\frac{x}{|x|} - y|x|\right|^n}\right) \equiv \frac{1-|x|^2}{\omega_n|x-y|^n}, \quad n \geq 2. \quad (40.11)$$

Exercise 40.7. Prove (40.11).

Theorem 40.10. *If $f \in L^1(S)$, then*

$$u(x) = \int_S P(x,y)f(y)\mathrm{d}\sigma(y), \quad x \in B,$$

is harmonic. If $f \in C(S)$, then u extends continuously to $\overline{B}$ and $u = f$ on S.

Proof. For each $x \in B$ (see (40.11)), $P(x,y)$ is a bounded function of $y \in S$, so $u(x)$ is well defined for $f \in L^1(S)$. It is also harmonic in B, because $P(x,y)$ is harmonic for $x \neq y$. Next, we claim that

$$\int_S P(x,y)\mathrm{d}\sigma(y) = 1. \tag{40.12}$$

Since P is harmonic in x, the mean value theorem implies ($y \in S$)

$$1 = \omega_n P(0,y) = \int_S P(ry',y)\mathrm{d}\sigma(y')$$

for all $0 < r < 1$. But

$$P(ry',y) = P(y,ry') = P(ry,y')$$

if $y,y' \in S$. The last formula follows from

$$|ry' - y|^2 = r^2 - 2r(y',y) + 1 = |ry - y'|^2.$$

We therefore conclude that

$$1 = \int_S P(ry',y)\mathrm{d}\sigma(y') = \int_S P(x,y')\mathrm{d}\sigma(y')$$

with $x = ry$. This proves (40.12). We claim also that for all $y_0 \in S$ and for a neighborhood $B_\sigma(y_0) \subset S$,

$$\lim_{r \to 1-} \int_{S \setminus B_\sigma(y_0)} P(ry_0,y)\mathrm{d}\sigma(y) = 0. \tag{40.13}$$

Indeed, for $y_0, y \in S$ and $0 < r < 1$,

$$|ry_0 - y| > r|y_0 - y|$$

and therefore

$$|ry_0 - y|^{-n} < (r|y_0 - y|)^{-n} \leq (r\sigma)^{-n}$$

if $y \in S \setminus B_\sigma(y_0)$, i.e., $|y - y_0| \geq \sigma$. Hence $|ry_0 - y|^{-n}$ is bounded uniformly for $r \to 1-$ and $y \in S \setminus B_\sigma(y_0)$. In addition, $1 - |ry_0|^2 \equiv 1 - r^2 \to 0$ as $r \to 1-$. This proves (40.13).

Now suppose $f \in C(S)$. Hence f is uniformly continuous, since S is compact. Hence for every $\varepsilon > 0$ there exists $\delta > 0$ such that

$$|f(x) - f(y)| < \varepsilon, \quad x,y \in S, |x - y| < \delta.$$

For all $x \in S$ and $0 < r < 1$, by (40.12),

$$\begin{aligned}
|u(rx) - f(x)| &= \left| \int_S (f(y) - f(x)) P(rx,y) \mathrm{d}\sigma(y) \right| \\
&\le \int_{|x-y|<\delta} |f(y) - f(x)| |P(rx,y)| \mathrm{d}\sigma(y) \\
&\quad + \int_{S \setminus B_\delta(x)} |f(y) - f(x)| |P(rx,y)| \mathrm{d}\sigma(y) \\
&\le \varepsilon \int_S |P(rx,y)| \mathrm{d}\sigma(y) + 2 \|f\|_\infty \int_{S \setminus B_\delta(x)} |P(rx,y)| \mathrm{d}\sigma(y) \\
&\le \varepsilon + 2 \|f\|_\infty \int_{S \setminus B_\delta(x)} P(rx,y) \mathrm{d}\sigma(y) \to 0,
\end{aligned}$$

as $\varepsilon \to 0$ and $r \to 1-$ by (40.13). Hence $u(rx) \to f$ uniformly as $r \to 1-$. □

Corollary 40.11. **(Without proof)** *If u is as in Theorem 40.10 and $f \in L^p(S)$, $1 \le p \le \infty$, then*

$$\|u(r\cdot) - f(\cdot)\|_p \to 0$$

as $r \to 1-$.

Exercise 40.8. Show that the Poisson kernel for the ball $B_R(x_0)$ is

$$P(x,y) = \frac{R^2 - |x - x_0|^2}{\omega_n R |x - y|^n}, \quad n \ge 2.$$

Exercise 40.9. ***(Harnack's inequality)*** Suppose $u \in C(\overline{B})$ is harmonic on B and $u \ge 0$. Show that for $|x| = r < 1$,

$$\frac{1-r}{(1+r)^{n-1}} u(0) \le u(x) \le \frac{1+r}{(1-r)^{n-1}} u(0).$$

Theorem 40.12. **(The *reflection principle*)** *Let $\Omega \subset \mathbb{R}^{n+1}$, $n \ge 1$, be open and satisfy the property that $(x,-t) \in \Omega$ if $(x,t) \in \Omega$. Let $\Omega_+ = \{(x,t) \in \Omega : t > 0\}$ and $\Omega_0 = \{(x,t) \in \Omega : t = 0\}$. If $u(x,t)$ is continuous on $\Omega_+ \cup \Omega_0$, harmonic in Ω_+, and $u(x,0) = 0$, then we can extend u to be harmonic on Ω by setting $u(x,-t) := -u(x,t)$.*

Proof. See [11, (2.68), p. 110]. □

Definition 40.13. If u is harmonic on $\Omega \setminus \{x_0\}$, $\Omega \subset \mathbb{R}^n$ open, then u is said to have a *removable singularity* x_0 if u can be defined at x_0 so as to be harmonic in Ω.

Theorem 40.14. *Suppose u is harmonic on $\Omega \setminus \{x_0\}$ and $u(x) = o\left(|x - x_0|^{2-n}\right)$ for $n > 2$ and $u(x) = o(\log|x - x_0|)$ for $n = 2$ as $x \to x_0$. Then u has a removable singularity at x_0.*

Proof. Without loss of generality we assume that $\Omega = B := B_1(0)$ and $x_0 = 0$. Since u is continuous on ∂B, by Theorem 40.10 there exists $v \in C(\overline{B})$ satisfying

$$\begin{cases} \Delta v = 0, & \text{in}\, B \\ v = u, & \text{on}\, S. \end{cases}$$

We claim that $u = v$ in $B \backslash \{0\}$, so that we can remove the singularity at $\{0\}$ by setting $u(0) := v(0)$. Indeed, given $\varepsilon > 0$ and $0 < \delta < 1$, consider the function

$$g_\varepsilon(x) = \begin{cases} u(x) - v(x) - \varepsilon(|x|^{2-n} - 1), & n > 2 \\ u(x) - v(x) + \varepsilon \log |x|, & n = 2 \end{cases}$$

in $B \backslash \overline{B_\delta(0)}$. These functions are real (as we can assume without loss of generality), harmonic, and continuous for $\delta \leq |x| \leq 1$. Moreover, $g_\varepsilon(x) = 0$ on ∂B and $g_\varepsilon(x) < 0$ on $\partial B_\delta(0)$ for all δ sufficiently small. By the maximum principle, $g_\varepsilon(x)$ is negative in $B \backslash \{0\}$. Letting $\varepsilon \to 0$, we see that $u - v \leq 0$ in $B \backslash \{0\}$. By the same arguments we may conclude that also $v - u \leq 0$ in $B \backslash \{0\}$. Hence $u = v$ in $B \backslash \{0\}$, and we can extend u to the whole ball by setting $u(0) = v(0)$. This proves the theorem. □

Chapter 41
Layer Potentials

In this chapter we assume that $\Omega \subset \mathbb{R}^n$, $n \geq 2$ is bounded and open, and that $S = \partial\Omega$ is a surface of class C^2. We assume also that both Ω and $\Omega' := \mathbb{R}^n \setminus \overline{\Omega}$ are connected.

Definition 41.1. Let $\nu(x)$ be a normal vector to S at x. Then

$$\partial_{\nu_-} u(x) := \lim_{t \to 0-} \nu(x) \cdot \nabla u(x + t\nu(x)),$$
$$\partial_{\nu_+} u(x) := \lim_{t \to 0+} \nu(x) \cdot \nabla u(x + t\nu(x)),$$

are called the *interior and exterior normal derivatives*, respectively, of u.

The interior Dirichlet problem (ID)

Given $f \in C(S)$, find $u \in C^2(\Omega) \cap C(\overline{\Omega})$ such that $\Delta u = 0$ in Ω and $u = f$ on S.

The exterior Dirichlet problem (ED)

Given $f \in C(S)$, find $u \in C^2(\Omega') \cap C(\overline{\Omega'})$ such that $\Delta u = 0$ in Ω' and at infinity and $u = f$ on S.

Definition 41.2. A function u is said to be *harmonic at infinity* if

$$|x|^{2-n} u\left(\frac{x}{|x|^2}\right) = \begin{cases} o(|x|^{2-n}), & n \neq 2 \\ o(\log|x|), & n = 2 \end{cases}$$

as $x \to 0$.

Remark 41.3. This definition implies the following behaviour of u at infinity

$$u(y) = \begin{cases} o(1), & n \neq 2 \\ o(\log|y|), & n = 2 \end{cases}$$

as $y \to \infty$.

V. Serov, *Fourier Series, Fourier Transform and Their Applications to Mathematical Physics*, Applied Mathematical Sciences 197,
DOI 10.1007/978-3-319-65262-7_41

The interior Neumann problem (IN)

Given $f \in C(S)$, find $u \in C^2(\Omega) \cap C(\overline{\Omega})$ such that $\Delta u = 0$ in Ω and $\partial_{\nu\_} u = f$ exists on S.

The exterior Neumann problem (EN)

Given $f \in C(S)$, find $u \in C^2(\Omega') \cap C(\overline{\Omega'})$ such that $\Delta u = 0$ in Ω' and at infinity and $\partial_{\nu_+} u = f$ exists on S.

Theorem 41.4. (Uniqueness)

(1) The solutions of (ID) and (ED) are unique.

(2) The solutions of (IN) and (EN) are unique up to a constant on Ω and Ω', respectively. When $n > 2$ this constant is zero on the unbounded component of Ω'.

Proof. If u solves (ID) with $f = 0$, then $u \equiv 0$, because this is just the uniqueness theorem for harmonic functions (see Corollary 39.14). If u solves (ED) with $f = 0$, we may assume that $\{0\} \notin \overline{\Omega'}$. Then $\widetilde{u} = |x|^{2-n} u\left(\frac{x}{|x|^2}\right)$ solves (ID) with $f = 0$ for the bounded domain $\widetilde{\Omega} = \left\{x : \frac{x}{|x|^2} \in \Omega'\right\}$. Hence $\widetilde{u} \equiv 0$, so that $u \equiv 0$, and part (1) is proved.

Exercise 41.1. Prove that if u is harmonic, then $\widetilde{u} = |x|^{2-n} u\left(\frac{x}{|x|^2}\right)$, $x \neq 0$, is also harmonic.

Concerning part (2), by Green's identity we have

$$\int_\Omega |\nabla u|^2 \mathrm{d}x = -\int_\Omega u \Delta u \mathrm{d}x + \int_S u \partial_{\nu\_} u \mathrm{d}\sigma(x).$$

Thus $\nabla u = 0$ in Ω, so that u is constant in Ω.

For (EN) let $r > 0$ be large enough that $\overline{\Omega} \subset B_r(0)$. Again by Green's identity we have

$$\begin{aligned}\int_{B_r(0) \setminus \overline{\Omega}} |\nabla u|^2 \mathrm{d}x &= -\int_{B_r(0) \setminus \overline{\Omega}} u \Delta u \mathrm{d}x + \int_{\partial B_r(0)} u \partial_r u \mathrm{d}\sigma(x) - \int_S u \partial_{\nu_+} u \mathrm{d}\sigma(x) \\ &= \int_{\partial B_r(0)} u \partial_r u \mathrm{d}\sigma(x),\end{aligned}$$

where $\partial_r u \equiv \frac{\mathrm{d}}{\mathrm{d}r} u$. Since for $n > 2$ and for large $|x|$ we have

$$u(x) = O\left(|x|^{2-n}\right), \quad \partial_r u(x) = O\left(|x|^{1-n}\right),$$

it follows that

$$\left|\int_{\partial B_r(0)} u \partial_r u \mathrm{d}\sigma(x)\right| \leq c r^{2-n} r^{1-n} \int_{\partial B_r(0)} \mathrm{d}\sigma(x) = c r^{3-2n} r^{n-1} = c r^{2-n} \to 0$$

as $r \to \infty$. Hence

$$\int_{\Omega'} |\nabla u|^2 \mathrm{d}x = 0.$$

This implies that u is constant on Ω' and $u = 0$ on the unbounded component of Ω', because for large $|x|$,

$$u(x) = O\left(|x|^{2-n}\right), \quad n > 2.$$

If $n = 2$ then $\partial_r u(x) = O\left(r^{-2}\right)$ for a function $u(x)$ that is harmonic at infinity.

Exercise 41.2. Prove that if u is harmonic at infinity, then u is bounded and $\partial_r u(x) = O\left(r^{-2}\right)$ as $r \to \infty$ if $n = 2$ and $\partial_r u(x) = O(|x|^{1-n})$, $r \to \infty$, if $n > 2$.

By Exercise 41.2 we obtain

$$\left| \int_{\partial B_r(0)} u \partial_r u \mathrm{d}\sigma(x) \right| \le c r^{-2} r = c r^{-1} \to 0, \quad r \to \infty.$$

Hence $\nabla u = 0$ in Ω' and u is constant in (each component of) Ω'. □

Remark 41.5. If Ω and Ω' are both simply connected, then the solution of (EN) for $n > 2$ is unique. This is a consequence of Theorem 41.4 for simply connected Ω'.

We now turn to the problem of finding the solutions (*existence problems*). Let us try to solve (ID) by setting

$$\widetilde{u}(x) := \int_S f(y) \partial_{\nu_y} K(x - y) \mathrm{d}\sigma(y), \tag{41.1}$$

where K is the (known) fundamental solution for Δ.

Remark 41.6. Note that (41.1) involves only the known fundamental solution and not the Green's function (which is difficult to find in general) as in the Poisson integral

$$w(x) = \int_S f(y) \partial_{\nu_y} G(x, y) \mathrm{d}\sigma(y). \tag{P}$$

We know that $\widetilde{u}(x)$ is harmonic in Ω, because $K(x - y)$ is harmonic for $x \in \Omega$, $y \in S$. It remains to verify the boundary conditions. Clearly $\widetilde{u}$ will not have the correct boundary values, but in a sense it is not far from correct. We shall prove it (very soon) that on S,

$$\widetilde{u} = \frac{f}{2} + Tf,$$

where T is a compact operator on $L^2(S)$. Thus, what we really want is to take

$$u(x) = \int_S \varphi(y) \partial_{\nu_y} K(x - y) \mathrm{d}\sigma(y), \quad x \notin S, \tag{41.2}$$

where φ is the solution of

$$\frac{1}{2}\varphi + T\varphi = f.$$

Similarly, we shall try to solve (IN) (and (EN)) in the form

$$u(x) = \int_S \varphi(y) K(x-y) \mathrm{d}\sigma(y), \quad x \notin S. \tag{41.3}$$

Definition 41.7. The functions $u(x)$ from (41.2) and (41.3) are called the *double and single layer potentials with moment (density)* φ, respectively.

Definition 41.8. Let $I(x,y)$ be continuous on $S \times S$, $x \neq y$. We call I a *continuous kernel* of order α, $0 \le \alpha < n-1$, $n \ge 2$, if

$$|I(x,y)| \le c|x-y|^{-\alpha}, \quad 0 < \alpha < n-1,$$

and

$$|I(x,y)| \le c_1 + c_2 |\log|x-y||, \quad \alpha = 0,$$

where $c > 0$ and $c_1, c_2 \ge 0$.

Remark 41.9. Note that a continuous kernel of order 0 is also a continuous kernel of order α, $0 < \alpha < n-1$.

We denote by $\widehat{I}$ the integral operator

$$\widehat{I}f(x) = \int_S I(x,y) f(y) \mathrm{d}\sigma(y), \quad x \in S,$$

with kernel I.

Lemma 41.10. *If I is a continuous kernel of order α, $0 \le \alpha < n-1$, then*

(1) $\widehat{I}$ is bounded on $L^p(S)$, $1 \le p \le \infty$.
(2) $\widehat{I}$ is compact on $L^2(S)$.

Proof. It is enough to consider $0 < \alpha < n-1$. Let us assume that $f \in L^1(S)$. Then

$$\begin{aligned}
\left\|\widehat{I}f\right\|_{L^1(S)} &\le \int_S \int_S |I(x,y)||f(y)| \mathrm{d}\sigma(y) \mathrm{d}\sigma(x) \\
&\le c \int_S |f(y)| \mathrm{d}\sigma(y) \int_S |x-y|^{-\alpha} \mathrm{d}\sigma(x) \\
&\le c \|f\|_{L^1(S)} \int_0^d r^{n-2-\alpha} \mathrm{d}r = c' \|f\|_{L^1(S)},
\end{aligned}$$

where $d = \operatorname{diam} S = \sup_{x,y \in S} |x-y|$.

If $f \in L^\infty(S)$, then

$$\left\|\widehat{I}f\right\|_{L^\infty(S)} \le c\,\|f\|_{L^\infty(S)} \int_0^d r^{n-2-\alpha}\mathrm{d}r = c'\,\|f\|_{L^\infty(S)}.$$

For $1 < p < \infty$ part (1) follows now by interpolation.

For part (2), let $\varepsilon > 0$ and set

$$I_\varepsilon(x,y) = \begin{cases} I(x,y), & |x-y| > \varepsilon, \\ 0, & |x-y| \le \varepsilon. \end{cases}$$

Since I_ε is bounded on $S \times S$, it follows that $\widehat{I_\varepsilon}$ is a Hilbert–Schmidt operator in $L^2(S)$, so that $\widehat{I_\varepsilon}$ is compact for each $\varepsilon > 0$.

Exercise 41.3. Prove that a Hilbert–Schmidt operator, i.e., an integral operator whose kernel $I(x,y)$ satisfies

$$\int_S \int_S |I(x,y)|^2 \mathrm{d}x\mathrm{d}y < \infty,$$

is compact in $L^2(S)$.

On the other hand, due to estimates for convolution,

$$\begin{aligned} \left\|\widehat{I}f - \widehat{I_\varepsilon}f\right\|_{L^2(S)} &\le c\left(\int_{|x-y|<\varepsilon} \left(\int |f(y)||x-y|^{-\alpha}\mathrm{d}\sigma(y)\right)^2 \mathrm{d}\sigma(x)\right)^{1/2} \\ &\le c\,\|f\|_{L^2(S)} \int_0^\varepsilon r^{n-2-\alpha}\mathrm{d}r \to 0, \quad \varepsilon \to 0. \end{aligned}$$

Thus, $\widehat{I}$ as the limit of $\widehat{I_\varepsilon}$ is also compact in $L^2(S)$. □

Lemma 41.11.

(1) If I is a continuous kernel of order α, $0 \le \alpha < n-1$, then $\widehat{I}$ transforms bounded functions into continuous functions.

(2) If $\widehat{I}$ is as in part (1), then $u + \widehat{I}u \in C(S)$ for $u \in L^2(S)$ implies $u \in C(S)$.

Proof. Let $|x-y| < \delta$. Then

$$\begin{aligned} |\widehat{I}f(x) - \widehat{I}f(y)| &\le \int_S |I(x,z) - I(y,z)||f(z)|\mathrm{d}\sigma(z) \\ &\le \int_{|x-z|<2\delta} (|I(x,z)| + |I(y,z)|)\,|f(z)|\mathrm{d}\sigma(z) \\ &+ \int_{S\setminus\{|x-z|<2\delta\}} |I(x,z) - I(y,z)||f(z)|\mathrm{d}\sigma(z) \\ &\le c\,\|f\|_\infty \int_{|x-z|<2\delta} \left(|x-z|^{-\alpha} + |y-z|^{-\alpha}\right)\mathrm{d}\sigma(z) \\ &+ \int_{S\setminus\{|x-z|<2\delta\}} |I(x,z) - I(y,z)||f(z)|\mathrm{d}\sigma(z) =: I_1 + I_2. \end{aligned}$$

Since $|z-y| \le |x-z| + |x-y|$, we have

$$I_1 \le c\|f\|_\infty \int_0^{3\delta} r^{n-2-\alpha} \mathrm{d}r \to 0, \quad \delta \to 0.$$

On the other hand, for $|x-y| < \delta$ and $|x-z| \ge 2\delta$ we have that

$$|y-z| \ge |x-z| - |x-y| > 2\delta - \delta = \delta.$$

So the continuity of I outside of the diagonal implies that

$$I(x,z) - I(y,z) \to 0, \quad x \to y,$$

uniformly in $z \in S \setminus \{|x-z| < 2\delta\}$. Hence, I_1 and I_2 will be small if y is sufficiently close to x. This proves the first claim.

For the second part, let $\varepsilon > 0$ and let $\varphi \in C(S \times S)$ be such that $0 \le \varphi \le 1$ and

$$\varphi(x,y) = \begin{cases} 1, & |x-y| < \varepsilon/2, \\ 0, & |x-y| \ge \varepsilon. \end{cases}$$

Write $\widehat{I}u = \widehat{\varphi I}u + \widehat{(1-\varphi)I}u =: \widehat{I_0}u + \widehat{I_1}u$. By the Cauchy–Bunyakovsky–Schwarz inequality we have

$$|\widehat{I_1}u(x) - \widehat{I_1}u(y)| \le \|u\|_2 \left(\int_S |I_1(x,z) - I_1(y,z)|^2 \mathrm{d}\sigma(z) \right)^{1/2} \to 0, \quad y \to x,$$

since I_1 is continuous (see the definition of φ). Now if we set

$$g := u + \widehat{I}u - \widehat{I_1}u \equiv u + \widehat{I_0}u,$$

then g is continuous for $u \in L^2(S)$ by the conditions of this lemma. Since the operator norm of $\widehat{I_0}$ can be made less that 1 on $L^2(S)$ and $L^\infty(S)$ (we can do this due to the choice of $\varepsilon > 0$ sufficiently small), then

$$u = \left(I + \widehat{I_0}\right)^{-1} g,$$

where I is the identity operator. Since g is continuous and the operator norm is less than 1, we have

$$u = \sum_{j=0}^{\infty} \left(-\widehat{I_0}\right)^j g.$$

This series converges uniformly, and therefore u is continuous. □

Let us consider now the double layer potential (41.2) with moment φ,

$$u(x)=\int_S \varphi(y)\partial_{\nu_y}K(x-y)\mathrm{d}\sigma(y),\quad x\in\mathbb{R}^n\backslash S.$$

First of all,

$$\partial_{\nu_y}K(x-y)=-\frac{(x-y,\nu(y))}{\omega_n|x-y|^n}. \tag{41.4}$$

Exercise 41.4. Prove that (41.4) holds for all $n\geq 1$.

It is clear also that (41.4) defines a harmonic function in $x\in\mathbb{R}^n\backslash S$, $y\in S$. Moreover, it is $O\left(|x|^{1-n}\right)$ as $x\to\infty$ ($y\in S$), so that u is also harmonic at infinity.

Exercise 41.5. Prove that (41.4) defines a harmonic function at infinity.

Lemma 41.12. *There exists $c>0$ such that*

$$|(x-y,\nu(y))|\leq c|x-y|^2,\quad x,y\in S.$$

Proof. It is quite trivial to obtain

$$|(x-y,\nu(y))|\leq|x-y||\nu(y)|=|x-y|.$$

But the latter inequality allows us to assume that $|x-y|\leq 1$. Given $y\in S$, by a translation and rotation of coordinates we may assume that $y=0$ and $\nu(y)=(0,0,\ldots,0,1)$. Hence $(x-y,\nu(y))$ transforms to x_n, and near y, S is the graph of the equation $x_n=\psi(x_1,\ldots,x_{n-1})$, where $\psi\in C^2(\mathbb{R}^{n-1})$, $\psi(0)=0$, and $\nabla\psi(0)=0$. Then

$$|(x-y,\nu(y))|=|x_n|\leq c|(x_1,\ldots,x_{n-1})|^2\leq c|x|^2=c|x-y|^2$$

by Taylor's expansion. □

We denote $\partial_{\nu_y}K(x-y)$ by $I(x,y)$.

Lemma 41.13. *I is a continuous kernel of order $n-2$, $n\geq 2$.*

Proof. If $x,y\in S$, then $I(x,y)$ is continuous for $x\neq y$; see (41.4). Hence

$$|I(x,y)|\leq\frac{c|x-y|^2}{\omega_n|x-y|^n}=c'|x-y|^{2-n}$$

by Lemma 41.12. □

Lemma 41.14.

$$\int_S I(x,y)\mathrm{d}\sigma(y)=\begin{cases}1, & x\in\Omega,\\ 0, & x\in\Omega',\\ \frac{1}{2}, & x\in S.\end{cases} \tag{41.5}$$

Proof. If $x \in \Omega'$, then $K(x-y)$ is harmonic in $x \notin S$, $y \in S$, and it is also harmonic in $y \in \Omega$, $x \in \Omega'$. Hence (see Exercise 39.2)

$$\int_S \partial_{\nu_y} K(x-y) d\sigma(y) = 0,$$

or

$$\int_S I(x,y) d\sigma(y) = 0, \quad x \in \Omega'.$$

If $x \in \Omega$, let $\delta > 0$ be such that $\overline{B_\delta(x)} \subset \Omega$. Denote $\Omega_\delta = \Omega \backslash \overline{B_\delta(x)}$ and $S_\delta = S \backslash (S \cap B_\delta(x))$, see Figure 41.1. Then $K(x-y)$ is harmonic in y in Ω_δ, and therefore by Green's identity,

$$\begin{aligned}
0 &= \int_{\Omega \backslash \overline{B_\delta(x)}} (1 \cdot \Delta_y K(x-y) - K(x-y) \Delta 1) \, dy \\
&= \int_S \partial_{\nu_y} K(x-y) d\sigma(y) - \int_{|x-y|=\delta} \partial_{\nu_y} K(x-y) d\sigma(y) \\
&= \int_S I(x,y) d\sigma(y) - \frac{\delta^{1-n}}{\omega_n} \int_{|x-y|=\delta} d\sigma(y) = \int_S I(x,y) d\sigma(y) - 1,
\end{aligned}$$

or

$$\int_S I(x,y) d\sigma(y) = 1.$$

Now suppose $x \in S$. In this case

$$\int_S I(x,y) d\sigma(y) = \lim_{\delta \to 0} \int_{S_\delta} I(x,y) d\sigma(y). \tag{41.6}$$

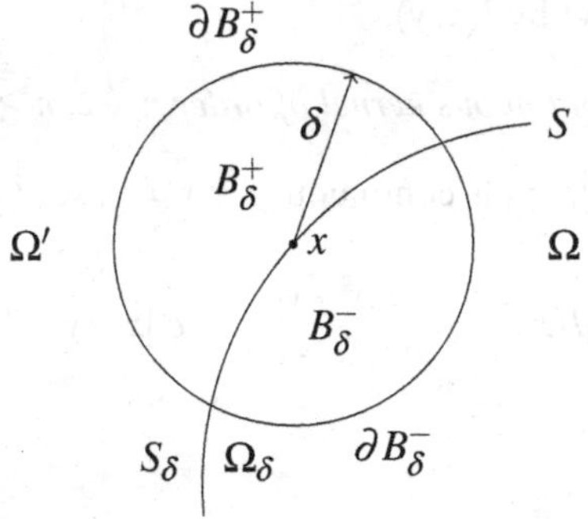

Fig. 41.1 Geometric illustration of the boundary near x.

If $y \in \Omega_\delta$, then for $x \in S$ we have that $x \neq y$. This implies that

$$0 = \int_{\Omega_\delta} \Delta_y K(x-y)\mathrm{d}y = \int_{S_\delta} \partial_{\nu_y} K(x-y)\mathrm{d}\sigma(y) - \int_{\partial B_\delta^-} \partial_{\nu_y} K(x-y)\mathrm{d}\sigma(y).$$

Therefore, see (41.4),

$$\begin{aligned}\lim_{\delta\to 0}\int_{S_\delta} \partial_{\nu_y} K(x-y)\mathrm{d}\sigma(y) &= \lim_{\delta\to 0}\int_{\partial B_\delta^-} \partial_{\nu_y} K(x-y)\mathrm{d}\sigma(y)\\ &= \lim_{\delta\to 0}\frac{\delta^{1-n}}{\omega_n}\int_{\partial B_\delta^-}\mathrm{d}\sigma(y)\\ &= \lim_{\delta\to 0}\frac{\delta^{1-n}}{\omega_n}\left(\delta^{n-1}\frac{\omega_n}{2}+o(\delta^{n-1})\right) = \frac{1}{2}.\end{aligned}$$

This means that the limit in (41.6) exists and (41.5) is satisfied. □

Lemma 41.15. *There exists* $c > 0$ *such that*

$$\int_S |\partial_{\nu_y} K(x-y)|\mathrm{d}\sigma(y) \le c, \quad x \in \mathbb{R}^n.$$

Proof. It follows from Lemma 41.13 that

$$\int_S |\partial_{\nu_y} K(x-y)|\mathrm{d}\sigma(y) \le \frac{c}{\omega_n}\int_S |x-y|^{2-n}\mathrm{d}\sigma(y) \le c_1, \quad x \in S.$$

Next, for $x \notin S$ define $\mathrm{dist}(x,S) = \inf_{y\in S}|x-y|$.

There are two possibilities now: if $\mathrm{dist}(x,S) \ge \delta/2$, then $|x-y| \ge \delta/2$ for all $y \in S$, and therefore

$$\int_S |\partial_{\nu_y} K(x-y)|\mathrm{d}\sigma(y) \le c\delta^{1-n}\int_S \mathrm{d}\sigma(y) = c', \tag{41.7}$$

where c' does not depend on $\delta > 0$ (because δ is fixed).

Suppose now that $\mathrm{dist}(x,S) < \delta/2$. If we choose $\delta > 0$ sufficiently small, then there is a unique $x_0 \in S$ such that

$$x = x_0 + t\nu(x_0), \quad t \in (-\delta/2, \delta/2).$$

Set $B_\delta = \{y \in S : |x_0 - y| < \delta\}$. We estimate the integrals of $|I(x,y)|$ over $S\backslash B_\delta$ and B_δ separately. If $y \in S\backslash B_\delta$, then

$$|x-y| \ge |x_0-y| - |x-x_0| > \delta - \delta/2 = \delta/2$$

and

$$|I(x,y)| \le c\delta^{1-n},$$

so that the integral over $S\backslash B_\delta$ satisfies (41.7), where again c' does not depend on δ.

To estimate the integral over B_δ, we note that (see (41.4))

$$|I(x,y)| = \frac{|(x-y,\nu(y))|}{\omega_n|x-y|^n} = \frac{|(x-x_0,\nu(y))+(x_0-y,\nu(y))|}{\omega_n|x-y|^n} \leq \frac{|x-x_0|+c|x_0-y|^2}{\omega_n|x-y|^n}. \tag{41.8}$$

The latter inequality follows from Lemma 41.12, since $x_0, y \in S$. Moreover, we have (due to Lemma 41.12)

$$\begin{aligned}
|x-y|^2 &= |x-x_0|^2+|x_0-y|^2+2(x-x_0,x_0-y)\\
&= |x-x_0|^2+|x_0-y|^2+2|x-x_0|\left(x_0-y,\frac{x-x_0}{|x-x_0|}\right)\\
&\geq |x-x_0|^2+|x_0-y|^2-2|x-x_0||(x_0-y,\nu(x_0))|\\
&\geq |x-x_0|^2+|x_0-y|^2-2c|x-x_0||x_0-y|^2\\
&\geq |x-x_0|^2+|x_0-y|^2-|x-x_0||x_0-y|,
\end{aligned}$$

if we choose $\delta > 0$ such that $|x_0-y| \leq \frac{1}{2c}$, where the constant $c > 0$ is from Lemma 41.12.

Since $|x-x_0||x_0-y| \leq \frac{1}{2}\left(|x-x_0|^2+|x_0-y|^2\right)$, we obtain finally

$$|x-y|^2 \geq \frac{1}{2}\left(|x-x_0|^2+|x_0-y|^2\right)$$

and (see (41.4) and (41.8))

$$\begin{aligned}
|I(x,y)| &\leq c\frac{|x-x_0|+|x_0-y|^2}{(|x-x_0|^2+|x_0-y|^2)^{n/2}}\\
&\leq c\frac{|x-x_0|}{(|x-x_0|^2+|x_0-y|^2)^{n/2}}+\frac{c}{|x_0-y|^{n-2}}.
\end{aligned}$$

This implies

$$\begin{aligned}
\int_{B_\delta}|I(x,y)|\mathrm{d}\sigma(y) &\leq c'\int_0^\delta\frac{|x-x_0|}{(|x-x_0|^2+r^2)^{n/2}}r^{n-2}\mathrm{d}r+c'\int_0^\delta\frac{r^{n-2}}{r^{n-2}}\mathrm{d}r\\
&\leq c'\delta+c'\int_0^\infty\frac{ar^{n-2}}{(a^2+r^2)^{n/2}}\mathrm{d}r,
\end{aligned}$$

where $a := |x-x_0|$. For the latter integral we have ($t = r/a$)

$$\int_0^\infty\frac{ar^{n-2}}{(a^2+r^2)^{n/2}}\mathrm{d}r = \int_0^\infty\frac{t^{n-2}}{(1+t^2)^{n/2}}\mathrm{d}t < \infty.$$

If we combine all estimates, then we may conclude that there is $c_0 > 0$ such that

$$\int_S |\partial_{\nu_y} K(x-y)| \mathrm{d}\sigma(y) \leq c_0, \quad x \in \mathbb{R}^n,$$

and this constant does not depend on x. □

Theorem 41.16. *Suppose $\varphi \in C(S)$ and u is defined by the double layer potential (41.2) with moment φ. Then for all $x \in S$,*

$$\lim_{t \to 0-} u(x + t\nu(x)) = \frac{\varphi(x)}{2} + \int_S I(x,y)\varphi(y) \mathrm{d}\sigma(y),$$
$$\lim_{t \to 0+} u(x + t\nu(x)) = -\frac{\varphi(x)}{2} + \int_S I(x,y)\varphi(y) \mathrm{d}\sigma(y)$$

uniformly on S with respect to x.

Proof. If $x \in S$ and $t < 0$, with $|t|$ sufficiently small, then $x_t := x + t\nu(x) \in \Omega$ and $u(x + t\nu(x))$ is well defined by

$$\begin{aligned} u(x + t\nu(x)) &= \int_S \varphi(y) I(x_t, y) \mathrm{d}\sigma(y) = \int_S (\varphi(y) - \varphi(x)) I(x_t, y) \mathrm{d}\sigma(y) + \varphi(x) \\ &\to \varphi(x) + \int_S \varphi(y) I(x,y) \mathrm{d}\sigma(y) - \varphi(x) \int_S I(x,y) \mathrm{d}\sigma(y) \\ &= \varphi(x) + \int_S \varphi(y) I(x,y) \mathrm{d}\sigma(y) - \varphi(x)/2, \quad t \to 0-. \end{aligned}$$

If $t > 0$, the arguments are the same except that

$$\int_S I(x_t, y) \mathrm{d}\sigma(y) = 0.$$

Uniform convergence follows from the fact that S is compact and $\varphi \in C(S)$. □

Corollary 41.17. *For $x \in S$,*

$$\varphi(x) = u_-(x) - u_+(x),$$

where $u_\pm = \lim_{t \to 0\pm} u(x_t)$.

We state without proof that the normal derivative of the double layer potential is continuous across the boundary in the sense of the following theorem.

Theorem 41.18. *Suppose $\varphi \in C(S)$ and u is defined by the double layer potential (41.2) with moment φ. Then for all $x \in S$,*

$$\lim_{t \to 0+} (\nu(x) \cdot \nabla u(x + t\nu(x)) - \nu(x) \cdot \nabla u(x - t\nu(x))) = 0$$

uniformly on S with respect to x.

Let us now consider the single layer potential

$$u(x) = \int_S \varphi(y)K(x-y)\mathrm{d}\sigma(y)$$

with moment $\varphi \in C(S)$.

Lemma 41.19. *The single layer potential u is continuous on $\mathbb{R}^n$.*

Proof. Since u is harmonic in $x \notin S$, we have only to show continuity for $x \in S$. Given $x_0 \in S$ and $\delta > 0$, let $B_\delta = \{y \in S : |x_0 - y| < \delta\}$. Then

$$\begin{aligned}|u(x) - u(x_0)| &\le \int_{B_\delta} (|K(x-y)| + |K(x_0-y)|)\,|\varphi(y)|\mathrm{d}\sigma(y) \\ &\quad + \int_{S\setminus B_\delta} |K(x-y) - K(x_0-y)||\varphi(y)|\mathrm{d}\sigma(y) \\ &\le c\delta(\text{or}\,\delta\log\frac{1}{\delta}\ \text{for}\, n=2) + \|\varphi\|_\infty \int_{S\setminus B_\delta} |K(x-y) - K(x_0-y)|\mathrm{d}\sigma(y) \to 0\end{aligned}$$

as $x \to x_0$ and $\delta \to 0$. □

Exercise 41.6. Prove that

$$\int_{B_\delta} (|K(x-y)| + |K(x_0-y)|)\,|\varphi(y)|\mathrm{d}\sigma(y) \le c\,\|\varphi\|_\infty \begin{cases} \delta, & n > 2, \\ \delta\log\frac{1}{\delta}, & n = 2. \end{cases}$$

Definition 41.20. Let us set

$$I^*(x,y) := \partial_{\nu_x} K(x-y) \equiv \frac{(x-y, \nu(x))}{\omega_n |x-y|^n}.$$

Theorem 41.21. *Suppose $\varphi \in C(S)$ and u is defined on $\mathbb{R}^n$ by the single layer potential* (41.3) *with moment φ. Then for $x \in S$,*

$$\lim_{t\to 0-} \partial_\nu u(x + t\nu(x)) = -\frac{\varphi(x)}{2} + \int_S I^*(x,y)\varphi(y)\mathrm{d}\sigma(y),$$

$$\lim_{t\to 0+} \partial_\nu u(x + t\nu(x)) = \frac{\varphi(x)}{2} + \int_S I^*(x,y)\varphi(y)\mathrm{d}\sigma(y).$$

Proof. We consider the double layer potential on $\mathbb{R}^n\setminus S$ with moment φ,

$$v(x) = \int_S \varphi(y)\partial_{\nu_y} K(x-y)\mathrm{d}\sigma(y),$$

and define the function f on a tubular neighborhood V of S by

$$f(x) = \begin{cases} v(x) + \partial_\nu u(x), & x \in V \setminus S, \\ \widehat{I}\varphi(x) + \widehat{I}^*\varphi(x), & x \in S, \end{cases} \tag{41.9}$$

where u is defined by (41.3).

Here the *tubular neighborhood* of S is defined as

$$V = \{x + t\nu(x) : x \in S, |t| < \delta\}.$$

We claim that f is continuous on V. It is clearly (see (41.9)) continuous on $V \setminus S$ and S, so it suffices to show that if $x_0 \in S$ and $x = x_0 + t\nu(x_0)$, then $f(x) - f(x_0) \to 0$ as $t \to 0\pm$. We have

$$\begin{aligned} f(x) - f(x_0) &= v(x) + \partial_\nu u(x) - \widehat{I}\varphi(x_0) - \widehat{I}^*\varphi(x_0) \\ &= \int_S I(x,y)\varphi(y)\mathrm{d}\sigma(y) + \int_S \varphi(y)\partial_{\nu_x} K(x-y)\mathrm{d}\sigma(y) \\ &\quad - \int_S I(x_0,y)\varphi(y)\mathrm{d}\sigma(y) - \int_S I^*(x_0,y)\varphi(y)\mathrm{d}\sigma(y) \\ &= \int_S (I(x,y) + I^*(x,y) - I(x_0,y) - I^*(x_0,y))\,\varphi(y)\mathrm{d}\sigma(y). \end{aligned}$$

We write this expression as an integral over $B_\delta = \{y \in S : |x_0 - y| < \delta\}$ plus an integral over $S \setminus B_\delta$. The integral over $S \setminus B_\delta$ tends uniformly to 0 as $x \to x_0$, because $|y - x| \geq \delta$ and $|y - x_0| \geq \delta$, so that the functions I and I^* have no singularities in this case.

On the other hand, the integral over B_δ can be bounded by

$$\|\varphi\|_\infty \int_{B_\delta} (|I(x,y) + I^*(x,y)| + |I(x_0,y) + I^*(x_0,y)|)\,\mathrm{d}\sigma(y).$$

Since

$$I(x,y) = -\frac{(x-y, \nu(y))}{\omega_n |x-y|^n}$$

and $\nu(x) = \nu(x_0)$ for $x = x_0 + t\nu(x_0) \in V$, we have

$$I^*(x,y) = I(y,x) = \frac{(x-y, \nu(x))}{\omega_n |x-y|^n} \equiv \frac{(x-y, \nu(x_0))}{\omega_n |x-y|^n}. \tag{41.10}$$

Hence

$$\begin{aligned} |I(x,y) + I^*(x,y)| &= \left| \frac{(x-y, \nu(x_0) - \nu(y))}{\omega_n |x-y|^n} \right| \leq \frac{|x-y||\nu(x_0) - \nu(y)|}{\omega_n |x-y|^n} \\ &\leq c \frac{|x-y||x_0-y|}{\omega_n |x-y|^n} \leq c' \frac{|x_0-y|}{|x_0-y|^{n-1}} = c'|x_0-y|^{2-n}, \end{aligned}$$

because $|x_0 - y| \leq |x_0 - x| + |x - y| \leq 2|x - y|$. Here we have also used the fact that $|\nu(x_0) - \nu(y)| \leq c|x_0 - y|$, since ν is C^1 (Figure 41.2).

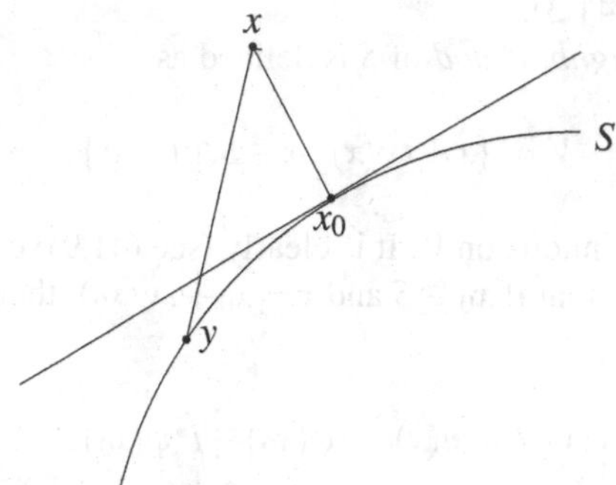

Fig. 41.2 Geometric illustration of the boundary at the point x_0.

This estimate allows us to obtain that the corresponding integral over B_δ can be dominated by

$$c \int_{|y-x_0|\leq\delta} |x_0 - y|^{2-n} \mathrm{d}\sigma(y) = c' \int_0^\delta r^{2-n} r^{n-2} \mathrm{d}r = c'\delta.$$

Thus $f := v + \partial_\nu u$ extends continuously across S. Therefore, for $x \in S$,

$$\widehat{I}\varphi(x) + \widehat{I}^*\varphi(x) = v_-(x) + \partial_{\nu_-} u(x) = \frac{1}{2}\varphi(x) + \widehat{I}\varphi(x) + \partial_{\nu_-} u(x).$$

It follows that

$$\partial_{\nu_-} u(x) = -\frac{\varphi(x)}{2} + \widehat{I}^*\varphi(x).$$

By similar arguments we obtain

$$\widehat{I}\varphi(x) + \widehat{I}^*\varphi(x) = v_+(x) + \partial_{\nu_+} u(x) = -\frac{1}{2}\varphi(x) + \widehat{I}\varphi(x) + \partial_{\nu_+} u(x)$$

and therefore

$$\partial_{\nu_+} u(x) = \frac{\varphi(x)}{2} + \widehat{I}^*\varphi(x).$$

This completes the proof. □

Corollary 41.22.

$$\varphi(x) = \partial_{\nu_+} u(x) - \partial_{\nu_-} u(x),$$

where u is defined by (41.3).

Lemma 41.23. *If $f \in C(S)$ and*

$$\frac{\varphi}{2}+\widehat{I}^*\varphi=f, \quad -\frac{\varphi}{2}+\widehat{I}^*\varphi=f,$$

then

$$\int_S \varphi \mathrm{d}\sigma=\int_S f \mathrm{d}\sigma, \quad \int_S f \mathrm{d}\sigma=0,$$

respectively.

Proof. It follows from (41.10) that

$$\begin{aligned}\int_S f(x)\mathrm{d}\sigma(x) &= \frac{1}{2}\int_S \varphi(x)\mathrm{d}\sigma(x)+\int_S \varphi(y)\mathrm{d}\sigma(y)\int_S I^*(x,y)\mathrm{d}\sigma(x)\\ &= \frac{1}{2}\int_S \varphi(x)\mathrm{d}\sigma(x)+\frac{1}{2}\int_S \varphi(y)\mathrm{d}\sigma(y)=\int_S \varphi(y)\mathrm{d}\sigma(y),\end{aligned}$$

where we have also used Lemma 41.14. □

Lemma 41.24. *Let $n=2$.*

(1) If $\varphi \in C(S)$, then the single layer potential u with moment φ is harmonic at infinity if and only if

$$\int_S \varphi(x)\mathrm{d}\sigma(x)=0,$$

and in that case $u \to 0$ as $|x| \to \infty$.

(2) Let $\varphi \in C(S)$ with

$$\int_S \varphi(x)\mathrm{d}\sigma(x)=0$$

and u as in part (1). If u is constant on $\overline{\Omega}$, then $u \equiv 0$.

Proof. Since $n=2$, we must have

$$\begin{aligned}u(x) &= \frac{1}{2\pi}\int_S \log|x-y|\varphi(y)\mathrm{d}\sigma(y)\\ &= \frac{1}{2\pi}\int_S (\log|x-y|-\log|x|)\,\varphi(y)\mathrm{d}\sigma(y)+\frac{1}{2\pi}\log|x|\int_S \varphi(y)\mathrm{d}\sigma(y).\end{aligned}$$

But $\log|x-y|-\log|x| \to 0$ as $x \to \infty$ uniformly for $y \in S$, and therefore, this term is harmonic at infinity (we have a removable singularity). Hence u is harmonic at infinity if and only if $\int_S \varphi(x)\mathrm{d}\sigma(x)=0$, and in this case $u(x)$ vanishes at infinity. This proves part (1).

In part (2), u is harmonic at infinity. If u is constant on $\overline{\Omega}$, then it solves (ED) with $f \equiv$ constant on S. But a solution of such a problem must be constant and vanish at infinity. Therefore, this constant is zero. Thus $u \equiv 0$. □

Remark 41.25. For $n>2$ the single layer potential u is a harmonic function at infinity without any additional conditions for the moment φ.

For solvability of the corresponding integral equations in the space $C(S)$ with integral operators $\widehat{I}$ and $\widehat{I^*}$ (see Theorems 41.16 and 41.21) we need the Fredholm alternative (see in addition Theorems 34.8 and 34.9).

Theorem 41.26 (First Fredholm theorem). *The null spaces of $\frac{1}{2}I-\widehat{I}$ and $\frac{1}{2}I-\widehat{I^*}$ have the same finite dimension*

$$\dim N\left(\frac{1}{2}I-\widehat{I}\right)=\dim N\left(\frac{1}{2}I-\widehat{I^*}\right)<\infty,$$

where I denotes the identity operator.

Proof. Since $\widehat{I}$ and $\widehat{I^*}$ are compact operators in $C(S)$, the null spaces $N\left(\frac{1}{2}I-\widehat{I}\right)$ and $N\left(\frac{1}{2}I-\widehat{I^*}\right)$ are closed subspaces of $C(S)$. If $\varphi\in N\left(\frac{1}{2}I-\widehat{I}\right)$ and $\psi\in N\left(\frac{1}{2}I-\widehat{I^*}\right)$, then

$$2\widehat{I}\varphi=\varphi,\quad 2\widehat{I^*}\psi=\psi,$$

i.e., $2\widehat{I}$ and $2\widehat{I^*}$ are identical on the corresponding null spaces. Since they are compact there, this is possible only when the corresponding null spaces are of finite dimension. The equality of these dimensions can be checked in the same manner as in the proof of Theorem 34.11. In this proof, part (2) of Lemma 41.11 must be taken into account. □

Theorem 41.27 (Second Fredholm theorem). *The ranges of the operators $\frac{1}{2}I-\widehat{I}$ and $\frac{1}{2}I-\widehat{I^*}$ on $C(S)$ can be described as*

$$R\left(\frac{1}{2}I-\widehat{I}\right)=\{f\in C(S):(f,\psi)_{L^2(S)}=0\,\textit{for any}\,\psi\in N\left(\frac{1}{2}I-\widehat{I^*}\right)\}$$

and

$$R\left(\frac{1}{2}I-\widehat{I^*}\right)=\{g\in C(S):(g,\varphi)_{L^2(S)}=0\,\textit{for any}\,\varphi\in N\left(\frac{1}{2}I-\widehat{I}\right)\}.$$

Proof. Let $f=\frac{\varphi}{2}-\widehat{I}\varphi$ for some $\varphi\in C(S)$. Then for all $\psi\in N(\frac{1}{2}I-\widehat{I^*})$ we have

$$\begin{aligned}(f,\psi)_{L^2(S)}&=(\frac{\varphi}{2}-\widehat{I}\varphi,\psi)_{L^2(S)}\\&=(\frac{\varphi}{2},\psi)_{L^2(S)}-(\varphi,\widehat{I^*}\psi)_{L^2(S)}=(\varphi,\frac{\psi}{2}-\widehat{I^*}\psi)_{L^2(S)}=0.\end{aligned}$$

Conversely, assume that $f\in C(S), f\neq 0$ satisfies $(f,\psi)_{L^2(S)}=0$ for all $\psi\in N(\frac{1}{2}I-\widehat{I^*})$. Assume to the contrary that there is no $\varphi\in C(S)$ such that $f=\frac{1}{2}\varphi-\widehat{I}\varphi$. Then f can be chosen to be orthogonal to all $\frac{1}{2}\varphi-\widehat{I}\varphi$, i.e.,

$$0 = (f, \frac{1}{2}\varphi - \widehat{I}\varphi)_{L^2(S)} = (f, \frac{1}{2}\varphi)_{L^2(S)} - (\widehat{I^*}f, \varphi)_{L^2(S)} = (\frac{1}{2}f - \widehat{I^*}f, \varphi)_{L^2(S)}.$$

This means that $f \in N(\frac{1}{2}I - \widehat{I^*})$. But at the same time, $f \perp N(\frac{1}{2}I - \widehat{I^*})$. Thus $f = 0$. This contradiction proves the opposite embedding. For the operator $\frac{1}{2}I - \widehat{I^*}$ the proof is the same, since $(\widehat{I^*})^* = \widehat{I}$. □

Since the ranges of $\frac{1}{2}I - \widehat{I}$ and $\frac{1}{2}I - \widehat{I^*}$ are closed due to Riesz's lemma (see Theorem 28.14) and due to part (2) of Lemma 41.11, we obtain the following result.

Theorem 41.28 (Fredholm alternative). *Either $\frac{1}{2}I - \widehat{I}$ and $\frac{1}{2}I - \widehat{I^*}$ are bijective or $\frac{1}{2}I - \widehat{I}$ and $\frac{1}{2}I - \widehat{I^*}$ have nontrivial null spaces with finite dimension*

$$\dim N\left(\frac{1}{2}I - \widehat{I}\right) = \dim N\left(\frac{1}{2}I - \widehat{I^*}\right) < \infty,$$

and the ranges of these operators are given by

$$R\left(\frac{1}{2}I - \widehat{I}\right) = \{f \in C(S) : (f, \psi)_{L^2(S)} = 0 \text{ for any } \psi \in N\left(\frac{1}{2}I - \widehat{I^*}\right)\}$$

and

$$R\left(\frac{1}{2}I - \widehat{I^*}\right) = \{g \in C(S) : (g, \varphi)_{L^2(S)} = 0 \text{ for any } \varphi \in N\left(\frac{1}{2}I - \widehat{I}\right)\}.$$

We will interpret this alternative as follows: either the integral equations

$$\frac{1}{2}\varphi - \widehat{I}\varphi = f, \quad \frac{1}{2}\psi - \widehat{I^*}\psi = g \tag{41.11}$$

have unique solutions φ and ψ for every given f and g from $C(S)$, or the corresponding homogeneous equations

$$\frac{1}{2}\varphi - \widehat{I}\varphi = 0, \quad \frac{1}{2}\psi - \widehat{I^*}\psi = 0$$

have the same number of linearly independent solutions $\varphi_1, \ldots, \varphi_m, \psi_1, \ldots, \psi_m$, and in this case equations (41.11) have solutions if and only if $f \perp \varphi_j$, $j = 1, 2, \ldots, m$, and $g \perp \psi_j$, $j = 1, 2, \ldots, m$, respectively.

In fact, it is possible to prove a stronger result (which is the analogue of Theorem 28.15) for the complete normed space $C(S)$; see [22] for a proof.

Theorem 41.29 (Riesz). *Let $A : C(S) \to C(S)$ be a compact linear operator. Then $I - A$ is injective if and only if it is surjective. If $I - A$ is injective (and therefore also bijective), then the inverse operator $(I - A)^{-1} : C(S) \to C(S)$ is bounded.*

Exercise 41.7. Show that

$$R\left(\frac{1}{2}I-\widehat{I}\right)=N\left(\frac{1}{2}I-\widehat{I^*}\right)^{\perp},\quad R\left(\frac{1}{2}I-\widehat{I^*}\right)=N\left(\frac{1}{2}I-\widehat{I}\right)^{\perp},$$

and then

$$C(S)=R\left(\frac{1}{2}I-\widehat{I}\right)\oplus N\left(\frac{1}{2}I-\widehat{I^*}\right)=R\left(\frac{1}{2}I-\widehat{I^*}\right)\oplus N\left(\frac{1}{2}I-\widehat{I}\right).$$

Now we are in a position to prove the main result of this chapter.

Theorem 41.30 (Main theorem). *Suppose Ω and Ω' are simply connected. Then*

(1) (ID) has a unique solution for every $f\in C(S)$.
(2) (ED) has a unique solution for every $f\in C(S)$.
(3) (IN) has a solution for every $f\in C(S)$ if and only if $\int_S f\mathrm{d}\sigma=0$. The solution is unique up to a constant.
(4) (EN) has a unique solution for every $f\in C(S)$ if and only if $\int_S f\mathrm{d}\sigma=0$.

Proof. We have already proved uniqueness (see Theorem 41.4) and the necessity of the conditions on f (see Exercise 39.2 and Lemma 41.24). So all that remains is to establish existence.

For (IN) and (EN) the function f must satisfy the condition

$$\int_S f\mathrm{d}\sigma=0,$$

or

$$(f,1)_{L^2(S)}=0.$$

Next, since

$$\int_S I(x,y)\mathrm{d}\sigma(y)=1/2,$$

we may conclude first that $1\in N(\frac{1}{2}I-\widehat{I})$, i.e., $\dim N(\frac{1}{2}I-\widehat{I})\ge 1$, and $\dim N(\frac{1}{2}I-\widehat{I^*})\le 1$ due to the fact that the single layer potential uniquely (up to a constant; see Theorem 41.4) solves (IN). Thus, due to Theorem 41.28 (Fredholm alternative), we have

$$\dim N\left(\frac{1}{2}I-\widehat{I}\right)=\dim N\left(\frac{1}{2}I-\widehat{I^*}\right)=1.$$

Using again this alternative, we see that the condition $(f,1)_{L^2(S)}=0$ is necessary and sufficient for the solvability of the equation $-\frac{1}{2}\varphi+\widehat{I^*}\varphi=f$, which solves (IN). For (EN) we can solve uniquely the equation $\frac{1}{2}\varphi+\widehat{I^*}\varphi=f$ if and only if $f\perp N(\frac{1}{2}I+\widehat{I})$. But since the solution of (ED) is unique, the null space of $\frac{1}{2}I+\widehat{I}$ consists only of the trivial solution (we have used here Lemma 41.24 for $n=2$). Therefore, the necessary and sufficient condition for f is automatically satisfied.

Concerning (ID) we consider the integral equation $\frac{1}{2}\varphi+\widehat{I}\varphi=f$. Let us set

$$\psi := f-\left(\frac{1}{2}\varphi+\widehat{I}\varphi\right),$$

so that $f=\psi+(\frac{1}{2}\varphi+\widehat{I}\varphi)$. Since $f\in C(S)$ and $\frac{1}{2}\varphi+\widehat{I}\varphi\in R(\frac{1}{2}I+\widehat{I})$, we can choose ψ uniquely (using Exercise 41.7) from $N(\frac{1}{2}I+\widehat{I^*})$, i.e.,

$$\frac{1}{2}\psi+\widehat{I^*}\psi=0.$$

If we consider now the single layer potential with moment ψ,

$$v(x) := \int_S \psi(y)K(x-y)\mathrm{d}\sigma(y),$$

then it is harmonic in $\Omega\cup\Omega'$ and

$$\partial_{\nu_+}v=\frac{1}{2}\psi+\widehat{I^*}\psi=0.$$

Hence $\partial_{\nu_+}v(x)\equiv 0$ for all $x\in\overline{\Omega'}$ (due to the uniqueness result), which means that $v\equiv$ constant in $\mathbb{R}^n\setminus\Omega$, and since v is required to be harmonic at infinity, this constant is equal to zero (see Theorem 41.4 and Lemma 41.24 for $n=2$). The final step is that a single layer potential v is continuous everywhere in the whole of $\mathbb{R}^n$ and $v_+\equiv 0$ on $\overline{\Omega'}$ including S. Thus $v\equiv 0$ in $\overline{\Omega}$ as well. The latter fact can be proved if we consider the Dirichlet boundary value problem

$$\begin{cases}\Delta v=0, & \text{in}\,\Omega,\\ v=0, & \text{on}\,S.\end{cases}$$

But Theorem 41.21 leads to $\partial_{\nu_+}v-\partial_{\nu_-}v=\psi$ or $\psi=0$. Therefore, we have proved that (ID) is uniquely solvable for all $f\in C(S)$. □

Exercise 41.8. Prove the second part of Theorem 41.30.

Chapter 42
Elliptic Boundary Value Problems

In this chapter we study the equation $Lu = f$ on some bounded domain $\Omega \subset \mathbb{R}^n$ with C^k-class (with appropriate $k \geq 1$) boundary $\partial\Omega = S$, where u is to satisfy certain boundary conditions on S. The object of interest is to prove existence, uniqueness, and regularity theorems. Our approach will be to formulate the problems in terms of sesquilinear forms and then to apply some Hilbert space theory (see Part III of this book). Here L will denote a differential operator in the divergence form

$$L(x,\partial) = \sum_{|\alpha|=|\beta|\leq m} (-1)^{|\alpha|}\partial^{\alpha}(a_{\alpha\beta}(x)\partial^{\beta}), \quad m = 1,2,\ldots, \tag{42.1}$$

satisfying the coercivity (generalized strong ellipticity) condition with $u \in C_0^{\infty}(\Omega)$,

$$\mathrm{Re} \sum_{|\alpha|=|\beta|=m} \int_{\Omega} a_{\alpha\beta}(x)\partial^{\beta}u \cdot \partial^{\alpha}\overline{u}\mathrm{d}x \geq \nu \sum_{|\alpha|=m} \int_{\Omega} |\partial^{\alpha}u|^2\mathrm{d}x, \tag{42.2}$$

where $\nu > 0$ is constant. It is easy to see that for the operator (42.1) with constant coefficients, the condition (42.2) is equivalent to

$$\mathrm{Re} \sum_{|\alpha|=|\beta|=m} a_{\alpha\beta}\xi^{\alpha}\xi^{\beta} \geq \nu \sum_{|\alpha|=m} |\xi^{\alpha}|^2.$$

The coefficients of L are assumed to be complex-valued (in general). We introduce the *Dirichlet form* $D(u,v)$ associated with this operator L as

$$D(u,v) = \sum_{|\alpha|=|\beta|\leq m} \int_{\Omega} a_{\alpha\beta}(x)\partial^{\beta}u \cdot \partial^{\alpha}\overline{v}\mathrm{d}x. \tag{42.3}$$

V. Serov, *Fourier Series, Fourier Transform and Their Applications to Mathematical Physics*, Applied Mathematical Sciences 197,
DOI 10.1007/978-3-319-65262-7_42

The *Dirichlet boundary value problem* for the operator L from (42.1) on the domain Ω can be formulated as follows: given $f \in L^2(\Omega)$, find a function u satisfying $Lu = f$ on Ω in the distributional sense, i.e., u is a distributional solution satisfying the boundary conditions

$$u|_S = g_0, \quad \partial_\nu u|_S = g_1, \ldots, \partial_\nu^{m-1} u|_S = g_{m-1}, \tag{42.4}$$

where S is a C^m-class surface, ν is an outward normal vector to S, and g_j, for $j = 0, 1, \ldots, m-1$, are from some appropriate space on S.

Remark 42.1 (This example is due to A. Bitsadze). The operator

$$L := \frac{1}{4}(\partial_x^2 + 2\mathrm{i}\partial_x\partial_y - \partial_y^2)$$

is elliptic on $\mathbb{R}^2$. The general solution of $Lu = 0$ is

$$u(x,y) = f(z) + \overline{z}g(z), \quad z = x + \mathrm{i}y,$$

where f and g are arbitrary holomorphic functions; see [11]. In particular, if we choose

$$g(z) = -zf(z),$$

where f is holomorphic on the unit disk B and continuous on $\overline{B}$, then $u(x,y) = (1 - |z|^2)f(z)$, and $u(x,y)$ solves the Dirichlet boundary value problem

$$\begin{cases} Lu = 0, & \text{in } B, \\ u = 0, & \text{on } \partial B. \end{cases}$$

Hence we have no uniqueness for this problem in B. The reason is that L is elliptic but not strongly elliptic.

Exercise 42.1. Show that L from Remark 42.1 is elliptic on $\mathbb{R}^2$ but not strongly elliptic in the sense of (42.2).

In order to investigate the solvability of the Dirichlet boundary value problem we need a regularity theorem for the operator L from (42.1). Here we define

$$\|u\|_s := \|u\|_{H^s(\Omega)}.$$

Theorem 42.2. *Suppose that the operator L from* (42.1) *satisfies the ellipticity condition* (42.2) *and the coefficients $a_{\alpha\beta}(x), |\alpha|, |\beta| \le m$ belong to the Sobolev space $W_\infty^k(\Omega)$, $k = 0, 1, 2, \ldots$. Then there is a constant $C > 0$ such that for all $u \in H_0^{m+k}(\Omega)$ we have*

$$\|u\|_{m+k} \leq C(\|Lu\|_{k-m} + \|u\|_{m+k-1}). \tag{42.5}$$

Proof. Since $C_0^\infty(\Omega)$ is dense in the Sobolev space $H_0^s(\Omega)$, $s \geq 0$, let us consider $u \in C_0^\infty(\Omega)$. Consider first the case $k = 0$. Then integration by parts leads to

$$\begin{aligned}
&|(Lu,u)_{L^2(\Omega)}| \\
&= \left| \sum_{|\alpha|=|\beta|=m} \int_\Omega a_{\alpha\beta}(x)\partial^\beta u \partial^\alpha \overline{u} \mathrm{d}x + \sum_{|\alpha|=|\beta|\leq m-1} \int_\Omega a_{\alpha\beta}(x)\partial^\beta u \partial^\alpha \overline{u} \mathrm{d}x \right| \\
&\geq \mathrm{Re} \sum_{|\alpha|=|\beta|=m} \int_\Omega a_{\alpha\beta}(x)\partial^\beta u \partial^\alpha \overline{u} \mathrm{d}x - C \sum_{|\alpha|=|\beta|\leq m-1} \int_\Omega |\partial^\beta u \partial^\alpha \overline{u}| \mathrm{d}x \\
&\geq \nu \sum_{|\alpha|=m} \int_\Omega |\partial^\alpha u|^2 \mathrm{d}x - C\|u\|_{m-1}^2,
\end{aligned}$$

where the constant $C > 0$ limits the norms of the coefficients $a_{\alpha\beta}(x)$ in $L^\infty(\Omega)$. Using now the Poincaré inequality (Theorem 20.23) and the properties of Sobolev space, we obtain from the latter inequality that

$$\|Lu\|_{-m} \|u\|_m \geq \nu_1 \|u\|_m^2 - C\|u\|_m \|u\|_{m-1},$$

that is, we have

$$\|Lu\|_{-m} \geq \nu_1 \|u\|_m - C\|u\|_{m-1}.$$

Thus (42.5) is proved for $k = 0$. Consider now the case $k = 1$. Since the coefficients of L belong to $W_\infty^1(\Omega)$, in that case we have $L(\partial_j u) = \partial_j Lu + \widetilde{L}u$, where $\widetilde{L}$ is again an elliptic operator in divergence form with coefficients from $L^\infty(\Omega)$. Hence,

$$\begin{aligned}
\|\partial_j u\|_m &\leq C(\|\partial_j Lu\|_{-m} + \|\widetilde{L}u\|_{-m} + \|\partial_j u\|_{m-1}) \\
&\leq C(\|Lu\|_{1-m} + \|u\|_m + \|u\|_m).
\end{aligned}$$

Here we have used the fact that $\|Lu\|_{-m} \leq C\|u\|_m$ by duality. Using the Poincaré inequality (Theorem 20.23), we obtain (42.5) for $k = 1$. The general case for k follows by induction. Thus, the theorem is proved. □

Corollary 42.3. *Suppose that $s \in \mathbb{R}$ and $s \geq m$. If the coefficients of the operator L from* (42.1)–(42.2) *belong to the Sobolev space $W_\infty^{s-m}(\Omega)$, then there is a constant $C > 0$ such that for all $u \in H_0^s(\Omega)$ we have*

$$\|u\|_s \leq C(\|Lu\|_{s-2m} + \|u\|_{s-1}). \tag{42.6}$$

Proof. The result follows by interpolation of Sobolev spaces $H_0^s(\Omega)$; see [39]. □

Theorem 42.4 (Regularity in Sobolev spaces). *Suppose that the coefficients of the operator L from* (42.1)–(42.2) *belong to $C^\infty(\Omega)$. Let u and f be distributions on Ω satisfying $Lu = f$. If $f \in H^s_{\text{loc}}(\Omega)$ for $s \geq 0$, then $u \in H^{s+2m}_{\text{loc}}(\Omega)$.*

Proof. Let $\varphi \in C_0^\infty(\Omega)$. Then the equation $\varphi Lu = \varphi f$ can be rewritten as

$$L(\varphi u) = \varphi f + L_1 u, \tag{42.7}$$

or

$$L(\widetilde{u}) = \widetilde{f} + L_1 u,$$

where the operator L_1 is of order $2m-1$ with $C_0^\infty(\Omega)$ coefficients and $\widetilde{f} \in H_0^s(\Omega)$. Our task is to show that $\widetilde{u}$ from (42.7) belongs to $H_0^{s+2m}(\Omega)$. We use first induction on $s \geq 0$ and $2m \geq 2$. Let us assume that $s = 0$ and $2m = 2$. Then $\widetilde{f} \in L^2(\Omega)$ with compact support and $\widetilde{u} \in L^2(\Omega)$ with compact support (we may assume this without loss of generality). Applying now Theorem 42.2 (see (42.5)) and using (42.7), we obtain

$$\begin{aligned} \|\widetilde{u}\|_1 &\leq C(\|L\widetilde{u}\|_{-1} + \|\widetilde{u}\|_0) \leq C(\left\|\widetilde{f}\right\|_{-1} + \|L_1 u\|_{-1} + \|\widetilde{u}\|_0) \\ &\leq C(\left\|\widetilde{f}\right\|_0 + \|\widetilde{u}\|_0 + \|\widetilde{u}\|_0) < \infty. \end{aligned}$$

That is, $\widetilde{u} \in H_0^1(\Omega)$. Applying again Theorem 42.2 with this $\widetilde{u} \in H_0^1(\Omega)$, we obtain that

$$\|\widetilde{u}\|_2 \leq C(\|L\widetilde{u}\|_0 + \|L_1 u\|_0 + \|\widetilde{u}\|_1) \leq C(\left\|\widetilde{f}\right\|_0 + \|\widetilde{u}\|_1 + \|\widetilde{u}\|_1).$$

Thus, $\widetilde{u} \in H_0^2(\Omega)$, and the starting point of induction is checked. Let us assume now that for every integer $s \geq 1$ and $2m \geq 2$ it is true that $f \in H^s_{\text{loc}}(\Omega)$ implies $u \in H^{s+2m}_{\text{loc}}(\Omega)$. Consider now $f \in H^{s+1}_{\text{loc}}(\Omega)$. Then $f \in H^s_{\text{loc}}(\Omega)$ as well, and we may apply the induction hypothesis, that is, the solution u of $Lu = f$ belongs to $H^{s+2m}_{\text{loc}}(\Omega)$. But then we have that (see (42.5) with $k = m+s+1$)

$$\begin{aligned} \|\widetilde{u}\|_{s+2m+1} &\leq C(\|L\widetilde{u}\|_{s+1} + \|\widetilde{u}\|_{2m+s}) \leq C(\left\|\widetilde{f}\right\|_{s+1} + \|L_1 u\|_{s+1} + \|\widetilde{u}\|_{2m+s}) \\ &\leq C(\left\|\widetilde{f}\right\|_{s+1} + \|\widetilde{u}\|_{s+2m} + \|\widetilde{u}\|_{2m+s}) < \infty, \end{aligned}$$

since L_1 is of order $2m-1$ and $u \in H^{s+2m}_{\text{loc}}(\Omega)$. The latter inequality means that this theorem has been proved for integer $s \geq 0$. For arbitrary $s \geq 0$ the result follows by interpolation of Sobolev spaces (see [39]). □

Corollary 42.5. *Suppose that the coefficients of the operator L from* (42.1)–(42.2) *belong to $W^m_\infty(\Omega)$. Then the distributional solution u of $Lu = f$ with $f \in L^2(\Omega)$ belongs to $H^{2m}_{\text{loc}}(\Omega)$.*

Proof. The result follows from the proof of Theorems 42.2 and 42.4. □

Theorem 42.6 (Gårding's inequality). *Suppose L from (42.1)–(42.2) has $L^\infty(\Omega)$ coefficients. Then for all $u \in H_0^m(\Omega)$ we have*

$$\operatorname{Re} D(u,u) \geq c_1 \|u\|_m^2 - c_2 \|u\|_0^2 \tag{42.8}$$

with some positive constants c_1 and c_2.

Proof. The proof is much the same as the proofs of previous theorems. Indeed, as before we can easily obtain that

$$\operatorname{Re} D(u,u) \geq \nu_1 \|u\|_m^2 - C \|u\|_{m-1}^2$$

with positive constants ν_1 and C. Since for all $u \in H_0^m(\Omega)$ we have

$$\|u\|_{m-1}^2 \leq \varepsilon \|u\|_m^2 + C_\varepsilon \|u\|_0^2$$

with arbitrary $\varepsilon > 0$, we obtain that

$$\operatorname{Re} D(u,u) \geq (\nu_1 - C\varepsilon) \|u\|_m^2 - CC_\varepsilon \|u\|_0^2.$$

This completes the proof. □

Exercise 42.2. Show that Theorem 42.6 does not hold for $L = \Delta^2$ on $\Omega \subset \mathbb{R}^n$ for any $u \in H^2(\Omega)$.

Exercise 42.3. Prove that the range of the operator $(-\Delta)^m + \mu I$, $\mu > 0$, considered on $H_0^m(B)$, B the unit ball in $\mathbb{R}^n$, is complete in $H^l(B)$ for all $l = 0, 1, 2, \ldots$. Hint: Prove the solvability of the equation $((-\Delta)^m + \mu I)u = (|x|^2 - 1)P(x)$ for a polynomial $P(x)$.

Returning now to the Dirichlet boundary value problem (42.4) for the operator (42.1)–(42.2), we must search for u in a space of functions for which these boundary conditions make sense. Based on Corollary 42.5, the natural candidate for this is $H^{2m}(\Omega)$. The trace formula for Sobolev spaces (see Part II of this book) says that g_j from (42.4) must be from the spaces $H^{2m-j-1/2}(\partial\Omega)$, $j = 0, 1, \ldots, m-1$, respectively. For these given functions g_j from (42.4) we may find a function $g \in H^{2m}(\Omega)$ such that

$$\left.\partial_\nu^j g\right|_{\partial\Omega} = g_j, \quad j = 0, 1, \ldots, m-1. \tag{42.9}$$

Then, setting $w := g - u$, we reduce our original problem (42.4) for L to solving

$$\begin{cases} Lw = Lg - f \text{ in } \Omega, \\ \partial_\nu^j w = 0 \text{ on } S, j = 0, 1, \ldots, m-1. \end{cases} \tag{42.10}$$

Since (42.9) and (42.10) imply that all derivatives of w of order strictly less than m vanish on S, we reformulate the Dirichlet boundary value problem (42.10) as follows: given $f \in L^2(\Omega)$, find $w \in H_0^m(\Omega)$ such that

$$D(w,v) = (Lg - f, v)_{L^2(\Omega)} \tag{42.11}$$

for all $v \in H_0^m(\Omega)$.

In order to solve the problem (42.11) we consider the properties of sesquilinear forms. Let H be a Hilbert space.

Definition 42.7. A complex-valued function $a(u,v)$ on $H \times H$ is said to be a *sesquilinear form* if

(1) $a(\lambda_1 u_1 + \lambda_2 u_2, v) = \lambda_1 a(u_1,v) + \lambda_2 a(u_2,v)$,
(2) $a(u, \mu_1 v_1 + \mu_2 v_2) = \overline{\mu_1} a(u,v_1) + \overline{\mu_2} a(u,v_2)$,

for all $u, v, u_1, u_2, v_1, v_2 \in H$ and for all $\lambda_1, \lambda_2, \mu_1, \mu_2 \in \mathbb{C}$.

Definition 42.8. A functional F on H is called a *conjugate linear functional* if

$$F(\mu_1 v_1 + \mu_2 v_2) = \overline{\mu_1} F(v_1) + \overline{\mu_2} F(v_2)$$

for all $v_1, v_2 \in H$ and for all $\mu_1, \mu_2 \in \mathbb{C}$.

Theorem 42.9 (Lax–Milgram). *Let $a(u,v)$ be a sesquilinear form on $H \times H$ such that*

(1) $|a(u,v)| \le M \|u\| \|v\|, u, v \in H$,
(2) $|a(u,u)| \ge \beta \|u\|^2, u \in H$,

where M and β are some positive constants. Then for every conjugate linear functional F on H there is a unique $u \in H$ such that $a(u,v) = F(v)$. Moreover,

$$\|u\| \le c_0 \|F\|_{H \to \mathbb{C}},$$

and the constant c_0 is independent of F.

Proof. For each fixed $u \in H$ the mapping $v \mapsto a(u,v)$ is a bounded conjugate linear functional on H. Hence, the Riesz–Fréchet theorem gives that there is a unique element $w \in H$ such that $\overline{a(u,v)} = (v,w)$ or $a(u,v) = (w,v)$. Thus we can define an operator $A : H \to H$ mapping u to w as

$$a(u,v) = (Au, v)$$

if and only if $w = Au$. It is clear that A is a bounded linear operator. The linearity follows from

$$a(\lambda_1 u_1 + \lambda_2 u_2, v) = \lambda_1 a(u_1,v) + \lambda_2 a(u_2,v) = \lambda_1 (w_1,v) + \lambda_2 (w_2,v)$$

if and only if $A(\lambda_1 u_1 + \lambda_2 u_2) = \lambda_1 A u_1 + \lambda_2 A u_2$. The boundedness follows from

$$\|Au\|^2 = (Au, Au) = a(u, Au) \le M \|u\| \, \|Au\|,$$

which implies $\|Au\| \le M \|u\|$. Let us note that $a(u,Au)$ is real and positive here. Next we show that A is one-to-one and that the range of A is equal to H. Indeed,

$$\beta \|u\|^2 \le |a(u,u)| = |(Au,u)| \le \|Au\| \, \|u\|$$

implies that $\beta \|u\| \le \|Au\| \le M \|u\|$. The first inequality implies that A is one-to-one and $\overline{R(A)} = R(A)$. Now we will show that in fact, $R(A) = H$. Let $w \in R(A)^{\perp}$. Then

$$\beta \|w\|^2 \le |a(w,w)| = |(Aw,w)| = 0,$$

and therefore $w = 0$, i.e., $\overline{R(A)} = H = R(A)$. Next, again due to the Riesz–Fréchet theorem for F we have that there is a unique $\widetilde{w} \in H$ such that

$$F(v) = (\widetilde{w}, v), \quad v \in H,$$

and $\|\widetilde{w}\| = \|F\|_{H \to \mathbb{C}}$. But since $R(A) = H$, we may find $u \in H$ such that $Au = \widetilde{w}$ if and only if

$$a(u,v) = (Au,v) = (\widetilde{w},v) = F(v),$$

which proves the solvability in this theorem. Furthermore,

$$\|u\| \le \frac{1}{\beta} \|Au\| = \frac{1}{\beta} \|\widetilde{w}\| = \frac{1}{\beta} \|F\|.$$

Finally, we need to show that this element u is unique. If there are two elements u_1 and u_2 such that

$$a(u_1,v) = F(v), \quad a(u_2,v) = F(v),$$

then $a(u_1 - u_2, v) = 0$ for all $v \in H$, and therefore

$$\beta \|u_1 - u_2\|^2 \le |a(u_1 - u_2, u_1 - u_2)| = 0.$$

This completes the proof. □

Remark 42.10. The Lax–Milgram theorem actually says that the operator A that was constructed there has a bounded inverse such that

$$\|A^{-1}\|_{H \to H} \le \frac{1}{\beta},$$

where β is the same as in condition (2).

Assume now that the sesquilinear form $a(u,v)$ satisfies all conditions of the Lax–Milgram theorem and that the sesquilinear form $b(u,v)$ is only bounded, i.e., there is a constant $M > 0$ such that

$$|b(u,v)| \leq M\,\|u\|\,\|v\|, \quad u,v \in H.$$

So, we may associate with $a(u,v)$ and $b(u,v)$ two operators A and B, respectively, such that A has bounded inverse and B is just bounded:

$$a(u,v) + b(u,v) = (Au,v) + (Bu,v).$$

In that case the problem of solving

$$a(u,v) + b(u,v) = F(v), \quad u,v \in H,$$

with a conjugate linear functional F can be reduced to solving

$$Au + Bu = w$$

with w from $F(v) = (w,v)_H$.

Theorem 42.11. *Let $A : H \to H$ be a bijective bounded linear operator with bounded inverse A^{-1} and let $B : H \to H$ be a compact linear operator. Then $A+B$ is injective if and only if it is surjective, and in this case $(A+B)^{-1} : H \to H$ is bounded.*

Proof. Write

$$A + B = A(I - (-A^{-1})B) =: A(I - K),$$

where K is compact, since B is compact. Applying Riesz's theorem (see Theorem 28.15), we obtain that $(A+B)^{-1} = (I-K)^{-1}A^{-1}$ is bounded. □

Corollary 42.12. *Let A be as in Theorem 42.11 and let B be bounded (not necessarily compact) with small norm, i.e., $\|B\| < \varepsilon$ with ε sufficiently small. Then $A+B$ is bijective and $(A+B)^{-1} : H \to H$ is bounded.*

Proof. The operator $A+B$ is bijective. Indeed, let $(A+B)u = 0$, or $u = -A^{-1}Bu$. Then

$$\|u\| \leq \left\|A^{-1}\right\| \|B\| \|u\| < \varepsilon \left\|A^{-1}\right\| \|u\|.$$

Thus $u = 0$ is the only possibility, and $A+B$ is injective. The operator A is surjective with $R(A) = H$. The same is true for $A+B$, since

$$A + B = A(I - (-A^{-1})B) = A(I - K)$$

with $\|K\| < 1$. Thus $A+B$ is injective with bounded inverse, since

$$(A+B)^{-1} = (I-K)^{-1}A^{-1}.$$

The corollary is proved. □

Theorem 42.13. *Let $D(u,v)$ be the Dirichlet form (42.3) corresponding to the operator L from (42.1)–(42.2) with $L^\infty(\Omega)$ coefficients. Then the following representation holds:*

$$D(u,v) = (Au,v)_{H^m(\Omega)} + (Bu,v)_{H^m(\Omega)}, \quad u,v, \in H_0^m(\Omega), \tag{42.12}$$

where $A : H_0^m(\Omega) \to H_0^m(\Omega)$ is a linear bounded operator with bounded inverse and $B : H_0^m(\Omega) \to H_0^m(\Omega)$ is compact.

Proof. The definition (42.3) of $D(u,v)$ allows us to write

$$D(u,v) = D_m(u,v) + D_{m-1}(u,v),$$

where $D_m(u,v)$ is a sesquilinear form that satisfies (see (42.2)) the conditions of the Lax–Milgram theorem (Theorem 42.9) and $D_{m-1}(u,v)$ has the form

$$D_{m-1}(u,v) = \sum_{|\alpha|=|\beta|\le m-1} \int_\Omega a_{\alpha\beta}(x)\partial^\beta u \partial^\alpha \overline{v} \mathrm{d}x. \tag{42.13}$$

Applying the Lax–Milgram theorem, we obtain that there is a bounded linear operator $A : H_0^m(\Omega) \to H_0^m(\Omega)$ with bounded inverse such that

$$D_m(u,v) = (Au,v)_{H^m(\Omega)}. \tag{42.14}$$

Concerning the sesquilinear form $D_{m-1}(u,v)$, we may say that (see (42.13)) since

$$\begin{aligned}|D_{m-1}(u,v)| &\le C \sum_{|\alpha|=|\beta|\le m-1} \int_\Omega |\partial^\beta u| \cdot |\partial^\alpha v| \mathrm{d}x \le C \|u\|_{H^{m-1}(\Omega)} \|v\|_{H^{m-1}(\Omega)} \\ &\le C \|u\|_{H^m(\Omega)} \|v\|_{H^m(\Omega)},\end{aligned}$$

there is a bounded linear operator $B : H_0^m(\Omega) \to H_0^m(\Omega)$ such that

$$D_{m-1}(u,v) = (Bu,v)_{H^m(\Omega)}. \tag{42.15}$$

We claim that B is compact in $H_0^m(\Omega)$. To see this, we first note that

$$\begin{aligned}\|Bu\|^2_{H^m(\Omega)} &= (Bu,Bu)_{H^m(\Omega)} = D_{m-1}(u,Bu) \\ &= \sum_{|\alpha|=|\beta|\le m-1} \int_\Omega a_{\alpha\beta}(x)\partial^\beta u \partial^u \overline{(Bu)} \mathrm{d}x \\ &\le C \|u\|_{H^{m-1}(\Omega)} \|Bu\|_{H^{m-1}(\Omega)} \le C \|u\|_{H^{m-1}(\Omega)} \|Bu\|_{H^m(\Omega)},\end{aligned}$$

or

$$\|Bu\|_{H^m(\Omega)} \leq C\|u\|_{H^{m-1}(\Omega)}. \tag{42.16}$$

Now let $u_j \in H_0^m(\Omega)$ be such that $\|u_j\|_{H^m(\Omega)}$ is bounded. Since $H_0^m(\Omega)$ is compactly embedded in $H_0^{m-1}(\Omega)$, we have that there is a subsequence u_{j_k} that converges strongly in $H_0^{m-1}(\Omega)$, i.e., u_{j_k} is a Cauchy sequence in $H_0^{m-1}(\Omega)$. Hence Bu_{j_k} is a Cauchy sequence in $H_0^m(\Omega)$; see (42.16). This means that Bu_{j_k} converges strongly in $H_0^m(\Omega)$, and thus B is compact in $H_0^m(\Omega)$. The theorem now follows from (42.14) and (42.15). □

Let us now return to the Dirichlet boundary value problem (42.10)–(42.11) for the elliptic differential operator L from (42.1)–(42.2). Theorem 42.13 allows us to rewrite (42.11) as

$$((A+B)w,v)_{H^m(\Omega)} = (Lg-f,v)_{L^2(\Omega)}, \tag{42.17}$$

and the task is to find $w \in H_0^m(\Omega)$ such that (42.17) holds for all $v \in H_0^m(\Omega)$. Since

$$F : H_0^m(\Omega) \ni v \mapsto (Lg-f,v)_{L^2(\Omega)}$$

is a conjugate linear functional, The Riesz–Fréchet theorem says that there is a unique $f_0 \in H_0^m(\Omega)$ such that

$$(Lg-f,v)_{L^2(\Omega)} = (f_0,v)_{H^m(\Omega)}.$$

Due to this fact, (42.17) can be rewritten in operator form as

$$(A+B)w = f_0. \tag{42.18}$$

Theorem 42.14 (Unique solvability). *The Dirichlet boundary value problem* (42.4) *for the operator L from* (42.1)–(42.2) *has a unique solution if and only if $\lambda = 0$ is not a point of the spectrum of L with homogeneous Dirichlet boundary conditions. Moreover, this unique solution u can be obtained as*

$$u = g + (A+B)^{-1} f_0, \tag{42.19}$$

where g satisfies (42.9) *and f_0 satisfies* (42.18).

Proof. Due to Theorem 42.11, it is enough to show that the operator $A+B$ is injective on $H_0^m(\Omega)$, i.e., $(A+B)u = 0$ implies $u = 0$. This is equivalent to the fact that

$$((A+B)u,v) = 0, \quad \text{for all } v \in H_0^m(\Omega),$$

implies $u = 0$, or $D(u,v) = 0$ for all $v \in H_0^m(\Omega)$ implies $u = 0$. But since $D(u,v) = (Lu,v)_{L^2(\Omega)}$, the statement of injectivity can be reformulated as

$$(Lu,v)_{L^2(\Omega)} = 0, \quad \text{for all } v \in H_0^m(\Omega),$$

or

$$\begin{cases} Lu = 0, \\ u \in H_0^m(\Omega) = 0, \end{cases}$$

or

$$\begin{cases} Lu = 0, & \text{in } \Omega, \\ u = 0, \partial_\nu u = 0, \ldots, \partial_\nu^{m-1} u = 0, & \text{on } \partial\Omega. \end{cases}$$

Formula (42.19) follows now from (42.18) and (42.9). The theorem is proved. □

Gårding's inequality (see (42.8)) allows us to get essential information about the kernel of the operator L and its adjoint. Let us define

$$W = \{u \in H_0^m(\Omega) : (Lu,v)_{L^2} = 0 \text{ if and only if } D(u,v) = 0 \text{ for all } v \in H_0^m(\Omega)\}$$

and

$$V = \{u \in H_0^m(\Omega) : (L^* u,v)_{L^2} = 0 \text{ if and only if } D(v,u) = 0 \text{ for all } v \in H_0^m(\Omega)\}.$$

Theorem 42.15. *Suppose L from* (42.1)–(42.2) *has $L^\infty(\Omega)$ coefficients. Then*

$$\dim W = \dim V < \infty.$$

Moreover, if $f \in L^2(\Omega)$, then there exists $u \in H_0^m(\Omega)$ such that $D(u,v) = (f,v)_{L^2}$ for all $v \in H_0^m(\Omega)$ if and only if $f \perp V$.

Proof. Gårding's inequality (42.8) implies that

$$|((L+c_2 I)u,u)_{L^2}| \geq \mathrm{Re}[D(u,u) + c_2(u,u)] \geq c_1 \|u\|_m^2, \quad c_1 > 0,$$

for all $u \in H_0^m(\Omega)$. Hence

$$\|(L+c_2 I)u\|_{L^2(\Omega)} \geq c_1 \|u\|_m \, .$$

The latter inequality means that the operator $L + c_2 I$ is invertible and its inverse $(L+c_2 I)^{-1}$ is compact as an operator in $L^2(\Omega)$, since the embedding $H_0^m(\Omega) \hookrightarrow L^2(\Omega)$ is compact. Moreover, the range $R((L+c_2 I)^{-1})$ is in $H_0^m(\Omega)$. Now we can see that $u \in W$ if and only if

$$((L+c_2 I)u,v)_{L^2(\Omega)} = c_2(u,v)_{L^2(\Omega)},$$

or

$$(L+c_2I)u = c_2u,$$

or

$$\left(\frac{1}{c_2}I - (L+c_2I)^{-1}\right)u = 0,$$

that is, u belongs to the kernel of the operator $\frac{1}{c_2}I - (L+c_2I)^{-1}$. We can say also that $u \in V$ if and only if u belongs to the kernel of the adjoint operator $\frac{1}{c_2}I - ((L+c_2I)^{-1})^*$. Therefore, the Fredholm alternative (see Theorem 34.11) gives us the statement of this theorem. □

Corollary 42.16. *Under the conditions of Theorem 42.15 we have that if in particular* $W = V = \{0\}$*, then the solution* u *of the problem* $D(u,v) = (f,v)_{L^2(\Omega)}$ *always exists and is unique for all* $f \in L^2(\Omega)$.

Proof. The result follows from the Fredholm alternative. □

Corollary 42.17. *Suppose that all conditions of Theorem 42.15 are satisfied. Assume in addition that* $L = L^*$*, i.e.,* L *is formally self-adjoint (this holds if and only if* $a_{\alpha\beta} = \overline{a_{\beta\alpha}}$ *for all* α, β*). Then there is an orthonormal basis* $\{u_k\}$ *for* $L^2(\Omega)$ *consisting of eigenfunctions that satisfy the Dirichlet boundary conditions* $\partial_\nu^j u_k = 0$ *on* $S = \partial\Omega$ *for* $0 \le j \le m-1$*. The corresponding eigenvalues are real and of finite multiplicity, and they accumulate only at infinity.*

Proof. The proof follows immediately from the Hilbert–Schmidt theorem (Corollary 28.11). □

Example 42.18. Let us consider the Dirichlet boundary value problem in a bounded domain $\Omega \subset \mathbb{R}^n, n \ge 2$ for the Schrödinger operator $-\Delta + q$ with complex-valued potential $q \in L^p(\Omega), n/2 < p \le \infty$: given $f \in H^{1/2}(\partial\Omega)$, find $u \in H^1(\Omega)$ such that

$$\begin{cases} (-\Delta + q - \lambda)u = 0, & \text{in } \Omega, \\ u = f, & \text{on } \partial\Omega, \end{cases} \tag{42.20}$$

where λ is real. Then Lemma 32.1 and Theorem 42.14 imply that (42.20) has a solution if and only if $\lambda = 0$ is not a point of the spectrum of this Schrödinger operator with homogeneous Dirichlet boundary conditions.

Example 42.19. Consider the Dirichlet boundary value problem in a bounded domain $\Omega \subset \mathbb{R}^n$, $n \ge 2$, for the Helmholtz operator $\Delta + k^2 n(x)$ with a complex-valued function $n(x)$ from $L^p(\Omega), n/2 < p \le \infty$: given $f \in H^{1/2}(\partial\Omega)$, find $u \in H^1(\Omega)$ such that

$$\begin{cases} (\Delta + k^2 n(x))u = 0, & \text{in } \Omega, \\ u = f, & \text{on } \partial\Omega, \end{cases} \tag{42.21}$$

where k is a real or complex number. The values k^2 for which there exists a nonzero function $u \in H_0^1(\Omega)$ satisfying (in the distributional sense)

$$\Delta u + k^2 n(x) u = 0, \quad \text{in } \Omega,$$

are called the Dirichlet eigenvalues of the Helmholtz operator, and the corresponding nonzero solutions are called the eigenfunctions for it. It is clear that $k^2 = 0$ is not an eigenvalue of this Helmholtz operator. The application of Lemma 32.1 and Theorem 42.11 lead to the solvability of (42.21). Namely, (42.21) has a unique solution if and only if k^2 is not a point of the spectrum (i.e., is not an eigenvalue) of this Helmholtz operator. In the case of a real-valued function $n(x)$, we may prove even more.

Theorem 42.20. *Assume that $n(x) \in L^p(\Omega)$, $n/2 < p \leq \infty$, is real-valued. Then there exists an orthonormal basis $\{u_k\}_{k=1}^{\infty}$ for $H_0^1(\Omega)$ consisting of eigenfunctions of the Helmholtz operator $-\Delta - \lambda n(x)$. The corresponding eigenvalues $\{\lambda_k\}_{k=1}^{\infty}$ are all real and accumulate only at infinity ($|\lambda_k| \to \infty$). If in addition $n(x) \geq 0$ ($n(x) \not\equiv 0$), then $\lambda_k > 0$ for all $k = 1, 2, \ldots$.*

Proof. We may rewrite the eigenvalue problem for (42.21) as (see Lemma 32.1 and Theorem 42.11)

$$(A - \lambda B)u = 0, \quad u \in H_0^1(\Omega), \lambda \neq 0, \tag{42.22}$$

where A is a bounded, self-adjoint, strictly positive linear operator in $H_0^1(\Omega)$ with bounded inverse, and B is a compact self-adjoint operator in $H_0^1(\Omega)$. Next, (42.22) can be rewritten as

$$\left(\frac{1}{\lambda} I - A^{-1/2} B A^{-1/2}\right) u = 0, \quad u \in H_0^1(\Omega). \tag{42.23}$$

This is an eigenvalue problem for the self-adjoint compact operator $A^{-1/2}BA^{-1/2}$. Using the Riesz–Schauder and Hilbert-Schmidt theorems (see Part III of this book), we may conclude that there exists a sequence $\{\frac{1}{\lambda_k}\}_{k=1}^{\infty}$ of eigenvalues of $A^{-1/2}BA^{-1/2}$ such that

$$\left|\frac{1}{\lambda_1}\right| \geq \left|\frac{1}{\lambda_2}\right| \geq \cdots \geq \left|\frac{1}{\lambda_k}\right| \geq \cdots \to 0$$

as $k \to \infty$ with corresponding eigenfunctions $\{u_k\}_{k=1}^{\infty}$ that form an orthonormal basis in $H_0^1(\Omega)$. Hence, the theorem is proved. □

Exercise 42.4. Prove that if $\{\varphi_j\}_{j=1}^{\infty}$ is an orthonormal basis in $H_0^k(\Omega)$, $k = 1, 2, \ldots$, then $\{\varphi_j\}_{j=1}^{\infty}$ is a basis (orthogonal) in $H_0^l(\Omega)$ for every integer $0 \leq l < k$. Show that the converse is not true.

Chapter 43
The Direct Scattering Problem for the Helmholtz Equation

In this chapter we will show that the scattering problem for an imperfect conductor in $\mathbb{R}^n$, $n \geq 2$, is well posed. More precisely, we consider a bounded domain $\Omega \subset \mathbb{R}^n$ (the conductor) containing the origin with connected complement such that $\partial\Omega$ is in the class C^2. Our aim is to show the existence of a unique solution $u \in C^2(\mathbb{R}^n \setminus \overline{\Omega}) \cap C(\mathbb{R}^n \setminus \Omega)$ of the *exterior impedance boundary value problem*

$$\begin{aligned}
&\Delta u + k^2 u = 0, \quad x \in \mathbb{R}^n \setminus \overline{\Omega}, \\
&u = u_0 + u_{\text{sc}}, \quad u_0 = \mathrm{e}^{\mathrm{i}k(x,\theta)}, \quad \theta \in \mathbb{S}^{n-1}, \\
&\lim_{r \to +\infty} r^{(n-1)/2} \left(\frac{\partial u_{\text{sc}}}{\partial r} - \mathrm{i}k u_{\text{sc}} \right) = 0, \quad r = |x|, \\
&\frac{\partial u}{\partial \nu} + \mathrm{i}\lambda u = 0, \quad x \in \partial\Omega,
\end{aligned} \tag{43.1}$$

where the boundary condition on $\partial\Omega$ is assumed in the sense of uniform convergence as $x \to \partial\Omega$, $\lambda(x) \in C(\partial\Omega)$, $\lambda(x) > 0$, and ν is the outward unit normal vector to $\partial\Omega$.

Theorem 43.1 (Representation formula for an exterior domain). *Let $u_{\text{sc}} \in C^2(\mathbb{R}^n \setminus \overline{\Omega}) \cap C(\mathbb{R}^n \setminus \Omega)$ be a solution of (43.1) such that $\frac{\partial u_{\text{sc}}}{\partial \nu}$ exists in the sense of uniform convergence as $x \to \partial\Omega$. Then for all $x \in \mathbb{R}^n \setminus \overline{\Omega}$ we have*

$$u_{\text{sc}}(x) = \int_{\partial\Omega} \left[u_{\text{sc}}(y) \partial_{\nu_y} G_k^+(|x-y|) - \partial_{\nu_y} u_{\text{sc}}(y) G_k^+(|x-y|) \right] \mathrm{d}\sigma(y), \tag{43.2}$$

where G_k^+ is defined in Chapter 32.

Proof. Let $x \in \mathbb{R}^n \setminus \overline{\Omega}$ be fixed and let $\varepsilon > 0$ be so small that $B_\varepsilon(x) \subset \mathbb{R}^n \setminus \overline{\Omega}$. Let $B_R(0)$ be a ball of radius R containing both Ω and $B_\varepsilon(x)$ (Figure 43.1).

V. Serov, *Fourier Series, Fourier Transform and Their Applications to Mathematical Physics*, Applied Mathematical Sciences 197,
DOI 10.1007/978-3-319-65262-7_43

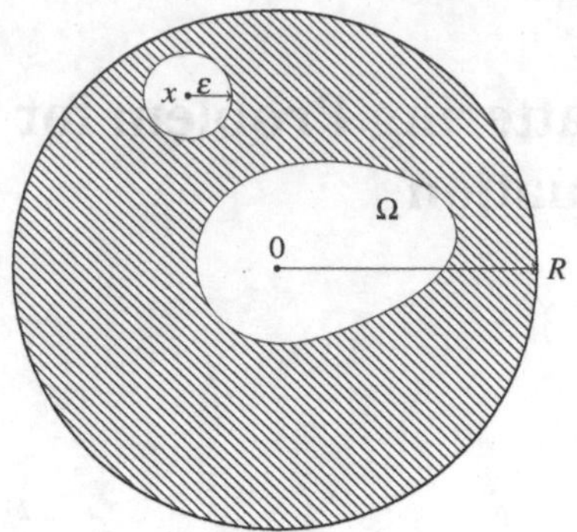

Fig. 43.1 Geometric illustration of $x \notin \Omega, B_\varepsilon(x)$ and $B_R(0)$.

Then from Green's identity we have that

$$0 = \int_{\partial\Omega \cup \partial B_\varepsilon(x) \cup \partial B_R(0)} \left[u_{\mathrm{sc}}(y)\partial_{\nu_y} G_k^+(|x-y|) - \partial_{\nu_y} u_{\mathrm{sc}}(y) G_k^+(|x-y|)\right] \mathrm{d}\sigma(y).$$

On the spheres $\partial B_\varepsilon(x)$ and $\partial B_R(0)$ we have

$$\partial_{\nu_y} = -\left.\frac{\partial}{\partial r}\right|_{r=\varepsilon}, \quad \partial_{\nu_y} = \left.\frac{\partial}{\partial r}\right|_{r=R},$$

and in addition, on $\partial B_\varepsilon(x)$ one has

$$\partial_{\nu_y}|x-y| = \nu_y \cdot \nabla_y |x-y| = -1.$$

The well known properties of Hankel functions (see, e.g., [23]) give that

$$\left(\rho^{-(n-2)/2} H^{(1)}_{(n-2)/2}(\rho)\right)'_\rho = -\rho^{-(n-2)/2} H^{(1)}_{n/2}(\rho).$$

Taking into account all these facts, we obtain that ($n \geq 2$)

$$\partial_{\nu_y} G_k^+(|x-y|) = \frac{\mathrm{i}}{4}\left(\frac{1}{2\pi}\right)^{(n-2)/2} k^{n/2}|x-y|^{-(n-2)/2} H^{(1)}_{n/2}(k|x-y|).$$

Using now the asymptotic behavior of $H^{(1)}_{n/2}(k|x-y|)$ for small arguments (see, e.g., [23])

$$H^{(1)}_{n/2}(k|x-y|) \sim \frac{-\mathrm{i}}{\pi}\left(\frac{k|x-y|}{2}\right)^{-n/2} \Gamma(n/2),$$

we obtain that on the sphere $\partial B_\varepsilon(x)$, one has

$$\partial_{\nu_y} G_k^+(|x-y|) \sim \frac{|x-y|^{1-n}}{\omega_n}, \quad n \geq 2.$$

Thus, letting $\varepsilon \to 0+$, we see that

$$\begin{aligned}
0 \sim & \int_{\partial\Omega} \left[u_{\mathrm{sc}}(y)\partial_{\nu_y} G_k^+(|x-y|) - \partial_{\nu_y} u_{\mathrm{sc}}(y) G_k^+(|x-y|)\right] \mathrm{d}\sigma(y) \\
& + \int_{|y|=R} \left[u_{\mathrm{sc}}(y)\frac{\partial}{\partial r} G_k^+(|x-y|) - \frac{\partial}{\partial r} u_{\mathrm{sc}}(y) G_k^+(|x-y|)\right] \mathrm{d}\sigma(y) \\
& - \int_{|x-y|=\varepsilon} \left[u_{\mathrm{sc}}(y)\frac{\varepsilon^{1-n}}{\omega_n} - \frac{\partial}{\partial r} u_{\mathrm{sc}}(y) G_k^+(\varepsilon)\right] \mathrm{d}\sigma(y),
\end{aligned}$$

and therefore (using the Sommerfeld radiation condition)

$$\begin{aligned}
u_{\mathrm{sc}}(x) = & \int_{\partial\Omega} \left[u_{\mathrm{sc}}(y)\partial_{\nu_y} G_k^+(|x-y|) - \partial_{\nu_y} u_{\mathrm{sc}}(y) G_k^+(|x-y|)\right] \mathrm{d}\sigma(y) \\
& + \int_{|y|=R} \left[u_{\mathrm{sc}}(y)\frac{\partial}{\partial r} G_k^+(|x-y|) - \frac{\partial}{\partial r} u_{\mathrm{sc}}(y) G_k^+(|x-y|)\right] \mathrm{d}\sigma(y) \\
= & \int_{\partial\Omega} \left[u_{\mathrm{sc}}(y)\partial_{\nu_y} G_k^+(|x-y|) - \partial_{\nu_y} u_{\mathrm{sc}}(y) G_k^+(|x-y|)\right] \mathrm{d}\sigma(y) \\
& + \int_{|y|=R} \left[u_{\mathrm{sc}}(y) o(1/R^{(n-1)/2}) - G_k^+(|x-y|) o(1/R^{(n-1)/2})\right] \mathrm{d}\sigma(y). \qquad (43.3)
\end{aligned}$$

In order to estimate the latter integral in (43.3) we need the following lemma.

Lemma 43.2. *For u_{sc} from* (43.1) *it is true that*

$$\lim_{R\to+\infty} \int_{|y|=R} |u_{\mathrm{sc}}(y)|^2 \mathrm{d}\sigma(y) < \infty.$$

Proof. The Sommerfeld radiation condition gives

$$\begin{aligned}
0 = & \lim_{R\to+\infty} \int_{|y|=R} \left|\frac{\partial u_{\mathrm{sc}}}{\partial r} - iku_{\mathrm{sc}}\right|^2 \mathrm{d}\sigma(y) \\
= & \lim_{R\to+\infty} \int_{|y|=R} \left(\left|\frac{\partial u_{\mathrm{sc}}}{\partial r}\right|^2 + k^2|u_{\mathrm{sc}}(y)|^2 + 2k\,\mathrm{Im}\left(u_{\mathrm{sc}}\frac{\partial \overline{u_{\mathrm{sc}}}}{\partial r}\right)\right) \mathrm{d}\sigma(y).
\end{aligned}$$

At the same time, Green's identity for $D_R := B_R(0) \setminus \overline{\Omega}$ implies

$$\begin{aligned}
\int_{|y|=R} u_{\mathrm{sc}}(y)\frac{\partial \overline{u_{\mathrm{sc}}}}{\partial r} \mathrm{d}\sigma(y) = & \int_{\partial\Omega} u_{\mathrm{sc}}(y)\partial_{\nu_y}\overline{u_{\mathrm{sc}}(y)} \mathrm{d}\sigma(y) - k^2 \int_{D_R} |u_{\mathrm{sc}}(y)|^2 \mathrm{d}y \\
& + \int_{D_R} |\nabla u_{\mathrm{sc}}(y)|^2 \mathrm{d}y.
\end{aligned}$$

This means that for R sufficiently large we have

$$\operatorname{Im} \int_{|y|=R} u_{sc}(y) \frac{\partial \overline{u_{sc}}}{\partial r} d\sigma(y) = \operatorname{Im} \int_{\partial\Omega} u_{sc}(y) \partial_{v_y} \overline{u_{sc}(y)} d\sigma(y).$$

We therefore have that

$$\lim_{R \to +\infty} \int_{|y|=R} \left(\left| \frac{\partial u_{sc}}{\partial r} \right|^2 + k^2 |u_{sc}(y)|^2 \right) d\sigma(y) = -2k \operatorname{Im} \int_{\partial\Omega} u_{sc}(y) \partial_{v_y} \overline{u_{sc}(y)} d\sigma(y).$$

This equality completes the proof of the lemma. □

Taking into account that

$$G_k^+(|x-y|) = O\left(\frac{1}{R^{(n-1)/2}}\right), \quad |y| = R \to \infty,$$

and Lemma 43.2, we can easily obtain from (43.3), by letting $R \to \infty$, the equality (43.2). Thus, the theorem is completely proved. □

Corollary 43.3 ***(Representation formula for an interior domain).*** *Let Ω be a bounded C^2 domain and $u \in C^2(\Omega) \cap C^1(\overline{\Omega})$ a solution of $(\Delta + k^2)u = 0$ in Ω. Then for all $x \in \Omega$ we have*

$$u(x) = \int_{\partial\Omega} \left[\partial_{v_y} u(y) G_k^+(|x-y|) - u(y) \partial_{v_y} G_k^+(|x-y|)\right] d\sigma(y). \tag{43.4}$$

Corollary 43.4. *Under the conditions of Corollary 43.3, $u(x)$ is real-analytic in $x \in \Omega$.*

Proof. Since $G_k^+(|x-y|)$ for $x \neq y$ is real-analytic in $x \in \Omega$, the representation formula (43.4) implies that the same is true for $u(x)$. □

Theorem 43.5. *Let $v \in C^2(\mathbb{R}^n \setminus \overline{\Omega}) \cap C(\mathbb{R}^n \setminus \Omega)$ be a solution of the Helmholtz equation $\Delta v + k^2 v = 0$ in $\mathbb{R}^n \setminus \overline{\Omega}$ satisfying the Sommerfeld radiation condition and the boundary condition $\partial_v v + i\lambda v = 0$, $\lambda(x) > 0$, on $\partial\Omega$. Then $v \equiv 0$.*

Proof. Let $B_R(0)$ contain Ω in its interior. Then Green's identity and the boundary conditions imply that ($\partial_v = \frac{\partial}{\partial r}$ on the spheres)

$$\begin{aligned}
&\int_{|y|=R} \left(\overline{v}(y) \frac{\partial v(y)}{\partial r} - v(y) \frac{\partial \overline{v}(y)}{\partial r} \right) d\sigma(y) \\
&\quad = \int_{\partial\Omega} \left(\overline{v}(y) \partial_{v_y} v(y) - v(y) \partial_{v_y} \overline{v}(y)\right) d\sigma(y) = -2i \int_{\partial\Omega} \lambda(y) |v(y)|^2 d\sigma(y).
\end{aligned}$$

But this is equivalent to the equalities

$$2\mathrm{i}\int_{|y|=R}\mathrm{Im}\left(\overline{v}(y)\frac{\partial v(y)}{\partial r}\right)\mathrm{d}\sigma(y)=2\mathrm{i}\int_{\partial\Omega}\mathrm{Im}\left(\overline{v}(y)\partial_{\nu_y}v(y)\right)\mathrm{d}\sigma(y)$$
$$=-2\mathrm{i}\int_{\partial\Omega}\lambda(y)|v(y)|^2\mathrm{d}\sigma(y).$$

Since $\lambda(y)>0$, these equalities lead to

$$\mathrm{Im}\int_{|y|=R}v(y)\frac{\partial\overline{v}(y)}{\partial r}\mathrm{d}\sigma(y)=\mathrm{Im}\int_{\partial\Omega}v(y)\partial_{\nu_y}\overline{v}(y)\mathrm{d}\sigma(y)$$
$$=\int_{\partial\Omega}\lambda(y)|v(y)|^2\mathrm{d}\sigma(y)\geq 0. \qquad (43.5)$$

We have proved in Lemma 43.2 that

$$\lim_{R\to+\infty}\int_{|y|=R}\left(\left|\frac{\partial v}{\partial r}\right|^2+k^2|v(y)|^2\right)\mathrm{d}\sigma(y)=-2k\,\mathrm{Im}\int_{\partial\Omega}v(y)\partial_{\nu_y}\overline{v(y)}\mathrm{d}\sigma(y). \qquad (43.6)$$

But the inequality (43.5) implies that

$$\lim_{R\to+\infty}\int_{|y|=R}\left(\left|\frac{\partial v}{\partial r}\right|^2+k^2|v(y)|^2\right)\mathrm{d}\sigma(y)=0$$

and

$$\mathrm{Im}\int_{\partial\Omega}v(y)\partial_{\nu_y}\overline{v(y)}\mathrm{d}\sigma(y)=0. \qquad (43.7)$$

Thus, we have from (43.5) and (43.7) (compare with Lemma 43.2)

$$\lim_{R\to+\infty}\int_{|y|=R}|v(y)|^2\mathrm{d}\sigma(y)=0$$

and

$$\int_{\partial\Omega}\lambda(y)|v(y)|^2\mathrm{d}\sigma(y)=0.$$

This implies that $v\equiv 0$ and $\partial_\nu v\equiv 0$ on $\partial\Omega$. These two facts and the representation formula (43.2) for the scattering solution provide that $v(x)\equiv 0$ in $\mathbb{R}^n\setminus\overline{\Omega}$. The theorem is proved. □

Corollary 43.6. *If the solution of the scattering problem* (43.1) *exists, then it is unique.*

Proof. If two solutions u_1 and u_2 of (43.1) exist, then their difference $v=u_1-u_2=u_{sc}^{(1)}-u_{sc}^{(2)}$ satisfies the hypothesis of Theorem 43.5. Hence $v\equiv 0$, i.e., $u_1=u_2$. □

Theorem 43.7 *(Rellich's lemma).* *Let $u\in C^2(\mathbb{R}^n\setminus\overline{\Omega})$ be a solution of the Helmholtz equation satisfying*

$$\lim_{R\to+\infty}\int_{|y|=R}|u(y)|^2 d\sigma(y)=0.$$

Then $u\equiv 0$ *in* $\mathbb{R}^n\setminus\overline{\Omega}$.

Proof. Let $r\gg 1$ and let

$$U(r)=\int_{\mathbb{S}^{n-1}}u(r\theta)\varphi(\theta)d\theta,$$

where u is a solution of the Helmholtz equation and φ is an arbitrary eigenfunction of the Laplacian Δ_S on the unit sphere, i.e.,

$$(\Delta_S+\mu^2)\varphi=0,\quad \mu^2=m(m+n-2),\quad m=0,1,2,\ldots,$$

where

$$\Delta=\frac{\partial^2}{\partial r^2}+\frac{n-1}{r}\frac{\partial}{\partial r}+\frac{1}{r^2}\Delta_S.$$

It follows that $U(r)$ satisfies the ordinary differential equation

$$U''(r)+\frac{n-1}{r}U'(r)+(k^2-\mu^2/r^2)U=0.$$

But the general solution of this equation is given by

$$U(r)=K_1r^{-(n-2)/2}H_\nu^{(1)}(kr)+K_2r^{-(n-2)/2}H_\nu^{(2)}(kr),$$

where K_1 and K_2 are arbitrary constants, $H_\nu^{(1,2)}$ are Hankel functions of order ν with $H_\nu^{(2)}=\overline{H_\nu^{(1)}}$ and $\nu^2=\mu^2+(\frac{n-2}{2})^2$. Since the hypothesis of this theorem implies that $U(r)=o(1/r^{(n-1)/2})$, we deduce that K_1 and K_2 are equal to zero and hence $U(r)\equiv 0$ for all $r\geq R_0$. The same is true for $u(r\theta)$ due to the completeness of the eigenfunctions of Δ_S on $\mathbb{S}^{n-1}$; see [34]. The claim follows now from the real analyticity of every solution of the Helmholtz equation. □

Theorem 43.8. *Let* $v\in C^2(\mathbb{R}^n\setminus\overline{\Omega})\cap C(\mathbb{R}^n\setminus\Omega)$ *be a solution of the Helmholtz equation satisfying the Sommerfeld radiation condition at infinity. Let* $\partial_\nu(u_0+v)$ *converge uniformly as* $x\to\partial\Omega$ *and let*

$$\operatorname{Im}\int_{\partial\Omega}v(y)\partial_{\nu_y}\overline{v(y)}d\sigma(y)\geq 0.$$

Then $v\equiv 0$ *in* $\mathbb{R}^n\setminus\Omega$.

Proof. This follows immediately from (43.6) and Rellich's lemma (Theorem 43.7). □

Remark 43.9. All the results of Theorems 43.1–43.8 and their corollaries remain true if we consider instead of $C^2(\mathbb{R}^n \setminus \overline{\Omega}) \cap C(\mathbb{R}^n \setminus \Omega)$ the Sobolev spaces H^s. Namely, we may assume that the problem (43.1) is considered in the space $H^2(\mathbb{R}^n \setminus \overline{\Omega}) \cap H^{3/2}(\mathbb{R}^n \setminus \Omega)$.

Here we have considered mostly uniqueness results for these boundary value problems. The solvability is provided using the results of Chapters 41 and 42 as follows. As we know, the single layer potential

$$u_{sc}(x) := \int_{\partial\Omega} \varphi(y) G_k^+(|x-y|) d\sigma(y), \quad x \in \mathbb{R}^n \setminus \partial\Omega, \tag{43.8}$$

with continuous density φ satisfies the Sommerfeld radiation condition at infinity, is a solution of the Helmholtz equation in $\mathbb{R}^n \setminus \partial\Omega$, is continuous in $\mathbb{R}^n$, and satisfies the discontinuity property (see Theorem 41.21)

$$\partial_{\nu\pm} u_{sc}(x) = \int_{\partial\Omega} \varphi(y) \partial_{\nu_x} G_k^+(|x-y|) d\sigma(y) \mp \varphi(x), \quad x \in \partial\Omega.$$

Let us note that these properties of the single layer potential are also valid for $\varphi \in H^{-1/2}(\partial\Omega)$, where the integrals are interpreted in the sense of duality pairing [22, 25]. Thus, (43.8) will solve the scattering problem (43.1), provided that

$$\varphi(x) - 2\int_{\partial\Omega} \varphi(y) \partial_{\nu_x} G_k^+(|x-y|) d\sigma(y) - 2\mathrm{i}\lambda(x) \int_{\partial\Omega} \varphi(y) G_k^+(|x-y|) d\sigma(y)$$
$$= 2(\partial_\nu u_0(x) + \mathrm{i}\lambda(x) u_0(x)), \quad x \in \partial\Omega, \tag{43.9}$$

where $u_0(x) = \mathrm{e}^{\mathrm{i}k(x,\theta)}$. Hence, to establish the existence for the problem (43.1), it suffices to show the existence of a solution to (43.9) in the normed space $C(\partial\Omega)$. To this end, we first recall that the integral operators in (43.9) are compact on $C(\partial\Omega)$ (see Theorem 34.9). Hence, by Riesz's theorem (see Theorem 41.29), it suffices to show that the homogeneous equation (corresponding to (43.9)) has only the trivial solution.

Let φ be a solution of this homogeneous equation. Then u_{sc} from (43.8) will be a solution of (43.9) with u_0 set equal to zero, and hence, by Theorem 43.5 and Corollary 43.6, we have that this $u_{sc}(x)$ is equal to zero for $x \in \mathbb{R}^n \setminus \overline{\Omega}$. By the continuity of (43.8) across $\partial\Omega$, $u_{sc}(x)$ is a solution of the Helmholtz equation in Ω as well, and $u_{sc}(x) = 0$ on the boundary $\partial\Omega$. If we assume now that $k^2 > 0$ is not a Dirichlet eigenvalue for $-\Delta$ in Ω, then $u_{sc}(x) \equiv 0$ in Ω, and by the discontinuity properties of the single layer potential we have that

$$0 = \partial_{\nu_-} u_{sc}(x) - \partial_{\nu_+} u_{sc}(x) = \varphi(x),$$

i.e., the homogeneous equation under consideration has only the trivial solution $\varphi = 0$. Hence by Riesz's theorem (Theorem 41.29), the corresponding inhomogeneous equation (43.9) has a unique solution that depends continuously on the right-hand side.

If $k^2 > 0$ is a Dirichlet eigenvalue of $-\Delta$ in Ω, then Riesz's theorem cannot be applied (it is not valid in this case), and the whole procedure mentioned above does not work. To obtain an integral equation that is uniquely solvable for all values of the number $k^2 > 0$ we need to modify the kernel of the representation (43.8). For the two-dimensional case such a modification is shown in [6, (3.51)]. The same idea can be considered for higher dimensions with the use of spherical harmonics. So, we have proved now the following solvability result, valid for all dimensions $n \geq 2$.

Theorem 43.10. *If $k^2 > 0$ is not a Dirichlet eigenvalue of $-\Delta$ in the domain $\Omega \subset \mathbb{R}^n$, then there exists a unique solution of the scattering problem* (43.1) *that depends continuously on $u_0(x) = \mathrm{e}^{\mathrm{i}k(x,\theta)}$.*

It is quite often necessary to consider the scattering problem (43.1) in Sobolev spaces H^s. In that case we define

$$H^1_{\mathrm{loc}}(\mathbb{R}^n \setminus \overline{\Omega}) := \{u : u \in H^1(B_R(0) \setminus \overline{\Omega})\}$$

with an arbitrary ball $B_R(0)$ of radius $R > 0$ centered at the origin and containing $\overline{\Omega}$. We recall that $H^{-s}(\partial\Omega), 0 \leq s < \infty$ is the dual space of $H^s(\partial\Omega)$. Then, for $f \in H^{-1/2}(\partial\Omega)$, a weak solution of

$$\begin{aligned} &\Delta u + k^2 u = 0, \quad x \in \mathbb{R}^n \setminus \overline{\Omega}, \\ &u = u_0 + u_{\mathrm{sc}}, \quad u_0 = \mathrm{e}^{ik(x,\theta)}, \quad \theta \in \mathbb{S}^{n-1}, \\ &\lim_{r \to +\infty} r^{(n-1)/2} \left(\frac{\partial u_{\mathrm{sc}}}{\partial r} - \mathrm{i}ku_{\mathrm{sc}} \right) = 0, \quad r = |x|, \\ &\partial_\nu u(x) + \mathrm{i}\lambda u(x) = 0, \quad x \in \partial\Omega, \end{aligned} \tag{43.10}$$

is defined as a function $u \in H^1_{\mathrm{loc}}(\mathbb{R}^n \setminus \overline{\Omega})$ such that

$$\int_{\mathbb{R}^n \setminus \overline{\Omega}} (\nabla u \cdot \nabla v - k^2 uv) \mathrm{d}x - \mathrm{i} \int_{\partial\Omega} \lambda uv \mathrm{d}\sigma(x) = - \int_{\partial\Omega} fv \mathrm{d}\sigma(x) \tag{43.11}$$

for all $v \in H^1(\mathbb{R}^n \setminus \overline{\Omega})$ that are identically equal to zero outside some ball $B_R(0)$ with radius $R > 0$ sufficiently large. In that case the analogue of Theorem 43.10 for Sobolev spaces can be proved (see [6, Chapter 8] for details).

Chapter 44
Some Inverse Scattering Problems for the Schrödinger Operator

The classical *inverse scattering problem* is to reconstruct the potential $q(x)$ from the knowledge of the far field data (scattering amplitude, see p. 232) $A(k,\theta',\theta)$, when k, θ', and θ are restricted to some given set.

If $q \in L^1(\mathbb{R}^n)$, then $q(y)u(y,k,\theta) \in L^1(\mathbb{R}^n)$ uniformly with respect to $\theta \in \mathbb{S}^{n-1}$ due to

$$q(y)u(y,k,\theta) = q(y)(\mathrm{e}^{\mathrm{i}k(\theta,y)} + u_{\mathrm{sc}}(y,k,\theta)) = q(y)\mathrm{e}^{\mathrm{i}k(\theta,y)} + |q|^{\frac{1}{2}} \cdot q_{\frac{1}{2}} u_{\mathrm{sc}}(y,k,\theta)$$

and Hölder's inequality. We may therefore conclude that the scattering amplitude $A(k,\theta',\theta)$ is well defined and continuous. Also, the following representation holds:

$$\begin{aligned} A(k,\theta',\theta) &= \int_{\mathbb{R}^n} \mathrm{e}^{-\mathrm{i}k(\theta',y)} q(y)(\mathrm{e}^{\mathrm{i}k(\theta,y)} + u_{\mathrm{sc}})\mathrm{d}y \\ &= \int_{\mathbb{R}^n} \mathrm{e}^{-\mathrm{i}k(\theta'-\theta,y)} q(y)\mathrm{d}y + R(k,\theta',\theta) \\ &= (2\pi)^{n/2}(\mathscr{F}q)(k(\theta'-\theta)) + R(k,\theta',\theta), \end{aligned}$$

where $R(k,\theta',\theta) \to 0$ as $k \to +\infty$ uniformly with respect to θ' and θ. This fact implies that

$$A(k,\theta',\theta) \approx (2\pi)^{n/2}(\mathscr{F}q)(k(\theta'-\theta)),$$

or

$$q(x) \approx (2\pi)^{-n/2}\mathscr{F}^{-1}(A(k,\theta',\theta))(x),$$

where the inverse Fourier transform must be understood in some special sense.

Let us introduce the cylinders $M_0 = \mathbb{R} \times \mathbb{S}^{n-1}$ and $M = M_0 \times \mathbb{S}^{n-1}$, and the measures μ_θ and μ on M_0 and M, respectively, as

V. Serov, *Fourier Series, Fourier Transform and Their Applications to Mathematical Physics*, Applied Mathematical Sciences 197,
DOI 10.1007/978-3-319-65262-7_44

$$\mathrm{d}\mu_\theta(k,\theta') = \frac{1}{4}|k|^{n-1}\mathrm{d}k|\theta-\theta'|^2\mathrm{d}\theta',$$

$$\mathrm{d}\mu(k,\theta',\theta) = \frac{1}{|\mathbb{S}^{n-1}|}\mathrm{d}\theta\mathrm{d}\mu_\theta(k,\theta'),$$

where $|\mathbb{S}^{n-1}| = \frac{2\pi^{n/2}}{\Gamma(\frac{n}{2})}$ is the area of the unit sphere $\mathbb{S}^{n-1}$, and $\mathrm{d}\theta$ and $\mathrm{d}\theta'$ denote the usual Lebesgue measures on $\mathbb{S}^{n-1}$. We shall define the inverse Fourier transform on M_0 and M as

$$(\mathscr{F}_{M_0}^{-1}\varphi_1)(x) = \frac{1}{(2\pi)^{n/2}}\int_{M_0} \mathrm{e}^{-ik(\theta-\theta',x)}\varphi_1(k,\theta')\mathrm{d}\mu_\theta,$$

$$(\mathscr{F}_M^{-1}\varphi_2)(x) = \frac{1}{(2\pi)^{n/2}}\int_M \mathrm{e}^{-ik(\theta-\theta',x)}\varphi_2(k,\theta',\theta)\mathrm{d}\mu.$$

If we write $\xi = k(\theta-\theta')$, then k and θ' are obtained by

$$k = \frac{|\xi|}{2(\theta,\widehat{\xi})}, \quad \theta' = \theta - 2(\theta,\widehat{\xi})\widehat{\xi}, \quad \widehat{\xi} = \frac{\xi}{|\xi|}. \tag{44.1}$$

Exercise 44.1. Let $u_\theta(k,\theta')$ be the coordinate mapping $M_0 \to \mathbb{R}^n$ given as

$$u_\theta(k,\theta') = k(\theta-\theta'),$$

where θ is considered a fixed parameter. Prove that

(1) the formulas (44.1) for k and θ' hold;
(2) the following is true:

$$\int_{M_0} \varphi \circ u_\theta(k,\theta')\mathrm{d}\mu_\theta(k,\theta') = \int_{\mathbb{R}^n}\varphi(x)\mathrm{d}x$$

if $\varphi \in S$ is even and

$$\int_M \varphi \circ u_\theta(k,\theta')\mathrm{d}\mu(k,\theta',\theta) = \int_{\mathbb{R}^n}\varphi(x)\mathrm{d}x$$

if $\varphi \in S$;
(3) in addition,

$$\mathscr{F}_{M_0}^{-1}(\varphi \circ u_\theta) = \mathscr{F}^{-1}\varphi$$

if $\varphi \in S$ is even and

$$\mathscr{F}_M^{-1}(\varphi \circ u_\theta) = \mathscr{F}^{-1}\varphi$$

if $\varphi \in S$. Here $\mathscr{F}^{-1}$ is the usual inverse Fourier transform in $\mathbb{R}^n$.

Exercise 44.2. Prove that

(1) $A(-k,\theta',\theta) = \overline{A(k,\theta',\theta)}$;
(2) $A(k,\theta',\theta) = A(k,-\theta,-\theta')$.

The approximation $q(x) \approx (2\pi)^{-\frac{n}{2}}\mathscr{F}^{-1}(A(k,\theta',\theta)(x))$ for all θ' and θ and for sufficiently large k allows us to introduce the following definitions.

Definition 44.1. The inverse Born approximations $q_{\mathrm{B}}^{\theta}(x)$ and $q_{\mathrm{B}}(x)$ of the potential $q(x)$ are defined by

$$q_{\mathrm{B}}^{\theta}(x) = (2\pi)^{-n/2}(\mathscr{F}_{M_0}^{-1}A)(x) = \frac{1}{(2\pi)^n}\int_{M_0} \mathrm{e}^{-ik(\theta-\theta',x)}A(k,\theta',\theta)\mathrm{d}\mu_{\theta}$$

and

$$q_{\mathrm{B}}(x) = (2\pi)^{-n/2}(\mathscr{F}_{M}^{-1}A)(x) = \frac{1}{(2\pi)^n}\int_{M} \mathrm{e}^{-ik(\theta-\theta',x)}A(k,\theta',\theta)\mathrm{d}\mu$$

in the sense of distributions.

Theorem 44.2 (Uniqueness). *Assume that the potential $q(x)$ belongs $L^p_{\mathrm{loc}}(\mathbb{R}^n)$, $\frac{n}{2} < p \le \infty$, $n \ge 3$, and has the special behavior $|q(x)| \le C|x|^{-\mu}$, $\mu > 2$, $|x| \to \infty$ at infinity. Then the knowledge of $q_{\mathrm{B}}^{\theta}(x)$ with θ restricted to an $(n-2)$-dimensional hemisphere determines $q(x)$ uniquely.*

Proof. It is not difficult to check that if $q(x)$ satisfies the conditions of the present theorem, then $q(x)$ will satisfy the conditions of Theorem 23.5:

$$q \in L^p(\mathbb{R}^n), \quad \frac{n}{2} < p \le \frac{n+1}{2},$$

or

$$q \in L^p_{\sigma}(\mathbb{R}^n), \quad \frac{n+1}{2} < p \le +\infty, \quad \sigma > 1 - \frac{n+1}{2p}.$$

Now we can represent $q_{\mathrm{B}}^{\theta}(x)$ in the form

$$\begin{aligned} q_{\mathrm{B}}^{\theta}(x) &= \frac{1}{(2\pi)^n}\int_{M_0} \mathrm{e}^{-ik(\theta-\theta',x)}A(k,\theta',\theta)\mathrm{d}\mu_{\theta}(k,\theta') \\ &= \frac{1}{(2\pi)^n}\int_{M_0} \mathrm{d}\mu_{\theta}(k,\theta')\int_{\mathbb{R}^n} \mathrm{e}^{-ik(\theta-\theta',x)}\mathrm{e}^{-ik(\theta',y)}q(y)u(y,k,\theta)\mathrm{d}y \\ &= \frac{1}{(2\pi)^n}\int_{M_0} \mathrm{d}\mu_{\theta}\int_{\mathbb{R}^n} \mathrm{e}^{-ik(\theta-\theta',x-y)}q(y)\mathrm{e}^{-ik(\theta,y)}u(y,k,\theta)\mathrm{d}y, \end{aligned}$$

where $u(y,k,\theta)$ is the solution of the Lippmann–Schwinger equation. Setting

$$v(y,k,\theta) := \mathrm{e}^{-ik(\theta,y)}u(y,k,\theta)$$

and making the change of variables $\xi = k(\theta - \theta')$, we obtain

$$q_{\mathrm{B}}^{\theta}(x) = \frac{1}{(2\pi)^n}\int_{\mathbb{R}^n} \mathrm{d}\xi \int_{\mathbb{R}^n} \mathrm{e}^{-\mathrm{i}(\xi,x-y)} q(y) v\left(y, \frac{|\xi|}{2(\theta,\widehat{\xi})}, \theta\right) \mathrm{d}y.$$

The usual Fourier transform of $q_{\mathrm{B}}^{\theta}(x)$ is equal to

$$\widehat{q_{\mathrm{B}}^{\theta}}(\xi) = \widehat{q}(\xi) + (2\pi)^{-n/2}\int_{\mathbb{R}^n} \mathrm{e}^{\mathrm{i}(\xi,y)} q(y)\left[v\left(y, \frac{|\xi|}{2(\theta,\widehat{\xi})}, \theta\right) - 1\right] \mathrm{d}y,$$

and it implies that

$$|\widehat{q_{\mathrm{B}}^{\theta}}(\xi) - \widehat{q}(\xi)| \le (2\pi)^{-n/2}\int_{\mathbb{R}^n} |q(y)| \left|v\left(y, \frac{|\xi|}{2(\theta,\widehat{\xi})}, \theta\right) - 1\right| \mathrm{d}y,$$

where the function $v(y,k,\theta)$ solves the equation

$$v(x,k,\theta) = 1 - \int_{\mathbb{R}^n} \mathrm{e}^{-ik(x,\theta)} G_k^+(|x-y|) \mathrm{e}^{ik(y,\theta)} q(y) v(y,k,\theta) \mathrm{d}y,$$

i.e.,

$$v = 1 - \widehat{\widetilde{G}}_k(qv),$$

where $\widetilde{G}_k = \mathrm{e}^{-ik(x-y,\theta)}G_k^+$. For k sufficiently large we obtain that

$$v = (I + \widehat{\widetilde{G}}_k q)^{-1}(1),$$

or

$$v = 1 - \widehat{\widetilde{G}}_q(q), \tag{44.2}$$

where $\widehat{\widetilde{G}}_q$ is an integral operator with kernel $\widetilde{G}_q = \mathrm{e}^{-ik(x-y,\theta)}G_q$, and the integral operator $\widehat{G}_q$ with this kernel also satisfies the equation $(H - k^2)\widehat{G}_q = I$. In order to prove (44.2), we recall that

$$\widehat{G}_q = \widehat{G}_k - \widehat{G}_k q \widehat{G}_q,$$

and therefore,

$$\widehat{\widetilde{G}}_q = \widehat{\widetilde{G}}_k - \widehat{\widetilde{G}}_k q \widehat{\widetilde{G}}_q,$$

or

$$\widehat{\widehat{G}}_q = (I + \widehat{\widehat{G}}_k q)^{-1} \widehat{\widehat{G}}_k.$$

The last equality implies that

$$\widehat{\widehat{G}}_q(q) = (I + \widehat{\widehat{G}}_k q)^{-1} \widehat{\widehat{G}}_k(q) = -(v-1),$$

because

$$(I + \widehat{\widehat{G}}_k q)^{-1} \widehat{\widehat{G}}_k(q) = -(v-1)$$

is equivalent to

$$\begin{aligned}\widehat{\widehat{G}}_k(q) &= -(I + \widehat{\widehat{G}}_k q)(v-1) = -(v-1) - (\widehat{\widehat{G}}_k q)(v) + (\widehat{\widehat{G}}_k q)(1) \\ &= -v + 1 - 1 + v + \widehat{\widehat{G}}_k(q) = \widehat{\widehat{G}}_k(q).\end{aligned}$$

We may therefore apply Theorem 23.5 to obtain

$$\|v-1\|_{L^{\frac{2p}{p-1}}_{-\sigma/2}(\mathbb{R}^n)} = \|\widehat{\widehat{G}}_q(q)\|_{L^{\frac{2p}{p-1}}_{-\sigma/2}} \le \frac{C}{k^\gamma} \|q\|_{L^{\frac{2p}{p+1}}_{\sigma/2}},$$

where γ, p, and σ are as in that theorem. It remains only to check that the potential $q \in L^p_{\text{loc}}(\mathbb{R}^n)$ with the special behavior at infinity belongs to $L^{\frac{2p}{p+1}}_{\sigma/2}(\mathbb{R}^n)$. But that is a very simple exercise. Hence, the latter inequality leads to

$$|\widehat{q^\theta_{\mathrm{B}}}(\xi) - \widehat{q}(\xi)| \le C \|q\|^2_{L^{\frac{2p}{p+1}}_{\sigma/2}(\mathbb{R}^n)} \left(\frac{|(\widehat{\xi}, \theta)|}{|\xi|} \right)^\gamma, \quad \xi \neq 0$$

with the same γ. If q_1 and q_2 are as q, then

$$\begin{aligned}|\widehat{q_1}(\xi) - \widehat{q_2}(\xi)| &= |\widehat{q_1}(\xi) - \widehat{q^\theta_{\mathrm{B}}} + \widehat{q^\theta_{\mathrm{B}}} - \widehat{q_2}(\xi)| \le |\widehat{q_1}(\xi) - \widehat{q^\theta_{\mathrm{B}}}| + |\widehat{q^\theta_{\mathrm{B}}} - \widehat{q_2}(\xi)| \\ &\le C \|q_1\|^2_{L^{\frac{2p}{p+1}}_{\sigma/2}(\mathbb{R}^n)} \left(\frac{|(\widehat{\xi}, \theta)|}{|\xi|} \right)^\gamma + C \|q_2\|^2_{L^{\frac{2p}{p+1}}_{\sigma/2}(\mathbb{R}^n)} \left(\frac{|(\widehat{\xi}, \theta)|}{|\xi|} \right)^\gamma = 0\end{aligned}$$

if $(\widehat{\xi}, \theta) = 0$. Thus, this theorem is proved, because $(\widehat{\xi}, \theta) = 0$ precisely as θ runs through an $(n-2)$-dimensional hemisphere see [31, 32]. □

Theorem 44.3 (Saito's formula). *Under the same assumptions for $q(x)$ as in Theorem 44.2,*

$$\lim_{k \to +\infty} k^{n-1} \int_{\mathbb{S}^{n-1}} \int_{\mathbb{S}^{n-1}} \mathrm{e}^{-ik(\theta - \theta', x)} A(k, \theta', \theta) \mathrm{d}\theta \mathrm{d}\theta' = \frac{(2\pi)^n}{\pi} \int_{\mathbb{R}^n} \frac{q(y)\mathrm{d}y}{|x-y|^{n-1}},$$

where the limit holds in the classical sense for $n < p \leq \infty$ and in the sense of distributions for $\frac{n}{2} < p \leq n$.

Proof. Let us consider only the case $n < p \leq \infty$. The proof for $\frac{n}{2} < p \leq n$ requires some changes.

By definition of the scattering amplitude,

$$\begin{aligned}
I &:= k^{n-1}\int_{\mathbb{S}^{n-1}}\int_{\mathbb{S}^{n-1}} A(k,\theta',\theta)\mathrm{e}^{-\mathrm{i}k(\theta-\theta',x)}\mathrm{d}\theta\mathrm{d}\theta' \\
&= k^{n-1}\int_{\mathbb{R}^n} q(y)\mathrm{d}y\int_{\mathbb{S}^{n-1}}\int_{\mathbb{S}^{n-1}} \mathrm{e}^{\mathrm{i}k(\theta-\theta',y-x)}\mathrm{d}\theta\mathrm{d}\theta' \\
&+ k^{n-1}\int_{\mathbb{R}^n} q(y)\mathrm{d}y\int_{\mathbb{S}^{n-1}}\int_{\mathbb{S}^{n-1}} \mathrm{e}^{-\mathrm{i}k(\theta',y)}R(y,k,\theta)\mathrm{e}^{-\mathrm{i}k(\theta-\theta',x)}\mathrm{d}\theta\mathrm{d}\theta' =: I_1 + I_2,
\end{aligned}$$

where $R(y,k,\theta)$ is given by

$$R(y,k,\theta) = -\int_{\mathbb{R}^n} G_k^+(|y-z|)q(z)u(z,k,\theta)\mathrm{d}z$$

and $u(z,k,\theta)$ is the solution of the Lippmann–Schwinger equation. Since

$$\begin{aligned}
\int_{\mathbb{S}^{n-1}}\int_{\mathbb{S}^{n-1}} \mathrm{e}^{\mathrm{i}k(\theta-\theta',y-x)}\mathrm{d}\theta\mathrm{d}\theta' &= \left|\int_{\mathbb{S}^{n-1}} \mathrm{e}^{\mathrm{i}k(\theta,y-x)}\mathrm{d}\theta\right|^2 \\
&= \frac{4\pi^{n-1}}{\Gamma^2(\frac{n-1}{2})}\left(\int_0^\pi \mathrm{e}^{\mathrm{i}k|y-x|\cos\psi}(\sin\psi)^{n-2}\mathrm{d}\psi\right)^2 \\
&= (2\pi)^n\frac{J^2_{\frac{n-2}{2}}(k|x-y|)}{(k|x-y|)^{n-2}},
\end{aligned}$$

we have that I_1 can be represented in the form

$$I_1 = (2\pi)^n k\int_{\mathbb{R}^n}\frac{q(y)}{|x-y|^{n-2}}J^2_{\frac{n-2}{2}}(k|x-y|)\mathrm{d}y.$$

We consider two cases: $k|x-y| < 1$ and $k|x-y| > 1$. In the first case, using Hölder's inequality the integral I_1' over $\{y : k|x-y| < 1\}$ can be estimated by

$$\begin{aligned}
|I_1'| &\leq Ck\int_{|x-y|<\frac{1}{k}}\frac{|q(y)|(k|x-y|)^{n-2}}{|x-y|^{n-2}}\mathrm{d}y \\
&\leq Ck^{n-1}\left(\int_{|x-y|<\frac{1}{k}}|q(y)|^p\mathrm{d}y\right)^{\frac{1}{p}}\left(\int_{|x-y|<\frac{1}{k}}1\cdot\mathrm{d}y\right)^{\frac{1}{p'}} \\
&= Ck^{n-1}k^{-\frac{n}{p'}}\left(\int_{|x-y|<\frac{1}{k}}|q(y)|^p\mathrm{d}y\right)^{\frac{1}{p}} = Ck^{\frac{n}{p}-1}\left(\int_{|x-y|<\frac{1}{k}}|q(y)|^p\mathrm{d}y\right)^{\frac{1}{p}} \to 0
\end{aligned}$$

as $k \to +\infty$, since $n < p \leq \infty$. This means that for every fixed x (or even uniformly with respect to x) I_1' approaches zero as $k \to \infty$. Hence, we have only to estimate the integral I_1'' over $\{y : k|x-y| > 1\}$. The asymptotic behavior of the Bessel function $J_\nu(\cdot)$ for large argument implies that

$$\begin{aligned}
I_1'' &= (2\pi)^n k \int_{|x-y|>\frac{1}{k}} \frac{q(y)}{|x-y|^{n-2}} \\
&\times \left[\sqrt{\frac{2}{\pi k|x-y|}} \cos\left(k|x-y| - \frac{n\pi}{4} + \frac{\pi}{4}\right) + O\left(\frac{1}{(k|x-y|)^{3/2}}\right)\right]^2 \mathrm{d}y \\
&= (2\pi)^n k \int_{|x-y|>\frac{1}{k}} \frac{q(y)}{|x-y|^{n-2}} \\
&\times \left[\frac{2\cos^2(k|x-y| - \frac{n\pi}{4} + \frac{\pi}{4})}{\pi k|x-y|} + O\left(\frac{1}{(k|x-y|)^2}\right)\right] \mathrm{d}y \\
&= \frac{(2\pi)^n}{\pi} \int_{|x-y|>\frac{1}{k}} \frac{q(y)\mathrm{d}y}{|x-y|^{n-1}} \\
&+ \frac{(2\pi)^n}{\pi} \int_{|x-y|>\frac{1}{k}} \frac{q(y)}{|x-y|^{n-1}} \cos\left(2k|x-y| - \frac{n\pi}{2} + \frac{\pi}{2}\right) \mathrm{d}y \\
&+ \frac{1}{k} \int_{|x-y|>\frac{1}{k}} \frac{|q(y)|O(1)}{|x-y|^n} \mathrm{d}y =: I_1^{(1)} + I_1^{(2)} + I_1^{(3)}.
\end{aligned}$$

It is clear that

$$\lim_{k\to+\infty} I_1^{(1)} = \frac{(2\pi)^n}{\pi} \int_{\mathbb{R}^n} \frac{q(y)\mathrm{d}y}{|x-y|^{n-1}}$$

and

$$\lim_{k\to+\infty} I_1^{(2)} = 0.$$

The latter fact follows from the following arguments. Since q belongs to $L^p(\mathbb{R}^n)$ for $p > n$ and has the special behavior at infinity, we may conclude that the L^1 norm of the function $\frac{q(y)}{|x-y|^{n-1}}$ is uniformly bounded with respect to x. Hence it follows from the Riemann–Lebesgue lemma that $I_1^{(2)}$ approaches zero uniformly with respect to x as $k \to +\infty$. For $I_1^{(3)}$ we have the estimate

$$|I_1^{(3)}| \leq \frac{C}{k^{1-\delta}} \int_{\mathbb{R}^n} \frac{|q(y)|\mathrm{d}y}{|x-y|^{n-\delta}}.$$

If we choose δ such that $1 > \delta > \frac{n}{p}$, then $\int_{\mathbb{R}^n} \frac{q(y)\mathrm{d}y}{|x-y|^{n-\delta}}$ will be uniformly bounded with respect to x. Therefore, $I_1^{(3)} \to 0$ as $k \to \infty$ uniformly with respect to x. If we collect all estimates, we obtain that

$$\lim_{k\to\infty} I_1 = \frac{(2\pi)^n}{\pi}\int_{\mathbb{R}^n}\frac{q(y)\mathrm{d}y}{|x-y|^{n-1}}.$$

Our next task is to prove that $I_2 \to 0$ as $k\to\infty$. Since

$$I_2 = k^{n-1}\int_{\mathbb{R}^n} q(y)\mathrm{d}y \int_{\mathbb{S}^{n-1}}\int_{\mathbb{S}^{n-1}} \mathrm{e}^{-\mathrm{i}k(\theta',y)}R(y,k,\theta)\mathrm{e}^{-\mathrm{i}k(\theta-\theta',x)}\mathrm{d}\theta\mathrm{d}\theta',$$

where

$$R(y,k,\theta) = -\int_{\mathbb{R}^n} G_k^+(|y-z|)q(z)u(z,k,\theta)\mathrm{d}z = -\widehat{G}_k(qu),$$

one can check that $R(y,k,\theta) = -\widehat{G}_q(q\mathrm{e}^{\mathrm{i}k(\theta,z)})$. Hence, I_2 can be represented as

$$\begin{aligned} I_2 &= -k^{n-1}\int_{\mathbb{R}^n} q(y)\mathrm{d}y\int_{\mathbb{S}^{n-1}} \mathrm{e}^{\mathrm{i}k(\theta',x-y)}\mathrm{d}\theta'\cdot\widehat{G}_q\left(q(z)\int_{\mathbb{S}^{n-1}}\mathrm{e}^{\mathrm{i}k(\theta,z-x)}\mathrm{d}\theta\right)\\ &= -k^{n-1}(2\pi)^n\int_{\mathbb{R}^n} q(y)\frac{J_{\frac{n-2}{2}}(k|x-y|)}{(k|x-y|)^{\frac{n-2}{2}}}\cdot\widehat{G}_q\left(q(z)\frac{J_{\frac{n-2}{2}}(k|x-z|)}{(k|x-z|)^{\frac{n-2}{2}}}\right)\mathrm{d}y\\ &= (2\pi)^n k\int_{\mathbb{R}^n} q_{\frac{1}{2}}(y)\frac{J_{\frac{n-2}{2}}(k|x-y|)}{(|x-y|)^{\frac{n-2}{2}}}\cdot\widehat{K}_q\left(|q(z)|^{\frac{1}{2}}\frac{J_{\frac{n-2}{2}}(k|x-z|)}{(|x-z|)^{\frac{n-2}{2}}}\right)\mathrm{d}y, \end{aligned}$$

where $\widehat{K}_q$ is an integral operator with kernel

$$K_q(x,y) = -|q(x)|^{\frac{1}{2}}G_q(k,x,y)q_{\frac{1}{2}}(y).$$

It follows from Theorem 23.5 that $\widehat{K}_q : L^2(\mathbb{R}^n)\to L^2(\mathbb{R}^n)$ with the norm estimate

$$\|\widehat{K}_q\|_{L^2\to L^2} \le \frac{C}{k^\gamma},$$

where γ is as in that theorem. We can therefore estimate I_2 using Hölder's inequality as

$$|I_2| \le \frac{C}{k^\gamma}k\int_{\mathbb{R}^n}|q(y)|\frac{J^2_{\frac{n-2}{2}}(k|x-y|)}{|x-y|^{n-2}}\mathrm{d}y.$$

By the same arguments as in the proof for I_1 we can obtain that

$$k\int_{\mathbb{R}^n}|q(y)|\frac{J^2_{\frac{n-2}{2}}(k|x-y|)}{|x-y|^{n-2}}\mathrm{d}y < \infty$$

uniformly with respect to x. This implies that

$$|I_2| \leq \frac{C}{k^\gamma} \to 0$$

as $k \to +\infty$. □

Remark 44.4. This proof holds also for $n = 2$. In dimension $n = 1$ there is an analogous result in which we replace the double integral on the left-hand side by the sum of four values of the integrand at $\theta = \pm 1$ and $\theta' = \pm 1$.

Theorem 44.5. *Let us assume that $n \geq 2$. Under the same assumptions for $q_1(x)$ and $q_2(x)$ as in Theorem 44.3 let us assume that the corresponding scattering amplitudes A_{q_1} and A_{q_2} coincide for some sequence $k_j \to \infty$ and for all $\theta', \theta \in \mathbb{S}^{n-1}$. Then $q_1(x) = q_2(x)$ in the sense of L^p for $n < p \leq \infty$ and in the sense of distributions for $\frac{n}{2} < p \leq n$.*

Proof. Saito's formula shows that we have only to prove that the homogeneous equation

$$\psi(x) := \int_{\mathbb{R}^n} \frac{q(y)dy}{|x-y|^{n-1}} = 0$$

has only the trivial solution $q(y) \equiv 0$. Let us assume that $n < p \leq \infty$. Introduce the space $S_0(\mathbb{R}^n)$ of all functions from the Schwartz space that vanish in some neighborhood of the origin. Due to the conditions for the potential $q(x)$ we may conclude (as before) that $\psi \in L^\infty(\mathbb{R}^n)$, and ψ defines a tempered distribution. Then for every function $\varphi \in S_0(\mathbb{R}^n)$ it follows that

$$0 = \langle \widehat{\psi}, \varphi \rangle = C_n \langle |\xi|^{-1} \widehat{q}(\xi), \varphi \rangle = C_n \langle \widehat{q}(\xi), |\xi|^{-1} \varphi \rangle.$$

Since $\varphi(\xi) \in S_0(\mathbb{R}^n)$, we have $|\xi|^{-1}\varphi \in S_0(\mathbb{R}^n)$ also. Hence, for every $h \in S_0(\mathbb{R}^n)$ the following equation holds:

$$\langle \widehat{q}, h \rangle = 0.$$

This means that the support of $\widehat{q}(\xi)$ is at most at the origin, and therefore $\widehat{q}(\xi)$ can be represented as

$$\widehat{q}(\xi) = \sum_{|\alpha| \leq m} C_\alpha D^\alpha \delta.$$

Hence, $q(x)$ is a polynomial. But due to the behavior at infinity we must conclude that $q \equiv 0$. This proves Theorem 44.5. □

Let us return now to the Born approximation of $q(x)$. Repeated use of the Lippmann–Schwinger equation leads to the following representation for the scattering amplitude $A(k, \theta', \theta)$:

$$A(k,\theta',\theta)=\sum_{j=0}^{m}\int_{\mathbb{R}^n}\mathrm{e}^{-ik(\theta',y)}q_{\frac{1}{2}}(y)\widehat{K}^j\cdot(|q|^{\frac{1}{2}}\mathrm{e}^{ik(x,\theta)})(y)\mathrm{d}y$$
$$+\int_{\mathbb{R}^n}\mathrm{e}^{-ik(\theta',y)}q_{\frac{1}{2}}(y)\widehat{K}^{m+1}(|q|^{\frac{1}{2}}(u(x,k,\theta))(y)\mathrm{d}y,$$

where $u(x,k,\theta)$ is the solution of the Lippmann–Schwinger equation and $\widehat{K}$ is an integral operator with kernel

$$K(x,y)=|q(x)|^{\frac{1}{2}}G_k^+(|x-y|)q_{\frac{1}{2}}(y).$$

The equality for A can be reformulated in the sense of integral operators in $L^2(\mathbb{S}^{n-1})$ as

$$\widehat{A}=\sum_{j=0}^{m}\Phi_0^*(k)\,\mathrm{sgn}\,q\widehat{K}^j\Phi_0(k)+\Phi_0^*(k)\,\mathrm{sgn}\,q\widehat{K}^{m+1}\Phi(k),$$

where Φ_0 and $\Phi(k)$ are defined by (23.9) and (23.10), and Φ_0^* is the L^2- adjoint of Φ_0.

Using this equality and the definition of Born's potential $q_{\mathrm{B}}(x)$, we obtain

$$q_{\mathrm{B}}(x)=\sum_{j=0}^{m}\mathscr{F}_M^{-1}\left[\Phi_0^*(k)\,\mathrm{sgn}\,q\widehat{K}^j\Phi_0(k)\right]+\mathscr{F}_M^{-1}\left[\Phi_0^*(k)\,\mathrm{sgn}\,q\widehat{K}^{m+1}\Phi(k)\right],$$

where the inverse Fourier transform is applied to the kernels of the corresponding integral operators. If we rewrite the latter formula as

$$q_{\mathrm{B}}(x)=\sum_{j=0}^{m}q_j(x)+\widetilde{q}_{m+1}(x),$$

then the term q_j has the form

$$q_j(x)=\mathscr{F}_M^{-1}\left(\int_{\mathbb{R}^n}|q(z)|^{\frac{1}{2}}\mathrm{e}^{-ik(z,\theta)}\mathrm{d}z\int_{\mathbb{R}^n}\mathrm{sgn}\,q(z)K^j(z,y,k)|q(y)|^{\frac{1}{2}}\mathrm{e}^{ik(\theta',y)}\mathrm{d}y\right)$$
$$=\mathscr{F}_M^{-1}\left(\Phi_0^*\,\mathrm{sgn}\,q\widehat{K}^j(|q|^{\frac{1}{2}}\mathrm{e}^{ik(\theta',y)})\right)$$
$$=\frac{1}{(2\pi)^n}\int_M\mathrm{e}^{-ik(\theta-\theta',x)}\mathrm{d}\mu(k,\theta',\theta)\left(\Phi_0^*\,\mathrm{sgn}\,q\widehat{K}^j(|q|^{\frac{1}{2}}\mathrm{e}^{ik(\theta',y)})\right),$$

and a similar formula holds for $\widetilde{q}_{m+1}$ with obvious changes.

In order to formulate the result about the reconstruction of singularities of the unknown potential $q(x)$, let us set $A(k,\theta',\theta)=0$ for $|k|\le k_0$, where $k_0>0$ is arbitrarily large.

Theorem 44.6. *Assume that the potential q belongs to $L^p_{2\delta}(\mathbb{R}^n)\cap L^1(\mathbb{R}^n)$ with $(3n-3)/2<p\le\infty$, $n\ge2$, and $2\delta>1-(n+1)/(2p)$. Then for all $j\ge2$ the terms*

$q_j(x)$ and $\widetilde{q}_j(x)$ in the Born series belong to the Hölder class $C^\alpha(\mathbb{R}^n)$ for all $\alpha \leq 1-(3n-3)/(2p)$.

Proof. For $x_1, x_2 \in \mathbb{R}^n$ we have (see [31])

$$q_j(x_1) - q_j(x_2) = C\int_{|k|\geq k_0} |k|^{n-1}\mathrm{d}k \int_{\mathbb{S}^{n-1}} \mathrm{d}\theta \int_{\mathbb{S}^{n-1}} \mathrm{d}\theta'(1-(\theta,\theta')) \times \Phi_0^* \operatorname{sgn}(q)\widehat{K}^j\Phi_0(k,\theta,\theta')(\mathrm{e}^{-\mathrm{i}k(\theta-\theta',x_1)} - \mathrm{e}^{-\mathrm{i}k(\theta-\theta',x_2)}). \tag{44.3}$$

For $l=1,2$ let us define $e_l(\theta) = \mathrm{e}^{\mathrm{i}k(\theta,x_l)} \in L^2(\mathbb{S}^{n-1})$ and $E_l(\theta) = e_l\theta \in (L^2(\mathbb{S}^{n-1}))^2$. Then the latter difference is equal to

$$C\int_{|k|\geq k_0} |k|^{n-1}\mathrm{d}k \Big((e_1, \Phi_0^* \operatorname{sgn}(q)\widehat{K}^j\Phi_0 e_1)_{L^2(\mathbb{S}^{n-1})} - (e_2, \Phi_0^* \operatorname{sgn}(q)\widehat{K}^j\Phi_0 e_2)_{L^2(\mathbb{S}^{n-1})} - (E_1, \Phi_0^* \operatorname{sgn}(q)\widehat{K}^j\Phi_0 E_1)_{L^2(\mathbb{S}^{n-1})} + (E_2, \Phi_0^* \operatorname{sgn}(q)\widehat{K}^j\Phi_0 E_2)_{L^2(\mathbb{S}^{n-1})}\Big). \tag{44.4}$$

Since $\|E_1 - E_2\|_{L^2(\mathbb{S}^{n-1})} = \|e_1 - e_2\|_{L^2(\mathbb{S}^{n-1})}$ and $\|e_l\|^2_{L^2(\mathbb{S}^{n-1})} = |\mathbb{S}^{n-1}|$, we obtain from (44.3)–(44.4) the estimate

$$|q_j(x_1) - q_j(x_2)| \leq C_n \int_{|k|\geq k_0} |k|^{n-1} \|e_1 - e_2\|_{L^2(\mathbb{S}^{n-1})} \left\|\Phi_0^* \operatorname{sgn}(q)\widehat{K}^j\Phi_0\right\| dk.$$

Note that (see [43])

$$\|e_1 - e_2\|^2_{L^2(\mathbb{S}^{n-1})} = \int_{\mathbb{S}^{n-1}} (2 - \mathrm{e}^{\mathrm{i}k(\theta,x_2-x_1)} - \mathrm{e}^{\mathrm{i}k(\theta,x_1-x_2)})\mathrm{d}\theta = 2\left(|\mathbb{S}^{n-1}| - (2\pi)^{n/2}\frac{J_{(n-2)/2}(|k||x_1-x_2|)}{(|k||x_1-x_2|)^{(n-2)/2}}\right),$$

where J_ν is the Bessel function of order ν. By Lemma 23.13 and Theorem 23.5 we get

$$\left\|\Phi_0^* \operatorname{sgn}(q)\widehat{K}^j\Phi_0\right\| \leq \frac{C}{|k|^{\gamma(j+1)+n-2}},$$

where $\gamma = 1-(n-1)/(2p)$.

If we set $r = |x_1 - x_2|$, we have to estimate the integral

$$\int_{k_0}^\infty \frac{k^{n-1}}{k^{\gamma(j+1)+n-2}} \left(|\mathbb{S}^{n-1}| - (2\pi)^{n/2}\frac{J_{(n-2)/2}(kr)}{(kr)^{(n-2)/2}}\right)^{1/2} dk.$$

We split this integral into two parts: over $1/r < k < \infty$ and over $k_0 < k < 1/r$. By the asymptotics of the Bessel functions for large argument, the first part can be estimated from above by

$$|\mathbb{S}^{n-1}|\int_{1/r}^{\infty}\frac{k^{n-1}}{k^{\gamma(j+1)+n-2}}\mathrm{d}k\leq C_n r^{\gamma(j+1)-2},$$

where j is chosen so large that $\gamma(j+1)>2$. For the second part of the integral we use the asymptotics of the Bessel functions for small argument [23], namely,

$$J_{(n-2)/2}(x)=\frac{x^{(n-2)/2}}{2^{(n-2)/2}\Gamma(n/2)}(1+O(x^2)),\quad x\to 0.$$

Since $|\mathbb{S}^{n-1}|=2\pi^{n/2}/\Gamma(n/2)$, we may estimate the second part from above by

$$Cr\int_{k_0}^{1/r}\frac{k^n\mathrm{d}k}{k^{\gamma(j+1)+n-2}}=Cr\int_{k_0}^{1/r}\frac{\mathrm{d}k}{k^{\gamma(j+1)-2}}\leq Cr^{\min(1,\gamma(j+1)-2)}.$$

To finish the proof we use the fact that $q_j\in L^\infty(\mathbb{R}^n)$, which holds since

$$|q_j(x)|\leq C\int_{k_0}^{\infty}\frac{k^{n-1}}{k^{\gamma(j+1)+n-2}}(\|e\|_{L^2(\mathbb{S}^{n-1})}+\|E\|_{L^2(\mathbb{S}^{n-1})})\mathrm{d}k<\infty$$

for $\gamma(j+1)>2$. The latter condition implies, for $\gamma=1-(n-1)/(2p)$ and $j\geq 2$, that $(3n-3)/2<p\leq\infty$. For $\widetilde{q}_j$ the proof is the same with obvious changes. This completes the proof. □

The first nonlinear term $q_1(x)$ can be rewritten as

$$q_1(x)=-\frac{\Gamma(n/2)}{2^{n+3}\pi^{3n/2}}\int_{\mathbb{R}^{2n}}G(y-x,z-x)q(y)q(z)\mathrm{d}y\mathrm{d}z,\tag{44.5}$$

where G is the tempered distribution

$$G(y,z)=\int_{-\infty}^{\infty}\int_{\mathbb{S}^{n-1}}\int_{\mathbb{S}^{n-1}}\widetilde{G_k^+}(|y-z|)\mathrm{e}^{\mathrm{i}k(\theta,z)-\mathrm{i}k(\theta',y)}|k|^{n-1}|\theta-\theta'|^2\mathrm{d}k\mathrm{d}\theta\mathrm{d}\theta'.$$

Lemma 44.7. *The $2n$-dimensional Fourier transform of G equals*

$$\widehat{G}(\xi,\eta)=-C_n\frac{(\xi,\eta)}{|\xi|^2|\eta|^2},$$

where C_n is a positive constant depending only on n.

Proof. The $2n$-dimensional Fourier transform of G becomes

$$\widehat{G}(\xi,\eta)=\int_{-\infty}^{\infty}\int_{\mathbb{S}^{n-1}}\int_{\mathbb{S}^{n-1}}|k|^{n-1}|\theta-\theta'|^2\mathrm{d}k\mathrm{d}\theta\mathrm{d}\theta'\int_{\mathbb{R}^n}\mathrm{e}^{-\mathrm{i}(k(\theta'-\theta)+\xi+\eta,y)}\mathrm{d}y$$
$$\times\int_{\mathbb{R}^n}\widetilde{G_k^+}(|s|)\mathrm{e}^{-\mathrm{i}(\eta-k\theta,s)}\mathrm{d}s.$$

Using the fact that

$$\widehat{G_k^+}(\cdot) = \lim_{\varepsilon \to 0} \frac{1}{|\cdot|^2 - k^2 - \mathrm{i}\varepsilon},$$

we may approximate $\widehat{G}$ in the topology of $S'(\mathbb{R}^{2n})$ by $\widehat{G_\varepsilon}$ given by

$$\widehat{G_\varepsilon} = \int_0^\infty \int_{\mathbb{S}^{n-1}} \int_{\mathbb{S}^{n-1}} k^{n-1} |\theta - \theta'|^2 \mathrm{d}k \mathrm{d}\theta \mathrm{d}\theta'$$
$$\times \int_{\mathbb{R}^n} \left(\frac{\mathrm{e}^{-\mathrm{i}(k(\theta'-\theta)+\xi+\eta, y)}}{|\eta|^2 - 2k(\theta,\eta) - \mathrm{i}\varepsilon} + \frac{\mathrm{e}^{\mathrm{i}(k(\theta'-\theta)-\xi-\eta, y)}}{|\eta|^2 + 2k(\theta,\eta) + \mathrm{i}\varepsilon} \right) \mathrm{d}y.$$

If we define the variable $\zeta = k(\theta - \theta')$ with the Jacobian $\frac{1}{2}k^{n-1}|\theta - \theta'|^2$, then $k\theta$ depends on ζ as

$$k\theta = \zeta - \frac{|\zeta|^2}{2(\zeta, \theta')}\theta'.$$

Since the Fourier transform of 1 equals $(2\pi)^n \delta$, it follows that

$$\widehat{G_\varepsilon}(\xi,\eta) = 2(2\pi)^n \int_{\mathbb{S}^{n-1}} \int_{\mathbb{R}^n}$$
$$\times \left(\frac{\delta(\zeta - \xi - \eta)}{|\eta|^2 - 2(\zeta,\theta') + \frac{|\zeta|^2(\eta,\theta')}{(\zeta,\theta')} - \mathrm{i}\varepsilon} + \frac{\delta(\zeta + \xi + \eta)}{|\eta|^2 + 2(\zeta,\theta') - \frac{|\zeta|^2(\eta,\theta')}{(\zeta,\theta')} + \mathrm{i}\varepsilon} \right) \mathrm{d}\zeta \mathrm{d}\theta'$$
$$= 2(2\pi)^n \int_{\mathbb{S}^{n-1}} (f,\theta') \left(\frac{1}{(g,\theta') - \mathrm{i}\varepsilon(f,\theta')} + \frac{1}{(g,\theta') + \mathrm{i}\varepsilon(f,\theta')} \right) \mathrm{d}\theta',$$

where $f = \xi + \eta$ and $g = |\xi|^2(\xi + \eta) - |\xi + \eta|^2 \xi$. This expression leads us to

$$\widehat{G}(\xi,\eta) = \lim_{\varepsilon \to 0} \widehat{G_\varepsilon}(\xi,\eta) = 4(2\pi)^n \text{p.v.} \int_{\mathbb{S}^{n-1}} \frac{(f,\theta')}{(g,\theta')} \mathrm{d}\theta' = C_n \frac{(f,g)}{|g|^2} = -C_n \frac{(\xi,\eta)}{|\xi|^2|\eta|^2},$$

where we have used the precise value of the principal value integral; see [30, proof of Lemma 2.4]. □

Lemma 44.8. *Assume that the potential q satisfies all conditions of Theorem 44.6. Then the first nonlinear term q_1 admits the representation*

$$q_1(x) = C_n \left| \int_{\mathbb{R}^n} \frac{x-y}{|x-y|^n} q(y) \mathrm{d}y \right|^2$$

with some positive constant C_n depending only on n.

Proof. The representation (44.5) and Lemma 44.7 imply that

$$q_1(x) = C_n \mathscr{F}_{2n}^{-1}\left(\frac{(\xi,\eta)}{|\xi|^2|\eta|^2}\widehat{q}(\xi)\widehat{q}(\eta)\right)(x,x) = C_n \mathscr{F}_{2n}^{-1}\left(\frac{\xi\widehat{q}(\xi)}{|\xi|^2}, \frac{\eta\widehat{q}(\eta)}{|\eta|^2}\right)(x,x),$$

where $\mathscr{F}_{2n}^{-1}$ denotes the $2n$-dimensional inverse Fourier transform. The claim follows now from $\mathscr{F}^{-1}(\xi/|\xi|^2) = C_n x/|x|^n$. □

Lemma 44.9. *Under the same assumptions on q as in Theorem 44.6 we have that*

(1) for $3(n-1)/2 < p < \infty$, q_1 belongs to $(W^1_{p,2\delta-1}(\mathbb{R}^n))^2$ with $1-(n+1)/(2p) < 2\delta < n - n/p$;

(2) for $p = \infty$, q_1 belongs to the Hölder space $C^1(\mathbb{R}^n)$.

Proof. We introduce the Riesz potential I^{-1} and Riesz transform R (see [37] and Chapter 21) as

$$I^{-1}f(x) = \mathscr{F}^{-1}\left(\frac{\widehat{f}(\xi)}{|\xi|}\right)(x), \quad Rf(x) = \mathscr{F}^{-1}\left(\frac{\xi\widehat{f}(\xi)}{|\xi|}\right)(x).$$

Note that $\nabla I^{-1} = R$ is bounded in $L^p(\mathbb{R}^n)$ for all $1 < p < \infty$; see [37]. From [27] we know that

$$I^{-1} : L^p_{\sigma+1}(\mathbb{R}^n) \to L^p_\sigma(\mathbb{R}^n)$$

for $-n/p < \sigma < n-1-n/p$. This proves (1). Part (2) can be proved like [31, Lemma 2.2]. □

The latter steps lead to the following main result.

Theorem 44.10 (Reconstruction of singularities). *Assume that the potential q belongs to $L^p_{2\delta}(\mathbb{R}^n) \cap L^1(\mathbb{R}^n)$ with p and δ as in Theorem 44.6. Then*

(1) for $\max(3(n-1)/2, n) < p \leq \infty$ the difference $q_{\mathrm{B}} - q$ is a continuous function in $\mathbb{R}^n$;

(2) for $3(n-1)/2 < p \leq \max(3(n-1)/2, n)$ the difference $q_{\mathrm{B}} - q - q_1$ is a continuous function in $\mathbb{R}^n$.

Proof. The proof of this theorem follows immediately from Lemmas 44.7, 44.8, and 44.9 and the Sobolev embedding theorem. □

The statement of Theorem 44.10 means that all singularities and jumps of the unknown potential can be recovered by the Born approximation. In particular, if the potential is the characteristic function of an arbitrary bounded domain, then this domain can be uniquely determined from the scattering data using a linear method.

Chapter 45
The Heat Operator

We turn our attention now to the *heat operator*

$$L = \partial_t - \Delta_x, \quad (x,t) \in \mathbb{R}^n \times \mathbb{R}.$$

The heat operator is a prototype of *parabolic operators*. These are operators of the form

$$\partial_t + \sum_{|\alpha| \le 2m} a_\alpha(x,t)\partial_x^\alpha,$$

where the sum satisfies the *strong ellipticity condition*

$$(-1)^m \sum_{|\alpha|=2m} a_\alpha(x,t)\xi^\alpha \ge \nu|\xi|^{2m},$$

for all $(x,t) \in \mathbb{R}^n \times \mathbb{R}$ and $\xi \in \mathbb{R}^n \setminus \{0\}$ with $\nu > 0$ constant.

We begin by considering the initial value problem

$$\begin{cases} \partial_t u - \Delta u = 0, & \text{in } \mathbb{R}^n \times (0,\infty), \\ u(x,0) = f(x). \end{cases}$$

This problem is a reasonable problem both physically and mathematically.

Assuming for the moment that $f \in S$, the Schwartz space, and taking the Fourier transform with respect to x only, we obtain

$$\begin{cases} \partial_t \widehat{u}(\xi,t) + |\xi|^2 \widehat{u}(\xi,t) = 0, \\ \widehat{u}(\xi,0) = \widehat{f}(\xi). \end{cases} \tag{45.1}$$

V. Serov, *Fourier Series, Fourier Transform and Their Applications to Mathematical Physics*, Applied Mathematical Sciences 197,
DOI 10.1007/978-3-319-65262-7_45

If we solve the ordinary differential equation (45.1), we obtain

$$\widehat{u}(\xi,t) = \mathrm{e}^{-|\xi|^2 t}\widehat{f}(\xi).$$

Thus (at least formally)

$$u(x,t) = \mathscr{F}^{-1}\left(\mathrm{e}^{-|\xi|^2 t}\widehat{f}(\xi)\right) = (2\pi)^{-n/2} f * \mathscr{F}^{-1}\left(\mathrm{e}^{-|\xi|^2 t}\right)(x,t) = f * K_t(x),$$

where

$$K_t(x) = (2\pi)^{-n/2}\mathscr{F}^{-1}\left(\mathrm{e}^{-|\xi|^2 t}\right) \equiv (4\pi t)^{-n/2}\mathrm{e}^{-\frac{|x|^2}{4t}}, \; t > 0, \tag{45.2}$$

is called the *Gaussian kernel*. We define $K_t(x) \equiv 0$ for $t \le 0$.

Exercise 45.1. Prove (45.2).

Let us first prove that

$$\int_{\mathbb{R}^n} K_t(x)\mathrm{d}x = 1.$$

Indeed, using polar coordinates, we have

$$\begin{aligned}
\int_{\mathbb{R}^n} K_t(x)\mathrm{d}x &= (4\pi t)^{-n/2}\int_{\mathbb{R}^n} \mathrm{e}^{-\frac{|x|^2}{4t}}\mathrm{d}x = (4\pi t)^{-n/2}\int_0^\infty r^{n-1}\mathrm{e}^{-\frac{r^2}{4t}}\mathrm{d}r\int_{|\theta|=1}\mathrm{d}\theta \\
&= \omega_n(4\pi t)^{-n/2}\int_0^\infty r^{n-1}\mathrm{e}^{-\frac{r^2}{4t}}\mathrm{d}r \\
&= \omega_n(4\pi t)^{-n/2}\int_0^\infty (4st)^{\frac{n-1}{2}}\mathrm{e}^{-s}\frac{1}{2}\sqrt{4t}\frac{\mathrm{d}s}{\sqrt{s}} \\
&= \frac{\omega_n}{2}\pi^{-n/2}\int_0^\infty s^{n/2-1}\mathrm{e}^{-s}\mathrm{d}s \\
&= \frac{\omega_n}{2}\pi^{-n/2}\Gamma(n/2) = \frac{1}{2}\frac{2\pi^{n/2}}{\Gamma(n/2)}\pi^{-n/2}\Gamma(n/2) = 1.
\end{aligned}$$

Theorem 45.1. *Suppose that $f \in L^\infty(\mathbb{R}^n)$ is uniformly continuous. Then $u(x,t) := (f * K_t)(x)$ satisfies $\partial_t u - \Delta u = 0$ and*

$$\|u(\cdot,t) - f(\cdot)\|_{L^\infty(\mathbb{R}^n)} \to 0$$

as $t \to 0+$.

Proof. For fixed $t > 0$,

$$\Delta_x K_t(x-y) = (4\pi t)^{-n/2}\mathrm{e}^{-\frac{|x-y|^2}{4t}}\left(\frac{|x-y|^2}{4t^2} - \frac{n}{2t}\right),$$

and for fixed $|x-y| \ne 0$,

$$\partial_t K_t(x-y) = (4\pi t)^{-n/2} e^{-\frac{|x-y|^2}{4t}} \left(\frac{|x-y|^2}{4t^2} - \frac{n}{2t} \right).$$

Therefore, $(\partial_t - \Delta_x) K_t(x-y) = 0$.

But we can differentiate (with respect to x and t) under the integral sign, since this integral will be absolutely convergent for all $t > 0$. We may therefore conclude that

$$\partial_t u(x,t) - \Delta_x u(x,t) = 0.$$

It remains only to verify the initial condition. We have

$$\begin{aligned} u(x,t) - f(x) = (f * K_t)(x) - f(x) &= \int_{\mathbb{R}^n} f(y) K_t(x-y) \mathrm{d}y - f(x) \\ &= \int_{\mathbb{R}^n} f(x-z) K_t(z) \mathrm{d}z - \int_{\mathbb{R}^n} f(x) K_t(z) \mathrm{d}z \\ &= \int_{\mathbb{R}^n} (f(x-z) - f(x)) K_t(z) \mathrm{d}z \\ &= \int_{\mathbb{R}^n} (f(x - \eta\sqrt{t}) - f(x)) K_1(\eta) \mathrm{d}\eta. \end{aligned}$$

The assumptions on f imply that

$$\begin{aligned} |u(x,t) - f(x)| \le &\sup_{x \in \mathbb{R}^n, |\eta| < R} |f(x - \eta\sqrt{t}) - f(x)| \int_{\mathbb{R}^n} K_1(\eta) \mathrm{d}\eta \\ &+ 2 \|f\|_{L^\infty(\mathbb{R}^n)} \int_{|\eta| \ge R} K_1(\eta) \mathrm{d}\eta < \varepsilon/2 + \varepsilon/2 \end{aligned}$$

for small t and for R large enough. So we can see that $u(x,t)$ is continuous (even uniformly continuous and bounded) for $(x,t) \in \mathbb{R}^n \times [0, \infty)$ and $u(x,0) = f(x)$. □

Corollary 45.2. $u(x,t) \in C^\infty(\mathbb{R}^n \times \mathbb{R}_+)$.

Proof. We can differentiate under the integral sign defining u as often as we please, because the exponential function increases at infinity faster than any polynomial. Thus, the heat equation takes arbitrary initial data (bounded and uniformly continuous) and smooths them out. □

Corollary 45.3. *Suppose $f \in L^p(\mathbb{R}^n)$, $1 \le p < \infty$. Then $u(x,t) := (f * K_t)(x)$ satisfies $\partial_t u - \Delta u = 0$ and*

$$\|u(\cdot,t) - f(\cdot)\|_{L^p(\mathbb{R}^n)} \to 0$$

as $t \to 0+$. And again $u(x,t) \in C^\infty(\mathbb{R}^n \times \mathbb{R}_+)$.

Theorem 45.4 (Uniqueness). *Suppose $u(x,t) \in C^2(\mathbb{R}^n \times \mathbb{R}_+) \cap C(\mathbb{R}^n \times \overline{\mathbb{R}_+})$ satisfies $\partial_t u - \Delta u = 0$ for $t > 0$ and $u(x,0) = 0$. If for every $\varepsilon > 0$ there exists $c_\varepsilon > 0$ such that*

$$|u(x,t)| \le c_\varepsilon e^{\varepsilon|x|^2}, \quad |\nabla_x u(x,t)| \le c_\varepsilon e^{\varepsilon|x|^2}, \tag{45.3}$$

then $u \equiv 0$.

Proof. For two smooth functions φ and ψ, it is true that

$$\varphi(\partial_t \psi - \Delta\psi) + \psi(\partial_t\varphi + \Delta\varphi) = \sum_{j=1}^n \partial_j(\psi\partial_j\varphi - \varphi\partial_j\psi) + \partial_t(\varphi\psi) = \nabla_{x,t}\cdot\vec{F},$$

where $\vec{F} = (\psi\partial_1\varphi - \varphi\partial_1\psi, \ldots, \psi\partial_n\varphi - \varphi\partial_n\psi, \varphi\psi)$. Given $x_0 \in \mathbb{R}^n$ and $t_0 > 0$, let us take

$$\psi(x,t) = u(x,t), \quad \varphi(x,t) = K_{t_0-t}(x-x_0).$$

Then

$$\begin{aligned} \partial_t\psi - \Delta\psi &= 0, \quad t > 0,\\ \partial_t\varphi + \Delta\varphi &= 0, \quad t < t_0. \end{aligned}$$

If we apply the divergence theorem in the region

$$\Omega = \{(x,t) \in \mathbb{R}^n \times \mathbb{R}_+ : |x| < r, 0 < a < t < b < t_0\},$$

we obtain

$$\begin{aligned} 0 = \int_{\partial\Omega} \vec{F}\cdot\nu \mathrm{d}\sigma &= \int_{|x|\le r} u(x,b)K_{t_0-b}(x-x_0)\mathrm{d}x - \int_{|x|\le r} u(x,a)K_{t_0-a}(x-x_0)\mathrm{d}x \\ &+ \int_a^b \mathrm{d}t \int_{|x|=r} \sum_{j=1}^n \left(u(x,t)\partial_j K_{t_0-t}(x-x_0) - K_{t_0-t}(x-x_0)\partial_j u(x,t)\right)\frac{x_j}{r}\mathrm{d}\sigma(x). \end{aligned}$$

Letting $r \to \infty$, the last sum vanishes by assumptions (45.3). We therefore have

$$0 = \int_{\mathbb{R}^n} u(x,b)K_{t_0-b}(x-x_0)\mathrm{d}x - \int_{\mathbb{R}^n} u(x,a)K_{t_0-a}(x-x_0)\mathrm{d}x.$$

Let us prove that

$$\lim_{b\to t_0-} \int_{\mathbb{R}^n} K_{t_0-b}(x-x_0)u(x,b)\mathrm{d}x = u(x_0,t_0)$$

and

$$\lim_{a\to 0+} \int_{\mathbb{R}^n} K_{t_0-a}(x-x_0)u(x,a)\mathrm{d}x = 0.$$

Since

$$\int_{\mathbb{R}^n} K_{t_0-b}(x-x_0)\mathrm{d}x = 1,$$

we have after a change of variables that

$$\int_{\mathbb{R}^n} K_{t_0-b}(x-x_0)u(x,b)\mathrm{d}x - u(x_0,t_0)$$
$$= \int_{\mathbb{R}^n} K_{t_0-b}(x-x_0)[u(x,b)-u(x_0,t_0)]\mathrm{d}x$$
$$= \int_{\mathbb{R}^n} K_\tau(z)[u(x_0+z,t_0+\tau)-u(x_0,t_0)]\mathrm{d}z.$$

We divide the latter integral into two parts: $|z|<\delta$ and $|z|>\delta$. The first part can be estimated from above by

$$\sup_{|z|<\delta} |u(x_0+z,t_0+\tau)-u(x_0,t_0)| \int_{|z|<\delta} K_\tau(z)\mathrm{d}z$$
$$\le \sup_{|z|<\delta} |u(x_0+z,t_0+\tau)-u(x_0,t_0)| \to 0$$

as $\tau \to 0$ and $\delta \to 0$ due to the continuity of $u(x,t)$ at the point (x_0,t_0). The second part can be estimated from above by (see (45.3))

$$c_\varepsilon \int_{|z|>\delta} K_\tau(z)\mathrm{e}^{\varepsilon|x_0+z|^2}\mathrm{d}z \le \frac{c'_\varepsilon}{(4\pi\tau)^{n/2}} \int_{|z|>\delta} \mathrm{e}^{-|z|^2/(4\tau)+\varepsilon|z|^2}\mathrm{d}z$$
$$= \frac{c'_\varepsilon}{(4\pi\tau)^{n/2}}\tau^{n/2} \int_{|y|>\delta/\sqrt{\tau}} \mathrm{e}^{-|y|^2/4+\varepsilon\tau|y|^2}\mathrm{d}y \to 0$$

as $\tau \to 0$. Thus the first limit is justified.

For the second limit we may first rewrite the integral as

$$\int_{\mathbb{R}^n} K_{t_0-a}(x-x_0)u(x,a)\mathrm{d}x = \int_{\mathbb{R}^n} K_{t_0-a}(z)u(x_0+z,a)\mathrm{d}z$$
$$= \int_{|z|\le R} K_{t_0-a}(z)u(x_0+z,a)\mathrm{d}z + \int_{|z|>R} K_{t_0-a}(z)u(x_0+z,a)\mathrm{d}z.$$

The first term in the latter sum can be estimated from above by

$$\sup_{|z|\le R} |u(x_0+z,a)| \int_{|z|\le R} K_{t_0-a}(z)\mathrm{d}z \le \sup_{|z|\le R} |u(x_0+z,a)| \to 0$$

as $a \to 0+$, since $u(x,t)$ is continuous up to the boundary ($t=0$) and therefore uniformly continuous on compact subsets there due to the fact that $u(x_0+z,0)=0$. The second term can be estimated from above by

$$\frac{c_\varepsilon}{(4\pi(t_0-a))^{n/2}} \int_{|z|>R} \mathrm{e}^{-|z|^2/(4(t_0-a))+\varepsilon|x_0+z|^2}\mathrm{d}z \to 0$$

as $R \to +\infty$ and $a \to 0$. □

Theorem 45.5. *The kernel $K_t(x)$ is a fundamental solution for the heat operator.*

Proof. Given $\varepsilon > 0$, set

$$K_\varepsilon(x,t) = \begin{cases} K_t(x), & t \ge \varepsilon, \\ 0, & t < \varepsilon. \end{cases}$$

Clearly $K_\varepsilon(x,t) \to K_t(x)$ as $\varepsilon \to 0$ in the sense of distributions. Even more is true, namely, $K_\varepsilon(x,t) \to K_t(x)$ pointwise as $\varepsilon \to 0$ and

$$\int_{\mathbb{R}^n} |K_\varepsilon(x,t)|\,\mathrm{d}x = \int_{\mathbb{R}^n} K_\varepsilon(x,t)\mathrm{d}x \le \int_{\mathbb{R}^n} K_t(x)\mathrm{d}x = 1.$$

We can therefore apply the dominated convergence theorem and obtain

$$\lim_{\varepsilon\to 0+} \int_{\mathbb{R}^n} K_\varepsilon(x,t)\mathrm{d}x = \int_{\mathbb{R}^n} K_t(x)\mathrm{d}x.$$

So it remains to show that as $\varepsilon \to 0$,

$$\partial_t K_\varepsilon(x,t) - \Delta_x K_\varepsilon(x,t) \to \delta(x,t),$$

or

$$\langle \partial_t K_\varepsilon - \Delta_x K_\varepsilon, \varphi\rangle \to \varphi(0), \quad \varphi \in C_0^\infty(\mathbb{R}^{n+1}).$$

Using integration by parts, we obtain

$$\begin{aligned}
\langle \partial_t K_\varepsilon - \Delta_x K_\varepsilon, \varphi\rangle &= \langle K_\varepsilon, -\partial_t\varphi - \Delta\varphi\rangle = \int_\varepsilon^\infty \mathrm{d}t \int_{\mathbb{R}^n} K_t(x)(-\partial_t - \Delta)\varphi(x,t)\mathrm{d}x \\
&= -\int_{\mathbb{R}^n} \mathrm{d}x \int_\varepsilon^\infty K_t(x)\partial_t\varphi(x,t)\mathrm{d}t \\
&\quad - \int_\varepsilon^\infty \mathrm{d}t \int_{\mathbb{R}^n} K_t(x)\Delta_x\varphi(x,t)\mathrm{d}x \\
&= \int_{\mathbb{R}^n} K_\varepsilon(x)\varphi(x,\varepsilon)\mathrm{d}x + \int_\varepsilon^\infty \mathrm{d}t \int_{\mathbb{R}^n} \partial_t K_t(x)\varphi(x,t)\mathrm{d}x \\
&\quad - \int_\varepsilon^\infty \mathrm{d}t \int_{\mathbb{R}^n} \Delta_x K_t(x)\varphi(x,t)\mathrm{d}x \\
&= \int_{\mathbb{R}^n} K_\varepsilon(x)\varphi(x,\varepsilon)\mathrm{d}x + \int_\varepsilon^\infty \mathrm{d}t \int_{\mathbb{R}^n} (\partial_t - \Delta)K_t(x)\varphi(x,t)\mathrm{d}x \\
&= \int_{\mathbb{R}^n} K_\varepsilon(x)\varphi(x,\varepsilon)\mathrm{d}x \to \varphi(0,0), \quad \varepsilon \to 0,
\end{aligned}$$

as we know from the proof of Theorem 45.1. □

Theorem 45.6. *If $f \in L^1(\mathbb{R}^{n+1})$, then*

$$u(x,t) := (f * K_t)(x,t) \equiv \int_{-\infty}^{t} \mathrm{d}s \int_{\mathbb{R}^n} K_{t-s}(x-y)f(y,s)\mathrm{d}y$$

is well defined almost everywhere and is a distributional solution of $\partial_t u - \Delta u = f$.

Exercise 45.2. Prove Theorem 45.6.

Let us now consider the heat operator in a bounded domain $\Omega \subset \mathbb{R}^n$ over a time interval $t \in [0,T]$, $0 < T \leq \infty$. In this case, it is necessary to specify the initial temperature $u(x,0)$, $x \in \Omega$, and also to prescribe a boundary condition on $\partial\Omega \times [0,T]$, see Figure 45.1.

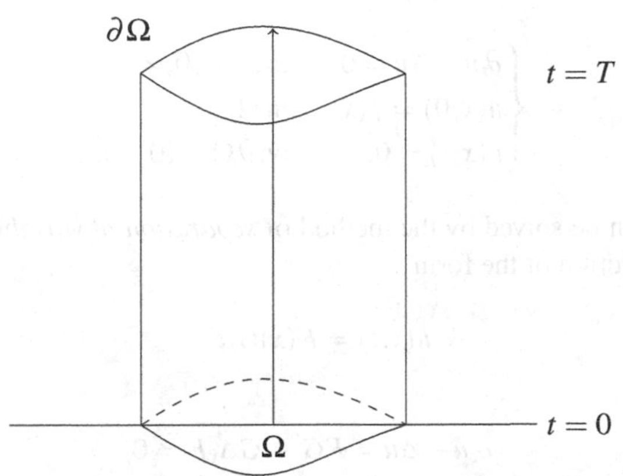

Fig. 45.1 Geometry of the boundary value problem for the Heat equation in Ω and $0 \leq t \leq T$.

The first basic result concerning such problems is the maximum principle.

Theorem 45.7. *Let* Ω *be a bounded domain in* $\mathbb{R}^n$ *and* $0 < T < \infty$. *Suppose* u *is a real-valued continuous function on* $\overline{\Omega} \times [0,T]$ *that satisfies* $\partial_t u - \Delta u = 0$ *in* $\Omega \times (0,T)$. *Then* u *assumes its maximum and minimum either on* $\Omega \times \{0\}$ *or on* $\partial\Omega \times [0,T]$.

Proof. Given $\varepsilon > 0$, set $v(x,t) := u(x,t) + \varepsilon|x|^2$. Then $\partial_t v - \Delta v = -2n\varepsilon$. Suppose $0 < T' < T$. If the maximum of v in $\overline{\Omega} \times [0,T']$ occurs at an interior point of $\Omega \times (0,T')$, then the first derivatives $\nabla_{x,t} v$ vanish there and the second derivative $\partial_j^2 v$ for all $j = 1,2,\ldots,n$ is nonpositive (consider $v(x,t)$ a function of one variable x_j, $j = 1,2,\ldots,n$). In particular, $\partial_t v = 0$ and $\Delta v \leq 0$, which contradicts $\partial_t v - \Delta v = -2n\varepsilon < 0$ and $\Delta v = 2n\varepsilon > 0$.

Likewise, if the maximum occurs in $\Omega \times \{T'\}$, then we have $\partial_t v(x,T') \geq 0$ and $\Delta v(x,T') \leq 0$, which contradicts $\partial_t v - \Delta v < 0$. Therefore,

$$\max_{\overline{\Omega}\times[0,T']} u \leq \max_{\overline{\Omega}\times[0,T']} v \leq \max_{(\Omega\times\{0\})\cup(\partial\Omega\times[0,T'])} u + \varepsilon \max_{\overline{\Omega}} |x|^2.$$

It follows that for $\varepsilon \to 0$ and $T' \to T$,

$$\max_{\overline{\Omega}\times[0,T]} u \leq \max_{(\Omega\times\{0\})\cup(\partial\Omega\times[0,T])} u \leq \max_{\overline{\Omega}\times[0,T]} u.$$

Replacing u by $-u$, we can obtain the same result for the minimum. □

Corollary 45.8 (Uniqueness). *There is at most one continuous function $u(x,t)$ in $\overline{\Omega}\times[0,T]$, $0<T<\infty$, that agrees with a given continuous function $f(x)$ in $\Omega\times\{0\}$, with $g(x,t)$ on $\partial\Omega\times[0,T]$ and satisfies $\partial_t u - \Delta u = 0$.*

Let us look now more closely at the following problem:

$$\begin{cases} \partial_t u - \Delta u = 0, & \text{in}\, \Omega \times (0,\infty), \\ u(x,0) = f(x), & \text{in}\, \Omega, \\ u(x,t) = 0, & \text{on}\, \partial\Omega \times (0,\infty). \end{cases} \tag{45.4}$$

This problem can be solved by the method of *separation of variables*. We begin by looking for a solution of the form

$$u(x,t) = F(x)G(t).$$

Then

$$\partial_t u - \Delta u = FG' - G\Delta_x F = 0$$

if and only if

$$\frac{G'}{G} = \frac{\Delta F}{F} := -\lambda^2,$$

or

$$G' + \lambda^2 G = 0, \quad \Delta F + \lambda^2 F = 0,$$

for some constant λ. The first equation has the general solution

$$G(t) = c e^{-\lambda^2 t},$$

where c is an arbitrary constant. Without loss of generality we assume that $c = 1$. It follows from (45.4) that

$$\begin{cases} \Delta F = -\lambda^2 F, & \text{in}\, \Omega, \\ F = 0, & \text{on}\, \partial\Omega, \end{cases} \tag{45.5}$$

because $u(x,t) = F(x)G(t)$ and $G(0) = 1$.

It remains to solve (45.5), which is an eigenvalue (spectral) problem for the Laplacian with Dirichlet boundary condition. It is known that the problem (45.5) has infinitely many solutions $\{F_j(x)\}_{j=1}^{\infty}$ with corresponding $\left\{\lambda_j^2\right\}_{j=1}^{\infty}$. The numbers $-\lambda_j^2$ are called *eigenvalues*, and the $F_j(x)$ are called *eigenfunctions* of the Laplacian. It is also known that $\lambda_j > 0$, $j = 1,2,\ldots$, $\lambda_j^2 \to \infty$, and $\{F_j(x)\}_{j=1}^{\infty}$ can be chosen as a complete orthonormal set in $L^2(\Omega)$ (or $\{F_j(x)\}_{j=1}^{\infty}$ forms an orthonormal basis of $L^2(\Omega)$). This fact allows us to represent $f(x)$ in terms of Fourier series:

$$f(x) = \sum_{j=1}^{\infty} f_j F_j(x), \tag{45.6}$$

where $f_j = (f, F_j)_{L^2(\Omega)}$ are called the Fourier coefficients of f with respect to $\{F_j\}_{j=1}^{\infty}$.

If we take now

$$u(x,t) = \sum_{j=1}^{\infty} f_j F_j(x) \mathrm{e}^{-\lambda_j^2 t}, \tag{45.7}$$

then we may conclude (at least formally) that

$$\partial_t u = -\sum_{j=1}^{\infty} f_j \lambda_j^2 F_j(x) \mathrm{e}^{-\lambda_j^2 t} = \sum_{j=1}^{\infty} f_j \Delta F_j(x) \mathrm{e}^{-\lambda_j^2 t} = \Delta u,$$

that is, $u(x,t)$ from (45.7) satisfies the heat equation and $u(x,t) = 0$ on $\partial\Omega \times (0,\infty)$. It remains to prove that $u(x,t)$ satisfies the initial condition and to determine for which functions $f(x)$ the series (45.6) converges and in what sense. This is the main question in the Fourier method.

It is clear that the series (45.6) and (45.7) (for $t \geq 0$) converge in the sense of $L^2(\Omega)$. It is also clear that if $f \in C^1(\Omega)$ vanishes at the boundary, then u will vanish on $\partial\Omega \times (0,\infty)$, and one easily verifies that u is a distributional solution of the heat equation ($t > 0$). Hence it is a classical solution, since $u(x,t) \in C^{\infty}(\Omega \times (0,\infty))$ (see Corollary 45.3).

Similar considerations apply to the problem

$$\begin{cases} \partial_t u - \Delta u = 0, & \text{in } \Omega \times (0,\infty), \\ u(x,0) = f(x), & \text{in } \Omega \\ \partial_\nu u(x,t) = 0, & \text{on } \partial\Omega \times (0,\infty). \end{cases}$$

This problem boils down to finding an orthonormal basis of eigenfunctions for the Laplacian with the Neumann boundary condition. Let us remark that for this problem, $\{0\}$ is always an eigenvalue and 1 is an eigenfunction.

Exercise 45.3. Prove that $u(x,t)$ of the form (45.7) is a distributional solution of the heat equation in $\Omega \times (0,\infty)$.

Exercise 45.4. Show that $\int_0^\pi |u(x,t)|^2 dx$ is a decreasing function of $t > 0$, where $u(x,t)$ is the solution of

$$\begin{cases} u_t - u_{xx} = 0, & 0 < x < \pi, t > 0, \\ u(0,t) = u(\pi,t) = 0, & t > 0. \end{cases}$$

Chapter 46
The Wave Operator

The *wave equation* is defined as

$$\partial_t^2 u(x,t) - \Delta_x u(x,t) = 0, \quad (x,t) \in \mathbb{R}^n \times \mathbb{R}. \tag{46.1}$$

The wave equation is satisfied exactly by the components of the classical electromagnetic field in vacuum.

The characteristic variety of (46.1) is

$$\text{char}_x(L) = \left\{(\xi,\tau) \in \mathbb{R}^{n+1} : (\xi,\tau) \neq 0, \tau^2 = |\xi|^2\right\},$$

and it is called the *light cone*. Accordingly, we call

$$\{(\xi,\tau) \in \text{char}_x(L) : \tau > 0\}$$

and

$$\{(\xi,\tau) \in \text{char}_x(L) : \tau < 0\}$$

the forward and backward light cones, respectively.

The wave operator is a prototype of *hyperbolic operators*. This means that the main symbol

$$\sum_{|\alpha|+j=k} a_\alpha(x,t)\xi^\alpha \tau^j$$

has k distinct real roots with respect to τ.

Theorem 46.1. *Suppose $u(x,t)$ is a C^2 function and that $\partial_t^2 u - \Delta u = 0$. Suppose also that $u = 0$ and $\partial_\nu u = 0$ on the ball $B = \{(x,0) : |x - x_0| \leq t_0\}$ in the hyperplane $t = 0$. Then $u = 0$ in the region $\Omega = \{(x,t) : 0 \leq t \leq t_0, |x - x_0| \leq t_0 - t\}$.*

V. Serov, *Fourier Series, Fourier Transform and Their Applications to Mathematical Physics*, Applied Mathematical Sciences 197,
DOI 10.1007/978-3-319-65262-7_46

Proof. By considering real and imaginary parts we may assume that u is real. Define $B_t = \{x : |x-x_0| < t_0 - t\}$. Let us consider the following integral:

$$E(t) = \frac{1}{2}\int_{B_t}\left((u_t)^2 + |\nabla_x u|^2\right)\mathrm{d}x,$$

which represents the energy of the wave in B_t at time t. Next,

$$\begin{aligned}E'(t) &= \int_{B_t}\left(u_t u_{tt} + \sum_{j=1}^n \partial_j u(\partial_j u)_t\right)\mathrm{d}x\\ &\quad - \frac{1}{2}\int_{\partial B_t}\left((u_t)^2 + |\nabla_x u|^2\right)\mathrm{d}\sigma(x) =: I_1 + I_2.\end{aligned}$$

Straightforward calculations using the divergence theorem show us that

$$\begin{aligned}I_1 &= \int_{B_t}\left(\sum_{j=1}^n \partial_j[(\partial_j u)u_t] - \sum_{j=1}^n (\partial_j^2 u)u_t + u_t u_{tt}\right)\mathrm{d}x\\ &= \int_{B_t} u_t(u_{tt} - \Delta_x u)\mathrm{d}x + \int_{\partial B_t}\sum_{j=1}^n (\partial_j u)\nu_j u_t \mathrm{d}\sigma(x)\\ &\le \int_{\partial B_t}|u_t||\nabla_x u|\mathrm{d}\sigma(x) \le \frac{1}{2}\int_{\partial B_t}\left(|u_t|^2 + |\nabla_x u|^2\right)\mathrm{d}\sigma(x) \equiv -I_2.\end{aligned}$$

Hence

$$\frac{\mathrm{d}E}{\mathrm{d}t} \le -I_2 + I_2 = 0.$$

But $E(t) \ge 0$ and $E(0) = 0$ due to the Cauchy data. Therefore, $E(t) \equiv 0$ if $0 \le t \le t_0$ and thus $\nabla_{x,t} u = 0$ in Ω. Since $u(x,0) = 0$, it follows that $u(x,t) = 0$ also in Ω. □

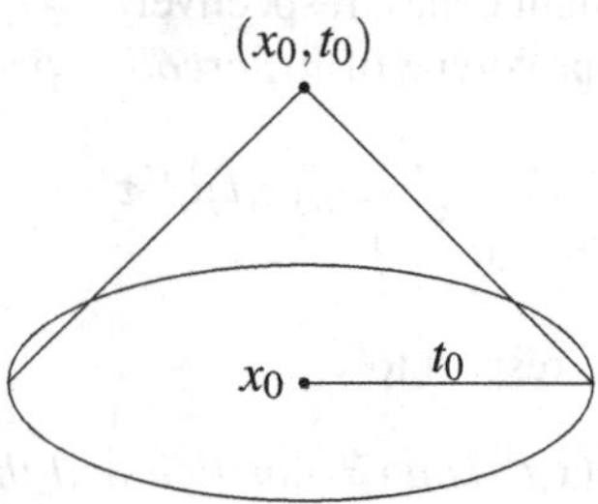

Fig. 46.1 Geometric illustration of the backward light cone at (x_0, t_0).

Remark 46.2. This theorem shows that the value of u at (x_0,t_0) depends only on the Cauchy data of u in the ball $\{(x,0) : |x-x_0| \le t_0\}$, see Figure 46.1.

Conversely, the Cauchy data in a region R in the initial ($t=0$) hyperplane influence only those points inside the forward light cones issuing from points of R. A similar result holds when the hyperplane $t=0$ is replaced by a spacelike hypersurface $S=\{(x,t): t=\varphi(x)\}$. A surface S is called *spacelike* if its normal vector $\nu=(\nu',\nu_0)$ satisfies $|\nu_0|>|\nu'|$ at every point of S, i.e., if ν lies inside the light cone. This means that $|\nabla\varphi|<1$.
Let us consider the Cauchy problem for the wave equation:

$$\begin{cases} \partial_t^2 u - \Delta u = 0, & x\in\mathbb{R}^n, t>0,\\ u(x,0)=f(x), & \partial_t u(x,0)=g(x). \end{cases} \tag{46.2}$$

Definition 46.3. If φ is a continuous function on $\mathbb{R}^n$ and $r>0$, we define the *spherical mean* $M_\varphi(x,r)$ as follows:

$$M_\varphi(x,r) := \frac{1}{r^{n-1}\omega_n}\int_{|x-z|=r}\varphi(z)\mathrm{d}\sigma(z) = \frac{1}{\omega_n}\int_{|y|=1}\varphi(x+ry)\mathrm{d}\sigma(y).$$

Lemma 46.4. *If φ is a C^2 function on $\mathbb{R}^n$, then $M_\varphi(x,0)=\varphi(x)$ and*

$$\Delta_x M_\varphi(x,r) = \left(\partial_r^2 + \frac{n-1}{r}\partial_r\right) M_\varphi(x,r).$$

Proof. It is clear that

$$M_\varphi(x,0) = \frac{1}{\omega_n}\int_{|y|=1}\varphi(x)\mathrm{d}\sigma(y) = \varphi(x).$$

For the second part we have by the divergence theorem that

$$\begin{aligned}\partial_r M_\varphi(x,r) &= \frac{1}{\omega_n}\int_{|y|=1}\sum_{j=1}^n y_j\partial_j\varphi(x+ry)\mathrm{d}\sigma(y) = \frac{1}{\omega_n}\int_{|y|\le 1} r\Delta\varphi(x+ry)\mathrm{d}y\\ &= \frac{1}{r^{n-1}\omega_n}\int_{|z|\le r}\Delta\varphi(x+z)\mathrm{d}z\\ &= \frac{1}{r^{n-1}\omega_n}\int_0^r \rho^{n-1}\mathrm{d}\rho\int_{|y|=1}\Delta\varphi(x+\rho y)\mathrm{d}\sigma(y).\end{aligned}$$

We therefore have

$$\partial_r\left(r^{n-1}\partial_r M_\varphi(x,r)\right) = \frac{r^{n-1}}{\omega_n}\int_{|y|=1}\Delta\varphi(x+ry)\mathrm{d}\sigma(y) \equiv r^{n-1}\Delta_x M_\varphi(x,r).$$

This implies that

$$(n-1)r^{n-2}\partial_r M_\varphi(x,r) + r^{n-1}\partial_r^2 M_\varphi(x,r) = r^{n-1}\Delta_x M_\varphi(x,r),$$

which proves the claim. □

Corollary 46.5. *Suppose $u(x,t)$ is a C^2 function on $\mathbb{R}^{n+1}$ and let*

$$M_u(x,r,t) = \frac{1}{r^{n-1}\omega_n}\int_{|x-z|=r} u(z,t)\mathrm{d}\sigma(z) = \frac{1}{\omega_n}\int_{|y|=1} u(x+ry,t)\mathrm{d}\sigma(y).$$

Then $u(x,t)$ satisfies the wave equation if and only if

$$\left(\partial_r^2 + \frac{n-1}{r}\partial_r\right) M_u(x,r,t) = \partial_t^2 M_u(x,r,t). \tag{46.3}$$

Lemma 46.6. *If $\varphi \in C^{k+1}(\mathbb{R})$, $k \geq 1$, then*

$$\partial_r^2 \left(\frac{1}{r}\partial_r\right)^{k-1} \left(r^{2k-1}\varphi(r)\right) = \left(\frac{\partial_r}{r}\right)^k (r^{2k}\varphi').$$

Proof. We employ induction on k. If $k=1$, then

$$\partial_r^2 \left(\frac{1}{r}\partial_r\right)^{k-1} \left(r^{2k-1}\varphi(r)\right) = \partial_r^2(r\varphi) = \partial_r(\varphi + r\varphi') = 2\varphi' + r\varphi''$$

and

$$\left(\frac{\partial_r}{r}\right)^k (r^{2k}\varphi') = \left(\frac{\partial_r}{r}\right)(r^2\varphi') = 2\varphi' + r\varphi''.$$

Assume that

$$\partial_r^2 \left(\frac{1}{r}\partial_r\right)^{k-1} \left(r^{2k-1}\varphi(r)\right) = \left(\frac{\partial_r}{r}\right)^k (r^{2k}\varphi').$$

Then

$$\begin{aligned}
\partial_r^2 \left(\frac{1}{r}\partial_r\right)^{k} (r^{2k+1}\varphi(r)) &= \partial_r^2 \left(\frac{1}{r}\partial_r\right)^{k-1} \left(\frac{\partial_r}{r}\left(r^{2k+1}\varphi\right)\right) \\
&= \partial_r^2 \left(\frac{1}{r}\partial_r\right)^{k-1} \left((2k+1)r^{2k-1}\varphi + r^{2k}\varphi'\right) \\
&= (2k+1)\partial_r^2 \left(\frac{1}{r}\partial_r\right)^{k-1} \left(r^{2k-1}\varphi\right) + \partial_r^2 \left(\frac{1}{r}\partial_r\right)^{k-1} \left(r^{2k}\varphi'\right) \\
&= (2k+1)\left(\frac{\partial_r}{r}\right)^k \left(r^{2k}\varphi'\right) + \left(\frac{\partial_r}{r}\right)^k (r^{2k}(r\varphi')')
\end{aligned}$$

$$
\begin{aligned}
&= \left(\frac{\partial_r}{r}\right)^k \left((2k+1)r^{2k}\varphi' + r^{2k}(r\varphi')'\right) \\
&= \left(\frac{\partial_r}{r}\right)^k \left((2k+1)r^{2k}\varphi' + r^{2k}\varphi' + r^{2k+1}\varphi''\right) \\
&= \left(\frac{\partial_r}{r}\right)^k \left((2k+2)r^{2k}\varphi' + r^{2k+1}\varphi''\right) \\
&= \left(\frac{\partial_r}{r}\right)^{k+1} \left(r^{2k+2}\varphi'\right).
\end{aligned}
$$

By the principle of induction, the proof is complete. □

Corollary 46.5 gives that if $u(x,t)$ is a solution of the wave equation (46.1) in $\mathbb{R}^n \times \mathbb{R}$, then $M_u(x,r,t)$ satisfies (46.3), i.e.,

$$\left(\partial_r^2 + \frac{n-1}{r}\partial_r\right) M_u = \partial_t^2 M_u,$$

with initial conditions

$$M_u(x,r,0) = M_f(x,r), \quad \partial_t M_u(x,r,0) = M_g(x,r), \tag{46.4}$$

since $u(x,0) = f(x)$ and $\partial_t u(x,0) = g(x)$.

Let us set

$$
\begin{aligned}
\widetilde{u}(x,r,t) &:= \left(\frac{\partial_r}{r}\right)^{\frac{n-3}{2}} \left(r^{n-2}M_u\right) \equiv TM_u, \\
\widetilde{f}(x,r) &:= TM_f, \quad \widetilde{g}(x,r) := TM_g,
\end{aligned} \tag{46.5}
$$

for $n = 2k+1$, $k = 1,2,\ldots$.

Lemma 46.7. *The following is true:*

$$
\begin{cases}
\partial_r^2 \widetilde{u} = \partial_t^2 \widetilde{u}, \\
\widetilde{u}|_{t=0} = \widetilde{f}, \quad \partial_t \widetilde{u}|_{t=0} = \widetilde{g},
\end{cases} \tag{46.6}
$$

where $\widetilde{u}$, $\widetilde{f}$, and $\widetilde{g}$ are defined in (46.5).

Proof. Since $n = 2k+1$, we have $\frac{n-3}{2} = k-1$ and $n-2 = 2k-1$. Hence we obtain from Lemmas 46.4 and 46.6 that

$$
\begin{aligned}
\partial_r^2 \widetilde{u} = \partial_r^2 TM_u = \partial_r^2 \left(\frac{\partial_r}{r}\right)^{k-1} \left(r^{2k-1}M_u\right) &= \left(\frac{\partial_r}{r}\right)^k \left(r^{2k}\partial_r M_u\right) \\
= \left(\frac{\partial_r}{r}\right)^{k-1} \left(2kr^{2k-2}\partial_r M_u + r^{2k-1}\partial_r^2 M_u\right)&
\end{aligned}
$$

$$= \left(\frac{\partial_r}{r}\right)^{k-1} \left(r^{2k-1}\left(\partial_r^2 M_u + \frac{n-1}{r}\partial_r M_u\right)\right) = \left(\frac{\partial_r}{r}\right)^{k-1} \left(r^{2k-1}\partial_t^2 M_u\right)$$
$$= \partial_t^2 \left(\frac{\partial_r}{r}\right)^{k-1} \left(r^{2k-1} M_u\right) = \partial_t^2 \widetilde{u}.$$

Moreover, the initial conditions are satisfied due to (46.4) and (46.5). □

But now, since (46.6) is a one-dimensional problem, we may conclude that $\widetilde{u}(x,r,t)$ from Lemma 46.7 is equal to

$$\widetilde{u}(x,r,t) = \frac{1}{2}\left\{\widetilde{f}(x,r+t) + \widetilde{f}(x,r-t) + \int_{r-t}^{r+t} \widetilde{g}(x,s)\mathrm{d}s\right\}. \tag{46.7}$$

Lemma 46.8. *If* $n = 2k+1, k = 1,2,\ldots,$ *then*

$$u(x,t) = M_u(x,0,t) = \lim_{r\to 0} \frac{\widetilde{u}(x,r,t)}{(n-2)!!r},$$

where $(n-2)!! = 1\cdot 3\cdot 5\cdots(n-2)$ *is the solution of* (46.2). *We have even more, namely,*

$$u(x,t) = \frac{1}{(n-2)!!}\left(\partial_r \widetilde{f}|_{r=t} + \widetilde{g}(x,t)\right). \tag{46.8}$$

Proof. By (46.5) we have

$$\widetilde{u}(x,r,t) = \left(\frac{\partial_r}{r}\right)^{k-1}\left(r^{2k-1}M_u\right) = \left(\frac{\partial_r}{r}\right)^{k-2}\left((2k-1)r^{2k-3}M_u + r^{2k-2}\partial_r M_u\right)$$
$$= (2k-1)(2k-3)\cdots 1\cdot M_u r + O(r^2), \quad r\to 0,$$

or

$$\frac{\widetilde{u}(x,r,t)}{(n-2)!!r} = M_u + O(r), \quad r\to 0.$$

Hence

$$M_u(x,0,t) = \lim_{r\to 0}\frac{\widetilde{u}(x,r,t)}{(n-2)!!r}.$$

But by definition of M_u we have that $M_u(x,0,t) = u(x,t)$, where $u(x,t)$ is the solution of (46.2). The initial conditions in (46.2) are satisfied due to (46.5). Next, since $\widetilde{u}(x,r,t)$ satisfies (46.7), we have

$$\lim_{r\to 0}\frac{\widetilde{u}(x,r,t)}{(n-2)!!r} = \frac{1}{2(n-2)!!}\lim_{r\to 0}\left(\frac{\widetilde{f}(x,r+t)+\widetilde{f}(x,r-t)}{r}+\frac{1}{r}\int_{r-t}^{r+t}\widetilde{g}(x,s)\mathrm{d}s\right)$$
$$= \frac{1}{2(n-2)!!}\left(\partial_r\widetilde{f}|_{r=t}+\partial_r\widetilde{f}|_{r=-t}+\widetilde{g}(x,t)-\widetilde{g}(x,-t)\right),$$

because $\widetilde{f}(x,t)$ and $\widetilde{g}(x,t)$ are odd functions of t. We therefore finally obtain

$$\lim_{r\to 0}\frac{\widetilde{u}(x,r,t)}{(n-2)!!r} = \frac{1}{(n-2)!!}\left(\partial_r\widetilde{f}|_{r=t}+\widetilde{g}(x,t)\right),$$

and the proof is complete. □

Now we are in a position to prove the main theorem for odd $n \geq 3$.

Theorem 46.9. *Suppose that $n \geq 3$ is odd. If $f \in C^{\frac{n+3}{2}}(\mathbb{R}^n)$ and $g \in C^{\frac{n+1}{2}}(\mathbb{R}^n)$, then*

$$u(x,t) = \frac{1}{(n-2)!!\omega_n}\left\{\partial_t\left(\frac{\partial_t}{t}\right)^{\frac{n-3}{2}}\left(t^{n-2}\int_{|y|=1}f(x+ty)\mathrm{d}\sigma(y)\right)\right. \left.+\left(\frac{\partial_t}{t}\right)^{\frac{n-3}{2}}\left(t^{n-2}\int_{|y|=1}g(x+ty)\mathrm{d}\sigma(y)\right)\right\} \tag{46.9}$$

solves the Cauchy problem (46.2).

Proof. Due to Lemmas 46.7 and 46.8, $u(x,t)$ given by (46.8) is the solution of the wave equation. It remains only to check that this u satisfies the initial conditions. But (46.9) gives us for small t that

$$u(x,t) = M_f(x,t)+tM_g(x,t)+O(t^2),$$

which implies that

$$u(x,0) = M_f(x,0) = f(x), \quad \partial_t u(x,0) = \partial_t M_f(x,0)+M_g(x,0) = g(x).$$

The last equality follows from the fact that $M_f(x,t)$ is even in t, and so its derivative vanishes at $t = 0$.

Remark 46.10. If $n = 3$, then (46.9) becomes

$$u(x,t) = \frac{1}{4\pi}\left\{\partial_t\left(t\int_{|y|=1}f(x+ty)\mathrm{d}\sigma(y)\right)+t\int_{|y|=1}g(x+ty)\mathrm{d}\sigma(y)\right\}$$
$$\equiv \frac{1}{4\pi}\left\{\int_{|y|=1}f(x+ty)\mathrm{d}\sigma(y)+t\int_{|y|=1}\nabla f(x+ty)\cdot y\mathrm{d}\sigma(y)\right.$$
$$\left.+\, t\int_{|y|=1}g(x+ty)\mathrm{d}\sigma(y)\right\}.$$

The solution of (46.2) for even n is readily derived from the solution for odd n by the "method of descent". This is just a trivial observation: if u is a solution of the wave equation in $\mathbb{R}^{n+1} \times \mathbb{R}$ that does not depend on x_{n+1}, then u satisfies the wave equation in $\mathbb{R}^n \times \mathbb{R}$. Thus to solve (46.2) in $\mathbb{R}^n \times \mathbb{R}$ with even n, we think of f and g as functions on $\mathbb{R}^{n+1}$ that are independent of x_{n+1}.

Theorem 46.11. *Suppose that n is even. If* $f \in C^{\frac{n+4}{2}}(\mathbb{R}^n)$ *and* $g \in C^{\frac{n+2}{2}}(\mathbb{R}^n)$*, then the function*

$$u(x,t) = \frac{2}{(n-1)!!\omega_{n+1}} \left\{ \partial_t \left(\frac{\partial_t}{t}\right)^{\frac{n-2}{2}} \left(t^{n-1} \int_{|y|\le 1} \frac{f(x+ty)}{\sqrt{1-y^2}} dy \right) + \left(\frac{\partial_t}{t}\right)^{\frac{n-2}{2}} \left(t^{n-1} \int_{|y|\le 1} \frac{g(x+ty)}{\sqrt{1-y^2}} dy \right) \right\} \tag{46.10}$$

solves the Cauchy problem (46.2).

Proof. If n is even, then $n+1$ is odd and $n+1 \ge 3$. We can therefore apply (46.9) in $\mathbb{R}^{n+1} \times \mathbb{R}$ to get that

$$u(x,t) = \frac{1}{(n-1)!!\omega_{n+1}} \times \left\{ \partial_t \left(\frac{\partial_t}{t}\right)^{\frac{n-2}{2}} \left(t^{n-1} \int_{y_1^2+\cdots+y_n^2+y_{n+1}^2=1} f(x+ty+ty_{n+1}) d\sigma(\tilde{y}) \right) + \left(\frac{\partial_t}{t}\right)^{\frac{n-2}{2}} \left(t^{n-1} \int_{y_1^2+\cdots+y_n^2+y_{n+1}^2=1} g(x+ty+ty_{n+1}) d\sigma(\tilde{y}) \right) \right\}, \tag{46.11}$$

where $\tilde{y} = (y, y_{n+1})$, solves (46.2) in $\mathbb{R}^{n+1} \times \mathbb{R}$ (formally). But if we assume now that f and g do not depend on x_{n+1}, then $u(x,t)$ does not depend on x_{n+1} either and solves (46.2) in $\mathbb{R}^n \times \mathbb{R}$. It remains only to calculate the integrals in (46.11) under this assumption. We have

$$\int_{|y|^2+y_{n+1}^2=1} f(x+ty+ty_{n+1}) d\sigma(\tilde{y}) = \int_{|y|^2+y_{n+1}^2=1} f(x+ty) d\sigma(\tilde{y}) = 2\int_{|y|\le 1} f(x+ty) \frac{dy}{\sqrt{1-|y|^2}},$$

because we have the upper and lower hemispheres of the sphere $|y|^2 + y_{n+1}^2 = 1$. Similarly for the second integral in (46.11). This proves the theorem. □

Remark 46.12. If $n = 2$, then (46.10) becomes

$$u(x,t)=\frac{1}{2\pi}\left\{\partial_t\left(t\int_{|y|\le 1}\frac{f(x+ty)}{\sqrt{1-y^2}}\mathrm{d}y\right)+t\int_{|y|\le 1}\frac{g(x+ty)}{\sqrt{1-y^2}}\mathrm{d}y\right\}.$$

Now we consider the Cauchy problem for the inhomogeneous wave equation

$$\begin{cases}\partial_t^2 u-\Delta_x u=w(x,t),\\ u(x,0)=f(x),\quad \partial_t u(x,0)=g(x).\end{cases}\tag{46.12}$$

We look for the solution $u(x,t)$ of (46.12) as $u=u_1+u_2$, where

$$\begin{cases}\partial_t^2 u_1-\Delta u_1=0,\\ u_1(x,0)=f(x),\quad \partial_t u_1(x,0)=g(x),\end{cases}\tag{A}$$

and

$$\begin{cases}\partial_t^2 u_2-\Delta u_2=w,\\ u_2(x,0)=\partial_t u_2(x,0)=0.\end{cases}\tag{B}$$

For the problem (B) we will use a method known as *Duhamel's principle.*

Theorem 46.13. *Suppose $w\in C^{[\frac{n}{2}]+1}(\mathbb{R}^n\times\mathbb{R})$. For $s\in\mathbb{R}$ let $v(x,t;s)$ be the solution of*

$$\begin{cases}\partial_t^2 v(x,t;s)-\Delta_x v(x,t;s)=0,\\ v(x,0;s)=0,\quad \partial_t v(x,0;s)=w(x,s).\end{cases}$$

Then

$$u(x,t):=\int_0^t v(x,t-s;s)\mathrm{d}s$$

solves (B).

Proof. By definition of $u(x,t)$ it is clear that $u(x,0)=0$. We also have

$$\partial_t u(x,t)=v(x,0;t)+\int_0^t \partial_t v(x,t-s;s)\mathrm{d}s.$$

This implies that $\partial_t u(x,0)=v(x,0;0)=0$. Differentiating once more in t, we get

$$\begin{aligned}\partial_t^2 u(x,t)&=\partial_t(v(x,0;t))+\partial_t v(x,0;t)+\int_0^t\partial_t^2 v(x,t-s;s)\mathrm{d}s\\ &=w(x,t)+\int_0^t \Delta_x v(x,t-s;s)\mathrm{d}s\\ &=w(x,t)+\Delta_x\int_0^t v(x,t-s;s)\mathrm{d}s=w(x,t)+\Delta_x u.\end{aligned}$$

Thus u solves (B), and the theorem is proved. □

Let us consider again the homogeneous Cauchy problem (46.2). Applying the Fourier transform with respect to x gives

$$\begin{cases} \partial_t^2 \widehat{u}(\xi,t) + |\xi|^2 \widehat{u}(\xi,t) = 0, \\ \widehat{u}(\xi,0) = \widehat{f}(\xi), \quad \partial_t \widehat{u}(\xi,0) = \widehat{g}(\xi). \end{cases}$$

But this ordinary differential equation with initial conditions can be easily solved to obtain

$$\widehat{u}(\xi,t) = \widehat{f}(\xi)\cos(|\xi|t) + \widehat{g}(\xi)\frac{\sin(|\xi|t)}{|\xi|} \equiv \widehat{f}(\xi)\partial_t\left(\frac{\sin(|\xi|t)}{|\xi|}\right) + \widehat{g}(\xi)\frac{\sin(|\xi|t)}{|\xi|}.$$

This implies that

$$\begin{aligned} u(x,t) &= \mathscr{F}^{-1}\left(\widehat{f}(\xi)\partial_t \frac{\sin(|\xi|t)}{|\xi|}\right) + \mathscr{F}^{-1}\left(\widehat{g}(\xi)\frac{\sin(|\xi|t)}{|\xi|}\right) \\ &= f * \partial_t\left((2\pi)^{-n/2}\mathscr{F}^{-1}\left(\frac{\sin(|\xi|t)}{|\xi|}\right)\right) + g * \left((2\pi)^{-n/2}\mathscr{F}^{-1}\left(\frac{\sin(|\xi|t)}{|\xi|}\right)\right) \\ &= f * \partial_t \Phi(x,t) + g * \Phi(x,t), \end{aligned} \tag{46.13}$$

where $\Phi(x,t) = (2\pi)^{-n/2}\mathscr{F}^{-1}\left(\frac{\sin(|\xi|t)}{|\xi|}\right)$.

The next step is to try to solve the equation

$$\partial_t^2 F(x,t) - \Delta_x F(x,t) = \delta(x)\delta(t).$$

By taking the Fourier transform in x we obtain

$$\partial_t^2 \widehat{F}(\xi,t) + |\xi|^2 \widehat{F}(\xi,t) = (2\pi)^{-n/2}\delta(t).$$

Therefore, $\widehat{F}$ must be a solution of $\partial_t^2 u + |\xi|^2 u = 0$ for $t \neq 0$, and so

$$\widehat{F}(\xi,t) = \begin{cases} a(\xi)\cos(|\xi|t) + b(\xi)\sin(|\xi|t), & t < 0, \\ c(\xi)\cos(|\xi|t) + d(\xi)\sin(|\xi|t), & t > 0. \end{cases}$$

To obtain the delta function at $t = 0$ we require that $\widehat{F}$ is continuous at $t = 0$, but $\partial_t \widehat{F}$ has a jump of size $(2\pi)^{-n/2}$ at $t = 0$. So we have

$$a(\xi) = c(\xi), \quad |\xi|(d(\xi) - b(\xi)) = (2\pi)^{-n/2}.$$

This gives two equations for the four unknown coefficients a,b,c,d. But it is reasonable to require $F(x,t) \equiv 0$ for $t < 0$. Hence, $a = b = c = 0$ and $d = (2\pi)^{-n/2}\frac{1}{|\xi|}$.

Therefore,

$$\widehat{F}(\xi,t) = \begin{cases} (2\pi)^{-n/2}\frac{\sin(|\xi|t)}{|\xi|}, & t>0, \\ 0, & t<0. \end{cases} \tag{46.14}$$

If we compare (46.13) and (46.14), we may conclude that

$$F(x,t) = (2\pi)^{-n/2}\mathscr{F}_{\xi}^{-1}\left(\frac{\sin(|\xi|t)}{|\xi|}\right), \quad t>0,$$

and

$$\Phi_{+}(x,t) = \begin{cases} \Phi(x,t), & t>0, \\ 0, & t<0, \end{cases}$$

is the fundamental solution of the wave equation, i.e., $F(x,t)$ with $t>0$.

There is one more observation. If we compare (46.9) and (46.10) with (46.13), then we may conclude that these three formulas are the same. Hence, we may calculate the inverse Fourier transform of

$$(2\pi)^{-n/2}\frac{\sin(|\xi|t)}{|\xi|}$$

in odd and even dimensions respectively with (46.9) and (46.10). In fact, the result is presented in these two formulas.

In solving the wave equation in the region $\Omega\times(0,\infty)$, where Ω is a bounded domain in $\mathbb{R}^n$, it is necessary to specify not only the Cauchy data on $\Omega\times\{0\}$ but also some conditions on $\partial\Omega\times(0,\infty)$ to tell the wave what to do when it hits the boundary. If the boundary conditions on $\partial\Omega\times(0,\infty)$ are independent of t, the method of separation of variables can be used.

Let us (for example) consider the following problem:

$$\begin{cases} \partial_t^2 u - \Delta_x u = 0, & \text{in } \Omega\times(0,\infty), \\ u(x,0) = f(x), \quad \partial_t u(x,0) = g(x), & \text{in } \Omega, \\ u(x,t) = 0, & \text{on } \partial\Omega\times(0,\infty). \end{cases} \tag{46.15}$$

We can look for a solution u in the form $u(x,t) = F(x)G(t)$ and get

$$\begin{cases} \Delta F(x) + \lambda^2 F(x) = 0, & \text{in } \Omega, \\ F(x) = 0, & \text{on } \partial\Omega, \end{cases} \tag{46.16}$$

and

$$G''(t) + \lambda^2 G(t) = 0, \quad 0<t<\infty. \tag{46.17}$$

The general solution of (46.17) is $G(t) = a\cos(\lambda t) + b\sin(\lambda t)$. Since (46.16) has infinitely many solutions $\{F_j\}_{j=1}^{\infty}$ with corresponding $\{\lambda_j^2\}_{j=1}^{\infty}$, $\lambda_j^2 \to +\infty$ $(\lambda_j > 0)$, and $\{F_j\}_{j=1}^{\infty}$ that can be chosen as an orthonormal basis in $L^2(\Omega)$, the solution $u(x,t)$ of (46.15) is of the form

$$u(x,t) = \sum_{j=1}^{\infty} F_j(x)\left(a_j\cos(\lambda_j t) + b_j\sin(\lambda_j t)\right). \tag{46.18}$$

At the same time, $f(x)$ and $g(x)$ have the $L^2(\Omega)$ representations

$$f(x) = \sum_{j=1}^{\infty} f_j F_j(x), \quad g(x) = \sum_{j=1}^{\infty} g_j F_j(x), \tag{46.19}$$

where $f_j = (f, F_j)_{L^2}$ and $g_j = (g, F_j)_{L^2}$. It follows from (46.15) and (46.18) that

$$u(x,0) = \sum_{j=1}^{\infty} a_j F_j(x), \quad u_t(x,0) = \sum_{j=1}^{\infty} \lambda_j b_j F_j(x). \tag{46.20}$$

Since (46.19) must be satisfied also, we obtain $a_j = f_j$ and $b_j = \frac{1}{\lambda_j} g_j$. Therefore, the solution $u(x,t)$ of (46.15) has the form

$$u(x,t) = \sum_{j=1}^{\infty} F_j(x)\left(f_j\cos(\lambda_j t) + \frac{1}{\lambda_j} g_j \sin(\lambda_j t)\right).$$

The series (46.18), (46.19), and (46.20) converge in $L^2(\Omega)$, because $\{F_j\}_{j=1}^{\infty}$ is an orthonormal basis in $L^2(\Omega)$. It remains only to investigate the convergence of these series in stronger norms (which depends on f and g, or more precisely, on their smoothness).

The Neumann problem with $\partial_\nu u(x,t) = 0$, $x \in \partial\Omega$, can be considered in a similar manner.

References

1. Adams R A and Fournier J J F, *Sobolev Spaces*, Elsevier, Oxford, 2003
2. Agmon S, *Spectral properties of Schrödinger operators and scattering theory*, Ann. Scuola Norm. Sup. Pisa Cl. Sci., 4 (1975), 151–218
3. Bergh J and Löfström J, *Interpolation Spaces: An Introduction*, Springer-Verlag, Berlin Heidelberg, 1976
4. Birman M S and Solomjak M Z, *Spectral Theory of Self-Adjoint Operators in Hilbert Space*, Kluwer, 1987
5. Blanchard P and Brüning E, *Mathematical Methods in Physics*, Birkhäuser, 1993
6. Cakoni F and Colton D, *A Qualitative Approach To Inverse Scattering Theory*, Springer-Verlag, 2014
7. Colton D and Kress R, *Integral Equation Methods in Scattering Theory*, Wiley, 1983
8. Colton D, *Partial Differential Equations: An Introduction*, Dover, 2004
9. Cycon H L, Froese R G, Kirsch W and Simon B, *Schrödinger Operators*, Academic Press, New-York, 1989
10. Evans L C, *Partial Differential Equations*, American Mathematical Society, 1998
11. Folland G B, *Introduction to Partial Differential Equations*, second edition, Princeton University Press, 1995
12. Folland G B, *Fourier Analysis and its Applications*, Brooks/Cole Publishing Company, 1992
13. Folland G B, *Real Analysis*, John Wiley, New York, 1984
14. Gårding L, *Singularities in Linear Wave Propagation*, Springer, 1987
15. Gårding L, *Some Points of Analysis and Their History*, American Mathematical Society, 1997
16. Gilbarg D and Trudinger N S, *Elliptic Partial Differential Equations of Second Order*, Springer, 2001
17. Gonzalez-Velasco E A, *Fourier Analysis and Boundary Value Problems*, Academic Press, San Diego, 1995
18. Grafakos L, *Classical and Modern Fourier Analysis*, Pearson Education Inc., 2004
19. Hörmander L, *The Analysis of Linear Partial Differential Operators I: Distribution Theory and Fourier Analysis*, second edition, Springer-Verlag, Berlin Heidelberg, 2003
20. Ikebe I, *Eigenfunction expansions associated with the Schroedinger operators and their applications to scattering theory*, Arch. Rational Mech. Anal., 5 (1960), 1–34
21. Kato T, *Growth properties of solutions of the reduced wave equation with a variable coefficient*, Comm. Pure Appl. Math., 12 (1959), 403–425
22. Kress, R, *Linear Integral Equations*, second edition, Springer-Verlag, New York, 1999
23. Lebedev N N, *Special Functions and Their Applications*, Dover, New York, 1972

V. Serov, *Fourier Series, Fourier Transform and Their Applications to Mathematical Physics*, Applied Mathematical Sciences 197,
DOI 10.1007/978-3-319-65262-7

24. Maurin K, *Methods of Hilbert Spaces*, Polish Scientific Publishers, Warsaw, 1967
25. McLean W, *Strongly Elliptic Systems and Boundary Integral Equations*, Cambridge University Press, 2000
26. Nikolskii S M, *Approximation of Functions of Several Variables and Imbedding Theorems*, Springer-Verlag, Berlin, Heidelberg, New York, 1975
27. Nirenberg L and Walker H, *Nullspaces of elliptic partial differential operators in* $\mathbb{R}^n$, J. Math. Anal. Appl., 42 (1973), 271–301.
28. Ouhabaz E M, *Analysis of Heat Equations on Domains*, Princeton University Press, 2005
29. Pinsky M A, *Introduction to Fourier Analysis and Wavelets*, Brooks/Cole, 2002
30. Päivärinta L and Serov V, *Recovery of singularities of a multidimensional scattering potential*, SIAM J. Math. Anal., 29 (1998), 697–711
31. Päivärinta L and Somersalo E, *Inversion of discontinuities for the Schrödinger equation in three dimensions*, SIAM J. Math. Anal., 22 (1991), 480–499
32. Saito Y, *Some properties of the scattering amplitude and inverse scattering problem*, Osaka, J. Math.19 (1982), 527–547
33. Saranen J and Vainikko G, *Periodic Integral & Pseudodifferential Equations with Numerical Approximation*, Springer, 2001
34. Shubin M A, *Pseudodifferential Operators and Spectral Theory*, Springer-Verlag, Berlin, 1987
35. Simon B, *A Comprehensive Course in Analysis*, American Mathematical Society, 2015
36. Stakgold I, *Green's Functions and Boundary Value Problems*, second edition, John Wiley, 1997
37. Stein E M, *Singular Integrals and Differentiability Properties of Functions*, Princeton University Press, Princeton, NJ, 1970
38. Stein E M and Shakarchi R, *Fourier Analysis: An Introduction*, Princeton University Press, 2003
39. Taylor M E, *Partial Differential Equations: Basic Theory*, Springer-Verlag, New York, 1996
40. Taylor M E, *Partial Differential Equations II: Qualitative Studies of Linear Equations*, Springer-Verlag, New York, 1996
41. Titchmarsh E C, *Eigenfunction Expansions Associated with Second-order Differential Equations*, vol. 1, Oxford University Press, Oxford, 1946
42. Triebel H, *Theory of Function Spaces*, Birkhäuser, Basel, 1983
43. Watson G N, *A Treatise on the Theory of Bessel Functions*, Cambridge University Press, Cambridge, UK, 1996
44. Wong M W, *Discrete Fourier Analysis*, Springer Basel AG, 2011
45. Vretblad A, *Fourier Analysis and Its Applications*, Springer-Verlag, New York, 2003
46. Zigmund A, *Trigonometric Series*, Vol. II, Cambridge University Press, 1959

Index

V. Serov, *Fourier Series, Fourier Transform and Their Applications to Mathematical Physics*, Applied Mathematical Sciences 197,
DOI 10.1007/978-3-319-65262-7

Printed in the United States
By Bookmasters